COMPOSITE STRUCTURES

2

Proceedings of the 2nd International Conference on Composite Structures, held at Paisley College of Technology, Scotland, from 14 to 16 September 1983, organised in association with the Scottish Development Agency and the National Engineering Laboratory.

COMPOSITE STRUCTURES

2

Edited by

I. H. MARSHALL

Department of Mechanical and Production Engineering,
Paisley College of Technology, Scotland

APPLIED SCIENCE PUBLISHERS
LONDON and NEW YORK

APPLIED SCIENCE PUBLISHERS LTD
Ripple Road, Barking, Essex, England

Sole Distributor in the USA and Canada
ELSEVIER SCIENCE PUBLISHING CO., INC.
52 Vanderbilt Avenue, New York, NY 10017, USA

British Library Cataloguing in Publication Data

International Conference on Composite
Structures (*2nd: 1983: Paisley College
of Technology*)
Composite structures 2.
1. Composite materials—Congresses
2. Composite construction—Congresses
I. Title   II. Marshall, I. H.
624.1'8   TA664

ISBN-13 : 978-94-009-6642-0   e-ISBN-13 : 978-94-009-6640-6
DOI : 10.1007 / 978-94-009-6640-6

WITH 62 TABLES AND 309 ILLUSTRATIONS

Preface

The papers contained herein were presented at the Second International Conference on Composite Structures (ICCS/2) held at Paisley College of Technology, Paisley, Scotland, in September 1983. The Conference was organised and sponsored by Paisley College of Technology in association with the Scottish Development Agency and the National Engineering Laboratory. It forms a natural progression from the highly successful First International Conference on Composite Structures (ICCS/1) held at Paisley in September 1981.

The last few decades have seen phenomenal advances in research and development of composite materials with new and exciting structural possibilities being unearthed on an almost daily basis. Composites have been rightly heralded as space-age materials of the future. However, along with the rather specialised aerospace applications a growing awareness of the wider potential of composites is also unmistakable. The extensive composite materials research programmes of the fifties and sixties are now yielding fruit in abundance, with composites being used in virtually every area of structural engineering from transportation to pressure vessels and so on. Although significant weight savings, paramount in transportation engineering, are possible, composites have gone far beyond being simply lighter than conventional materials. They offer real structural advantages with almost unbounded potential. The ability to tailor a particular matrix material to suit prevailing environmental conditions whilst maintaining adequate reinforcement to withstand applied loading is unquestionably an attractive proposition.

There is growing evidence that today's advances in developing new composite materials are finding almost immediate practical applications in structural engineering. Composites can no longer be considered as rather specialised esoteric materials, only of interest to researchers in materials science. Rather, they have firmly established themselves as real and viable materials of construction with wide-ranging applications. A measurement of the rate of advance in structural engineering can be found by comparing the present volume with that pertaining to ICCS/1. In a space of only two years, tremendous strides are evident. With this in mind and with a conscious need for dissemination of knowledge between users, manufacturers, designers and researchers involved in composite structural engineering, the present series of international conferences was organised.

Authors from fourteen countries combine with delegates from virtually every major industrial nation in the world to make this conference, once again, a truly international gathering of specialists in an ever-expanding technology. Topics under discussion range from performance studies in natural fibre composite structures to thermal control of composite structures in outer space.

An international conference can only succeed in making a contribution to knowledge through the considerable efforts of a number of enthusiastic and willing individuals. In particular, thanks are due to the following:

The Conference Steering Committee

Professor J. Anderson	Paisley College of Technology
Dr W. S. Carswell	National Engineering Laboratory
J. Gleave	Scottish Development Agency
C. L. Phillips	Scott-Bader Ltd
Dr J. Rhodes	University of Strathclyde
Dr E. J. Smith	Pilkington Brothers Ltd
J. A. Wylie	Paisley College of Technology

The International Advisory Panel

Dr E. Anderson	Battelle Laboratories, Geneva (Switzerland)
Dr W. M. Banks	University of Strathclyde (UK)
Professor A. R. Bunsell	Ecole des Mines de Paris (France)
Professor T. Hayashi	Chuo University, Tokyo (Japan)
Professor R. M. Jones	Virginia Polytechnic Institute and State University (USA)
L. N. Phillips, OBE	Royal Aircraft Establishment (UK)
Professor S. W. Tsai	Air Force Materials Laboratory, Ohio (USA)

The Local Organising Committee
G. Macaulay
J. S. Paul
J. Kirk
F. J. Allan
The Conference Secretary, Mrs C. MacDonald.

Grateful thanks are due to many other individuals who contributed to the success of this event. A final thanks to Nan, Simon and Louise for their support during the conference.

I. H. MARSHALL

Contents

ix

Session V: Finite Element Studies
(*Chairman:* E. ANDERSON, *Battelle Laboratories, Geneva, Switzerland*)

Session VI: Structural Analysis: Structural Systems
(*Chairman:* G. J. TURVEY, *University of Lancaster, England*)

Session VII: Research and Development
(*Chairman:* A. R. BUNSELL, ¡*Ecole des Mines de Paris, France*)

Session VIII: Design
(*Chairman:* W. M. BANKS, *University of Strathclyde, Glasgow, Scotland*)

Session IX: Experimental Investigations
(*Joint Chairmen:* H. F. BRINSON, *Virginia Polytechnic Institute and State University, Blacksburg, U.S.A.* and F. L. MATTHEWS, *Imperial College, London, England*)

1

The Monitoring of Damage in Carbon Fibre Composite Structures by Acoustic Emission

A. R. Bunsell

*Ecole Nationale Supérieure des Mines de Paris,
Centre des Matériaux, BP 87, 91003 Evry Cédex, France*

ABSTRACT

Acoustic emission monitoring provides a means of following the accumulation of damage in carbon fibre reinforced resin structures. The identification of the sources of emissions presents difficulties and amplitude analysis of the emissions has been suggested as a means of doing so. The acoustic activity recorded during steady loading is shown to be reproducible and a model of damage accumulation which explains the behaviour is proposed. This approach offers the possibility of calculating minimum lifetimes or of developing proof testing techniques for certain structures in carbon fibre reinforced resin.

INTRODUCTION

Carbon fibre reinforced epoxy resin is being used in an increasingly wide variety of structures as it gains acceptance in design offices and amongst engineers. The cost of the composite remains however relatively high, although it is falling, which means that carbon fibre reinforced plastics (CFRP) are used mainly for high performance structures for which high reliability is required. In cases where the composite is used in primary structures such as in aircraft or rotating machinery unforeseen failure would be disastrous. Despite the increased experience and increasing confidence with the composite there remains the suspicion that an unforeseen failure could occur after prolonged periods of steady or cyclic

1

loading. Studies have shown that CFRP has very good fatigue properties[1,2] and that creep when loaded in the direction parallel to the fibres is undetectable;[3] however it is known that the occasional sudden failure can occur when the composite is subjected to prolonged high loading.[4] There is therefore a need to be able to detect the progressive deterioration of the composite and predict failure or a minimum safe lifetime. An associated problem is the need to develop proof tests for CFRP structures so that they can be subsequently used with confidence.

Carbon fibres are extremely fine having diameters of about 7 μm so that any section of a CFRP structure is likely to reveal hundreds of thousands, if not millions, of fibres packed into the resin matrix. The number and fineness of the fibres combined with the opacity of the composite prevents any direct means of monitoring internal damage accumulation. In addition failure in these structures is often associated with dispersed damage and not with the development of one major flaw or crack, as is often the case with metals. For this reason the usual proof test for metal structures involving an overload seems unsuitable for composite structures. An overload, if it does not produce failure, causes plastic deformation around the tip of a crack in a metal so that further use at lower loads does not induce crack growth, the cracks being arrested by the plastic zones. There exists therefore physical reasons why such a proof test should be applicable to metal structures but there are no such arguments for applying the same test to composite materials. In the absence of geometrical constraints producing localised stress concentrations, damage in carbon fibre reinforced composites can be expected to be of a global rather than local nature, particularly if the fibres are positioned so as to dominate composite behaviour and control failure, as is usually the case in filament wound structures. Failure of this type has been modelled by Rosen[5] and Zweben.[6]

Conventional techniques such as extensometry having revealed themselves as being incapable of detecting the evolution of internal damage in CFRP, particularly when loaded parallel to the fibres, and faced with the knowledge that delayed failure can occur, the acoustic emission technique presents itself as an indirect means of doing just that.

THE ACOUSTIC EMISSION TECHNIQUE

Processes which produce vibrations in the composite can be detected by one or more transducers, coupled to the specimen surface, which respond to the surface waves which are created. The transducers can be of a variety of types, wire or piezoelectric strain gauges, piezoelectric resonant transducers, laser and infrared techniques.[7,8] The role of the transducer is to convert the vibrations created by the internal mechanisms into an electrical signal which can then be processed and analysed.

Before arriving at the transducer the stress wave may have undergone considerable modification due to attenuation or reflection inside the specimen as well as the anisotropic response of the material. The response of the transducer is usually not perfect so that the signal which is processed cannot be assumed to be a true replica of the original stress wave. These changes to the acoustic signal pose real problems for source identification and despite many claims to the contrary it remains unlikely that positive identification of failure mechanisms can be made by the technique. Because of this difficulty and the indirect nature of the technique it would be preferable if acoustic emission monitoring was accompanied by other means of damage detection. In the case of transparent composites such as glass reinforced resin optical methods may be used but no such technique seems applicable to CFRP. Acoustic emission is an imperfect technique for monitoring damage in CFRP structures but the best we have at present.

The most common acoustic emission system employs a PZT–piezoelectric ceramic transducer which is coupled to the structure to be monitored with silicone grease or similar acoustic coupling agent. These transducers respond mechanically to the vibrations at the surface and develop electric charges on their opposing faces. The voltage thus produced, usually of the order of a few microvolts, can be monitored by associated electronic equipment and the signal subsequently analysed. The mechanical response of the transducer depends on the physical properties of the ceramic, the constraints on its free deformation due to the surrounding support and its geometry. The most commonly used transducer has a nominal mechanical resonant frequency around 150 kHz and so responds to the vibration by ringing like a bell.

The electrical signal produced is then amplified by a high input impedance preamplifier with a gain of about 40 dB placed close to the transducer to avoid capacitive pick up, attenuation and interference of the original signal. The electrical signal then passes through a spectral filter which cuts off low frequency signals (e.g. < 100 kHz) which may be due to machine noise. The upper band of the filter cuts off high frequency signals (e.g. > 1 MHz) to eliminate interference spikes. The signal passes to an amplifier with a variable gain, normally up to 100 dB. A discriminator then produces pulses for all signals exceeding the set threshold level and these can then be summed by a counter. The output can be either in the form of

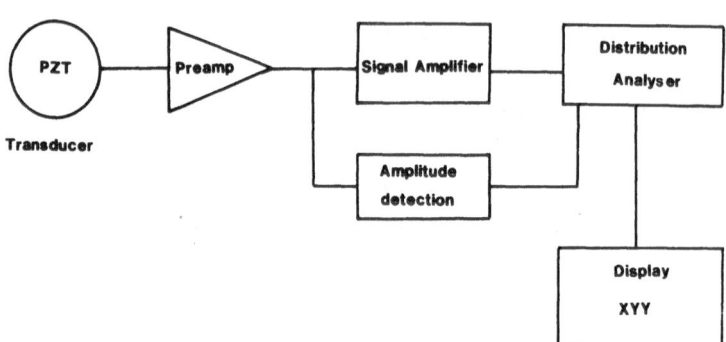

FIG. 1.   Schematic representation of an acoustic emission system.

an accumulative count or a rate of counting by summation over fixed periods and periodic resetting to zero of the counter. Figure 1 gives a schematic view of an acoustic emission system, and Fig. 2 shows the response of the transducer.

There are other means of processing the acoustic signal, for example the energy of each signal can be estimated by recording the square of the signal amplitude. The emissions can be analysed to obtain the distribution of signal amplitudes obtained during a test and also spectral analysis by fast Fourier transformation of the signals frequencies. These techniques have been discussed by a number of authors.[9-12] After initial enthusiasm for spectral or frequency analysis of acoustic emissions this has been superseded by amplitude analysis as being easier to interpret.

The amplitudes of the emissions recorded during tests on composites vary by several orders of magnitude. The amplitude distributions can be described in two ways, as a differential distribution $g(v)$ defined as the

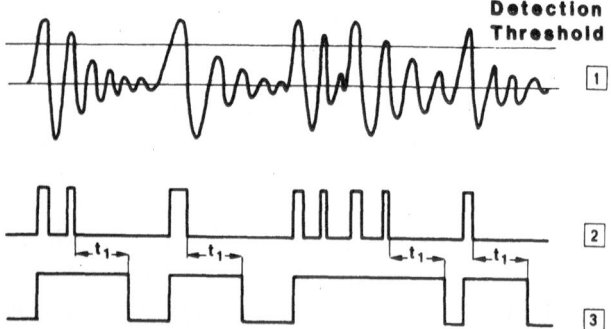

FIG. 2.   (1) Electrical signal obtained from piezoelectric transducer after amplification. (2) Digitised signal used for emission counting. (3) Signal recorded at the output of the envelope generator with a dead time t_1.

number of events having amplitudes of the value V (measured in dB or volts), or as an accumulative distribution $f(V)$ defined as the number of events which are greater than V. The two functions are related by:

$$g(v) = \frac{df(V)}{dV} \tag{1}$$

Pollock[13] has proposed an empirical relationship:

$$f(V) = N_0 \left(\frac{V}{V_0}\right)^{-b} \tag{2}$$

where V_0 is the smallest detectable amplitude for the monitoring system employed. N_0 is the total number of emissions exceeding the threshold V_0 and b is a parameter characterising the distribution function.

Two useful pieces of equipment which help in understanding the acoustic emission activity are a video tape recorder, although this is limited to fairly short tests and is restricted to a 4 MHz band width, and a transient recorder with a high digitising speed and long time window coupled to a digital computer enabling fast Fourier transforms.[7,8]

The use of several transducers allows sources of emissions to be located by analysing the time of flight of the emission to each transducer. This technique has been extensively and successfully used in fault detection in many types of pressure vessel.[14] Similarly guard transducers placed at either side of the regions of the structure under scrutiny can eliminate the effects of extraneous vibrations coming from outside that zone.

SOURCES OF EMISSIONS

In a useful review on acoustic emission monitoring of fibre composite materials and structures Williams and Samson[15] suggest the following possible sources for emissions:

(I)   fibre and matrix fracture;
(II)   interfacial debonding;
(III) relaxation of the fibres after failure;
(IV)   fibre pull-out;
(V)   structural flaws—intra- and interlaminar cracks;
(VI)   stress concentrations associated with specimen geometry.

Whilst the above list would not be disputed by other authors most restrict their considerations to fibre fracture, matrix cracking and interfacial

debonding.[16–18] Others, perhaps realising the difficulties of separating
mechanisms write only of fibre fracture and matrix failure.[19,20]

Although it is important to understand the possible sources of emissions
from composite structures it is paramount to realise that emissions may be
monitored which can come from outside the structure, such as grip or
machine noises and emissions generated by the material which arise from
processes which do not effect ultimate strength or behaviour. Fuwa *et al.*[11]

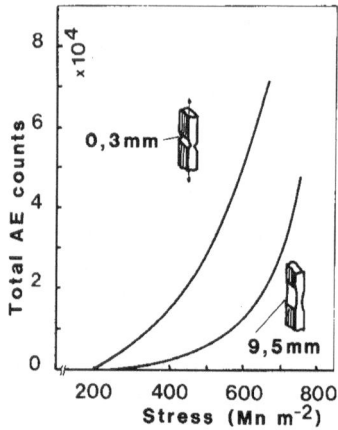

FIG. 3.   Acoustic emission from notched unidirectional CFRP specimens. The dimensions
are those of the radii of curvatures of the notches.

report on tensile tests of waisted unidirectional CFRP specimens. The
waisting of the specimens was in the thickness direction. Two radii of
curvature were used and as can be seen from Fig. 3 each type of specimen
produced distinctly different acoustic emission. The specimens with the
sharper notch started to emit earlier and produced more emissions when
compared to the activity recorded from the other type of specimen.
However the two types of specimen had indistinguishable tensile strengths
and the strengths obtained were the same as that found with unnotched
specimens of cross section equal to the minimum section of the waisted
specimens. Closer inspection revealed that many emissions were being
generated at the root of the notches, by shear failure of the composite
producing cracks parallel to the fibre direction. These cracks relieved the
stress concentration at the crack tip and the large fracture surfaces
produced parallel to the fibres generated large numbers of emissions.
Longitudinal cracking occurred earlier with the sharper notches because of
the higher stress concentration. The strength of this type of specimen is

notch insensitive so that despite very different acoustic emission patterns the strengths were similar. The emission produced by delamination masked the emissions coming from the failure mechanisms which determined the ultimate failure of the composite.

This last observation is important as there is considerable controversy in the literature as to the relative strengths of emissions coming from fibre breaks and matrix failure. It has been shown that delamination during fatigue produces a lot of emissions due to crack propagation in the matrix and that the acoustic emission can be used to characterise the damage which is produced.[21-23] Some fundamental fracture mechanisms which occur in advanced filament reinforced composites were described by Mullin *et al.*[24] They concluded that when the composite was loaded in the fibre direction it failed because of mechanisms controlled by the fibres. Three possibilities were envisaged, all induced by loading of the fibre by shear deformation of the matrix:

(1) With sufficiently high bond and matrix strength the fibre can fail in tension producing a disc-shaped crack in a plane normal to the fibre axis and passing through the point of fracture.

(2) If the bond is insufficient to load the filament to failure it may unzip at the interface from the ends of the fibre which become progressively unbonded as loading continues.

(3) If the bond strength is sufficient to transfer the load to the filament and the matrix is weak in tension a tensile crack may propagate in the high shear load transfer region in the matrix near the fibre ends.

Two possibilities are considered for the first type of failure depending on the response of the matrix. On fracture the fibre releases elastic energy which can be absorbed by the matrix in the immediate vicinity of the break or it can create the disc-shaped crack. Low strength fibres or an extensible, crack insensitive matrix would be expected to produce low energy pulses whereas strong fibres producing disc cracks in the matrix would be expected to produce high energy pulses. The effect of close fibre packing could be expected to modify the generation of matrix cracks as would loading rate. A high strain rate would be more likely to produce matrix cracking as at low strain rates the matrix would have more time to accommodate the changing stress field induced by the fracture.

The original work described by Mullin *et al.*[24] referred to studies on single or several boron fibres embedded in the matrix. A later study[25] found analogous behaviour in CRFP. The type of cracking produced depended on the Young's modulus of the fibres used and the rigidity of the matrix.

Failure of the specimens was seen to be initiated by fibre breakage and both disc cracks, normal to the fibre, and debonding were observed.

Most authors state that the most energetic emission originates from fibre failure and that matrix cracking is of an inherently lower energy[16,20] but it should be noted when considering CFRP that most of the evidence for this comes from studies on boron and glass fibre composites. Boron fibres have a cross section which is 400 times greater than that of carbon fibres, their usual respective diameters being 140 μm and 7 μm so that the break of a boron fibre is inevitably a major event. Glass fibres usually have rather greater diameter (10–20 μm) and have a much lower Young's modulus and a greater strain to failure when compared to carbon fibres. The elastic strain energy release rate due to fibre failure is given by:[26]

$$G_c = 0.32(\sigma_f^2 V_f / 2E_f) \tag{3}$$

$$\sigma_f^2 = (G_m / E_f d_f^2)[8/(1/V_f^{1/2} - 1)]$$

where G_c is the critical energy release rate; σ_f the fibre breaking stress; E_f Young's modulus of the fibre; G_m shear modulus of the matrix; d_f fibre diameter; and V_f fibre volume fraction.

The energy release rate for a crack running parallel to the fibres is given by:

$$G_c = K_c^2 \left(\frac{S_{11}S_{22}}{2}\right)^{1/2} \left[\left(\frac{S_{11}}{S_{22}}\right)^{1/2} + \frac{2S_{12} + S_{66}}{2S_{22}}\right]^{1/2} \tag{4}$$

where K_c is the critical energy release rate and S_{ij} the lamina's compliances $(i, j = 1, 2, 6)$.

Equation (3) shows that the energy release rate on fibre breakage is strongly dependent on fibre diameter and the smaller it is the less energetic will be the event. The energy release rate for cracks parallel to the fibres is independent of the fibre diameter. In short it cannot be assumed that high amplitude emissions from CFRP originate at fibre breaks even if it is the case for other composites.

A comparison of acoustic emission obtained from three lots of specimens all containing the same number of fibres but with different matrix conditions suggested strongly that the major source of emissions from unidirectional CFRP is the fracture of fibres.[7] Specimens with which the resin had been removed so as to leave only the fibre bundle produced emission early during tensile loading and the only possible source was the failure of fibres. Those specimens which were of uncured prepreg produced a similar pattern of emission and as the matrix was viscous the only source

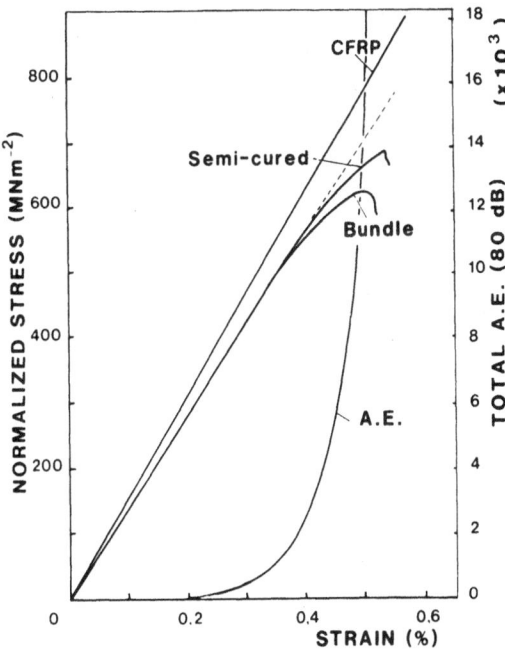

FIG. 4. Typical stress–strain curves of cured and semi-cured CFRP and bundles of high modulus carbon fibre with the acoustic emission curve obtained with the curved specimens.

of emissions was again fibre breaks. The cured specimens exhibited similar acoustic emission so that by extension from what had been observed with the bundle and uncured specimens it was concluded that fibre failure was the most important source of emission. Figure 4 shows the tensile behaviour of each type of specimen.

A study of failure mechanism discrimination in unidirectional and crossplied CFRP specimens gave some encouragement to signal amplitude analysis.[28] Unidirectional specimens loaded at different angles with respect to the fibre direction will fail by one of three mechanisms depending on the angle. When loaded parallel to the fibres it is their failure which dominates, at angles up to about 45° shear failure parallel to the fibres occurs and at greater angles the matrix fails due to the tensile forces normal to the fibres. Figure 5 shows the cumulative amplitude distribution of the acoustic emission obtained at different angles. The parameter *b* is the gradient of the curves and it can be seen to decrease with increase in angle. Figure 6 shows that the variation of *b* reflects quite well the anisotropic mechanical behaviour of the unidirectional specimens. Also in Fig. 6 is the variation in *b* for (0°, 90°) crossplied specimens. It is interesting that the curve is not

A. R. Bunsell

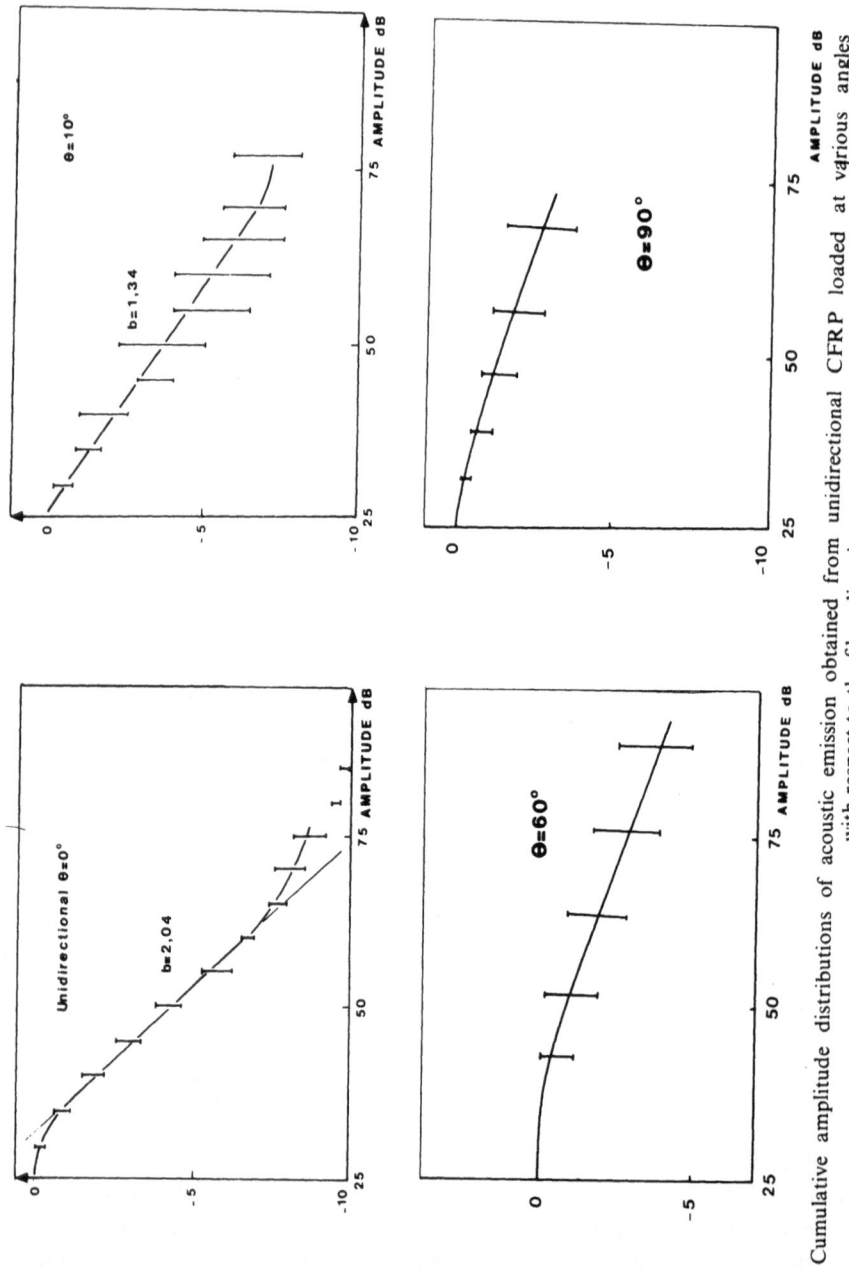

Fig. 5. Cumulative amplitude distributions of acoustic emission obtained from unidirectional CFRP loaded at various angles with respect to the fibre direction.

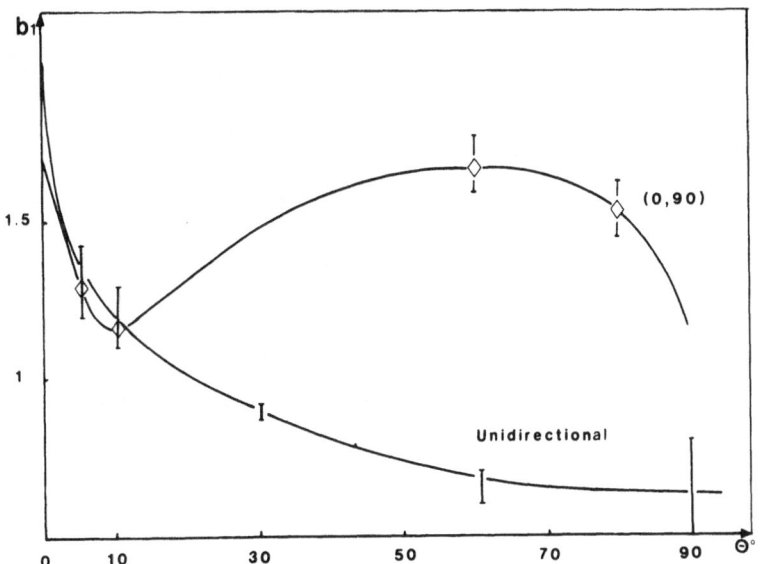

FIG. 6.   Variation of the coefficient *b* for unidirectional and (0°, 90°) crossply specimens.

symmetrical, which it would be expected to be at first sight. The mechanisms which dominate the failure of unidirectional specimens at different angles are known. It can therefore be seen that fibre breakage gave rise to low amplitude emissions and that the larger amplitude emission became relatively more important as the loading angle increased, revealing that they originated at matrix cracking parallel to the fibres. The behaviour of the crossplied specimens can be understood when it is remembered that at 0° the outer layers of the specimen contained fibres aligned parallel to the applied load whereas at 90° those layers parallel to the loading direction were sandwiched between layers in which the fibres were at right angles to the load. At 0° interlaminar shear did not play a role, tensile loading of the matrix in the central layers with fibres at 90° did exist but was suppressed by the high modulus fibres in the outer layers. Loading of the 0° direction for the crossplied specimens was therefore very similar to the unidirectional case and was dominated by fibre failure. At 90° the outer layers containing fibres at right angles to the load were less restrained, and tensile failure of the matrix could occur so producing the asymmetric curve in Fig. 6.

It has been shown that strain rate could be expected to influence the acoustic emission activity and that matrix cracking would be particularly sensitive to loading rate.[24] Rotem[29] has shown that whereas unidirectional glass fibre reinforced epoxy produces emission which is strain rate sensitive

A. R. Bunsell

this is not the case for unidirectional CFRP. It is therefore concluded that matrix cracking plays a much greater role in the failure of glass fibre composites than in CFRP in which fibre failure dominates. The difference in the behaviour of the two composites is attributed to the greater breaking strain of the glass fibres inducing high shear stresses at failure in the surrounding resin. As carbon fibres with increasingly high strains to failure are being produced it is conceivable that a change in acoustic emission response may be detected although the fineness of the fibres probably will mitigate against the effect. Otsuka and Scarton[30] reported few emissions from CFRP specimens compared to identical glass fibre composite specimens which agrees with the results mentioned above. These authors suggested however that the difference may have been due to limitations in their equipment which was not fast enough to detect all of the emissions coming from the CFRP.

THE KAISER EFFECT AND FAILURE PROCESSES

Many materials when reloaded a second time generate no acoustic emission until the previous maximum applied load has been exceeded. This is known as the Kaiser effect and there has been some discussion in the literature as to its applicability to carbon fibre reinforced epoxy composites.[31-33] The debate has been settled with general agreement that the Kaiser effect is nearly observed but not quite. Reloading unidirectional CFRP and also many filament wound CFRP structures results in re-emission just before the previously applied load is reached.[4,34]

If the CFRP structure is repeatedly loaded the number of emissions which are detected in each cycle falls regularly and the load at which emissions are recorded approaches more closely to the maximum. It seems therefore that a stabilisation or shake-down process is occurring although it seems that complete stabilisation of the structure does not occur. Figure 7 shows that after a period of constant load amplitude cycling an increased pressure in a filament wound structure does not result in an immediate increase of activity. At stresses slightly higher than the maximum previously applied no emissions are recorded but at still higher stresses the original acoustic emission curve is rejoined. Exactly analogous behaviour is seen after a period of steady loading[35] and these observations have considerable significance when considering the processes of failure in CFRP.

A model of failure in simple carbon fibre reinforced epoxy resin specimens has been suggested which describes the observed acoustic

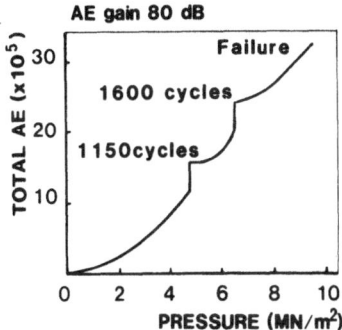

FIG. 7. Acoustic emission obtained from CFRP pressure vessels. Cyclic pressures result in emission near the maximum pressure each cycle and vertical traces for the curve. The numbers refer to the number of cycles conducted. Increasing the pressure after cycling produces no immediate emissions but at higher pressures the original curve is rejoined.

activity.[36] A unidirectional CFRP specimen loaded in the fibre direction is considered to behave as a fibre bundle and the role of the matrix serves to isolate individual fibre breaks in a narrow section of the composite. This fibre bundle chain model was first proposed by Rosen.[5] Each section or link of the chain acts as a short fibre bundle. It has been shown[36] that as fibres are broken in a bundle the load which can be applied to it without producing further breaks is described by:

$$P = (N_0^{\cdot} - N_f)f_0 \left(\ln \left(\frac{N_0}{N_0 - N_f} \right)^{1/\delta} \right) \tag{5}$$

where P is the load supported by the bundle; N_0 the total number of fibres in the bundle; N_f the number of broken fibres; f_0 a constant; and δ the Weibull shape parameter for the fibre strength distribution.

Equation (5) describes the curve shown in Fig. 8 and it can be seen that as damage increases the load which can be safely applied to the bundle passes through a maximum. Damage accumulation under steady loading conditions produces the vertical curve A_1A_2CB and, if the fibres are perfectly elastic, this damage must be due to stress redistribution induced by the viscoelastic properties of the matrix. The composite becomes unstable and fails when the damage reaches the level B. At loads less than the tensile breaking load of the composite it can sustain greater damage without failure than is produced up to just before failure in a simple tensile test. If loading is increased when the accumulated damage has reached the point A_2 then no emission will be recorded until the load P_2 is reached. This is the behaviour observed with many CFRP structures.

A. R. Bunsell

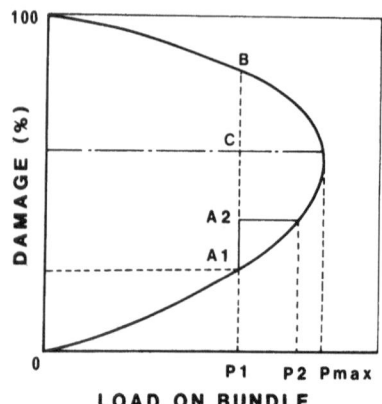

FIG. 8.   As fibres are damaged in a bundle the load which can be applied to it without producing unstable failure passes through a maximum. At a load $P_1 < P_{max}$ the damage which the bundle can sustain without breaking, B, is greater than that at P_{max}, C.

It is often assumed that acoustic activity should accelerate just before final failure[37] but the above model does not support this assumption unless other mechanisms such as splitting occur just before failure. The ultimate fracture of the composite occurs when one section becomes sufficiently weakened to fail and as there are many such sections or links any increased activity in one link will not be significant at the level of the whole structure. Laroche[35] linked an Instron tensile machine to the acoustic emission apparatus such that the loading rate was adjusted so as to maintain the acoustic emission activity constant as loading increased. As loading increased the rate of loading dropped dramatically as would be expected but failure still occurred abruptly with no acceleration of activity being detected.

LIFE PREDICTION AND PROOF TESTING

It has been suggested that the monitoring of acoustic emission could be a means of predicting the lifetimes or the proof testing of composite structures.[34,35–40] The representation of damage in CFRP shown in Fig. 8 suggests a way for predicting lifetimes but requires that the master enveloping curve of damage as a function of load be known for the composite and that the rate of damage accumulation be also known. It is not generally possible to obtain all of the master curve but a simple test to failure gives the first part of the curve from no damage up to damage

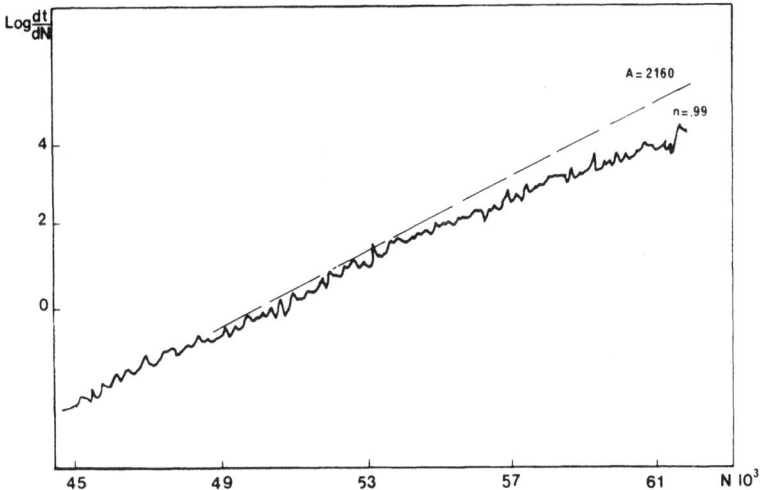

FIG. 9. The acoustic emission recorded from a unidirectional CFRP specimen loaded in the fibre direction follows an almost perfect logarithmic function of time so that $\ln(dt/dN)$ is nearly a straight line.

accumulated at fracture. The damage at failure can therefore be used as a conservative estimate of damage at failure for all loads lower than the ultimate strength of the specimen so permitting a minimum lifetime to be calculated.

Damage accumulation as shown by acoustic emission has been seen to obey the equation:

$$\frac{dN}{dt} = \frac{A}{(t+\tau)^n} \tag{6}$$

where N is the number of emissions; t the time; τ a time constant; A a parameter which depends only on the applied stress; n a power less than but nearly equal to unity.

For a unidirectional specimen $n = 0{\cdot}99$ and if we put $n = 1$ we can obtain:

$$\ln\left(\frac{dt}{dN}\right) = \ln\left(\frac{\tau}{A}\right) + \frac{N}{A} \tag{7}$$

so that $\ln(dt/dN)$ is a linear function of the total number of emissions. Figure 9 shows that eqn. (7) is obeyed by a loaded unidirectional CFRP specimen. The parameter A was found to be well represented by the equation:

$$A = \lambda e^{k\sigma} \tag{8}$$

where λ and k are constants and σ is the applied stress.

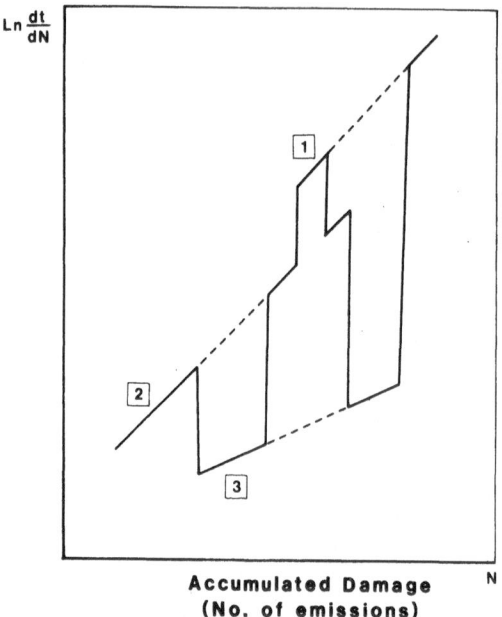

FIG. 10. Schematic representation of the effect on $\ln(dt/dN)$ of varying the load level between load 1, the lowest load, to level 3, the highest. Activity is seen to be a function of accumulated damage and independent of loading history. This is observed for load changes of the order of 30% of the breaking load (e.g. variations between 60% and 90% σ_c).

The value of A therefore varies with the applied stress and Fig. 10 shows the effect on $\ln(dt/dN)$ as a function of N, as the applied load was varied. When the load is increased there is an increase in acoustic activity so that $\ln(dt/dN)$ falls and a new straight line is obtained at the higher load. Returning to the previous lower load produces just the activity which would have been expected at the value of the total number of emissions accrued if the load had not been changed. This means that a period at a higher load was equivalent to an accelerated mechanical ageing. Increasing or reducing the applied load within 10 or 15% and then returning to the original load produces activity which is an extrapolation as a function of the accumulated emission of the initial behaviour.

It seems that from the observed behaviour a minimum lifetime and proof testing technique could be determined if preliminary tests were conducted on similar vessels in order to obtain the first part of the general master curve. Steady loading for periods at progressively higher loads would reveal the variation in A and permit the exact measurement of the parameter n which is found to vary with different fibre lay ups.[41] The use of eqn. (6)

would allow the time to accumulate any given number of emissions to be calculated.

The scenario of failure where large load variations occurs may be more complex than that which has been observed under steady loading conditions. This is because not only the viscoelastic nature of the matrix may be controlling the damage process but also its plastic deformation may contribute. The observed behaviour under cyclic loading is however at least analogous to that observed under steady loading.[4]

As the behaviour which has been described depends on the viscoelastic behaviour of the matrix, changes of the matrix or of the conditions of test which influence its properties can be expected to modify the acoustic activity. This has been found to be the case.[42]

CONCLUSION

The acoustic emission technique is an indirect means of monitoring the internal damage occurring in composite materials and does so where other techniques fail. Quantifying and interpreting the emissions remain difficult however and many studies confine themselves to a qualitative description or comparison of acoustic activity with the activity obtained from standard and supposedly perfect structures. Wherever the technique is employed all possible sources of spurious emissions must be identified. Carbon fibre reinforced plastic lends itself better to acoustic emission analysis than do some other composites as carbon fibres can be considered to be both perfectly elastic and brittle, and they do not creep or fail in fatigue. When loaded under steady loads the acoustic emission activity from many CFRP structures is reproducible and can be described mathematically. Confirmation of the sources of emissions in CFRP is generally not possible and although reasonable speculation is possible in some cases, proof is usually not. It seems reasonable however to believe that for certain structures proof testing or minimum life prediction techniques based on acoustic emission monitoring are feasible.

The acoustic emission signals recorded from a composite structure may be analysed and transformed in an ever increasing number of ways but due to the inherent uncertain nature of the emissions and their sources doubts must be raised as to the ultimate use of these procedures. Whilst relatively simple acoustic emission monitoring and analysis is of undoubted, particularly technological, use, it remains true that a structure which is not emitting represents the best situation which could be interpreted as it would mean that most probably no damage of any serious nature would be occurring. As it is, a conservative approach is required which assumes that all emissions are generated by failure mechanisms and that if activity is continuing the structure is approaching closer to its critical failure level.

REFERENCES

1. OWEN, M. J. and MORRIS, S., Fatigue resistance of carbon fibre RP, *Modern Plastics*, April, 1970.
2. DHARAN, C. H. K., Fatigue failure in graphite fibre and glass fibre polymer composites, *J. Mat. Sci.*, **10**, 1975, 1665.
3. STURGEON, J. B., BUTT, R. I. and LARKE, L. W., Creep of carbon fibre reinforced plastics, R.A.E. Tech. Report 76168, 1976.
4. FUWA, M., HARRIS, B. and BUNSELL, A. R., Acoustic emission during cyclic loading of carbon fibre reinforced plastics, *J. Phys.*, D, **8**, 1975, 1460.
5. ROSEN, B. W., Tensile failure of fibrous composites, *AIAA Journal*, **2**, 1964, 1985.
6. ZWEBEN, K., Tensile failure of fibre composites, *AIAA Journal*, **6**, 1968, 2325.
7. WHITE, R. G. and TRÉTOUT, H., Acoustic emission detection using a piezoelectric strain gauge for failure mechanisms identification in cfrp, *Composites*, **10**, 1979, 101.
8. GREEN, R. E., Jr, Basic wave analysis of acoustic emission, in *Mechanics of Nondestructive Testing*, Blacksburg, Va, 10–12 Sept., Plenum Press, pp. 55–76, 1980.
9. POLLOCK, A. A., Stress wave emission in nondestructive testing, *Nondestructive Testing*, **2**, 1969, 178.
10. SPEAKE, J. H. and CURTIS, G. J., Characterisation of the fracture processes in CFRP using spectral analysis of the acoustic emission arising from the amplification of stress, *Int. Conf. on Carbon Fibres, Their Place in Modern Technology*, London, 1974, Paper 29.
11. FUWA, M., BUNSELL, A. R. and HARRIS, B., An evaluation of acoustic emission techniques applied to carbon fibre composites, *J. Phys.*, D, **9**, 1976, 363.
12. GUILD, F. J., WATTON, D., ADAMS, R. D. and SHORT, S., The application of acoustic emission to fibre reinforced composite materials, *Composites*, **7**, 1976, 173.
13. POLLOCK, A. A., Acoustic emission amplitudes, *Nondestructive testing*, **6**(5), October 1973, 63.
14. RYDER, J. T. and WADIN, J. R., Acoustic emission monitoring of a quasi-isotropic graphite/epoxy laminate under fatigue loading, *ANST Spring Conference*, San Diego, 1979, 9.
15. WILLIAMS, J. H. and SAMSON, S. L., Acoustic emission monitoring of fibre composite materials and structures, *J. Comp. Mat.*, **12**, 1978, 348.
16. MEHAN, R. L. and MULLIN, J. V., Analysis of composite failure mechanisms using acoustic emission, *J. Comp. Mat.*, **5**, 1971, 266.
17. LIPTAI, R. G., Acoustic emission from composite materials, *Composite Materials Testing and Design*, ASTM-STP 497, 1973, 285.

18. SWINDLEHURST, W. E. and ENGEL, C., A model for acoustic emission generation in composite materials, *Fibre Science and Technology*, **11**, 1978, 463.

19. ALTUS, E. and ROTEM, A., The characteristics of acoustic emission pulse from fibre reinforced composite, *Israel J. Technol.*, **15**, 1977, 79.

20. BECHT, J., SCHWALBE, H. J. and EISENBLAETTER, J., Acoustic emission as an aid for investigating the deformation and fracture of composite materials, *Composites*, **7**, 1976, 245.

21. DE CHARENTENAY, F. X. and BENZEGGAH, M., Fracture mechanisms of mode I delamination in composite materials, in *Advances in Composite Materials*, Vol. 1, Bunsell, A. R. *et al.* (eds), Pergamon Press, 1980.

22. DE CHARENTENAY, F. X., KAMIMURA, K. and LEMASÇON, A., Fatigue delamination in unidirectional carbon-epoxy composites, *Materials, Experimentation and Design in Fatigue*, 199, Westbury House, 1981.

23. FLITCROFT, J. E. and ADAMS, R. D., A study of shear crack propagation in glass and carbon fibre reinforced plastics using acoustic emission monitoring, *J. Phys.*, D, **15**, 1982, 991.

24. MULLIN, J. V., BERRY, J. M. and GATTI, A., Some fundamental fracture mechanisms applicable to advanced filament composites, *J. Comp. Mat.*, **2**(1), 1968, 82.

25. MULLIN, J. V. and MAZZIO, V. F., A comparative study of tensile fracture mechanisms, *J. Comp. Mat.*, **6**, 1972, 268.

26. ROTEM, A. and ALTUS, E., Fracture modes and acoustic emission of composite materials, *J. Testing Eval.*, **7**(1), 1979, 33.

27. FUWA, M., BUNSELL, A. R. and HARRIS, B., Tensile failure mechanisms in carbon fibre reinforced plastics, *J. Mat. Sci.*, **10**, 1975, 2062.

28. VALENTIN, D., BONNIAU, P. and BUNSELL, A. R., Failure mechanisms discrimination in carbon fibre reinforced epoxy, To be published in *Composites*.

29. ROTEM, A., Effect of strain rate on acoustic emission from fibre composites, *Composites*, **9**, 1978, 53.

30. OTSUKA, H. and SCARTON, H. A., Variations in acoustic emission between graphite and glass epoxy composites, *J. Comp. Mat.*, **5**, 1981, 591.

31. KIM, H. C., NETO, R. and STEPHENS, R. W. B., Some observations on acoustic emission during continuous tensile cycling of a carbon fibre epoxy composite, *Nature Physical Science*, **273**, 1972, 78.

32. STONE, E. W. and DINGWALL, P. F., The Kaiser effect in stress wave emission testing of carbon fibre composites, *Nature Physical Science*, **241**, 1973, 68.

33. KIM, H. C., NETO, R. and STEPHENS, R. W. B., Reply by Kim, Neto and Stephens, *Nature Physical Science*, **241**, 1973, 70.

34. BUNSELL, A. R., Acoustic emission for proof testing of carbon fibre reinforced plastics, *NDT International*, 1977, 21.

35. LAROCHE, D. and BUNSELL, A. R., Stress and time dependent damage in carbon fibre reinforced plastics, *Advances in Composite Materials*, Vol. 2, Bunsell, A. R. *et al.* (eds), 985, Pergamon Press, 1980.

36. BUNSELL, A. R., LAROCHE, D. and VALENTIN, D., Damage and failure in carbon fibre reinforced epoxy resin, *ASTM-STP Long Term Behavior of Composites*, to be published.

37. HOLT, J. and WORTHINGTON, P. J., A comparison of fatigue damage detection

in carbon and glass fibre-epoxy composite materials by acoustic emission, *Int. J. Fatigue*, 1981, 31.

38. HAMSTAD, M. A., Acceptance testing of graphite-epoxy composite parts with acoustic emission, *NDT International*, Dec. 1982, 307.

39. VALENTIN, D. and BUNSELL, A. R., A study of damage accumulation in carbon fibre reinforced epoxy resin during mechanical loading monitored by acoustic emission, *J. Reinforced Plastics and Composites*, to be published.

40. HAGEMAIER, D. J., McFAUL, H. J. and MOON, D., Non destructive testing of graphite fibre composite structures, *Materials Evaluation*, 1971, 133.

41. VALENTIN, D. and BUNSELL, A. R., Damage in carbon fibre reinforced epoxy resin produced during cyclic loading and monitored by acoustic emission, in *Fatigue and Creep of Composite Materials*, Lilholt, H. and Talreja, R. (eds), RISØ, 1982.

42. VALENTIN, D. and BUNSELL, A. R., The modelling of failure processes and the role of the matrix in the failure of carbon fibre reinforced epoxy resin, in *Progress in Science and Engineering of Composites*, Hayashi, T., Kawata, K. and Umekawa, S. (eds), Proceedings ICCM/4, Tokyo, Japan. Oct. 1982.

2

Biaxial Failure of GRP—Mechanisms, Modes and Theories

M. J. Owen

Department of Mechanical Engineering, University of Nottingham,
Nottingham NG7 2RD, England

ABSTRACT

Numerous failure theories have been proposed for GRP. Experimental observations under biaxial loading reveal scatter, multiple failure mechanisms and failure modes, which depend on material type and stress conditions. Biaxial failure theories can be represented as surfaces whose shape depend on both failure theory and the choice of single valued characteristic strengths. Experimental results suggest surfaces which are quadratic functions of the stresses and strengths and which often lie well inside the maximum normal stress boundaries. Failure theories which use a complex stress test to evaluate an interaction coefficient are generally unacceptable because the resulting surfaces are so sensitive to small changes in strength data. It is necessary to consider different classes of reinforcement (unidirectional, woven fabric, and random mat) separately in proposing failure theories. For unidirectional materials quadratic theories only appear to be well defined under tension–tension–shear conditions. For mats and fabrics adaptations of the early Norris theories fitted separately in each stress octant appear to be satisfactory. Failure theories only predict material failure as distinct from structural failure and should be treated with caution when applied to design.

INTRODUCTION

The mechanical properties of materials are usually determined from specimens tested under simple loading states from which the elastic constants, the onset of damage, and the ultimate strength can be determined. If enough tests are conducted the variability of these parameters can be established. Under repeated or sustained loading, fatigue or creep properties can also be established. GRP consist of arrays of closely spaced elastic fibres in a solidified resin matrix. The properties of such a material are highly directional. However, in most applications the reinforcements are used in the form of multi-layered laminates, woven fabrics, or random mats to meet complex loading conditions. For the purposes of analysis GRP are assumed to be homogeneous with directional and possibly layered properties. Standard methods of test in the tensile, compressive, or flexural modes use narrow specimens usually cut parallel to a principal material axis. Thus they are usually fibre controlled, but for the coarser types of reinforcement are of questionable homogeneity.

The designer is concerned to predict the deformation and strength of structures from such properties, possibly with supplementary data from tests on coupons (representative elements) or structural subassemblies. Real structures are subjected to loading which produces multi-axial stressing throughout the material usually of a sustained nature and with varying magnitudes. GRP structures are usually thin and as a first approximation may be considered to be subjected to plane stresses which can be transformed to the principal material axes as normal stresses σ_1, σ_2, and in-plane shear stress σ_6. This approximation glosses over the existence of interlaminar shear stresses σ_4 and σ_5, arising due to local variations in bending moments or due to strain compatibility requirements between layers at cut edges. Occasionally the third normal stress σ_3 is also present.

The prediction of deformation and stress usually assumes linear elasticity and is based on the orthotropic form of Hooke's Law and laminate analysis.[1] Comparisons between finite element analysis and strain gauge analysis usually give reasonable agreement, although it may be necessary to allow for non-linear elasticity in shear.[2] The prediction of strength is more complex and it is necessary to define 'failure' and to distinguish between material failure and structural failure. Structural failure may be caused by excessive deflection, by buckling, by localized damage (peeling, delamination, etc.) or some form of crack propagation. Material failure is regarded as due to the onset of damage and separation in a recognizable tensile, compressive or in-plane shear mode although the separate stress components may produce interactive effects under complex loading. This paper concentrates on material failure, i.e. on the prediction of material strength under biaxial stress conditions.

FAILURE THEORIES

Failure theories are functions of the stresses and strengths of the material which are assumed to represent failure under all loading conditions without regard to failure mechanism or failure mode. For isotropic materials there are three well-known strength theories, maximum principal stress, maximum shear stress (Tresca), and distortional energy (Mises–Hencky). In each case a function of the stresses is equated with a single parameter, the tensile yield strength, or the fatigue strength of the material. Recognition of the fact that anisotropic materials have more than one strength parameter has led to numerous proposals for failure theories. More than 40 such theories have been proposed for anisotropic metals, wood, reinforced plastics, etc. There have been a number of reviews of the available theories.[3-6] Although most of the theories have at some time been applied to glass reinforced plastics, their application needs to be treated with great caution, especially when judging general validity from a few experiments.

Table 1 shows a small selection of the available theories specialized for plane stress. Group 1 do not require data from a complex stress test. Group 2 require complex stress data to evaluate an interaction coefficient. The latter include tensor theories which permit the strength parameters to be transformed to the stress axes. Many of the theories are quadratic functions of the stresses and strengths. Failure theories for plane stress involve σ_1, σ_2 and σ_6, and can be represented as surfaces in three-dimensional space (Fig. 1). Thorough investigation of a failure theory involves comparison of experimental data for complex stress loading with the chosen failure surface. This requires an experimental facility capable of changing the relative values of σ_1, σ_2 and σ_6 over a wide range.

All failure theories involve single-valued strength parameters. Standard tensile test methods are reasonably consistent but decisions have to be made on the damage state defining failure (rupture is the simplest), and the statistical measure chosen to represent scattered data, e.g. the mean or the A-allowable strength.[14] Compressive and shear test methods are far less consistent and the damage states and the mode of failure are different from tensile loading. The distribution of strength values will almost certainly be different from the tensile values. Thus the shape of the failure surface will depend both on the test method and the strength values chosen to represent scattered data. With the Group 2 failure theories, it is also assumed that a single combined stress test resulting in a single strength value will uniquely define the failure surface. In addition to the scatter problem, the resulting

TABLE 1
Failure theories

Key letter	Group
	I (do not require biaxial data)
A	*Maximum Stress*[8]

$$\sigma_1 = X \text{ or } X'$$
$$\sigma_2 = Y \text{ or } Y'$$
$$\sigma_6 = S$$

| B | *Azzi and Tsai*[9] |

$$\left(\frac{\sigma_1}{X}\right)^2 - \frac{\sigma_1\sigma_2}{X^2} + \left(\frac{\sigma_2}{Y}\right)^2 + \left(\frac{\sigma_6}{S}\right)^2 = 1$$

| C | *Norris and McKinnon*[10] |

$$\frac{\sigma_1^2}{X^2} + \frac{\sigma_2^2}{Y^2} + \frac{\sigma_6^2}{S^2} = 1$$

| D | *Norris Distortional Energy*[11] |

$$\frac{\sigma_1^2}{X^2} - \frac{\sigma_1\sigma_2}{XY} + \frac{\sigma_2^2}{Y^2} + \frac{\sigma_6^2}{S^2} = 1 \qquad \frac{\sigma_1^2}{X^2} = 1 \qquad \frac{\sigma_2^2}{Y^2} = 1$$

| | **II (require biaxial data)** |
| F | *Modified Marin*[12] |

$$\left(\frac{X'-X}{XX'}\right)\sigma_1 + \left(\frac{Y'-Y}{YY'}\right)\sigma_2 + \left(\frac{\sigma_1^2}{XX'}\right) - \frac{\bar{k}_2\sigma_1\sigma_2}{XX'} + \left(\frac{\sigma_2^2}{YY'}\right) + \frac{\sigma_6^2}{S^2} = 1$$

| G | *Gol'denblat and Kopnov (Tsai and Wu*[13] *interpretation)* |

$$\frac{1}{2}\left(\frac{1}{X} - \frac{1}{X'}\right)\sigma_1 + \frac{1}{2}\left(\frac{1}{Y} - \frac{1}{Y'}\right)\sigma_2 + \left\{\frac{1}{4}\left(\frac{1}{X} + \frac{1}{X'}\right)^2 \sigma_1^2\right.$$
$$\left. + \frac{1}{4}\left(\frac{1}{Y} + \frac{1}{Y'}\right)^2 \sigma_2^2 + 2F_{XY}\sigma_1\sigma_2 + \left(\frac{\sigma_6^2}{S^2}\right)\right\}^{1/2} = 1$$

| H | *Tsai and Wu*[13] |

$$\left(\frac{1}{X} - \frac{1}{X'}\right)\sigma_1 + \left(\frac{1}{Y} - \frac{1}{Y'}\right)\sigma_2 + \left(\frac{\sigma_1^2}{XX'}\right) + 2F_{XY}\sigma_1\sigma_2 + \left(\frac{\sigma_2^2}{YY'}\right) + \left(\frac{\sigma_6^2}{S^2}\right) = 1$$

$$\left(\text{and } \frac{1}{XX'} \cdot \frac{1}{YY'} - F_{XY}^2 \geq 0\right)$$

stability criterion

failure surface tends to depend on the stress ratio chosen to evaluate an interaction coefficient. It has been realized for many years that failure surfaces are very sensitive to the choice of combined stress data. Tsai and Wu[13] included a stability criterion to ensure that the failure surface is closed. A quadratic surface could be a cylinder or open hyperbolic surface. Intuitively a closed surface is expected so that failure is defined for all combinations of the stress components.

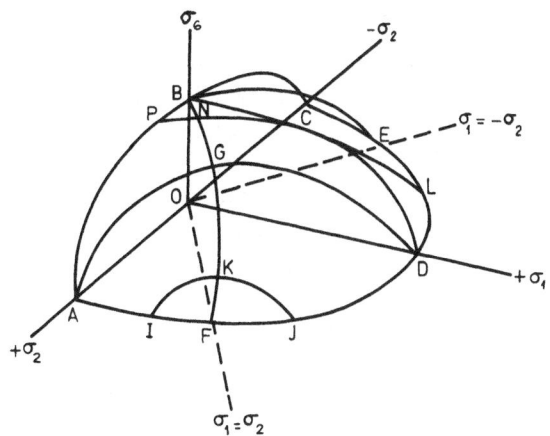

FIG. 1.   Failure surface in σ_1, σ_2, σ_6 space.[23]

TEST METHODS

Only three test methods have been seriously used to evaluate failure theories for composite materials: off-axis tension or compression,[3,24] thin-walled cylinders subjected to internal pressure and axial loading[3-5] or torsion and axial loading,[13] and flat cruciform specimens.[6] All three methods involve experimental difficulties. Off-axis specimens must be sufficiently long or have pin attached grips to allow shear deformation to occur without applying a moment. Thin-walled cylinders involve gripping problems especially diametral constraint at the ends, and they are also expensive to make. With prepared mat or fabric there are problems with reinforcement joints.[4] Cruciform specimens also require special test facilities and there have been difficulties with premature failures in the arms.[6]

UNIDIRECTIONAL GRP

Many high performance GRP laminates are made from hot pressed unidirectional prepreg sheet. Each layer of the resulting laminate is usually regarded as being subjected to plane stress and the designer's problem is that of predicting first ply failure. The characteristic tensile strengths of the fundamental lamina measured parallel and perpendicular to the fibres differ by almost two orders of magnitude, and the in-plane shear strength is generally of the same magnitude as the transverse tensile strength. The compressive strengths are less well defined. Based on the tensile and shear strengths, the failure surface must be a long slender rod-like shape and there are two basic failure modes—fibre failure and matrix/interface failure arising from a combination of transverse tension and shear.[15] Compressive modes of failure will be observed by structural failures such as buckling.

Tsai and Hahn[1] drew attention to the fact that the interaction coefficients are very difficult to establish for Group 2 failure theories. They therefore proposed to make the Tsai and Wu theory[13] consistent with the distortional energy theory of failure for isotropic materials, by assuming a value for the interaction coefficient.

$$F_{XX}\sigma_X^2 + 2F_{XY}\sigma_X\sigma_Y + F_{YY}\sigma_Y^2 + F_{SS}\sigma_S^2 + F_X\sigma_X + F_Y\sigma_Y = 1$$

Five of the six coefficients can be established from simple uniaxial test data (tension and compression in the two principal directions and in-plane shear). Hence

$$F_{XX} = \frac{1}{X} \cdot \frac{1}{X'} \qquad F_{YY} = \frac{1}{Y} \cdot \frac{1}{Y'} \qquad F_X = \frac{1}{X} - \frac{1}{X'}$$

$$F_Y = \frac{1}{Y} - \frac{1}{Y'} \qquad F_{SS} = \frac{1}{S^2}$$

where X, X' = longitudinal tensile and compressive strengths;
Y, Y' = transverse tensile and compressive strengths;
S = in-plane shear strength.

The Tsai and Wu stability criterion[13] is put in the form

$$F_{XY}^* = F_{XY}/(F_{XX} \cdot F_{YY})^{1/2} = -\tfrac{1}{2} \qquad \text{to define } F_{XY}$$

The Tsai and Hahn approach side-steps the issue of determining the shape of the failure surface. There seems to be no published evidence of fatigue investigations along these lines probably because aerospace designers appear to work on the basis of static design using residual strength values.[14]

Hashin and Rotem,[15] Rotem and Hashin,[16,17] and Rotem[18] simplified the approach of Puck and Schneider[19] and accepted from the outset that failure took one of two possible modes, fibre failure or matrix failure.

Under static loading failure is defined on an either/or basis. Either $\sigma_A = \sigma_A^s$ (fibre failure) or $[(\sigma_T/\sigma_T^s)^2 + (\tau/\tau^s)^2] = 1$ (matrix failure) where σ_A, σ_T, τ are the axial, transverse and shear stresses relative to the fibre directions. σ_A^s, σ_T^s and τ^s are the corresponding static strengths (single valued). For fatigue loading σ_A, σ_T, and τ are cyclic stresses and σ_A^u, σ_T^u and τ^u are fatigue strengths used in place of the static strengths. The fatigue stresses and strengths in practice must all be at the same stress ratio.

Hashin and Rotem[15] stated that the theory could be used for compression by putting in the appropriate values of strength but they have not done this in any of their publications and have not considered what the failure modes might be.

Hashin and Rotem[15] showed that transverse failure would occur for off-axis angles exceeding $1.76°$ and showed that the results for off-axis tests on a lamina gave good agreement with their theory of failure. They extended their work to fatigue failure essentially by fitting straight lines to the S–N curves and using data from tests at two different off-axis angles to establish a fatigue function for in-plane shear and transverse tension, thus avoiding experimental difficulties of scatter at these two particular conditions. Rotem and Hashin[16,17] extended the work to angle-plied laminates both under static and fatigue loading. They predicted failure using laminate theory and noted that for off-axis angles greater than 45°, clean fractures occurred which were accurately predicted. For off-axis angles less than 45°, failure occurred by delamination, i.e. in another mode, caused by interlaminar shear stress at the edges. Both Rotem[18] and Hashin[20] separately have tried to extend the failure theory to this mode. Sims and Brogdon[21] carried out similar work to Rotem and Hashin[17] but used the Tsai–Hill (Azzi and Tsai[9]) criterion. They also concluded that first-ply failure could be adequately predicted under both static and fatigue loading.

For unidirectional materials the failure modes are simple, and in laminates are only complicated by the intervention of interlaminar shear failure. In practice the long rod-like failure surfaces and the domination of the transverse mode failure seem to make the predictions relatively insensitive to which quadratic failure theory is used. The evidence points to fatigue failure surfaces being non-intersecting with static failure surfaces, provided scatter is adequately dealt with.

FABRIC REINFORCED GRP

The majority of fabric and woven roving fabric laminates have principal tensile and compressive strengths which are approximately equal, and have in-plane shear strengths which lie between one quarter and one half of the UTS. The failure surface (Fig. 1) is therefore likely to be ellipsoidal with an aspect ratio near one. The characteristic bundle of fibres occurs in fabrics with superimposed twist, crimp, and resin-rich areas. Although the transverse mode of failure characteristic of unidirectional material can still occur within the bundle, it does not lead to specimen failure (rupture) or even cause a significant reduction in strength. Additional forms of damage occur at the fibre cross-overs (shear and compression) and in or adjacent to the resin windows (Fig. 2). With some types of specimen, structural failure

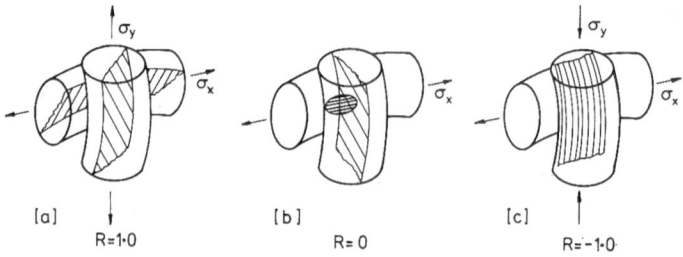

FIG. 2.   Cross-over damage.[5]

also becomes involved. The strength and failure of thin-walled cylinders are affected by reinforcement overlaps,[22] and by bulging (Fig. 3). Cruciform specimens tend to fail at one or other of the arms.[6] There is difficulty in correlating strength data between the larger cylinder or cruciform specimens and conventional tensile or compressive specimens.[23]

Owen and Found[24] reported off-axis tension and compression specimens subjected both to static and fatigue loading. They used a woven fabric and polyester resin system and attempted to fit both Group 1 and Group 2 failure theories (Table 1) for the mean strengths. Conventional presentations of the failure theories against off-axis strength and angle

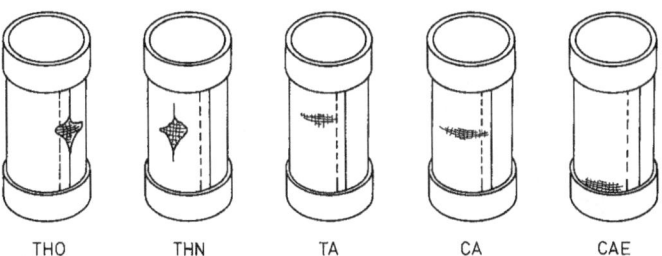

FIG. 3A.   Failure modes in tubes.[5] THO: Tensile failure perpendicular to the hoop stress at the overlap. THN: Tensile failure perpendicular to the hoop stress not at the overlap. TA: Tensile failure perpendicular to the axial stress. CA: Compressive failure perpendicular to the axial stress not at the end (bulging). CAE: Compressive failure perpendicular to the axial stress at the end (bulging).

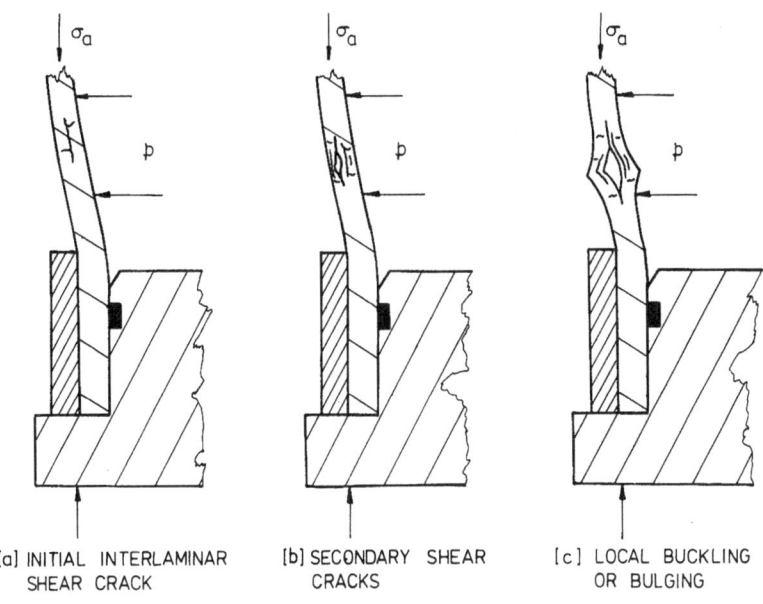

[a] INITIAL INTERLAMINAR SHEAR CRACK   [b] SECONDARY SHEAR CRACKS   [c] LOCAL BUCKLING OR BULGING

FIG. 3B.   Compressive failure at the tube ends (CAE).[5]

appear to show good agreement between the Group 2 theories and experimental data (Fig. 4). However when the fit of the whole failure surface was examined the fit was found to be chaotic (Fig. 5) resulting in intersecting failure surfaces depending on the choice of off-axis data, and fatigue and onset of damage failure surface which intersected the rupture surfaces. However the S–N curves implied non-intersecting failure surfaces (Fig. 6).

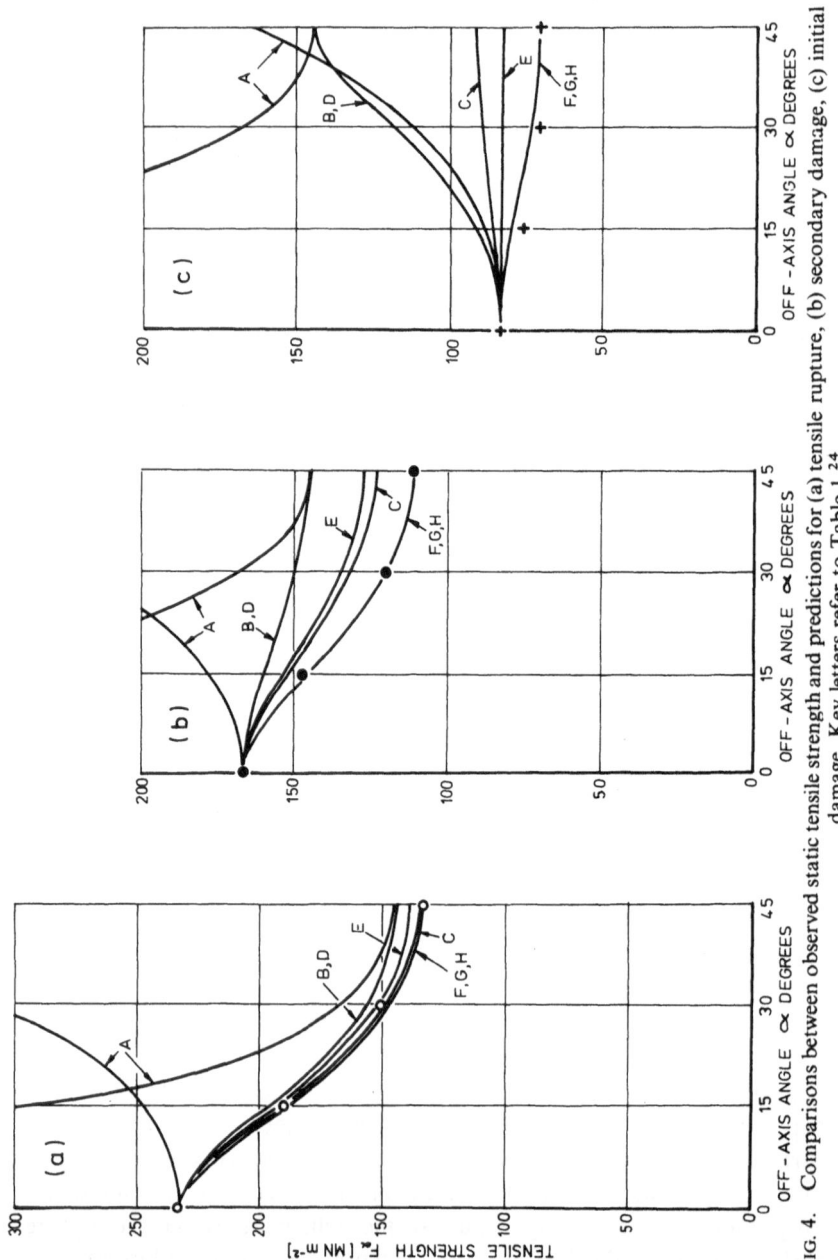

FIG. 4. Comparisons between observed static tensile strength and predictions for (a) tensile rupture, (b) secondary damage, (c) initial damage. Key letters refer to Table 1.[24]

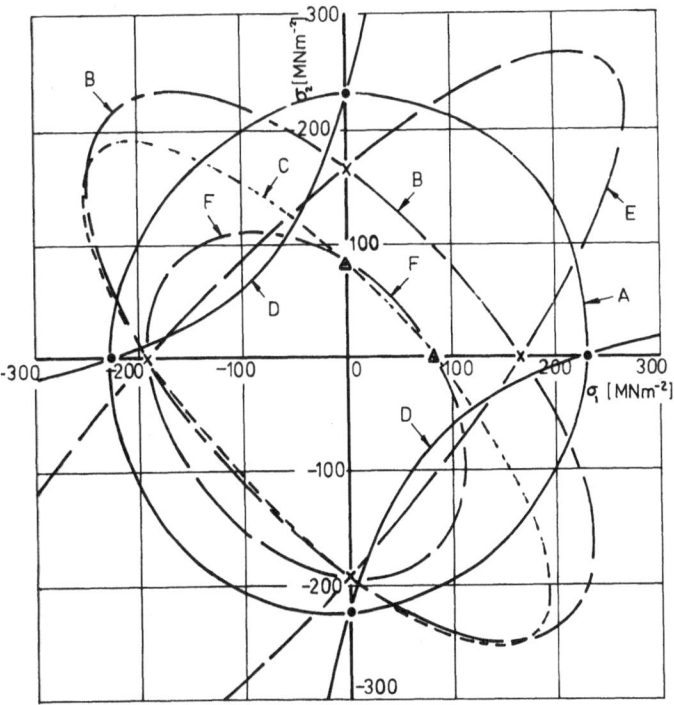

FIG. 5. Intersection of computed failure surfaces with $\sigma_6 = 0$ plane for static failure. Curves A, B, C based on 45° off-axis tensile results for rupture, secondary damage, and initial damage respectively. Curves D, E, F based on 45° off-axis compression results for rupture, secondary damage, and initial damage respectively.[24]

Owen and Griffiths[23] used the same material system in the form of thin-walled cylinders to provide a more comprehensive experimental examination of the failure surface. The same problems were encountered in fitting surfaces to the results but it was further revealed that the scatter was substantial (Fig. 7), there were inconsistencies between cylinder data (at $R = 0$) and uniaxial data, and that there was a major difficulty in the interpretation of the shear strength (Fig. 8).

Rice,[5] and Owen and Rice[7] examined four combinations of a woven roving fabric, a woven fabric, and two terephthalic polyester resins confining their attention to the base plane ($\sigma_6 = 0$) of the failure surface. In addition to comparing the behaviour of the material systems, they compared the static and fatigue strengths and mapped failure modes for all specimens. The failure modes are summarized in Fig. 3 and a typical map of the failure modes is shown in Fig. 9. It was found that there were

M. J. Owen

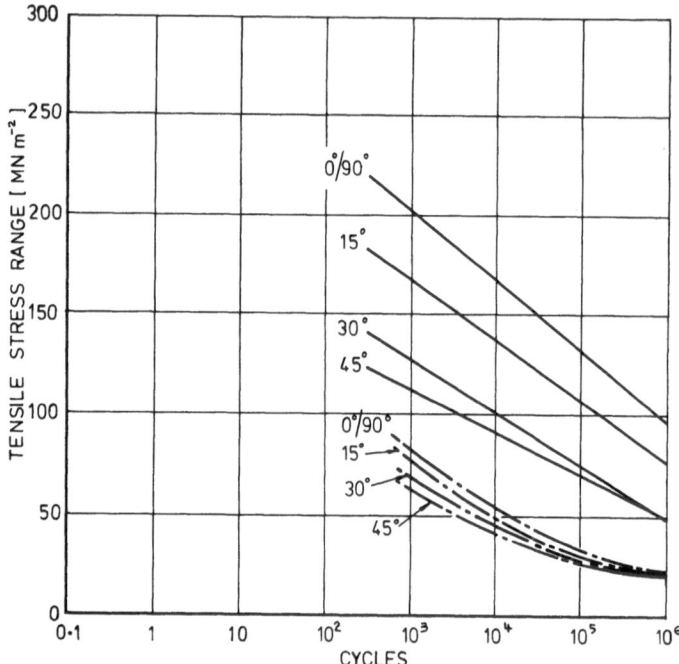

FIG. 6. Fatigue curves for various off-axis angles: zero-tension stress. —— rupture; —·— initial damage.[24]

significant differences in failure modes according to the material system and it was also noted that the improvement in strength normally associated with flexibilized resin systems was of very haphazard benefit when examined under biaxial loading conditions.

From a comparison of all the available results for fabric reinforced materials, the following features emerge:

(1) Scatter is significant and should be treated statistically.
(2) There are numerous failure mechanisms and failure modes which vary according to material combination, stress conditions and specimen design.
(3) Fatigue and static failure surfaces ought to be non-intersecting.
(4) With certain types of specimen, structural behaviour partially obscures material behaviour.
(5) There are particular difficulties in defining the in-plane shear strength.
(6) The experimental results fall well inside a maximum stress boundary and are broadly of quadratic form.

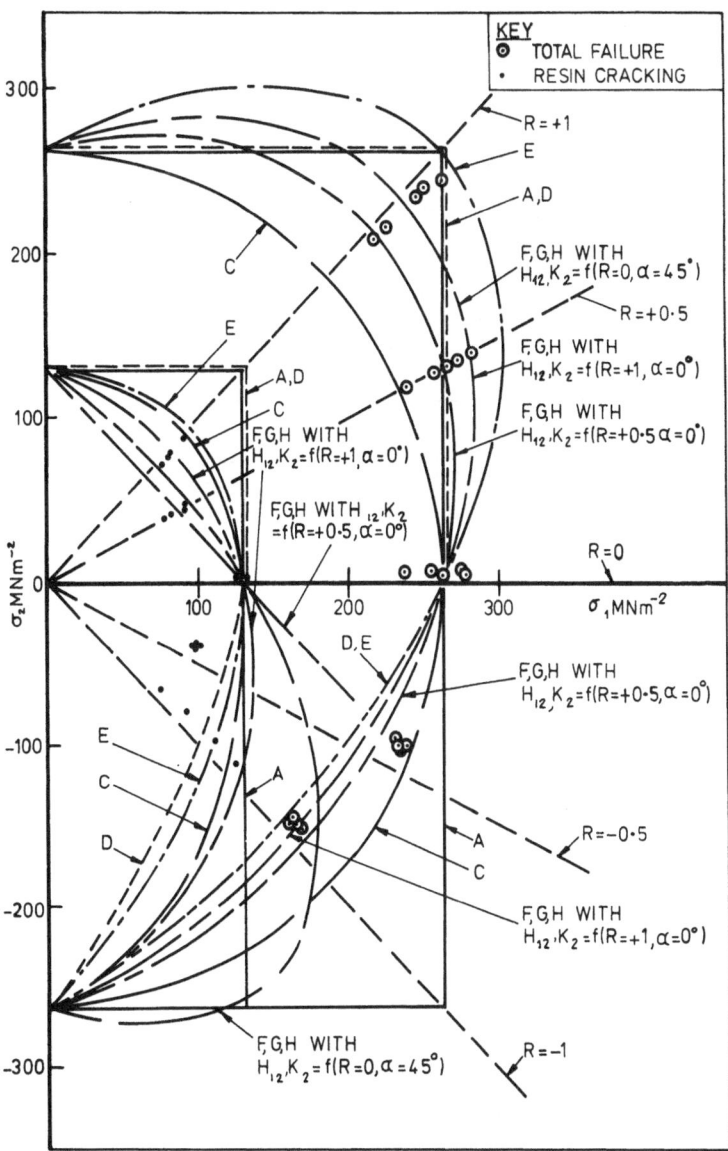

FIG. 7. Static results for tubes with $\alpha = 0°$ and R taking values from $+1$ to -1 (corresponding to curve FJDLE in Fig. 1). For key to curves see Table 1.[23]

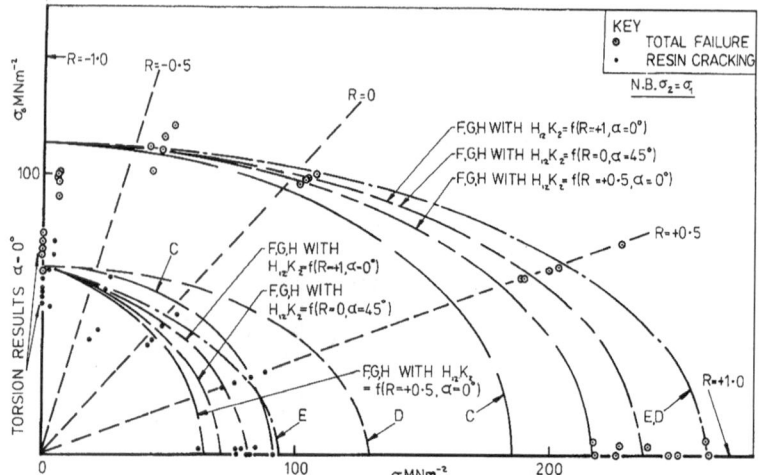

FIG. 8.   Static results for tubes with $\alpha = 45°$ and $|R|$ taking values from $+1$ to -1 (corresponding to curve FB in Fig. 1). For key to curves see Table 1.[23]

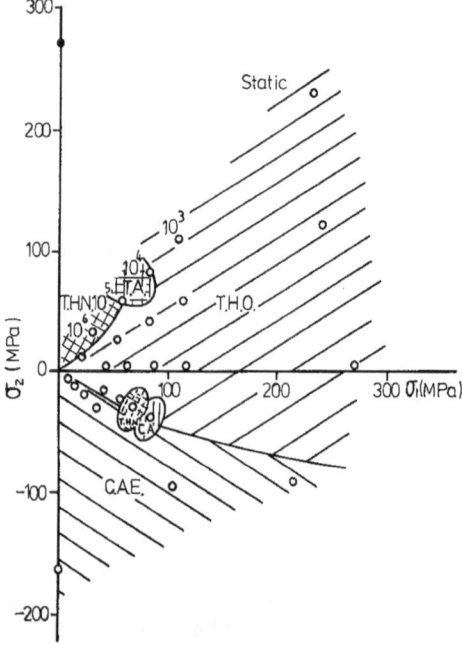

FIG. 9.   Failure modes for cylinders with ICI Impolex T500P resin and Turner and Newall ECK10 woven roving fabric.[7]

(7) The only economically acceptable complex stress test is the off-axis tension or compression test.

(8) The off-axis test is unsatisfactory for the determination of interaction coefficients.

(9) The interaction coefficients need to be specified to conform with 6–8 above, but the Tsai and Hahn proposal if unacceptable for woven fabrics.

(10) It appears to be necessary to fit a failure theory separately in each stress octant.

(11) Modifications of the early theories due to Norris[10,11] would appear to be the most suitable.

(12) It is proposed that the following equations should be used to allow for differences in tensile and compressive strength.

$$\left(\frac{\sigma_1}{X}\right)^2 + \left(\frac{\sigma_2}{Y}\right)^2 + \left(\frac{\sigma_6}{S}\right)^2 = 1 \qquad \text{(tension–tension shear octant)}$$

$$\left(\frac{\sigma_1}{X'}\right)^2 + \left(\frac{\sigma_2}{Y'}\right)^2 + \left(\frac{\sigma_6}{S}\right)^2 = 1 \qquad \text{(compression–compression shear octant)}$$

$$\left(\frac{\sigma_1}{X}\right)^2 - \frac{\sigma_1\sigma_2}{|XY'|} + \left(\frac{\sigma_2}{Y'}\right)^2 + \left(\frac{\sigma_6}{S}\right)^2 = 1 \qquad \text{(tension–compression shear octant)}$$

$$\left(\frac{\sigma_1}{X'}\right)^2 - \frac{\sigma_1\sigma_2}{|X'Y|} + \left(\frac{\sigma_2}{Y}\right)^2 + \left(\frac{\sigma_6}{S}\right)^2 = 1 \qquad \text{(compression–tension shear octant)}$$

X' and Y' are treated as positive, but σ_1, σ_2 should be treated as negative if compressive.

X, Y, X' and Y' should be established from conventional tensile or compressive tests parallel to the principal material axes, and a value for a stated probability of failure calculated. The suggested method for the determination of S *for fabric reinforcements only* is to use the 45° off-axis specimen for which $\sigma_1 = \sigma_2 = \sigma_6 = 0.5\sigma$ and to use 0.5σ at failure as the appropriate measure of the in-plane shear stress. Since the tensile and compressive values are likely to differ, it may be appropriate to take the lower value. This approach permits straightforward extension to fatigue conditions for constant R by substituting fatigue strengths for static strengths and cyclic stresses for σ_1, σ_2 and σ_6.

(13) In order to avoid intersecting failure surfaces, the values of static strength used for the determination of the static failure surface should represent the same probability of failure or survival as the fatigue strengths.

(14) A similar approach could be used at the onset of damage.

CHOPPED STRAND MAT GRP

Chopped strand mat GRP can be regarded as plane isotropic although there is usually a small difference in strength between the roll and cross-roll directions. Nevertheless it is usually possible to work in terms of the principal stresses without regard to material axes. Owen and Found[25] and Owen, Griffiths and Found[26] reported the results of thin-walled cylinder tests subjected to internal pressure, and axial load, material batch differences, and internal reinforcement joints were all significant problems in the interpretation of results. Figure 10 represents mean static and fatigue strengths at rupture. The boundary at rupture falls well inside the well-known failure theories for isotropic materials in the tension–tension quadrant. However, it is obvious that the failure behaviour could be represented by a quadratic type of failure theory and that the fatigue failure surfaces are non-intersecting although non-concentric. It is proposed that failure theories of the following type could be used.

$$\frac{\sigma_1^2}{T^2} + \frac{\sigma_2^2}{T^2} = 1 \qquad (T\text{–}T \text{ quadrant})$$

or

$$\frac{\sigma_1^2}{C^2} + \frac{\sigma_2^2}{C^2} = 1 \qquad (C\text{–}C \text{ quadrant})$$

or

$$\frac{\sigma_1^2}{T^2} - \frac{\sigma_1\sigma_2}{|TC|} + \frac{\sigma_2^2}{C^2} = 1 \qquad (T\text{–}C \text{ quadrant})$$

or

$$\frac{\sigma_1^2}{C^2} - \frac{\sigma_1\sigma_2}{|CT|} + \frac{\sigma_2^2}{T^2} = 1 \qquad (C\text{–}T \text{ quadrant})$$

In the foregoing equations, σ_1 and σ_2 are principal stresses and T and C are the tensile and compressive strengths of the laminate for a stated

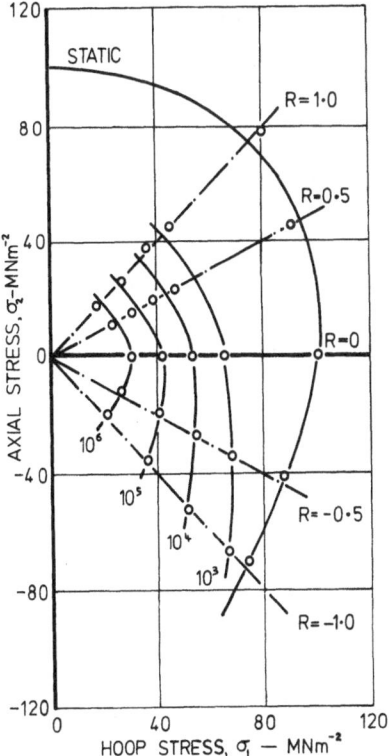

FIG. 10.   Constant life curves for chopped strand mat/polyester resin cylinders at rupture.[25]

probability of failure. Because of the possibility of slight anisotropy of the reinforcement it is suggested that the lower strength with due allowance for scatter should be used. Fatigue strength at an appropriate life-time and R ratio could be used to replace the static strengths. Whilst a similar approach could be applied at definable damage states such as the onset of debonding and the onset of resin cracking, it has been noted that the tension–tension state has a particularly severe effect on the onset of debonding under fatigue loading.[25]

CONCLUSIONS

Biaxial stress failure theories for GRP need to be applied with considerable caution. Experimental results indicate that failure under complex stress conditions occurs well inside the maximum stress boundaries. Failure theories which define failure in a manner which exceeds the maximum stress boundaries are particularly dangerous. Suitable quadratic functions of the stresses and strengths have been defined for mat and fabric laminates. Because of the long rod-like nature of the failure surface for unidirectional materials, and the dominant matrix mode of failure, similar equations would probably be useful for these materials.

ACKNOWLEDGEMENTS

The author is indebted to the Science and Engineering Research Council and to the Ministry of Defence (Navy) for financial support for work on this subject over many years. The contributions of former Research Assistants in the University of Nottingham, Dr D. J. Rice, Dr J. R. Griffiths, and Dr M. S. Found, are gratefully acknowledged.

REFERENCES

1. TSAI, S. W. and HAHN, H. T., *Introduction to Composite Materials*, Westport CT, Technomic Publishing Co., 1980.
2. OWEN, M. J. and BISHOP, P. T., The significance of microdamage in glass reinforced plastics at macroscopic stress concentrators, *J. Phys. (D): appl. Phys.*, 5 (1972), 1621–1636.
3. FOUND, M. S., *Biaxial Stress Fatigue of Glass Reinforced Plastics*, Ph.D. Thesis, University of Nottingham, May 1972.
4. GRIFFITHS, J. R., *Fatigue of Glass Reinforced Plastics under Complex Stresses*, Ph.D. Thesis, University of Nottingham, Oct. 1974.
5. RICE, D. J., *Fatigue and Failure Mechanisms in Glass Reinforced Plastics under Complex Stresses*, Ph.D. Thesis, University of Nottingham, May 1981.
6. SMITH, E. W., *Cyclic Biaxial Deformation and Failure of a Glass-fibre Reinforced Composite*, Ph.D. Thesis, Cambridge University, Dec. 1976.
7. OWEN, M. J. and RICE, D. J., Biaxial strength behavior of glass-reinforced polyester resins, in: *Composite Materials: Testing and Design* (Daniel, I. M., ed.) (6th Conference), ASTM STP 787, American Society for Testing and Materials, 1982, pp. 124–144.
8. STOWELL, E. Z. and LIU, T. S., On the mechanical behaviour of fibre reinforced crystalline materials, *J. Mech. Phys. Solids*, 9 (1961), 242.
9. AZZI, V. D. and TSAI, S. W., Anisotropic strength of composites, *Exp. Mech.*, 5 (1965), 283–288.
10. NORRIS, C. B. and MCKINNON, P. F., Compression, tension and shear tests on yellow-poplar plywood panels of sizes that do not buckle with tests made at various angles to the face grain, *U.S. Forest Products Laboratory Report*, No. 1328, 1946.

11. NORRIS, C. B., Strength of orthotropic materials subjected to combined stresses, *U.S. Forest Products Laboratory Report*, No. 1816, 1951.
12. FRANKLIN, H. G., Classic theories of failure of anisotropic materials, *Fibre Sci. Tech.*, 1 (1969), 137–150.
13. TSAI, S. W. and WU, E. M., A general theory of strength for anisotropic materials, *J. Comp. Mater.*, 5 (1971), 58–80.
14. GUYETT, R. P. and CARDRICK, A. W., The certification of composite airframe structures, *Aeronaut. J.* (1980), 188–203.
15. HASHIN, Z. and ROTEM, A., A fatigue failure criterion for fiber reinforced materials, *J. Comp. Mater.*, 7 (1973), 448–464.
16. ROTEM, A. and HASHIN, Z., Failure modes of angle ply laminates, *J. Comp. Mater.*, 9 (1975), 191–206.
17. ROTEM, A. and HASHIN, Z., Fatigue failure of angle ply laminates, *AIAA J.*, 14 (1976), 868–872.
18. ROTEM, A., Fatigue failure of multidirectional laminate, *AIAA J.*, 17 (1979), 271–277.
19. PUCK, A. and SCHNEIDER, W., On failure mechanisms and failure criteria of filament wound glass-fibre/resin composites, *Plastics Polym.*, Feb. (1969), 33–44.
20. HASHIN, Z., Fatigue failure criteria for unidirectional fiber composites, *Trans ASME, J. appl. Mech.*, 48 (1981), 846–852.
21. SIMS, D. F. and BROGDON, V. H., Fatigue behavior of composites under different loading modes, in: *Fatigue of Filamentary Composite Materials* (Reifsnider, K. L. and Lauraitis, K. N., eds), ASTM STP 636, American Society for Testing and Materials, 1977, pp. 185–205.
22. OWEN, M. J. and GRIFFITHS, J. R., Internal reinforcement joints in grp under static and fatigue loading, *Composites*, April (1979), 89–94.
23. OWEN, M. J. and GRIFFITHS, J. R., Evaluation of biaxial stress fatigue failure surfaces for a glass reinforced polyester resin under static and fatigue loading, *J. Mater. Sci.*, 13 (1978), 1521–1537.
24. OWEN, M. J. and FOUND, M. S., The fatigue behaviour of a glass fabric reinforced polyester resin under off-axis loading, *J. Phys. D.: appl. Phys.*, 8 (1975), 480–497.
25. OWEN, M. J. and FOUND, M. S., Static and fatigue failure of glass fibre reinforced polyester resins under complex stress conditions, *Faraday Special Discussions of the Chemical Society*, No. 2 (1972), 77–89.
26. OWEN, M. J., GRIFFITHS, J. R. and FOUND, M. S., Biaxial stress fatigue testing of thin-walled GRP cylinders, in: *Proc. 1975 Int. Conf. Composite Materials* (Scala, E., Anderson, E., Toth, I. and Noton, B. R., eds), Vol. 2, New York, Metallurgical Society of the AIME, 1976, pp. 917–941.

3

Damage Detection in Carbon Fibre Epoxy Structures Using Acoustic Emission

D. Valentin and A. R. Bunsell

Ecole Nationale Supérieure des Mines de Paris,
Centre des Matériaux, BP 87, 91003 Evry Cédex, France

ABSTRACT

A comparative study of damage accumulation during steady and cyclic loading in carbon fibre epoxy tubes has been possible by using a failure model developed for flat specimens. Those differences which were found between the plate and tube specimens can be explained by micrographical observations. The failure model, developed initially for steady loading conditions, has revealed the effect of regular unloading on the speed of damage accumulation in these structures. As variation in applied stress produces damage which is additional to that produced under steady loading however the rate of damage accumulation can be described in a similar manner when macroscopic damage does not occur.

INTRODUCTION

The increasing use of carbon fibre reinforced plastics (CFRP) is largely due to the benefits it affords in reducing energy consumption of structures during their manufacture and because of its low weight during their use. However there are insufficient means of predicting long term behaviour of these structures when they are subjected to mechanical loading. As little is known about damage accumulation in these materials it is necessary to invoke safety factors which are much greater than the optimum.

This study has been concerned with the accumulation of damage in CFRP structures in the form of tubes and has employed the acoustic emission technique to monitor the failure processes. In those cases where the principal failure process is fibre fracture a failure model exists for steady loading which was first developed for unidirectional plate specimens[1] and then applied to crossplied specimens.[2] The model has been used in this study of the behaviour of tubes first under static conditions and then during regular reductions of pressure.

ACOUSTIC EMISSION AND DAMAGE

Under conditions with which acoustic emission is primarily generated by fibre breaks it has been shown that under steady loading the rate of emission is given by:

$$\frac{\mathrm{d}N}{\mathrm{d}t} = \frac{A}{(t+\tau)^n} \tag{1}$$

where t is time, τ a time constant, n a power less than 1 depending on the structure tested. A depends only on the applied stress such that:

$$A = \lambda \, e^{k\sigma} \tag{2}$$

where λ and k are constants.

Table 1 shows the value of n for several different types of flat specimen including crossplied specimen $(0°, 90°, T)$ which included a satin weave cloth layer. These values of n were found to be independent of the applied stress except for the $(\pm 30°)_s$ specimens for which n was found to decrease with increasing applied stress. This results in a decreasing tendency for the structure to stabilise and stop emitting and suggests a second process generating emissions such that:

$$\frac{\mathrm{d}N}{\mathrm{d}t} = \frac{A_1}{t+\tau_1} + \frac{A_2}{t+\tau_2} \tag{3}$$

for applied stresses greater than about 70% of the failure stress and with $A_2 \gg A_1$ and $\tau_2 \gg \tau_1$.

TABLE 1
Values of n for each type of specimen

Specimen	$0°$	$(0°, 90°)_s$	$(\pm 15°)_s$	$(0°, 90°, T)$	$(\pm 30°)_s$
n	0·99	0·98	0·05	0·9	$0·7 < n < 1$

The type of behaviour seen at high stresses with the $(\pm 30°)_s$ specimens was observed at all stress levels with the $(\pm 45°)_s$ specimens. This behaviour must be due to the interlaminar shear forces as the additional failure process is inactive during a relaxation test and the acoustic emission is once again described by eqn. (1).

In the cases of unidirectional and $(0°, 90°)_s$ specimens the value of n is very nearly 1 and if n is put equal to 1 in eqn. (1) we obtain:

$$N = A \log \left(\frac{t + \tau}{\tau} \right) \tag{4}$$

and hence

$$\log \frac{dt}{dN} = \frac{N}{A} + \log \frac{\tau}{A} \tag{5}$$

so that $\log dt/dN$ is a linear function of the number emissions N. Figure 1 shows the straight line curves obtained at different stress levels and it is important to note that changes of load level within a range of 20 or 30 % and then returning to the original load produces a simple extrapolation of the original curve. This implies that for a given stress the rate of damage accumulation is a unique function of the number of emissions and is dependent on load history.

The goal of the present study was to see if similar results to those found with plates were to be found with tubes when they were subjected to similar loading and also to examine the influence of cyclic loading.

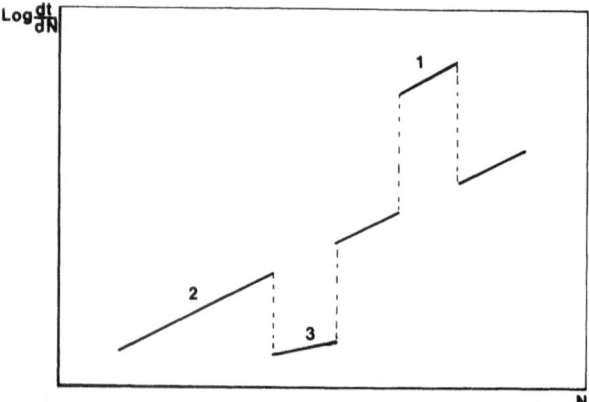

FIG. 1. Schematic representation of the effect on $\ln(dt/dN)$ of varying the load level between load 1, the lowest load, and level 3, the highest. Activity is seen to be a function of accumulated damage and independent of loading history.

Experimental Details

The carbon fibre filament wound tubes were made by the S.E.P. and Toray T6K carbon fibres were used as they were in the plate specimens. The epoxy resin employed had the code FM 702 and was very similar to the Ciba Geigy 914. The tubes had the following dimensions: internal diameter 96 mm, length 330 mm, wall thickness 2 mm, and they consisted of ten crossplied layers, five in each direction. The ($\pm 88°$) tubes had in addition one $0°$ layer positioned between the fourth and fifth layer in order to increase longitudinal bending rigidity. A ($\pm 88°$) tube containing a satin, carbon fibre cloth layer orientated at $45°$ and positioned between the fifth and sixth layers was also tested. The tubes were pressurised internally and were held in such a way as to allow complete freedom of deformation. The testing apparatus is described in greater detail elsewhere.[2]

The acoustic emission was recorded using a Dunegan-Endevco D-140 B piezoelectric transducer which had a resonant frequency of about 200 kHz and was connected to a preamplifier with a pass band of 100 kHz to 300 kHz and a fixed gain of 40 dB. A 3000 series system was then used to amplify the signals using a total gain of 80 dB. The rate of emission was recorded with a logarithmic time base reset to zero after a chosen fixed number of emissions and was load controlled as the time base was stopped at a pressure lower than the maximum pressure applied during the cycle. In this way it was possible to obtain directly the logarithmic emission rate ($\log dt/dN$), which corresponded to the accumulated periods of steady loading, as a function of N. The gradient of the curve which was obtained gave $1/A$. The applied pressure cycles were trapezoidal and at a low frequency of four cycles per hour consisting of 12 min at the maximum pressure and 3 min at the minimum pressure.

Experimental Results

Burst tests were conducted on each type of tube and allowed the failure pressure and the acoustic activity during a simple pressure test to be determined. The number of tubes available for the tests was limited so that only one burst test for each type of tube was conducted. The results are shown in Table 2. In all cases the tubes emitted in a similar fashion to that observed with unidirectional specimens[3] which is to say that the numbers of emissions increased exponentially with pressure.

The ($\pm 88°$) and ($\pm 75°$) tubes were seen to almost respect the Kaiser effect as had been seen for the plate specimens, see Fig. 2. This observation showed that the pressurisation apparatus did not generate spurious noise. However the ($\pm 45°$) tube did not show the same behaviour and it is thought

TABLE 2
Breaking stress of tube specimen

Tube	Pressure at rupture (bars)	Stress at rupture (MPa)
($\pm 88°$)	500	1 200
($\pm 75°$)	290	700
($\pm 45°$)	140	336

probable that was due to damage produced by the shear forces at the fibre matrix interfaces and possibly because the structure was much less rigid than the other tubes. The greater displacements produced during its testing may have generated spurious noise from the apparatus.

The different types of failure produced with each type of tube are shown in Fig. 3. The ($\pm 88°$) and ($\pm 75°$) tubes reveal mainly fibre failure along the generator of the cylinder. The ($\pm 45°$) tubes on the other hand showed interlaminar shear failure.

With the exception of the ($\pm 45°$) tubes the creep tests showed behaviour which was in every respect similar to that seen with the plate specimens, as shown by Fig. 4. The curve of $\log(\mathrm{d}t/\mathrm{d}N)$ as a function of N was described for the ($\pm 88°$) tubes by putting $n = 0.92$ and the value of n for the ($\pm 75°$) tubes was found to be very nearly 1. Successive periods at steady pressure

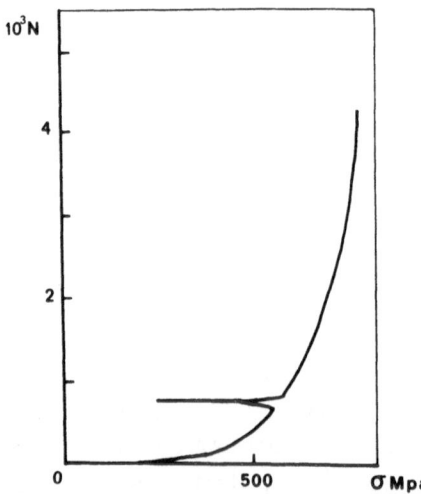

FIG. 2.   Cumulative acoustic emission curve during a burst test of a (± 88) tube showing the Kaiser effect.

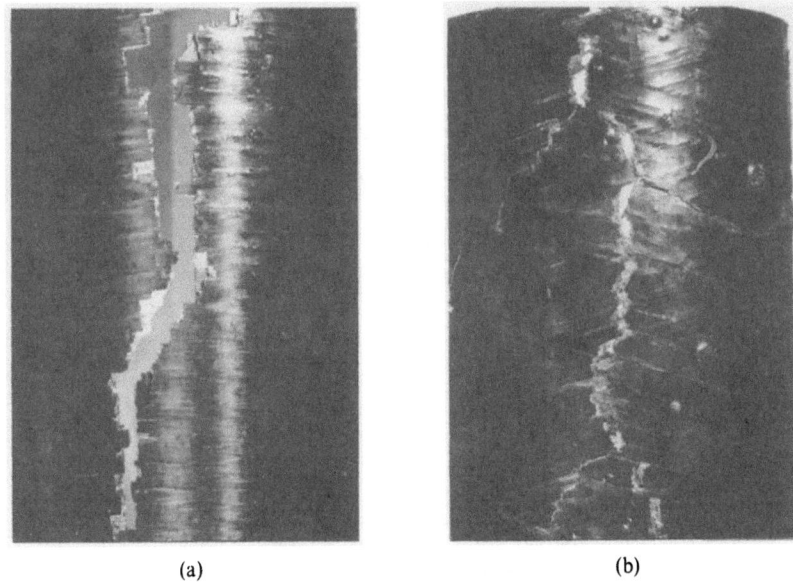

(a)   (b)

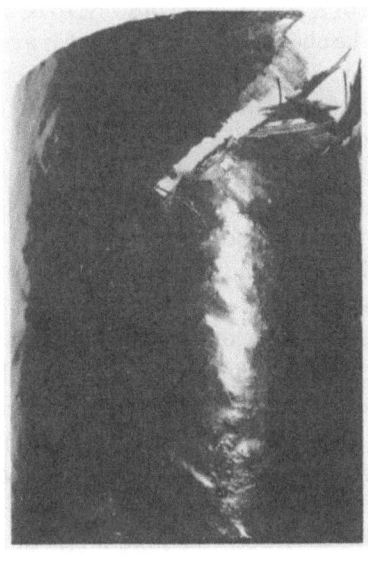

(c)

FIG. 3.   Fibre dominated failure of ($\pm 88°$) (a) and ($\pm 75°$) (b) tubes and typical delamination
dominated failure of a ($\pm 45°$) tube (c).

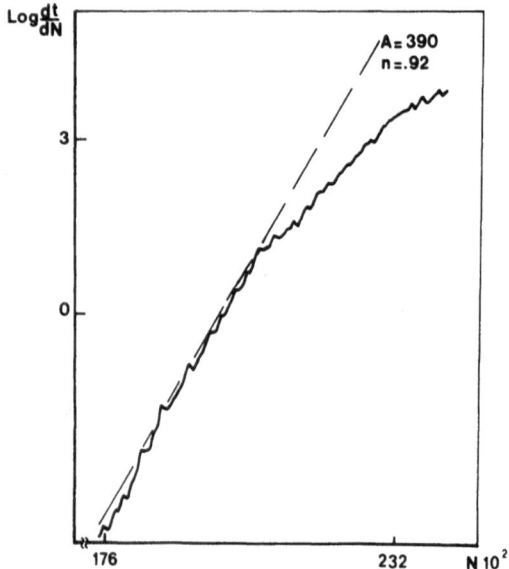

FIG. 4.   Acoustic emission rate for a ($\pm 88°$) tube during a creep test at an applied stress of 840 MPa (pressure = 350 bars).

allowed the variation of A as a function of stress to be determined and again eqn. (2) was found to be obeyed. The coefficient k is a constant for each type of structure tested by λ and was found to vary considerably between different tubes. The latter parameter is a function of the acoustic coupling between the composite and the transducer. The values of k were $4.8 \times 10^{-3}\,\mathrm{MPa}^{-1}$ and $10.6 \times 10^{-3}\,\mathrm{MPa}^{-1}$ for the ($\pm 88°$) and ($\pm 75°$) tubes respectively. It can be seen from Fig. 5 that the variations of the parameter A as a function of the stress ratio for the ($\pm 88°$) and ($\pm 75°$) tubes were very similar. This lends support to the supposition that similar failure processes were involved.

The effect of an additional cloth layer in the ($\pm 88°$, T) tube was not noticeable until a stress of about 800 MPa above which another failure mechanism was detected. The behaviour was then similar to that seen with the ($\pm 30°$)$_s$ and ($\pm 45°$)$_s$ plate specimens and the emission rate obeyed eqn. (3) showing a second asymptote. Micrographs of the failure of these tubes suggest that the relative increase in acoustic activity was due to cracking at the interface between the fibre layers at $\pm 88°$ and the satin cloth, as can be seen in Fig. 6a. Cracks were also seen in the ($\pm 88°$) tubes without a cloth layer but these were far fewer in number and confined to the $0°$ layer as shown in Fig. 6b. No such large scale cracking was detected in

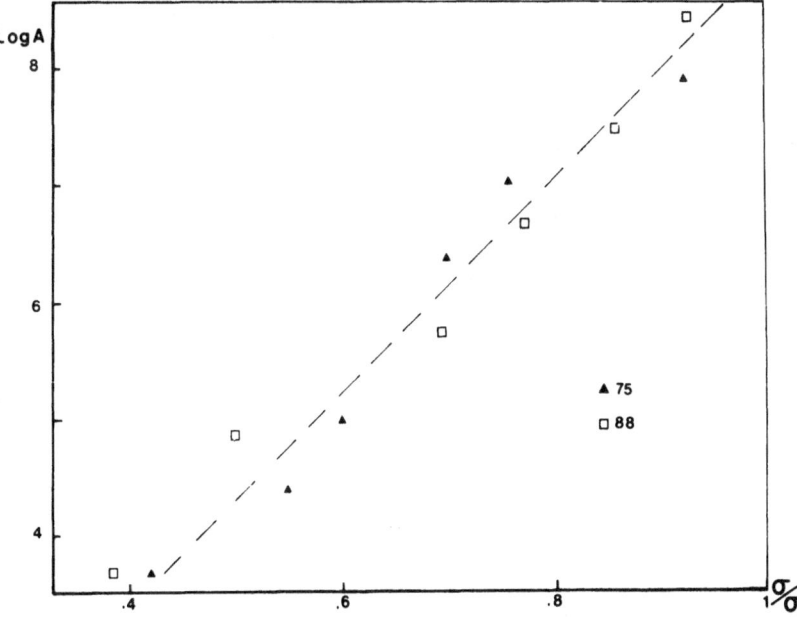

FIG. 5. Variation of the parameter A as a function of stress ratio for ($\pm 88°$) and ($\pm 75°$) tubes.

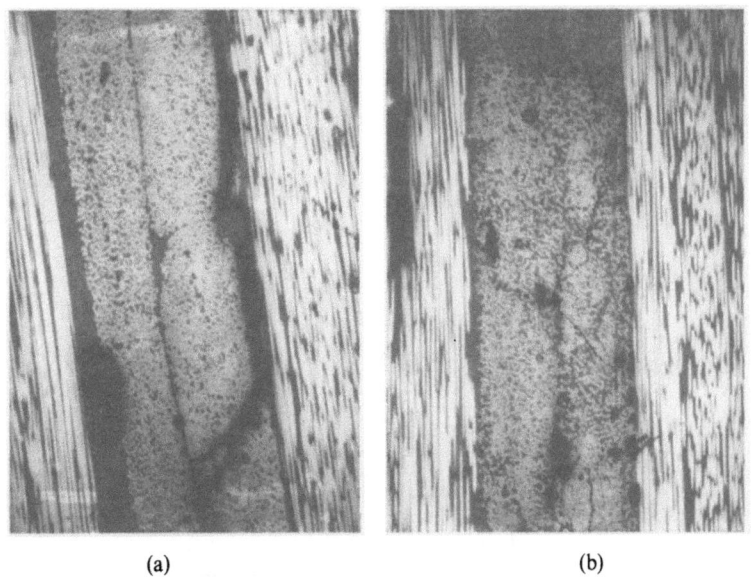

(a)   (b)

FIG. 6. (a) Photomicrograph showing the cracking at the $\pm 88°$ fiber layers and the satin cloth interfaces of a ($\pm 88°$, T) tube. (b) Microcracking of the $0°$ layer of a ($\pm 88°$) tube. Magnification: $\times 85$.

the ($\pm 75°$) tubes. These observations explain the anomaly seen with the ($\pm 88°$) and ($\pm 75°$) tubes when compared to the plate specimens and which suggested that the value of n for the ($\pm 88°$) tubes should have been nearer to unity than the value observed for the ($\pm 75°$) tubes. The greater rate of activity in the ($\pm 88°$) tubes has been revealed to be due to the failure of the additional $0°$ layer.

Although it has been shown that the failure processes producing emissions were not only fibre breaks, it is considered that the concepts described in the failure model developed for unidirectional specimens, which relate the accumulation of internal damage to eventual failure, can still be considered to be applicable. In particular the observation that an overload of a unidirectional specimen is equivalent to an acceleration of time has been verified for the ($\pm 88°$) and ($\pm 75°$) tubes, as shown in Fig. 7. This effect was also seen with the ($\pm 88°$, T) tubes even though the additional failure process was predominant, as Fig. 8 shows.

Cyclic pressurisation of the ($\pm 88°$) and ($\pm 75°$) tubes was found to produce similar behaviour to that seen under steady pressures leading towards stabilisation if the pressure variations were not great ($R > 0.7$) and if the maximum pressure was less than 75 % of the nominal burst pressure.

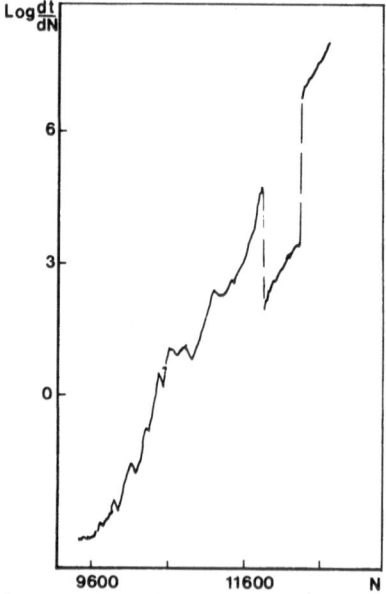

FIG. 7.   The influence of an overload of 15 bars during a creep test performed at 300 bars on a ($\pm 75°$) tube.

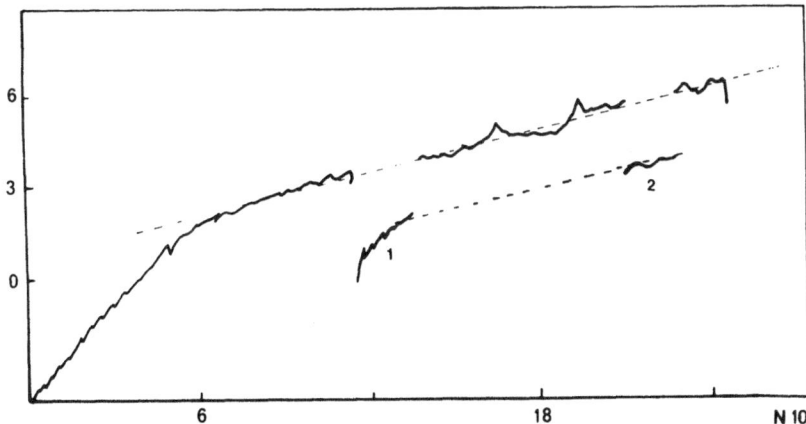

FIG. 8.   The influence of an overload (1, 2) of 15 bars (36 MPa) during a creep test performed at 430 bars (1032 MPa) on a (±88°, T) tube.

However it was found that for a given maximum pressure cyclic loading produced a greater rate of emission than was found under steady pressure (*A* cyclic > *A* creep). If steady pressurisation was followed by a period of cyclic pressurisation and then the steady pressure again applied the acoustic activity was found to be intermediate between the original creep activity and that produced during cycling (A_1 creep > A_2 creep) as shown in Fig. 9. This behaviour was observed with both the (±88°) and (±75°) tubes and has been reported for unidirectional and crossplied plate specimens.[4] The

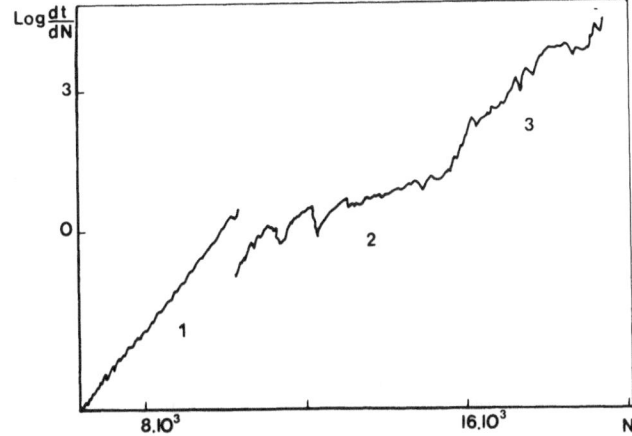

FIG. 9.   The effect of successive steady (1, 3) and cyclic loading (2) on the acoustic emission on a (±88°) tube. The maximum applied load was 400 bars. $R = 0.8$.

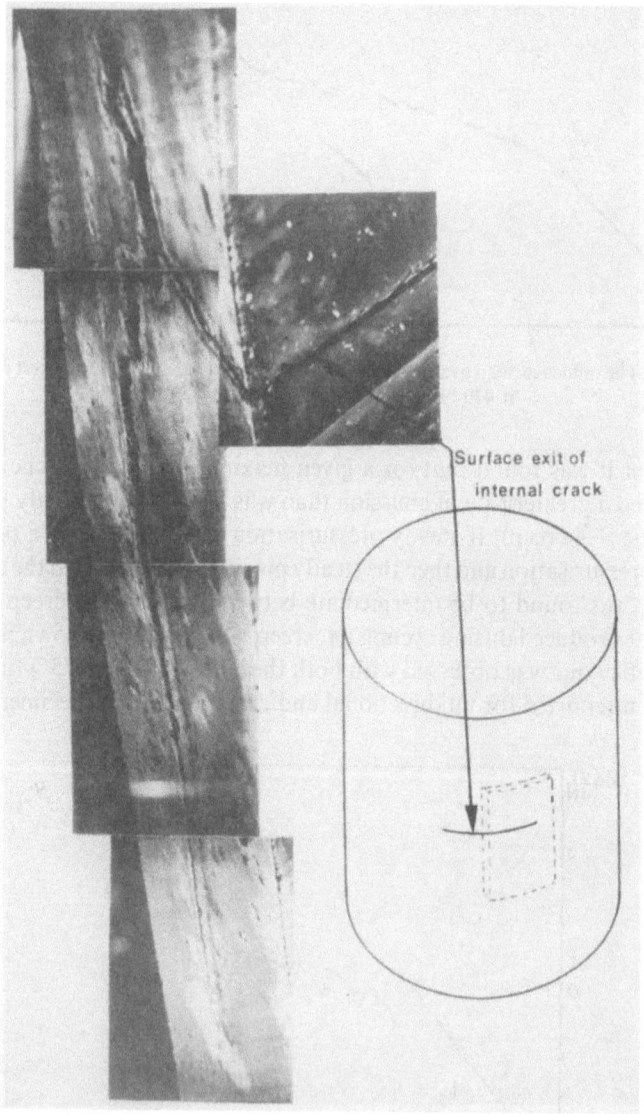

Fig. 10. Photomicrograph showing macroscopic damage of a ($\pm 88°$) tube during cyclic loading ($\sigma = 1075$ MPa, $R = 0.5$). Magnification: $\times 40$.

second creep curve in Fig. 8 is no longer a simple extrapolation of the first as in the case of a simple overload and we must conclude that the scenario of damage accumulation has been modified.

Outside the above cyclic conditions ($R < 0.7$, $\sigma/\sigma_r < 0.75$) large scale damage of the tubes prevented stabilisation of the acoustic activity. The microcracks found after steady pressurisation of the ($\pm 88°$) tubes combined during cyclic pressurisation producing major cracking at the $0°$, $88°$ interface. These cracks could sometimes be seen at the tube surface, as shown in Fig. 10.

Several cracks were found at the ends of the ($\pm 75°$) tubes which could have been the origin of emissions due to rubbing of the fracture surfaces during cyclic pressurisation.

CONCLUSION

The acoustic emission technique has again been shown to be useful for detecting damage in CFRP structures. Except for the ($\pm 45°$) tubes in which interlaminar shear failure occurred the other tubes ($\pm 88°$) and ($\pm 75°$) produced emissions during steady pressurisation which were quantifiable. The mechanisms which produced emissions in these tubes were irreversible and there was an equivalence of time and applied stress for their mechanical ageing. Cyclic pressurisation revealed more complex acoustic emission behaviour although analogous acoustic activity to that seen under creep conditions was seen under some cyclic conditions. The accelerated activity generated by cyclic loading was not found to produce a simple time-applied stress equivalence.

The failure model already shown to apply to plate specimens was seen to be applicable to tube specimens although its applicability under cyclic pressurisation is limited to pressure conditions with which macroscopic damage is not produced.

REFERENCES

1. BUNSELL, A. R., LAROCHE, D. and VALENTIN, D., Damage and failure in carbon fibre reinforced epoxy resin, ASTM STP to be published.
2. VALENTIN, D. and BUNSELL, A. R., A study of damage accumulation in carbon fibre reinforced epoxy resin structures during mechanical loading monitored by acoustic emission, *J. Reinforced Plastics and Composites*, to be published.

3. FUWA, M., BUNSELL, A. R. and HARRIS, B., Tensile failure mechanisms in carbon fibre reinforced plastics, *J. Mat. Sci.*, **10** (1975), 2062.
4. VALENTIN, D. and BUNSELL, A. R., Damage in carbon fibre reinforced epoxy resin produced during cyclic loading and monitored by acoustic emission, in *Fatigue and Creep of Composite Materials*, H. Lilholt and R. Talreja (eds) (1982), 329, RISØ.

4

Characterization of Composite Materials by Means of the Ultrasonic Stress Wave Factor

J. C. DUKE JR, E. G. HENNEKE, W. W. STINCHCOMB
and K. L. REIFSNIDER

Materials Response Group,
Engineering Science and Mechanics Department,
Virginia Polytechnic Institute and State University,
Blacksburg, Virginia 24061-4899, USA

ABSTRACT

The usual approach to nondestructively evaluating a composite structure involves inspection and mechanical analysis of the inspection results. Such an approach has met with only limited success. On the other hand, the ultrasonic stress wave factor technique directly evaluates the material. Despite requiring access to only one surface of the material, the technique interrogates the material in the directions of applied load. Using the stress wave factor technique it is possible to determine the failure location in the material. The correlation of the stress wave factor with stiffness is shown. In addition, the use of the technique for determining the strength or life of composite material structures is discussed.

INTRODUCTION

Damage, change in the condition of the material properties, which reduces either the stiffness, strength, or life of composite materials, is of major importance with regard to the design and use of structures of such materials. Numerous nondestructive techniques and methodologies exist which have as their objective the detection and physical description of material imperfections, matrix cracks, delaminations, porosity, fiber or matrix rich regions, etc. In such instances it is necessary to analytically establish the individual and, or, collective significance of these imperfections as regards service performance. Finally the success of this analysis must be validated. Alternatively it is possible to directly evaluate the material condition nondestructively. A technique which possesses this capability is the ultrasonic stress wave factor technique.[1] It involves the measurement of the response of the material to an ultrasonic stress wave pulse in directions which lie in the laminate plane.

TECHNIQUE DESCRIPTION

In general, fiber reinforced laminated composite materials are fabricated in forms that limit access for nondestructive examination. Most often the material is examined in a direction perpendicular to that in which the load is applied, using either ultrasonic, or X-ray radiographic methods. If access is further limited to only one of these surfaces only pulse-echo ultrasonic inspection is generally considered. A new method of utilizing ultrasound for nondestructively examining composite material subject to such constraints has been developed by Vary.[1] The technique employs both a sending and a receiving ultrasonic transducer that are in contact with the same surface of the object being examined (Fig. 1). Ultrasonic pulses are repeatedly (repetition rate, $1/P$) introduced into the specimen and the oscillations of the receiving transducer caused by these pulses are counted above a selected threshold (number of oscillations, C) for a period of time (time, T). As a

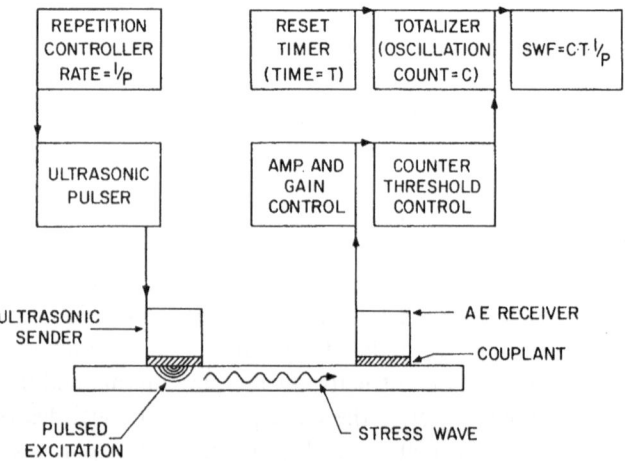

FIG. 1.   Schematic diagram of the Stress Wave Factor nondestructive examination method.

means of quantifying this measurement the stress wave factor (SWF) has been defined $SWF \equiv CT(1/P)$. The unique feature of this method is its sensitivity to the structural configuration of the object. That is to say that the material's internal structure, lamina interfaces, lamina orientation, and boundaries cause the stress wave to propagate out into the material to a point where the receiver is located. This feature makes it possible to interrogate the material in the same directions in which stresses resulting from applied load would act. Interrogation in these directions, as such, would be expected to find regions of the material responding in a fashion which would be peculiar to their condition, as related to mechanical performance. For example, the region which has a poor performance regarding stress wave energy propagation would be expected to test as a region of low SWF. If this number is the lowest in the object, its performance would be considered to be the poorest. Consequently, if all the regions were subject to an identical state of stress, the deformation in this region might be expected to be the most severe, and the location at which failure initiates.

DETERMINATION OF FAILURE LOCATION

By repeating the measurement at different locations it is possible to determine a relative evaluation of the material. Considerable care must however be taken to reproducibly perform this examination, otherwise the comparison is pointless. Special attention has been given to adequately characterize the parameters associated with the performance of this technique. Of particular significance is the type and amount of couplant material, the transducer attachment method, the threshold level and period of counting, and the procedure used for associating SWF values with a region of the material.[2] Figure 2 displays the results of examining a $[0, 90_3]_s$ E glass laminated composite. The transducers were maintained at a separation of 3·125 cm and were translated in increments of 0·63 cm in order to examine the entire specimen. The value of SWF obtained has been plotted in two ways: the solid line displays the SWF value for the increment at the point directly between the two transducers, and the broken line records the data that is the average of all SWF values obtained while a 0·63 cm region remains between the two transducers. The latter procedure was developed to try to associate the SWF value with a small region (point) of the specimen. The schematic below the graph depicts the specimen: the grid indicates the various transducer positions and the shaded area the

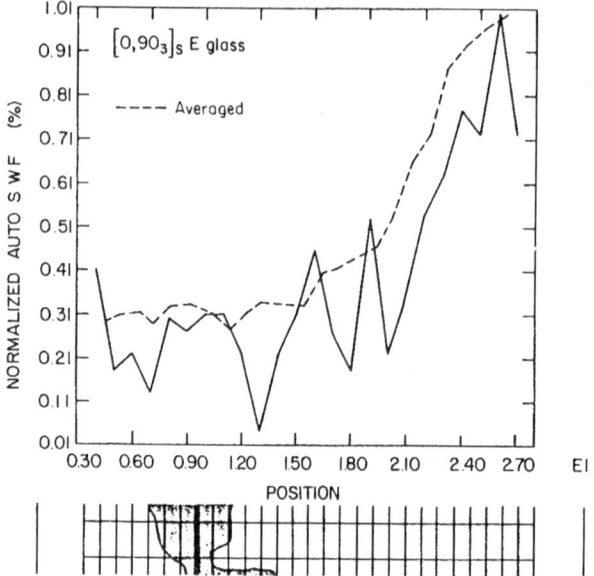

FIG. 2.   Plot of the variation of Stress Wave Factor along the length of a $[0, 90_3]_s$ laminate of E glass epoxy. The location of the failure resulting from quasi-static tensile loading is also shown.

location of failure resulting from quasi-static tensile loading. The region of low SWF correlates quite well with the failure location.

DETERMINATION OF STIFFNESS

Classically stiffness is considered to be a structure-insensitive property. However, because the very nature of fiber reinforced laminated composite material is that of a structure, the property of stiffness must of necessity be structure sensitive; this is not to say that the stiffnesses of the constituents of the composite material, i.e. the fiber and the matrix are structure sensitive. Nevertheless it is necessary to recognize that, since laminated composite materials may in general continue to support load even after sustaining considerable damage and may exhibit property variations from point to point in even the undamaged condition, because of problems during manufacture, experimental determination of stiffness is inextricably linked with a gauge length.

Several experiments were performed that were directed at comparing the stiffness of the material to the measured SWF. The technique of moiré interferometry was utilized to obtain the full field in-plane axial displacement during quasi-static tensile loading. Since the interference patterns obtained may be interpreted as displacement, it is possible to determine the stiffness at every point along the length of a straight sided coupon specimen, subjected to tensile load. Figure 3 displays a plot of measured SWF versus the axial displacement. For each region for which the SWF measurement was made, the axial displacement measured over the corresponding region has been plotted; each region measured 3·125 cm in length. The displacement values used were obtained at a load for which no detectable damage to the composite material resulted (76 MPa).

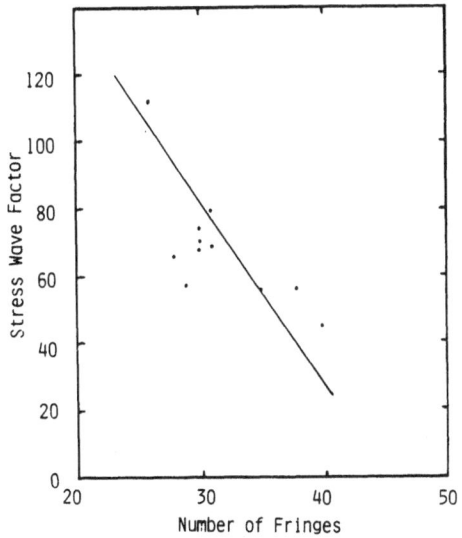

FIG. 3. Plot of Stress Wave Factor versus the number of interference fringes. The fringes correspond to the axial displacement of a $[0, 90_3]_s$ laminate of E glass epoxy at a nominal stress of 76 MPa. A moiré interferometric technique was used to provide this measure of full field displacement. The length over which individual measurements of both have been made was 3·125 cm.

It is clear from the figure that these two measures are correlated. Since the stress wave used to make the measurement is not constrained to propagate solely in the axial direction, some variation due to nonuniform strains in the transverse directions might be expected. Nevertheless, within the experimental accuracy of the two measurement techniques, the correlation is quite good. Since the deformation at the loads employed is essentially linear and elastic, and the regions of material involved in the measurement are identical, the correlation of SWF with longitudinal stiffness is the same.

However, the exact nature of the correlation is material and laminate dependent (Fig. 4). As a result, in order to apply this technique to determining the stiffness from point to point in a structure, the correlation of the SWF with stiffness for the particular laminate and composite system would need to be known. In addition, the correlation of SWF with stiffness for damaged conditions is expected to be different. This question is presently the subject of extensive investigation in the Materials Response Laboratory at Virginia Polytechnic.

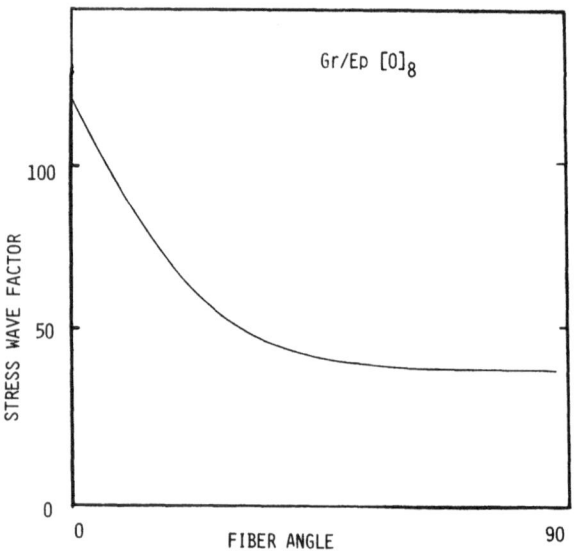

FIG. 4.   Plot of the variation of the Stress Wave Factor with fiber angle. The laminate used for this evaluation was $[0]_8$ graphite epoxy.

It is possible to obtain the same longitudinal stiffness for damage conditions of composite materials that result from different load histories. In certain instances those damage conditions are different, consequently the SWF would be expected to be sensitive to the difference. Of course the stiffness is in reality a tensorial quantity and its complete determination for different damage conditions would reflect this difference. However, it is not possible to determine this experimentally unless the difference in the damage condition occurs on a scale larger than the scale of the strain measurement.

DETERMINATION OF STRENGTH/LIFE

Although good agreement is seen between measured low values of SWF and failure location, some question exists regarding the exact relationship of the measured SWF and strength/life. In this regard, a fundamental difficulty exists because of the terminal nature of strength/life determining tests; the potential strength/life of other regions cannot be determined. It is important to recognize at this point that, in addition, the strength/life is dependent on the applied load history and future. That is to say, for a particular piece of composite material, the strength/life of any region is not unique. Consequently it is somewhat unrealistic to expect a relationship to exist in general between SWF and strength/life. Perhaps a particular relation may exist between the SWF, strength/life, and a specific load history. To fully investigate this aspect of SWF technique, additional clarification of the issue is required. First, as a composite material is subjected to mechanical loading the condition of the material becomes more and more damaged. Figure 5 demonstrates this for a specimen that has been cyclically loaded. Similarly such progressive damage development occurs for quasi-static, sustained or spectral loading. In general, a SWF measurement cannot be uniquely related to the material strength since as

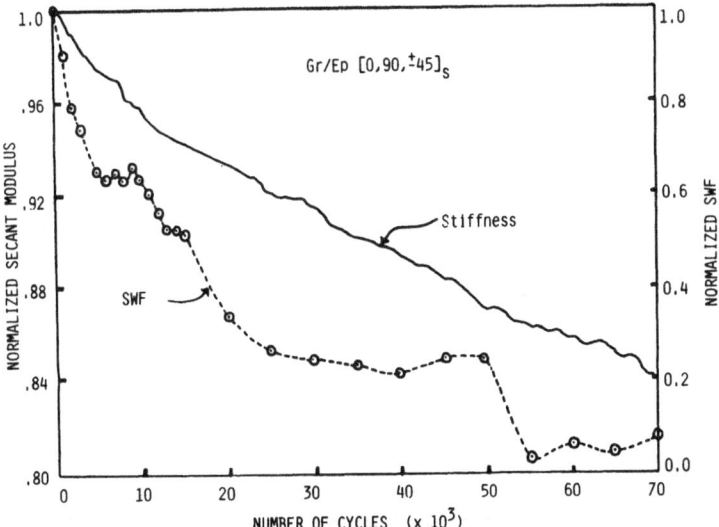

FIG. 5. Plot of the variation of the Stress Wave Factor (normalized) and longitudinal secant modulus (normalized) resulting from cyclic loading ($R = 0 \cdot 1$) of a [0, 90, ± 45]$_s$ laminate of graphite epoxy.

the material condition changes so does the SWF value; there would be only one value of strength.

Two areas of investigation need to be considered: one directed at establishing the correlation between SWF and strength when the material condition being evaluated has resulted from cyclic loading; and the other is to determine the correlation between SWF and remaining life.

CONCLUSIONS

The Stress Wave Factor:

(1)   seeks to directly evaluate the mechanical performance of composite materials;

(2)   correlates with the longitudinal elastic stiffness for a given laminate and composite system;

(3)   is capable of determining the failure location;

(4)   is not uniquely related to the strength;

(5)   is capable of detecting differences in damage conditions that are characterized by similar stiffness;

(6)   may be used to nondestructively evaluate the mechanical performance of a composite material structure.

ACKNOWLEDGEMENTS

The authors wish to acknowledge the support of this work through NASA Grant NAG 3-172 and 3-323; special appreciation in this regard is expressed to Alex Vary, the technical monitor. In addition, the efforts of Anil Govada in support of this work are gratefully acknowledged.

REFERENCES

1. VARY, A. and BOWLES, K. J., An ultrasonic-acoustic technique for non-destructive evaluation of fiber composite quality, *Polymer Engng Sci.*, **19** (1979), 373–377.

2. HENNEKE, E. G., II, DUKE, J. C., Jr, STINCHCOMB, W. W., GOVADA, A. and LEMASCON, A., A study of the stress wave factor technique for the characterization of composite materials, Interim Report of NASA NAG 3-172, June 1982.

5

Acoustic Emission (AE) as a Tool for Use on Composite Structures

P. T. COLE

*Dunegan/Endevco Division, Endevco UK Ltd,
Melbourn, Royston, Herts SG8 6AQ, England*

ABSTRACT

Interest in and use of AE as a tool for both development and testing of composite structures have increased considerably in recent years. Monitoring of GRP (glassfibre reinforced plastic) manlift booms is now routine and last year the Society of the Plastics Industry released its 'Recommended Practice for AE Testing of Fibreglass Tanks/Vessels'. based on experience of testing more than 1400 tanks and 50 pressure vessels.

This paper outlines the practical application of AE to a variety of structures, with observations on assessing its suitability for use in different applications and reports on the Dunegan/Endevco Field Test experience of testing 40 tanks using the SPI (CARP Committee) practice since its release in 1982.

INTRODUCTION

Acoustic emission (AE) is the term used to describe the resulting acoustic stress waves when strain energy is released rapidly due to the occurrence of microstructural changes in a material.

In a composite material, AE is released when matrix crazing, fibre breakage, debonding or any other microstructural failure occurs. In addition, AE may be released due to fretting (or rubbing) at previously damaged areas. Interpretation of AE signals in conjunction with 'background' information such as:

—the loading conditions of the structure,
—the material,
—previous history,

will give invaluable information about the progression of damage and the suitability of the structure for further use. The 'background' information relating to the structure is equally as important as the AE data in making decisions and will affect the feasibility of using AE in the first place as well as the way in which an AE test is conducted.

At present, AE monitoring is most successfully used on structures where their 'background' is reasonably well defined, since this reduces the number of variables and results in more reliable interpretation of the AE data.

Acoustic emission is very different from conventional NDT tools in that it is telling about the dynamic growth of defects rather than their static presence. AE has a jargon of its own and for this reason, the paper is split into three sections:

(1) 'basics' underlines the fundamental technique and factors to be considered when using AE;
(2) 'established applications' describes two well known and specific applications which are straightforward and lend themselves ideally to the use of AE as a structural test;
(3) 'high technology applications' looks at techniques being developed for use in the military and aerospace areas where computer technology now plays a large part in reducing masses of data to a usable form and in a time scale which makes its use economic.

BASICS OF AE APPLICATION TO COMPOSITE STRUCTURES

—The presence of acoustic emission indicates damage is occurring (or has occurred in the case of fretting).
—Acoustic signals travel omnidirectionally from the damaged area (source) at a speed of between 1000 metres per second and 10 000 metres per second (dependent upon construction and fibre orientation—most GRP structures show a velocity of around 2500 metres per second). These signals are essentially broad-band and so can be detected by monitoring at a range of frequencies (usually 150 kHz is used on composites).

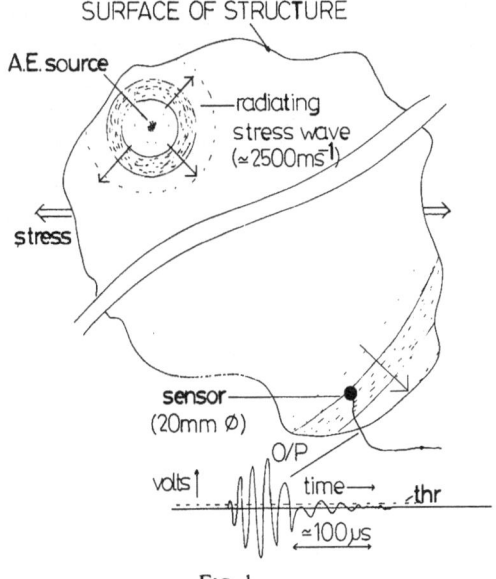

FIG. 1.

—Each signal is detectable in the structure only for a short time (the structure disperses and absorbs the acoustic energy). High frequency components are absorbed more rapidly, and naturally bigger signals are detectable for longer. This time can vary from several millionths of a second to seconds, depending upon the cause and structural acoustics of the particular structure. Each signal is called an 'event'.

—Signals are detected using a surface mounted piezoelectric sensor which converts the acoustic wave into an electrical signal (they are designed to be sensitive to structure-borne acoustic signals and to reject airborne noise).

Figure 1 is a simplified illustration of the process of AE generation and detection in a structure.

Analysis of AE Data from Composite Structures

A single emission contains relatively little analytical information since its size and shape have been changed as it travels through the structure. Its source could be any of the possible fracture mechanisms or even fretting at a damage site. The information which is useful from it is:

—its presence and location tell you that something is happening and where it is;

—its peak amplitude (maximum voltage at the sensor output) is related to the strain energy released (and rate of release) at source. Obviously a large peak amplitude indicates a more significant fracture event. The range of detectable peak amplitudes which can come from a composite structure is enormous (energy is proportional to amplitude squared so covers a greater range still). Using a sensor of average sensitivity, peak amplitudes range from less than 20 dB (10 microvolts) to more than 100 dB (one tenth of a volt!). This covers the range (in terms of damage) from minor matrix crazing being the source of AE (or individual low strength fibres breaking), to major structural fractures such as bundles of fibres fracturing at high stress, major delaminations forming or rapid fracture of resin-rich areas. The amplitude detected at the sensor will depend upon how far the acoustic signal has had to travel to reach it, and the monitoring frequency. This is illustrated in Fig. 2 (attenuation data from a polyester resin chemical tank).

It is clear that unless the exact source to sensor distance is known, the detected amplitudes can only give an indication of the extent of the damage. The one certainty is that high detected amplitudes indicate serious damage (probably close to the sensor). A horizontal dotted line on Fig. 2 (level with $100\,\mu\mathrm{V}$ amplitude) shows the detection threshold usually used for monitoring composite structures such as tanks and booms. This level has been found by experience to be a good compromise between rejecting the very minor emissions which often occur when a composite structure is loaded, and providing a reasonable area of coverage for detecting any serious damage.

Other characteristics of the AE signal have been used successfully; for instance, CARP* report that long duration events are an indication of bond failure in particular types of pipe joint. This has been borne out by our experience in field testing of structures when secondary bonds have been failing.

The most useful and straightforward analysis of AE data is possible when there are a number of AE events, and the location and occurrence of these is correlated with the loading or use of the structure:

(1)   Emission continuing under constant load conditions indicates that damage is occurring due to creep. Whilst emission continues, there is always a danger of the structure failing (failure is usually

* Committee on Acoustic Emission from Reinforced Plastics (CARP), a section of the Corrosion Resistant Structures working group of the Society of the Plastics Industry (SPI).

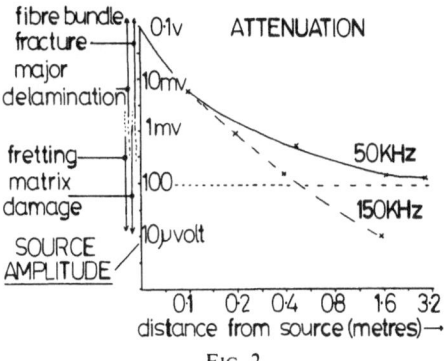

FIG. 2.

immediately preceded by a dramatic increase in the rate of emission). For structures used under long term constant load conditions (such as storage tanks), it is obviously vital that creep is not occurring.[1] AE monitoring is the perfect tool for ensuring this. Conversely, if there is no AE at a particular constant load level, providing conditions do not change, the structure should last indefinitely.

(2) Emission during load increase indicates that damage is occurring during this particular load cycle, or may have occurred previously and be due to 'fretting' at the damage site. Quantification is the key here and several methods can be used to gain further information:

—Comparison of emission density from various parts of the structure.

—Comparison with other identical structures.

—Unload and reload the structure to see if the emission is a 'once only' occurrence (due to relief of local areas of high stress) or will continue at every loading (indicating fatigue damage or fretting).

—Does the rate of emission (i.e. total per unit load increase) increase with load? If it does, then progressive damage is indicated.

—Are there any sudden increases in activity? This will be obvious; most composite structures show a dramatic increase in activity at 30–60 % of failure load (in a continuous load to failure test). This increase may be more than 100 times in terms of AE activity per unit load increase.

Quantification of AE

Emissions ('events') can be counted—hence total events, event rate, etc.

Emissions can be measured for amplitude (as described previously)—hence number of events above a particular amplitude. One of the most useful (and conventional) measures of AE is the AE 'count' (*not* to be confused with event counts!). This is the number of times the AE waveform crosses the threshold of detection (Fig. 1: thr), hence a single event can have one count or a thousand or more counts. Counts are amplitude dependent (bigger signals give more counts), so give a useful single measure of AE.

ESTABLISHED APPLICATIONS

Aerial Man-Lift Booms

In the USA, fibreglass is used as a structural insulating boom in 'cherry pickers' which are used by the power companies when working on elevated electrical equipment. In the absence of AE testing, these booms were replaced on a time expired basis (regardless of condition) and even this did not prevent catastrophic failures occurring. In 1975 Dunegan/Endevco, in conjunction with Georgia Power Company, developed an AE test for assessing the structural integrity of these booms.[2] There are now a number of companies specialising in providing a testing service for this application alone. The basis of the test is very straightforward:

—Booms have a rated maximum working load.
—It was found that routine use often resulted in the boom seeing higher loads, up to three times the rated level.
—The booms should remain structurally sound so that this misuse will not result in a failure.

Acoustic emission is used as follows.

—Sensors are attached along the boom and at critical areas such as hydraulic ram mountings.
—The boom is progressively loaded to three times its rated working load under controlled conditions whilst monitoring for AE.
—Provided there is no indication (by AE) of serious damage, the boom is considered fit for further use. The AE testing interval may be annual or sooner if a problem is suspected or the boom known to have been misused.

The test is cheap and quick, so routine use has resulted in enormous economic savings as well as increased safety, since only damaged booms are removed from service (previously they were replaced every two years irrespective of condition).

Fibreglass Petrochemical Tanks

About the same time that the test was developed for the booms, work started to establish methods of testing fibreglass chemical tanks. When designed, manufactured, shipped, installed and operated correctly, they are a cheap solution to containment and processing of corrosive liquids. Unfortunately, shortcomings with any of these design and use factors may lead to total collapse with little obvious warning (often due to strain corrosion slowly eating away the structure where the corrosion barrier has failed). Apart from being able to 'hear' active strain corrosion as it is occurring, AE offers a simple method of assessing structural integrity. The test is based as follows:

—Tanks should have a safety factor of between 5 and 10 (depending on what specification they are built to), this being the ratio of maximum rated working load to failure load.

—Tests to failure on GRP tanks show that[1] AE starts at about 10–20 % of ultimate failure and shows a dramatic upturn when serious structural damage is occurring, usually between 30 and 60 % of failure load.

—Sensors are attached to all critical areas of the tank[3] and the tank is filled to maximum operating load in stages (preferably with process fluid).

—Serious structural problems are immediately obvious due to the large amount of AE from them. This means the test can be terminated without danger of total structural failure.

—Minor amounts of AE indicate localised structural damage. If found early enough, this enables repairs to take place.

—Damage occurring due to creep is identified by AE occurring during constant load.

Figure 3 illustrates the testing of a chemical tank. The CARP document 'Recommenced practice for acoustic emission testing of fibre glass

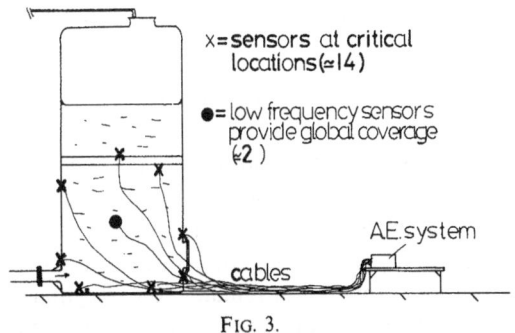

FIG. 3.

tanks/vessels'[3] published by the Society of the Plastics Industry gives full operational details of the test procedure, based on tests of more than 1400 tanks.

Amongst the tanks tested by Dunegan/Endevco UK Field Test Service—more than 40 in total, ranging in size from 1 to 10 metres high—the following illustrate typical problems identified by AE:

(a) In-service GRP neutraliser (5 × 4 metres): suspected deterioration due to unforeseen vibration problems in service and high operating temperature.

   Results of AE test: high AE activity (counts quantity and high amplitudes) from nearly all areas of tank, even at half working load. This means the structure is rapidly degrading even at low loads and is in danger of collapse.

(b) In-service GRP mixing tank (2 m high): no suspected problems.

   Results of AE test showed little general damage but one very high amplitude and long duration burst of activity from a nozzle proved to be due to secondary bond failure of a nozzle gusset where it joined the tank (insufficiently supported attached pipeline).

(c) In-service GRP sodium bisulphide storage vessel: widespread AE activity and creep damage at below 50 % of working load. It was considered that so much damage was occurring to the tank that it would be unwise to load it to maximum working load. On close visual inspection one of the most active areas appeared 'patchy' when viewed with a strong light shining through the wall (to carry out the visual examination, it was necessary to shut down and clean out the tank).

(d) GRP tank moved to a new site: AE monitoring during the first filling after the move identified damage occurring to the base supporting structure (several high amplitude emissions from one area) due to excess stress on a hold-down which had been tightened prior to initial filling.

Deciding on a course of action after an AE test is simple if the tank either has not emitted or is very active. Some tanks fall into a category where failure processes are just initiating, usually indicated by slight AE during hold (i.e. at constant level indicating creep damage) at high loads (it may only be a dozen emissions in the final 30-min hold period). The AE result is clear in that damage *is* propagating, but the tank may last for a year or more if the rate of propagation remains low. Regular AE testing can monitor the rate of deterioration. Using the tank at reduced loads may extend its life, but AE testing *does not* predict the future and the tank is being used beyond its design safety factor if creep damage is occurring at all, so the only fully safe course of action is to repair or replace it.

AE testing does not always 'reject' tanks. There are several instances where AE may confirm as structurally sound a tank which has failed visual inspection. With new tanks, air entrapment in low stress areas is a typical example; with in-service tanks, usually something external to the tank's normal use is the cause for concern. One example which we monitored was a tank which had been sprayed *externally* with a corrosive chemical due to a nearby fractured pipe. The outside of the tank had a 'furry' appearance and visual inspection would have no choice but to fail it. AE monitoring showed that the structure was not degrading—the effect of the chemical had been purely temporary and cosmetic.

Used with common sense, simple AE monitoring such as in these examples provides reliable, low cost information on the condition of GRP structures and can be used by personnel with a minimum of training in AE techniques (courses are run by a Society of the Plastics Industry sub-group specifically for tank testing).

HIGH TECHNOLOGY APPLICATIONS

The applications discussed so far have revolved around the simple ability to tell when and in what general area activity is occurring. If an AE signal reaches three or more sensors placed on the same structure, it is possible to locate its origin far more accurately by calculating its location from the time differences in arrival at each sensor (this is how the epicentre of an earthquake is located). The main assumption with this approach is that the signals can reach a sufficient number of sensors to be located. This may not occur in the case of small signals, large sensor spacings or too high a monitoring frequency (because of signal attenuation as it travels in the structure). This approach is too costly for routine use on large, low-cost structures like GRP tanks due to the large number of sensors needed and the complexity of the processing instrumentation. These drawbacks are not important in the military and aerospace application area where the value of the information gained during development testing far outweighs the cost. Once the ability to do planar location efficiently is available, it is possible to:

—identify individual damage sites;
—compensate for signal attenuation in the structure (this gives better quantification of AE sources);

—study the behaviour of damage progression in both static loading and fatigue tests;

—study the effect of deliberately introduced defects without risking total failure of the structure.

Problems which limited the usefulness of older 'computerised' AE systems included 'unstable' location algorithms (this meant that errors in measurement of time arrival due to poor acoustics in the structure gave enormous and variable errors of location). This is largely overcome with modern methods of computing location. Also, modern systems make more efficient use of available sensor channels (older systems required four sensors to be struck to locate the source and had restrictions in the positioning of the sensors). Modern systems allow random positioning of sensors, can display an 'unwrapped' planar view of the structure and are able to locate from only three sensors being struck. In addition, they provide instant 'zonal location' analysis of the low amplitude data (which has reached less than three sensors), so you have complete monitoring, as it happens, of damage occurring anywhere in the structure and are making the most efficient use of the information available.

Success in using AE to do planar location with the resulting analysis ability depends upon the correct choice of operating frequency and sensor location. Figure 4 illustrates what is possible in good circumstances. The data was acquired during loading of a small GRP structure (5 mm wall thickness) which had an artificial defect introduced (saw cut 1·5 mm maximum depth at 45° to vertical axis). Defects are not always 'drawn out'

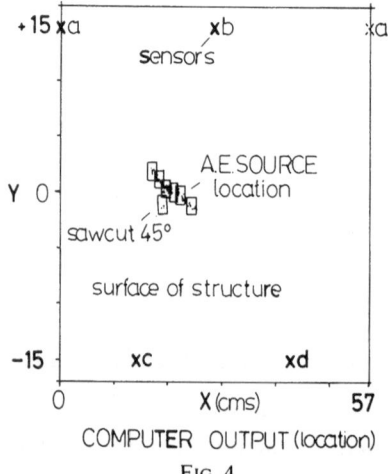

COMPUTER OUTPUT (location)

FIG. 4.

this precisely. The resolution obtained in practice depends upon the acoustics of the structure and individual test set up.

Recent work by D/E has shown the ability when using these methods to discriminate defect types to a limited extent and to find defects which are usually considered 'static' (such as inclusions in a winding). This is most likely a result of the defect changing the local stress field and causing minor damage propagation (this is useful information itself, since you may then be able to study the effect of defect types on the structural integrity under different loading conditions). Defects (deliberately introduced) which have shown this behaviour include:

— metallic inclusions trapped in the windings of a hybrid CFRP structure;
— localised lack of bonding between metallic and composite sections of a hybrid structure;
— small cracks in the metallic part of a hybrid structure which result in damage propagation in the composite at very low loads.

Lack of impregnation results in non-localised (scattered) activity at lower loads than would be expected for a good structure.

The next few years will see a rapid growth in the use of AE techniques on 'high technology' composite applications which, together with simplifications of system operation and increase in analysis speed will enable routine use by far more people.

REFERENCES

1. FOWLER, T. J. and GRAY, E., *Development of an Acoustic Emission Test for FRP Equipment*, American Society of Civil Engineers' convention, April 1979.
2. WADIN, J. R., *Listening to...'cherry picker' booms*, San Juan Capistrano, Dunegan/Endevco, March 1977.
3. COMMITTEE ON ACOUSTIC EMISSION FROM REINFORCED PLASTICS (CARP), C. Howard Adams, *Recommended Practice for Acoustic Emission Testing of Fibreglass Tanks/Vessels*, 37th Annual Conference, Reinforced Plastics/ Composites Institute, The Society of the Plastics Industry, Inc., January 1982.

6

Axisymmetric Elastic Large Deflection Behaviour of Stiffened Composite Plates

G. J. TURVEY

*Department of Engineering, University of Lancaster,
Bailrigg, Lancaster LA1 4YR, England*

ABSTRACT

Governing equations for the axisymmetric, elastic large deflection behaviour of ring-stiffened circular plates are presented. Some of the limitations/problems associated with a Dynamic Relaxation (DR) solution of the finite-difference approximations to these equations are briefly described. Numerical results are presented in dimensionless graphical format for several uniformly loaded, ring-stiffened, composite circular plates with either simply supported or clamped boundaries. The results illustrate the effectiveness of the stiffener in suppressing deflections, etc., for different material combinations and also the reduction in the stiffening efficiency as the lateral pressure increases.

NOTATION

A	$(= Eh(1 - v^2)^{-1})$: Plate in-plane stiffness.
b_s	Width of ring stiffener.
D	$(= Ah^2/12)$: Plate flexural stiffness.
e_r, e_θ	Plate mid-plane radial and tangential strain components.
$e_s, e_{\theta s}$	Ring stiffener eccentricity and axial strain.
E, E_s	Plate and stiffener elastic moduli.
F_r, F_z	Radial and normal plate body forces due to ring stiffener.
h_0, h_s	Plate thickness and ring stiffener depth.
k_r, k_θ	Plate mid-plane radial and tangential curvature components.

72

$k_{\theta s}$ Ring stiffener curvature.

M_r, M_θ Radial and tangential stress couples.

$\bar{M}_r, \bar{M}_\theta$ $(= M_r r_0^2 E_s^{-1} h_0^{-4}, M_\theta r_0^2 E_s^{-1} h_0^{-4})$: Dimensionless radial and tangential stress couples.

M_s Ring stiffener bending moment.

N_r, N_θ Radial and tangential stress resultants.

$\bar{N}_r, \bar{N}_\theta$ $(= N_r r_0^2 E_s^{-1} h_0^{-3}, N_\theta r_0^2 E_s^{-1} h_0^{-3})$: Dimensionless radial and tangential stress resultants.

N_s Ring stiffener axial force.

q Lateral pressure.

q_1 Modified lateral pressure at finite-difference nodes adjacent to ring stiffener.

$\bar{q}$ $(= q r_0^4 E_s^{-1} h_0^{-4})$: Dimensionless lateral pressure.

r_0, r_s Plate and ring stiffener radii.

r, θ, z Polar co-ordinate directions.

T_r Rotational plate body force due to ring stiffener.

u Radial in-plane displacement component.

w Plate deflection.

$\bar{w}$ $(= w h_0^{-1})$: Dimensionless plate deflection.

δr Radial mesh interval.

ν Poisson's ratio.

$(\ )$ Total derivative with respect to r.

Subscripts/Superscripts

c   Refers to the plate centre.

e   Refers to the plate edge.

s   Refers to the stiffened plate.

u   Refers to the unstiffened plate.

1. INTRODUCTION

The elastic small deflection response of stiffened isotropic material plates — particularly stiffened metallic plates — has been the subject of considerable research effort over the past four decades. Much of this effort has been summarised and discussed in the treatise by Troitsky.[1] During the same period a relatively smaller number of studies have been concerned with the elastic and elasto-plastic large deflection response of this type of structural element (*see*, for example, Refs 2–6), though the number of such studies being reported is now rapidly increasing. The advent of modern composite materials has aroused interest in composite stiffened plates. As yet, interest has been focused on buckling (see, for example, Refs 7 and 8) rather than flexural response and, moreover, the small number of works concerned with the latter type of response have been restricted to small deflection considerations. Several types of composite material stiffened plates are readily identified: (1) metal–composite stiffened plates (metal plate with a high modulus composite stiffener), (2) composite stiffened plates (single composite material for both plate and stiffener) and (3) hybrid–composite stiffened plates (one composite material for the plate and another for the stiffener). This categorisation is neither complete nor universally accepted; it merely serves the purpose of the present paper. All three types of composite stiffened plate have a potentially wide range of engineering structural applications, but before they may be realised and properly exploited much basic research into the principal factors governing their behaviour must be undertaken.

The primary purpose of this paper is to make a small contribution to this much needed research effort by carrying out a preliminary investigation of the axisymmetric, non-linear behaviour of circular plates with eccentric rectangular section ring stiffeners subjected to lateral pressure loading. The plate–stiffener combination is modelled using discretely stiffened plate theory, i.e. the effect of the stiffener is incorporated into the plate analysis via statically equivalent local body forces. A computer program based on the Dynamic Relaxation (DR) algorithm has been developed for the numerical solution of the finite difference approximations to the large deflection stiffened circular plate equations. The program has been used to carry out a series of analyses in order to gain a preliminary appreciation of composite ring-stiffened plate response. A stiffened composite plate requires a large number of parameters for its precise definition. It is not feasible, or desirable, to attempt to vary systematically all of these parameters in a preliminary investigation of this type. Accordingly, all of the numerical results have been obtained for one stiffener geometry and one stiffener location. The study is particularly concerned with the effect of different plate and stiffener material combinations on the stiffened plate response. Thus, five plate materials, viz. steel, aluminium, BFRP (Boron Fibre Reinforced Plastic), GFRP (Glass Fibre Reinforced Plastic) and CFRP (Carbon Fibre Reinforced Plastic), in combination with one stiffener material, viz. CFRP, are considered in studying the flexural response of uniformly loaded, simply supported and clamped, ring-stiffened circular plates. A dimensionless graphical format has been used for the presentation of the numerical results in order to illustrate more vividly the effects of the stiffener in reducing the plate deflection, etc., for each material combination and also to demonstrate the loss of effectiveness of the stiffener as the lateral pressure increases.

2. STIFFENED PLATE GEOMETRY, MATERIALS AND RELATIVE WEIGHTS

Figure 1 illustrates a typical diametral section through the eccentric ring-stiffened circular plate. For ease of reference the principal parameters defining the stiffened plate geometry are also marked on Fig. 1 together with the positive co-ordinate system.

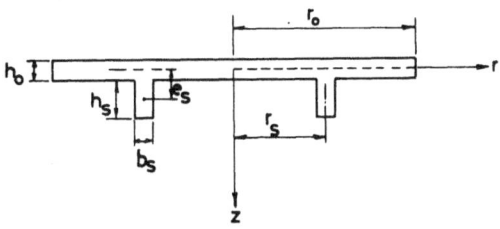

FIG. 1. Diametral section through a ring-stiffened circular plate.

The relevant elastic, etc., properties of the plate and stiffener materials are listed in Table 1. It is evident that the Young's modulus of aluminium is equal to that of quasi-isotropic CFRP and is only marginally smaller than that of quasi-isotropic BFRP. By contrast, from Table 1 it is apparent that the Young's modulus of random mat GFRP (assumed isotropic) is nearly

TABLE 1
Plate and stiffener material properties

Material	Specific gravity (ρ)	Elastic (fibre-direction) modulus (kN/mm^2)	Poisson's ratio (v)
Steel	7·8	200·0	0·30
Aluminium	2·7	70·0	0·30
BFRP*	2·0	78·7	0·32
GFRP**	1·5	7·0	0·30
CFRP*	1·6	70·0	0·30
CFRP***	1·6	180·0	0·28

* Quasi-isotropic; ** random mat; *** uni-directional.

G. J. Turvey

TABLE 2

Geometric properties and relative weights of stiffened and unstiffened composite circular plates

Materials	Plate type	Plate geometry				Relative wt (%)
		$r_0 h_0^{-1}$	$r_s r_0^{-1}$	$h_s h_0^{-1}$	$b_s h_0^{-1}$	
Steel	Unstiffened	50	—	—	—	100·00
Steel–CFRP**	Ring-stiffened	50	0·466 7	4	1	101·64
Aluminium	Unstiffened	50	—	—	—	34·62
Aluminium–CFRP**	Ring-stiffened	50	0·466 7	4	1	36·26
BFRP*	Unstiffened	50	—	—	—	25·64
BFRP*–CFRP**	Ring-stiffened	50	0·466 7	4	1	27·28
GFRP*	Unstiffened	50	—	—	—	19·23
GFRP*–CFRP**	Ring-stiffened	50	0·466 7	4	1	20·87
CFRP*	Unstiffened	50	—	—	—	20·51
CFRP*–CFRP**	Ring-stiffened	50	0·466 7	4	1	22·15

* Isotropic/quasi-isotropic; ** uni-directional.

thirty times smaller than that of steel. These similarities/disparities in elastic moduli were found to have implications for the numerical computations (*see* Section 4).

Details of the geometric properties and the relative weights of each of the unstiffened plates considered in the paper are listed in Table 2. A two-material descriptor is used for each of the stiffened plates. The first material appearing in this descriptor is that of the plate and the second is that of the stiffener. The relative weight of each plate is expressed as a percentage of the weight of the unstiffened steel plate. It is, therefore, evident from Table 2 that the lightest plates are about five times lighter than the unstiffened steel plate.

3. STIFFENED PLATE EQUATIONS (AXISYMMETRIC DEFORMATIONS)

The discretely stiffened plate analysis presented here, albeit in outline only, belongs to the category which represents the stiffener in terms of equivalent body forces in the plate equilibrium equations. Several sets of such stiffened plate equations have been developed, though almost exclusively within the context of a rectangular Cartesian co-ordinate system. Polar forms of these equations appear, in the author's experience, to be rare. The present axisymmetric ring-stiffened plate equations have been derived as degenerate forms of more general non-axisymmetric stiffened plate equations. The derivation of the latter equations employs assumptions and procedures similar to those used by Basu *et al.*[2] in the derivation of their discretely stiffened rectangular plate theory. Space restrictions do not allow a detailed derivation of the ring-stiffened plate equations to be presented here. Instead, only the final forms of the plate and stiffener equilibrium, compatibility and constitutive equations are separately presented.

3.1. Equilibrium Equations
(*a*) *Plate equilibrium equations*

The in-plane and out-of-plane plate equilibrium equations may be expressed as follows:

$$N_r^{\cdot} + r^{-1}(N_r - N_\theta) + \boxed{\delta r^{-1} F_r} = 0$$

$$M_r^{\cdot} + r^{-1}(2M_r^{\cdot} - M_\theta^{\cdot}) + N_r w^{\cdot\cdot} + N_\theta r^{-1} w^{\cdot} + q + \boxed{\delta r^{-1} F_z} = 0 \tag{1}$$

In eqns (1) the terms in the dashed-line boxes represent the body force contributions of the ring stiffener to the plate equilibrium. A third body force—the couple, T_r—exists due to the rotational deformation of the ring-stiffener, but is not shown explicitly in eqns (1). This body force is introduced by modifying the q-term in the second of eqns (1) to:

$$q_1 = q \pm \tfrac{1}{2}\delta r^{-2} T_r \qquad (2)$$

at the nodes immediately adjacent to the stiffener node in the finite-difference approximations of the plate equilibrium equations.

(b) Stiffener equilibrium equations

The stiffener equilibrium equations may be expressed as follows:

$$F_r = -r^{-1} N_s$$
$$F_z = N_s r^{-1} w^{\cdot}$$
$$T_r = r^{-1}(M_s + N_s e_s) \qquad (3)$$

3.2. Compatibility Equations
(a) Plate compatibility equations

These equations take their usual form for axisymmetric, non-linear deformations:

$$e_r = u^{\cdot} + \tfrac{1}{2}(w^{\cdot})^2 .$$
$$e_\theta = r^{-1} u$$
$$k_r = -w^{\cdot\cdot}$$
$$k_\theta = -r^{-1} w^{\cdot} \qquad (4)$$

(b) Stiffener compatibility equations

For axisymmetric deformations these relationships simplify to:

$$e_{\theta s} = r^{-1}(u - e_s w^{\cdot})$$
$$k_{\theta s} = k_\theta \qquad (5)$$

3.3. Constitutive Equations
(a) Plate constitutive equations

These equations assume the following forms for isotropic plate materials:

$$N_r = A(e_r + v e_\theta)$$
$$N_\theta = A(v e_r + e_\theta)$$
$$M_r = D(k_r + v k_\theta)$$
$$M_\theta = D(v k_r + k_\theta) \qquad (6)$$

(*N.B.* The form of eqns (6) implies that any in-plane–out-of-plane coupling arising from the use of quasi-isotropic lay-ups in the composite material plates has been ignored.)

(*b*) *Stiffener constitutive equations*

The rectangular section ring-stiffener constitutive equations may be expressed as follows:

$$N_s = E_s h_s b_s e_{\theta s}$$

$$M_s = (E_s h_s^3 b_s / 12) k_{\theta s} \tag{7}$$

3.4. Boundary Conditions

For ring-stiffened circular plates the question of stiffener boundary conditions does not arise. Only two sets of plate edge boundary conditions are considered in the paper, viz. simply supported and clamped. In both cases complete in-plane fixity is assumed. The plate boundary conditions imply that the following restrictions on the problem variables exist around the edge:

(*a*) *Simply supported edge* (in-plane fixed)

$$u = w = M_r = 0 \tag{8a}$$

(*b*) *Clamped edge* (in-plane fixed)

$$u = w = w^{\cdot} = 0 \tag{8b}$$

At the plate centre a mixed set of conditions must be satisfied:

$$u = w^{\cdot} = 0$$

$$N_r = N_\theta$$

$$M_r = M_\theta \tag{8c}$$

In addition to eqns (8c) the variables: w, N_r, N_θ, M_r and M_θ exhibit point symmetry and u exhibits point antisymmetry about the plate centre.

4. COMMENTS ON COMPUTATIONAL PROCEDURE

As mentioned earlier in the paper, the DR-algorithm was used to solve the finite-difference approximations to the governing equations of ring-stiffened circular plates, i.e. eqns (1)–(8). The DR procedure has been adequately documented elsewhere[9-11] and, therefore, a detailed description is out of place here. Instead, only the following comments are offered.

All of the calculations were carried out on two relatively coarse ($7\frac{1}{2}$-interval), uniform, interlacing finite-difference meshes. The use of relatively coarse, uniform finite-difference meshes not only limits the solution accuracy, but also severely restricts the available locations for the ring-stiffener. Whilst this practice may be justified for the preliminary study described here, the use of graded meshes offers a more flexible approach.

Some convergence difficulties were experienced with the DR-iterative procedure, particularly in the case of GFRP-CFRP stiffened plates. Indeed, no converged solutions were obtained for these plates. It is thought that the high ratio, stiffener elastic modulus: plate elastic modulus ($\simeq 26$), is the primary cause of these computational problems and that a series of special measures (including, for example, the use of damping factors at the stiffener attachment nodes which are different from those at the other plate nodes) would have to be employed to achieve satisfactory solution convergence. However, these thoughts belong to the realms of conjecture until time permits these difficulties to be investigated and resolved.

5. DISCUSSION OF STIFFENED PLATE PARAMETER STUDY RESULTS

As is clear from Table 2 all of the computed results have been obtained for one stiffened plate geometry (see also Fig. 1) in which the radius:thickness ratio (r_0/h_0) is 50, the ring stiffener:plate radius ratio (r_s/r_0) is 0·4667 and the stiffener depth:plate thickness (h_s/h_0) and stiffener width:plate thickness (b_s/h_0) ratios are respectively 4 and 1. The choice of these particular geometric parameters was arbitrary but, nevertheless, convenient for the present study in which attention is focused on the large deflection response of ring-stiffened circular plates for different composite material combinations. Although five different plate materials are listed in Table 2, the computational effort has been somewhat less than would normally be associated with this number of plate materials. Because the elastic properties of aluminium and quasi-isotropic CFRP are identical (see Table 2) and, moreover, because material coupling is ignored in eqns (6) identical results are computed for aluminium-CFRP and CFRP-CFRP ring-stiffened plates. It should, therefore, be appreciated that all of the results for aluminium-CFRP plates shown in Figs 2–5 apply equally well to

CFRP–CFRP plates but that the latter are about one-third lighter in weight. Also, for the reasons stated under Section 4 the computations for GFRP–CFRP ring-stiffened plates had to be abandoned.

The main results of the parameter study which illustrate the axisymmetric large deflection response of composite stiffened circular plates are presented in Figs 2–4. The load-plate centre deflection response for simply supported and clamped ring-stiffened plates are shown in Figs 2(a) and 2(b) respectively. On these graphs the full-line curves represent the stiffened plate results and the dashed-line curves represent the results for a similar plate but with the stiffener absent. It is evident from these figures that the CFRP ring stiffener causes a reduction in the plate centre deflection. The difference between unstiffened and stiffened plate centre deflections increases as the plate elastic modulus reduces and reduces as the lateral pressure increases. A few results are denoted by crosses on Figs 2(a) and 2(b). They are the results computed for BFRP–CFRP ring-stiffened circular plates and they fall, as might have been anticipated, only slightly below the corresponding aluminium–CFRP (or CFRP–CFRP) stiffened plate values. Comparing Figs 2(a) and 2(b) it appears that the presence of a ring stiffener is much more effective in reducing the plate centre deflection, i.e. increasing the plate lateral stiffness, than changing the plate edge condition from simply supported to clamped.

The plate centre stress couples versus lateral pressure are plotted in Figs 3(a) and 3(b) for simply supported and clamped edges respectively. In general these stress couples are larger in the clamped than in the simply supported edge plates. This is because the membrane stresses are more fully developed (for a given value of lateral pressure) in the latter plates. It is interesting to observe that in steel–CFRP plates the effect of the stiffener is to reduce the magnitude of the stress couple over the whole of the pressure range considered, whereas for aluminium–CFRP plates sustaining high lateral pressures the central stress couple exceeds the unstiffened value. Again, the central stress couples for simply supported and clamped BFRP–CFRP ring-stiffened circular plates are denoted by crosses on these two figures. Also on each of Figs 3(a) and 3(b) a third dashed-line curve is shown. These two curves correspond respectively to simply supported and clamped, unstiffened GFRP plates and serve to confirm that the lateral load is being supported predominantly by membrane action.

Figures 4(a) and 4(b) depict the variation of the central stress resultants with lateral pressure for simply supported and clamped, stiffened and unstiffened, circular plates respectively. It is obvious from these figures that these stress resultants are substantially reduced when the plate is stiffened

G. J. Turvey

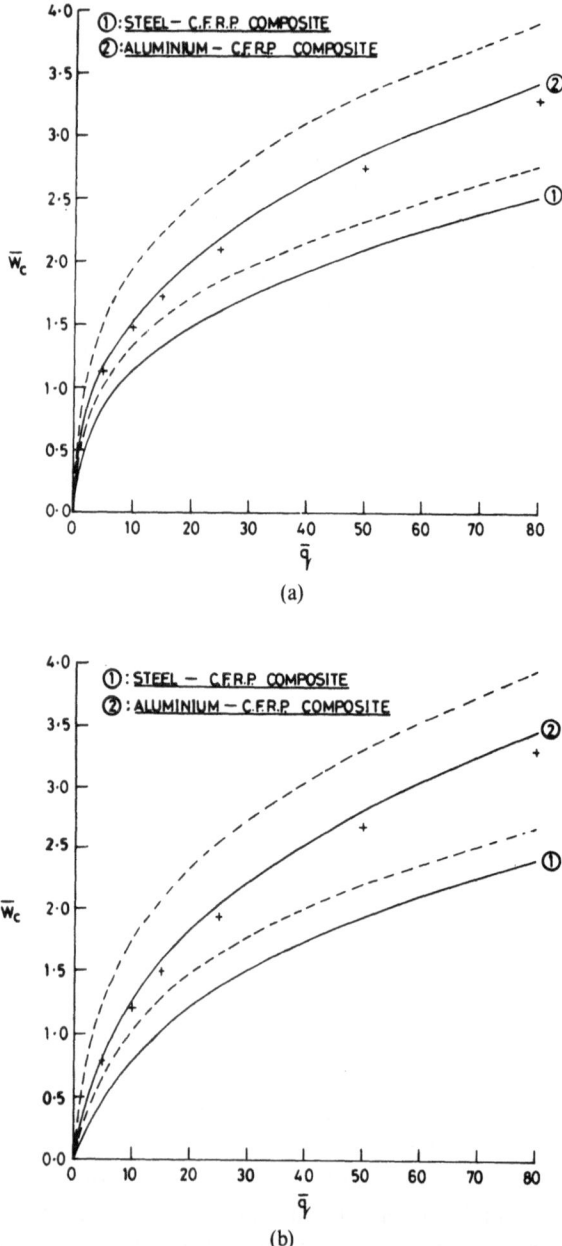

(a)

(b)

Fig. 2. Plate centre deflection versus lateral pressure for uniformly loaded ring-stiffened composite circular plates: (a) simply supported edge; (b) clamped edge.

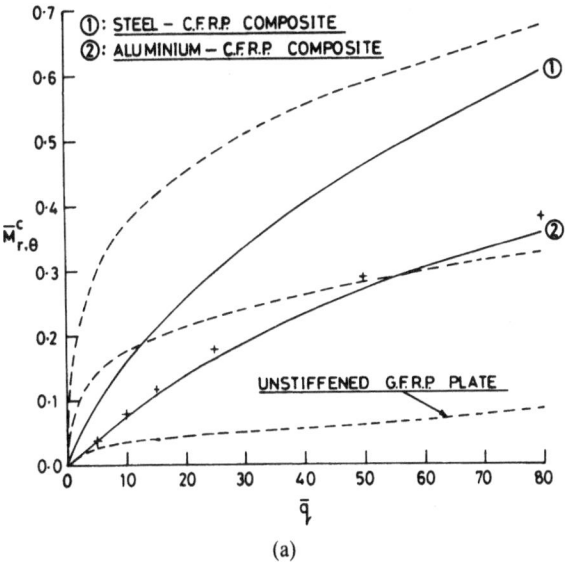

(a)

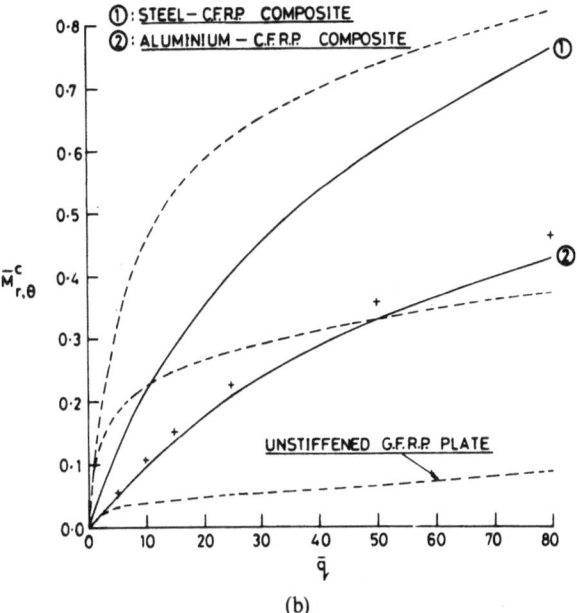

(b)

FIG. 3. Central stress couple versus lateral pressure for uniformly loaded ring-stiffened composite circular plates: (a) simply supported edge; (b) clamped edge.

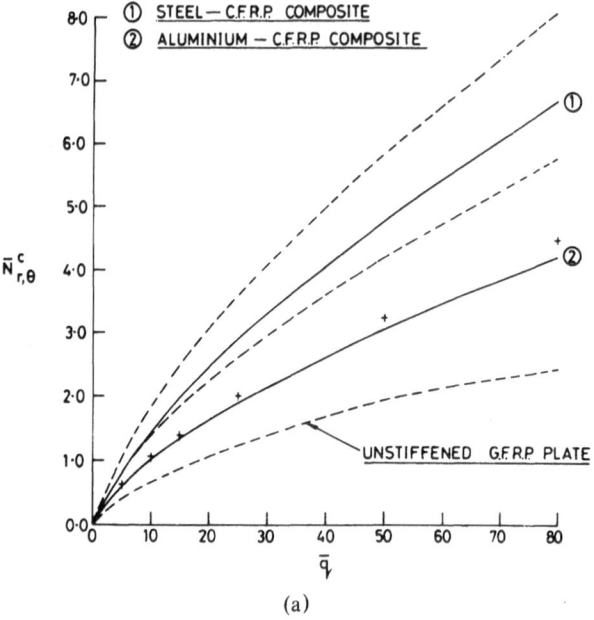

(a)

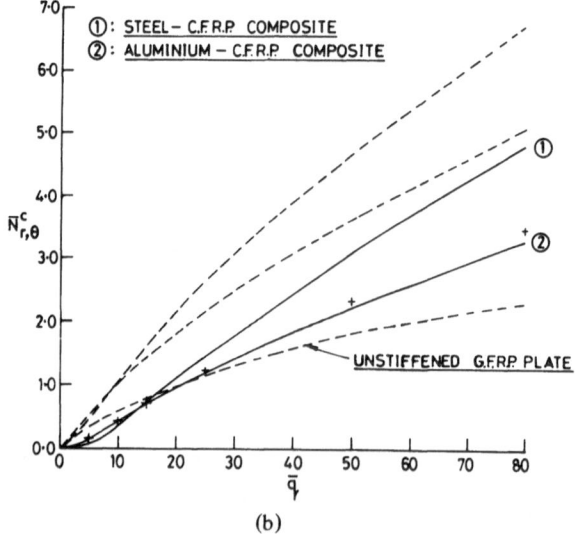

(b)

FIG. 4.   Central stress resultant versus lateral pressure for uniformly loaded ring-stiffened composite circular plates: (a) simply supported edge; (b) clamped edge.

by a ring stiffener—this is particularly so when the plate edge is clamped. The BFRP–CFRP stiffened plate results are indicated on these figures by crosses and the third dashed-line curve on each figure represents the unstiffened GFRP plates results. These latter central stress resultants serve to corroborate the earlier statement that membrane action is the dominant load supporting mode for the unstiffened GFRP plates.

The final set of results, which are simply the stiffened plate results expressed as percentages of the unstiffened plate results and plotted against the lateral pressure, are presented in Figs 5(a)–(c). The results of the parameter study are presented in this alternative format in an attempt to quantify the efficiency/effectiveness of the CFRP ring stiffener in its stiffening role as the lateral pressure increases. Thus, referring to Fig. 5(a), it is apparent that at low pressures the lateral stiffness is improved by the CFRP stiffener by between 15 % and 29 % for steel–CFRP plates and by between 22 % and 32 % for aluminium–CFRP plates depending on whether the plate edge is simply supported or clamped. However, at high lateral pressures the improvement in lateral stiffness due to the CFRP stiffener is much less—varying between 9 % and 12 % for steel–CFRP and 10 % and 12 % for aluminium–CFRP plates according to the edge support condition. In some design situations it is beneficial to keep in mind that even though the efficiency of the ring stiffener is load dependent, significant stiffness improvements may accrue with only a small increase in overall weight due to the presence of a stiffener. According to Table 2 the weight increase penalty incurred by stiffening the steel plate is about 1·6 % and for the aluminium plate it is about 4·7 %.

The effectiveness of the ring stiffener in reducing the plate centre stress couples as the lateral pressure increases is illustrated in Fig. 5(b). It is evident that the plate edge conditions only have a marginal influence on the loss of effectiveness of the ring stiffener as the lateral pressure increases. Thus, at low pressures the presence of the ring stiffener causes a reduction in the central stress couples to between 20 % and 40 % of the unstiffened plate value depending on the type of plate material. At higher pressures, however, the reduction is only marginal—about 10 %—in aluminium–CFRP plates, whereas in steel–CFRP plates the central stress couples are actually about 10 % greater than those in similar unstiffened steel plates. It is believed that these differences in the stiffened and unstiffened plate centre stress couples may be explained by slower development of membrane action in the stiffened plates due to the presence of the ring stiffener.

Finally, in Fig. 5(c) the effectiveness of the ring stiffener in limiting/changing the radial edge stress couple in clamped circular plates is

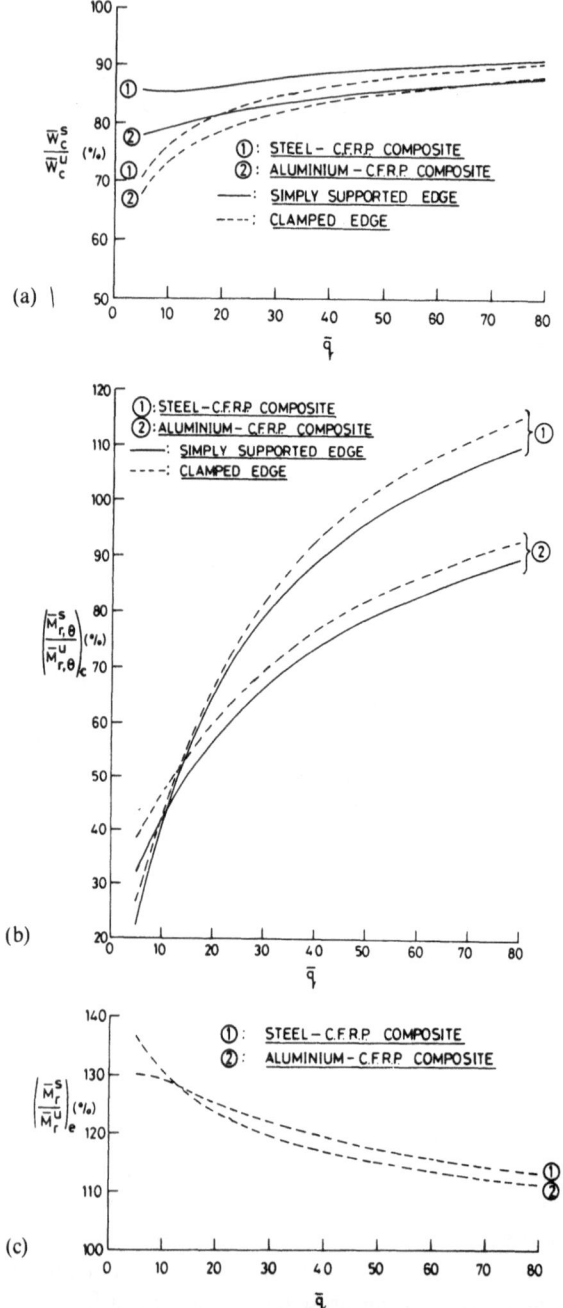

FIG. 5. Stiffened: unstiffened plate parameter ratios versus lateral pressure for uniformly loaded, simply supported and clamped, composite circular plates: (a) central deflection ratio; (b) central stress couple ratio; (c) edge radial stress couple ratio (clamped plate only).

illustrated. The type of plate material appears only to have a marginal influence. At low loads the presence of the ring stiffener tends to produce an increase in the radial edge stress couple of about 30 % compared with the unstiffened plate value. However, as the load increases the effect of the ring stiffener changes so that there is only about a 12 % amplification in the stress couple.

6. CONCLUSIONS

The governing equations of ring-stiffened circular plates undergoing axisymmetric, elastic large deformations have been presented. Solutions to the finite-difference approximations to these equations have been obtained using the DR-iterative procedure. The procedure was found to converge well, i.e. without recourse to special procedures, when the ratio, stiffener elastic modulus: plate elastic modulus, was small but not when it was large ($\simeq 26$) as with GFRP–CFRP ring-stiffened plates. Numerical results have been presented for the primary quantities of interest, viz. deflections, etc., for several stiffened composite material circular plates with either simply supported or clamped edges and subjected to uniform lateral pressure. The results demonstrate, for example, that a ring stiffener may be a more effective means of reducing plate centre deflection than changing the edge support condition from simply supported to clamped. Also, some quantification of the loss of effectiveness of the ring stiffener in reducing deflections, etc., as the lateral pressure increases is provided.

ACKNOWLEDGEMENTS

The author wishes to express his appreciation to the Department of Engineering for providing computing facilities and to his father, Mr G. Turvey, for preparing tracings of the figures.

REFERENCES

1. TROITSKY, M. S., *Stiffened Plates—Bending, Stability and Vibrations*, Elsevier Scientific Publishing Company, 1976.
2. BASU, A. K., DJAHANI, P. and DOWLING, P. J., *Elastic Post-Buckling Behaviour of Discretely Stiffened Plates*, Int. Coll. on Stability of Steel Structures, Liege, April 1977.

3. WEGMULLER, A., Full-range analysis of eccentrically stiffened plates, *ASCE J. Struct. Div.*, **100** (1974), 143–159.
4. CRISFIELD, M. A., *Large-Deflection Elasto-Plastic Buckling Analysis of Eccentrically Stiffened Plates Using Finite Elements*, TRRL Report 725, 1976.
5. WEBB, S. E. and DOWLING, P. J., Large-deflexion elasto-plastic behaviour of discretely stiffened plates, *Proc. ICE*, **69** (1980), 375–401.
6. PUTHLI, R. S., Collpase analysis of three dimensional assemblages of eccentrically stiffened hot rolled steel plates and shallow shells, *Heron*, **25** (1980), 1–45.
7. TURVEY, G. J. and WITTRICK, W. H., The influence of orthotropy on the stability of some multi-plate structures in compression, *Aero. Quart.*, **24** (1973), 1–8.
8. WILLIAMS, J. G. and STEIN, M., Buckling behaviour and structural efficiency of open-section stiffened composite compression panels, *AIAA J.*, **14** (1976), 1618–1626.
9. DAY, A. S., An introduction to dynamic relaxation, *The Engineer*, **219** (1965), 218–221.
10. CASSELL, A. C. and HOBBS, R. E., Numerical stability of dynamic relaxation analysis of non-linear structures, *Int. J. Num. Meth. Eng.*, **10** (1976), 1407–1410.
11. TURVEY, G. J., Large deflection of tapered annular plates by dynamic relaxation, *ASCE J. Eng. Mech. Div.*, **104** (1978), 351–366.

7

Vibration of Web-Stiffened Foam Sandwich Panel Structures

C. C. Chao, C. C. Wang and C. Y. Chan

*Department of Power Mechanical Engineering,
National Tsing Hua University, Hsinchu, Taiwan*

ABSTRACT

This paper presents free vibration of rectangular web-stiffened foam sandwich panels simply supported at all four edges. The sandwich structure is composed of orthotropic facing and a number of equally spaced identical stiffeners embedded in the polyurethane core. Transverse shear deformations and rotatory inertia are considered in the thick laminate formulation and the free vibration is treated by way of the panel segment method. Equations of motion of stiffeners are solved in connection with plate–beam interface continuity and panel boundary conditions. Natural frequencies and mode shapes are obtained as a result of non-trivial solution of a system of homogeneous equations.

INTRODUCTION

Sandwich structures with composite facing are noted for high strength and high stiffness at lower weight in aerospace engineering. To alleviate the high cost of the honeycomb core material and to facilitate the joint design of structural elements, the web-stiffened foam sandwich panel seems to be a solution to the problem in composite structural engineering. It is composed of the upper and lower faces and a number of equally spaced identical web-stiffeners embedded in the polyurethane (PU) filled core as shown in Fig. 1. The simply supported rectangular sandwich panel may be conceived as a composite beam–plate combination with isotropic core material and orthotropic facing and web-stiffeners.

Vibration studies of isotropic stiffened plates using the composite beam–plate theory were attributed to Wah[1] and Long.[2] In 1980, Chao and Lee[3] performed an analysis on the vibration of eccentrically stiffened laminates. On the other hand, investigation of web-stiffened sandwich structures has been made by Chen et al.,[4,5] while the foam-filled sandwich panels were analysed by Chong and Hartsock.[6,7] The present research on the vibration behavior of web-stiffened foam sandwich panel structure is a combined effort of them all. The problem is treated through the panel segment method in a way similar to Ref. 3 except that transverse shear deformation is included in addition.

The process involves simultaneous solution of equations of motion of the stiffeners in connection with all the beam–plate interface continuity and boundary conditions. Assuming a set of appropriate displacement functions, a system of homogeneous equations in the undetermined mode shape parameters is derived. Condition for the existence of non-trivial solution that requires vanishing of the coefficient determinant gives the natural frequencies and the corresponding mode shapes. Results for the first four modes as obtained by the classical lamination and present theories are compared for the foam sandwich panels with and without web-stiffeners.

FORMULATION OF PLATE EQUATIONS

In the present study the plate equations are derived in accordance with thick plate theory with transverse shear deformation taken into account. All plate segments are identical in geometry and material properties, and so are the stiffeners. As shown in Fig. 1 the local co-ordinates x, y and z are for the plate segments and $\bar{x}$, $\bar{y}$ and $\bar{z}$ for the web-stiffeners. The corresponding displacements of the mid-plane are denoted by u_0, v_0, and w for the plate segments and $\bar{u}$, $\bar{v}$ and $\bar{w}$ for the stiffeners. The displacements in the plate segments with transverse shear deformation can be given by

$$u = u_0(x, y, t) + z\psi_x(x, y, t)$$
$$v = v_0(x, y, t) + z\psi_y(x, y, t)$$
$$w = w(x, y, t) \tag{1}$$

where ψ_x and ψ_y are rotations of the cross-sections perpendicular to the x- and y-axis, respectively.

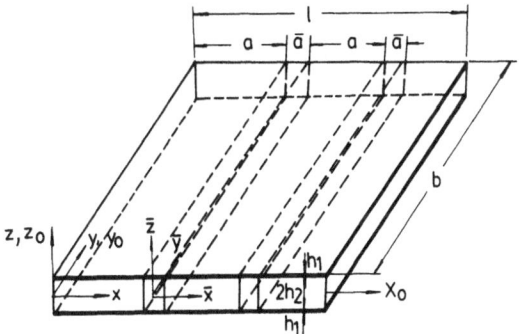

FIG. 1.   A typical web-stiffened foam sandwich panel.

From the strain–displacement relations, the plate deformation is obtained in terms of the mid-plane strains and curvatures

$$\{\varepsilon\} = \{\varepsilon^0\} + z\{\kappa\}$$ (2)

with

$$\varepsilon_x^0 = u_{0,x} \qquad \varepsilon_y^0 = v_{0,y} \qquad \gamma_{xy}^0 = u_{0,y} + v_{0,x}$$

$$\gamma_{yz}^0 = w_{,y} + \psi_y \qquad \gamma_{xz}^0 = w_{,x} + \psi_x$$

$$\kappa_x = \psi_{x,x} \qquad \kappa_y = \psi_{y,y} \qquad \kappa_{xy} = \psi_{x,y} + \psi_{y,x}$$ (3)

For orthotropic material as in the present study of the web-stiffened foam sandwich structure, the stress–strain relations can be expressed in the form:

$$\begin{Bmatrix} \sigma_{xx} \\ \sigma_{yy} \\ \sigma_{xy} \\ \sigma_{yz} \\ \sigma_{xz} \end{Bmatrix} = \begin{bmatrix} \bar{Q}_{11} & \bar{Q}_{12} & \bar{Q}_{16} & 0 & 0 \\ \bar{Q}_{12} & \bar{Q}_{22} & \bar{Q}_{26} & 0 & 0 \\ \bar{Q}_{16} & \bar{Q}_{26} & \bar{Q}_{66} & 0 & 0 \\ 0 & 0 & 0 & \bar{Q}_{44} & \bar{Q}_{45} \\ 0 & 0 & 0 & \bar{Q}_{45} & \bar{Q}_{55} \end{bmatrix} \left(\begin{Bmatrix} \varepsilon_x^0 \\ \varepsilon_y^0 \\ \gamma_{xy}^0 \\ \gamma_{yz}^0 \\ \gamma_{xz}^0 \end{Bmatrix} + Z \begin{Bmatrix} \kappa_x \\ \kappa_y \\ \kappa_{xy} \\ 0 \\ 0 \end{Bmatrix} \right)$$ (4)

in which $\bar{Q}_{ij}$, with $i, j = 1, 2, 6$, are the conventional reduced in-plane stiffnesses for the state of plane stress;[8,9] and those with $i, j = 4, 5$ account for the transverse shear deformation.[10] The resultant forces N_x, N_y, N_{xy}, resultant moments M_x, M_y, M_{xy}, and resultant transverse shear forces Q_y, Q_x are related to the plate deformation variables in the form

$$\begin{Bmatrix} N \\ M \end{Bmatrix} = \begin{bmatrix} A & B \\ \hline B & D \end{bmatrix} \begin{Bmatrix} \varepsilon^0 \\ \kappa \end{Bmatrix}$$ (5)

and

$$\{Q\} = [H]\{\gamma\}$$

where

$$\{\gamma\} = \begin{Bmatrix} \gamma_{yz}^0 \\ \gamma_{xz}^0 \end{Bmatrix} \quad \text{and} \quad [H] = \begin{bmatrix} A_{44} & A_{45} \\ A_{45} & A_{55} \end{bmatrix} \tag{6}$$

Considering the stress and moment resultants, the plate stress equations of motion can be written as

$$N_{x,x} + N_{xy,y} = \bar{\rho}\ddot{u}_0 + R\ddot{\psi}_x$$

$$N_{xy,x} + N_{y,y} = \bar{\rho}\ddot{v}_0 + R\ddot{\psi}_y$$

$$M_{x,x} + M_{xy,y} - Q_x = R\ddot{u}_0 + I\ddot{\psi}_x$$

$$M_{xy,x} + M_{y,y} - Q_y = R\ddot{v}_0 + I\ddot{\psi}_y$$

$$Q_{x,x} + Q_{y,y} + q = \bar{\rho}\ddot{w} \tag{7}$$

where q is the transverse load and

$$(\bar{\rho}, R, I) = \sum_{i=1}^{N} \rho_i \int_{z_{i-1}}^{z_i} (1, Z, Z^2)\, \mathrm{d}Z \tag{8}$$

with ρ_i as mass density of the ith layer, $Z_i - Z_{i-1}$ its thickness.

For the web-stiffened foam sandwich structures made up of symmetrical specially orthotropic laminated composite plates, it is known that all $B_{ij} = 0$ and $R = 0$. Also, $A_{16} = A_{26} = A_{45} = 0$, $D_{16} = D_{26} = 0$. The displacement equations of motion for each sandwich plate segments are obtained as

$$A_{11}u_{0,xx} + A_{66}u_{0,yy} + (A_{12} + A_{66})v_{0,xy} = \bar{\rho}\ddot{u}_0$$

$$(A_{12} + A_{66})u_{0,xy} + A_{66}v_{0,xx} + A_{22}v_{0,yy} = \bar{\rho}\ddot{v}_0$$

$$D_{11}\psi_{x,xx} + D_{66}\psi_{x,yy} + (D_{12} + D_{66})\psi_{y,xy} - A_{55}(w_{,x} + \psi_x) = I\ddot{\psi}_x$$

$$(D_{12} + D_{66})\psi_{x,xy} + D_{66}\psi_{y,xx} + D_{22}\psi_{y,yy} - A_{44}(w_{,y} + \psi_y) = I\ddot{\psi}_y$$

$$A_{55}w_{,xx} + A_{44}w_{,yy} + A_{55}\psi_{x,x} + A_{44}\psi_{y,y} + q = \bar{\rho}\ddot{w} \tag{9}$$

Subject to the simply supported boundary conditions at the opposite edges $y = 0$ and b

$$w = 0 \qquad u = 0 \qquad \psi_x = 0$$

$$M_y = D_{12}\psi_{x,x} + D_{22}\psi_{y,y} = 0$$

$$N_y = A_{12}u_{0,x} + A_{22}v_{0,y} = 0 \tag{10}$$

FREE VIBRATION OF SANDWICH PLATE SEGMENTS

Let us consider the problem of free vibration of the sandwich plate segments. To satisfy the equations of motion and the boundary conditions, we assume

$$u_0 = U(x) \sin \alpha y \, e^{i\omega t} \qquad v_0 = V(x) \cos \alpha y \, e^{i\omega t}$$

$$w = W(x) \sin \alpha y \, e^{i\omega t} \qquad \psi_x = \Psi_x(X) \sin \alpha y \, e^{i\omega t}$$

$$\psi_y = \Psi_y(X) \cos \alpha y \, e^{i\omega t} \tag{11}$$

in which $\alpha = m\pi/b$.

Substituting eqn. (11) into eqn. (9) yields

$$A_{11} U'' - (A_{66}\alpha^2 - \bar{\rho}\omega^2)U - (A_{12} + A_{66})\alpha V' = 0 \tag{12a}$$

$$A_{66} V'' - (A_{22}\alpha^2 - \bar{\rho}\omega^2)V + (A_{12} + A_{66})\alpha U' = 0 \tag{12b}$$

$$D_{11}\Psi_x'' - (D_{22}\alpha^2 + A_{55} - I\omega^2)\Psi_x - (D_{12} + D_{66})\alpha\Psi_y' - A_{55}W' = 0 \tag{12c}$$

$$D_{66}\Psi_y'' - (D_{22}\alpha^2 + A_{44} - I\omega^2)\Psi_y + (D_{12} + D_{66})\alpha\Psi_x' - A_{44}\alpha W = 0 \tag{12d}$$

$$A_{55} W'' - (A_{44}\alpha^2 - \bar{\rho}\omega^2)W + A_{55}\Psi_x' - A_{44}\alpha\Psi_y = 0 \tag{12e}$$

The prime denotes differentiation with respect to x and the transverse load $q = 0$ for the consideration of free vibration analysis.

It is seen that equations of $U(x)$ and $V(x)$ are uncoupled with those of $W(x)$, $\Psi_x(x)$ and $\Psi_y(x)$. For the existence of a non-trivial solution, the determinant of the coefficient matrix must vanish for each group. For the in-plane part, we define a discriminant

$$D = \begin{Bmatrix} D_1 \\ D_2 \end{Bmatrix} = \frac{-c \pm d}{2A_{11}A_{66}}$$

with

$$c = -[A_{11}(A_{22}\alpha^2 - \bar{\rho}\omega^2) + A_{66}(A_{66}\alpha^2 - \bar{\rho}\omega^2) - (A_{12} + A_{66})^2\alpha^2]$$

$$d = [c^2 - 4A_{11}A_{66}(A_{22}\alpha^2 - \bar{\rho}\omega^2)(A_{66}\alpha^2 - \bar{\rho}\omega^2)]^{1/2}$$

Then, it is found that for $D \geq 0$:

$$U(x) = A_1 \cosh \lambda_1 x + A_2 \sinh \lambda_1 x + A_3 \cosh \lambda_2 x + A_4 \sinh \lambda_2 x$$
$$V(x) = \beta_1 A_1 \sinh \lambda_1 x + \beta_1 A_2 \cosh \lambda_1 x + \beta_2 A_3 \sinh \lambda_2 x$$
$$+ \beta_2 A_4 \cosh \lambda_2 x \tag{13}$$

for $D_1 \geq 0 > D_2$:

$$U(x) = A_1 \cosh \lambda_1 x + A_2 \sinh \lambda_1 x + A_3 \cos \lambda_3 x + A_4 \sin \lambda_3 x$$
$$V(x) = \beta_1 A_1 \sinh \lambda_1 x + \beta_1 A_2 \cosh \lambda_1 x - \beta_3 A_3 \sin \lambda_3 x$$
$$+ \beta_3 A_4 \cos \lambda_3 x \tag{14}$$

and for $D < 0$:

$$U(x) = A_1 \cos \lambda_3 x + A_2 \sin \lambda_3 x + A_3 \cos \lambda_4 x + A_4 \sin \lambda_4 x$$
$$V(x) = -\beta_3 A_1 \sin \lambda_3 x + \beta_3 A_2 \cos \lambda_3 x - \beta_4 A_3 \sin \lambda_4 x$$
$$+ \beta_4 A_4 \cos \lambda_4 x \tag{15}$$

where

$$\begin{Bmatrix} \lambda_1 \\ \lambda_2 \end{Bmatrix} = \left(\frac{-c \pm d}{2 A_{11} A_{66}} \right)^{1/2} \qquad \begin{Bmatrix} \lambda_3 \\ \lambda_4 \end{Bmatrix} = \left(\frac{c \pm d}{2 A_{11} A_{66}} \right)^{1/2}$$
$$\beta_i = [A_{11}\lambda_i - (A_{66}\alpha^2 - \bar{\rho}\omega^2)]/(A_{12} + A_{66})\alpha\lambda_i \qquad i = 1, 2, 3, 4$$

Equations (12c)–(12e) are three coupled homogeneous ordinary differential equations of order six. Assuming the solution of Ψ_x, Ψ_y and W in the exponential form e^{nx} yields a third order characteristic equation in n^2. In the present study, three real roots are obtained for n^2, i.e. two unequal positive and one negative, of which the square roots are δ_1, δ_2 and δ_3. The general solution can be obtained in the form:

$$W(x) = C_1 \cosh \delta_1 x + C_2 \sinh \delta_1 x + C_3 \cosh \delta_2 x + C_4 \sinh \delta_2 x$$
$$+ C_5 \cos \delta_3 x + C_6 \sin \delta_3 x$$
$$\Psi_x(x) = \gamma_1 C_2 \cosh \delta_1 x + \gamma_1 C_1 \sinh \delta_1 x + \gamma_3 C_4 \cosh \delta_2 x$$
$$+ \gamma_3 C_3 \sinh \delta_2 x + \gamma_5 C_6 \cos \delta_3 x - \gamma_5 C_5 \sin \delta_3 x$$
$$\Psi_y(x) = \gamma_2 C_1 \cosh \delta_1 x + \gamma_2 C_2 \sinh \delta_1 x + \gamma_4 C_3 \cosh \delta_2 x$$
$$+ \gamma_4 C_4 \sinh \delta_2 x + \gamma_6 C_5 \cos \delta_3 x + \gamma_6 C_6 \sin \delta_3 x \tag{16}$$

in which C_1, C_2, C_3, C_4, C_5, C_6 are undetermined coefficients and

$$\begin{Bmatrix} \gamma_1 \\ \gamma_3 \\ \gamma_5 \end{Bmatrix} = A_{55}D_{66} \begin{Bmatrix} \delta_1^3/\Delta_1 \\ \delta_2^3/\Delta_2 \\ -\delta_3^3/\Delta_3 \end{Bmatrix}$$

$$+ [A_{44}(D_{12} + D_{66})\alpha^2 - A_{55}(D_{22}\alpha^2 + A_{44} - I\omega^2)] \begin{Bmatrix} \delta_1/\Delta_1 \\ \delta_2/\Delta_2 \\ \delta_3/\Delta_3 \end{Bmatrix}$$

$$\begin{Bmatrix} \gamma_2 \\ \gamma_4 \\ \gamma_6 \end{Bmatrix} = [A_{44}D_{11} - A_{55}(D_{12} + D_{66})]\alpha \begin{Bmatrix} \delta_1^2/\Delta_1 \\ \delta_2^2/\Delta_2 \\ -\delta_3^2/\Delta_3 \end{Bmatrix}$$

$$- A_{44}(D_{66}\alpha^2 + A_{55} - I\omega^2)\alpha \begin{Bmatrix} 1/\Delta_1 \\ 1/\Delta_2 \\ 1/\Delta_3 \end{Bmatrix}$$

$$\begin{Bmatrix} \Delta_1 \\ \Delta_2 \\ \Delta_3 \end{Bmatrix} = \left[D_{11} \begin{Bmatrix} \delta_1^2 \\ \delta_2^2 \\ -\delta_3^2 \end{Bmatrix} - (D_{66}\alpha^2 + A_{55} - I\omega^2) \right]$$

$$\times \left[D_{66} \begin{Bmatrix} \delta_1^2 \\ \delta_2^2 \\ -\delta_3^2 \end{Bmatrix} - (D_{22}\alpha^2 + A_{44} - I\omega^2) \right] + (D_{12} + D_{66})^2\alpha^2 \begin{Bmatrix} \delta_1^2 \\ \delta_2^2 \\ -\delta_3^2 \end{Bmatrix}$$

$$\tag{17}$$

We note that the frequency parameter ω appearing in the above constants plays an important role in the evaluation of the displacements, and rotations of the sandwich panel segment. It is to be determined through the force, moment, and displacement continuities at the beam–plate interfaces and the boundary conditions of the whole sandwich panel.

STIFFENER VIBRATION

A stiffener is considered to be composed of a web-beam along with two facing strips of the beam width, to which the beam is attached. Different materials may be considered. In view of the adjacent foam sandwich structure, the stiffener equations are derived in connection with the thick plate theory. Let $\bar{u}$, $\bar{v}$ and $\bar{w}$ be the neutral axis displacements, and $\bar{\psi}_x$ and $\bar{\psi}_y$ the beam cross section rotations, all of which are lengthwise functions of $\bar{y}$ only. Then for an arbitrary point in the web-stiffener, the respective displacements in the $\bar{x}$, $\bar{y}$ and $\bar{z}$ directions can be approximated as

$$U_1 = \bar{u}(\bar{y}, t) + \bar{z}\bar{\psi}_x(\bar{y}, t)$$
$$\bar{U}_2 = \bar{v}(\bar{y}, t) + \bar{z}\bar{\psi}_y(\bar{y}, t)$$
$$\bar{U}_3 = \bar{w}(\bar{y}, t) \tag{18}$$

From the strain–displacement relations, we obtain all the strain components for the web-stiffener,

$$\bar{\varepsilon}_1 = \bar{\varepsilon}_3 = 0 \qquad \bar{\varepsilon}_2 = \bar{v}_{,\bar{y}} + \bar{z}\bar{\psi}_{y,\bar{y}}$$

$$\bar{\gamma}_{.2} = \bar{u}_{,\bar{y}} + \bar{z}\bar{\psi}_{x,\bar{y}} \qquad \bar{\gamma}_{13} = \bar{\psi}_x \qquad \bar{\gamma}_{23} = \bar{w}_{,\bar{y}} + \bar{\psi}_y \tag{19}$$

and the stress–strain relations can be written in a way similar to eqn. (4), in short

$$\{\bar{\sigma}\} = [\bar{\bar{Q}}]\{\bar{\varepsilon}\} \tag{20}$$

where $[\bar{\bar{Q}}_{ij}]$ is the material stiffness of the beam. For materials in the present research, the force and moment resultants including transverse forces acting in its cross section are obtained as

$$\bar{N}_y = \bar{A}_{22}\bar{v}_{,\bar{y}} \qquad \bar{M}_y = \bar{D}_{22}\bar{\psi}_{y,\bar{y}} \qquad \bar{N}_{xy} = \bar{A}_{66}\bar{u}_{,\bar{y}}$$
$$\bar{M}_{xy} = \bar{D}_{66}\bar{\psi}_{x,\bar{y}} \qquad \bar{Q}_x = \bar{A}_{55}\bar{\psi}_x \qquad \bar{Q}_y = \bar{A}_{44}(\bar{w}_{,\bar{y}} + \bar{\psi}_y) \tag{21}$$

in which $(\bar{A}_{ij}, \bar{D}_{ij}) = \bar{a}\sum_k \int \bar{Q}_{ij}^{(k)}(1, Z^2)\,\mathrm{d}z$, with $\bar{a}$ being width of the web.

The resultant forces and moments exerted by the plating on a stiffener element are shown in Fig. 2. On the left hand side, forces and moments exerted by the (ith) plate segment along edge $x = a$ are denoted by the subscript i, and on the right hand side, forces and moments exerted by the (jth) plate segment along edge $x = 0$ are denoted by the subscript j. The plate segment equations of motion are given by

$$\bar{N}_{xy,\bar{y}} + (f_j - f_i) = \rho_B\ddot{\bar{u}} \qquad \bar{N}_{y,\bar{y}} + (S_j - S_i) = \rho_B\ddot{\bar{v}}$$
$$\bar{Q}_{y,\bar{y}} + (Q_j - Q_i) + q = \rho_B\ddot{\bar{w}} \qquad \bar{M}_{y,\bar{y}} - (T_j - T_i) - \bar{Q}_y = J_x\ddot{\bar{\psi}}_y$$

$$\bar{M}_{xy,\bar{y}} + (m_j - m_i) - \bar{Q}_x - \frac{\bar{a}}{2}(Q_j + Q_i) = J_y\ddot{\bar{\psi}}_x \tag{22}$$

where J_x and J_y are beam polar moment of inertia about the $\bar{x}$- and $\bar{y}$-axes, respectively. Because f, S, Q, T and m are the stress resultants in the foam

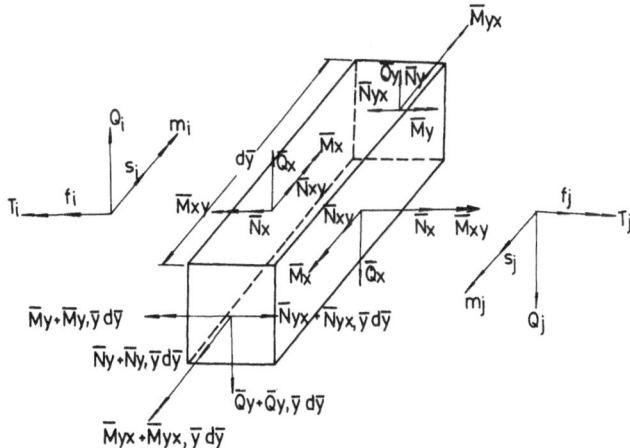

FIG. 2.   Force and moment resultants on a stiffener element.

sandwich plate segment, they can be expressed in terms of plate segment displacement along the edge $x = a$ in the ith plate segment

$$S_i = A_{66}(u_{0,y} + v_{0,x})|_{x=a} \qquad f_i = (A_{11}u_{0,x} + A_{12}v_{0,y})|_{x=a}$$

$$Q_i = A_{55}(w_{,x} + \psi_x)|_{x=a} \qquad m_i = (D_{11}\psi_{x,x} + D_{12}\psi_{y,y})|_{x=a}$$

$$T_i = D_{66}(\psi_{x,y} + \psi_{y,x})|_{x=a} \tag{23}$$

Similarly, f_j, S_j, Q_j, T_j and m_j can be written for the jth plate segment along $x = 0$.

Since the stiffeners are also simply supported at the ends $\bar{y} = 0$ and $\bar{y} = b$, a displacement field can be assumed similar to eqns (11) as for the plate segments except that the stiffener parameters $\bar{U}$, $\bar{V}$, $\bar{W}$, $\bar{\Psi}_x$ and $\bar{\Psi}_y$ are now constants for the individual stiffeners. Substitution of the above relations for each of the R beam stiffeners yields $5R$ homogeneous equations, in the stiffener parameters $\bar{U}$, $\bar{V}$, $\bar{W}$, $\bar{\Psi}_x$ and $\bar{\Psi}_y$ along with the related plate segment variables U, V, W, Ψ_x and Ψ_y.

BEAM–PLATE DISPLACEMENT CONTINUITY

Displacements of each plate segment and its contiguous stiffeners must be continuous at the interface, i.e. for

Segment i	Segment j

$$\begin{array}{ll} U(a) = \bar{U} \qquad V(a) = \bar{V} & \qquad U(0) = \bar{U} \qquad V(0) = \bar{V} \\ W(a) = W + \tfrac{1}{2}\bar{a}\bar{\Psi}_x & \qquad W(0) = \bar{W} - \tfrac{1}{2}\bar{a}\bar{\Psi}_x \\ \bar{\Psi}_x(a) = \bar{\Psi}_x \qquad \Psi_y(a) = \bar{\Psi}_y & \qquad \Psi_x(0) = \bar{\Psi}_x \qquad \Psi_y(0) = \bar{\Psi}_y \end{array} \tag{24}$$

When eqns (24) are applied to edges of all plate segments which terminate at a stiffener, $10R$ simultaneous equations can be obtained for the R stiffeners.

BOUNDARY CONDITIONS

At the extreme right and left edges of the web-stiffened foam sandwich panel, the S_2 type boundary conditions are assumed, i.e. at $x = 0$ for the first segment and at $x = a$ for the last segment, we have 10 more equations

$$w = v = 0 \qquad N_x = A_{11}u_{0,x} + A_{12}v_{0,y} = 0$$

$$M_x = D_{11}\psi_{x,x} + D_{12}\psi_{y,y} = 0$$

$$\psi_y = 0 \tag{25}$$

Reduction can be made by use of eqns (11).

NUMERICAL SOLUTION

In the free vibration problem of an R web-stiffened foam sandwich panel structure, a total of $15R + 10$ homogeneous equations are provided for the solution of the natural frequencies and the corresponding mode shapes. There are $5R$ equations for the R web-stiffeners with $5R$ unknowns in $\bar{U}, \bar{V}, \bar{W}, \bar{\Psi}_x, \bar{\Psi}_y$ and $10(R + 1)$ equations for the $R + 1$ panel segments with $10(R + 1)$ unknowns in A_1, A_2, A_3, A_4 and $C_1, C_2, \ldots, C_6$. Non-trivial solution requires the determinant of the coefficient matrix of the system of homogeneous equations to vanish. In the computation, global co-ordinates x_0 and y_0 are used for the web-stiffened foam sandwich panels. In search for the natural frequency which appears in the coefficient matrix as a yet unknown parameter, the golden section method is used to find an highly accurate ω in every interval where the determinant changes sign. Natural frequencies of the first four modes are obtained as f_{11}, f_{12}, f_{21} and f_{22} in that the first and second index denotes the vibration modes in the global y_0 and x_0 direction respectively. Also, the corresponding normalized mode shapes are calculated by using the original system of simultaneous equations. Results are presented in Figs 3 and 4.

RESULTS AND DISCUSSION

In the present study, we have taken the transverse shear effects of a thick plate into full consideration, which is a necessity for foam sandwich panels.

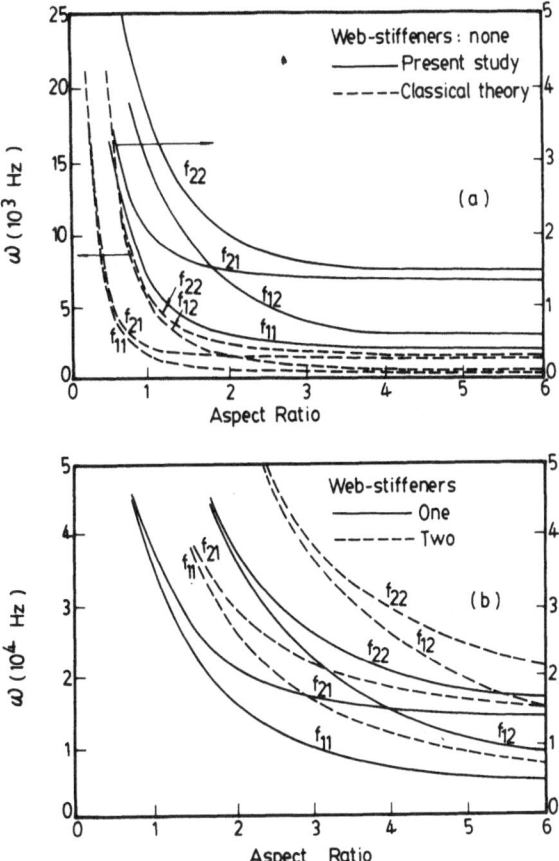

FIG. 3. Natural frequencies of foam sandwich panels.

Five variables u, v, w, ψ_x and ψ_y are introduced with equations and appropriate boundary and interface conditions hand in hand. Geometry and material properties are as follows:

Facing to core thickness ratio = 0·02, core thickness to panel length ratio = 0·1, total web width to panel length ratio = 0·025.

For the graphite/epoxy facing and web material, $E_1 = 17·5 \times 10^6$ psi, $E_2 = 1·15 \times 10^6$ psi, $G_{12} = G_{23} = G_{13} = 0·8 \times 10^6$ psi, $v_{12} = 0·30$, and $\rho_f = 1·48 \times 10^{-4}$ lb-sec^2/in^4.

For the PU foam, $E_c = 625$ psi, $G_c = 246$ psi, $v_c = 0·267$, and $\rho_c = 2·85 \times 10^{-6}$ lb-sec^2/in^4. The panel width is fixed at 10 in as its length may change.

In an attempt to verify the validity of the foam sandwich theory,

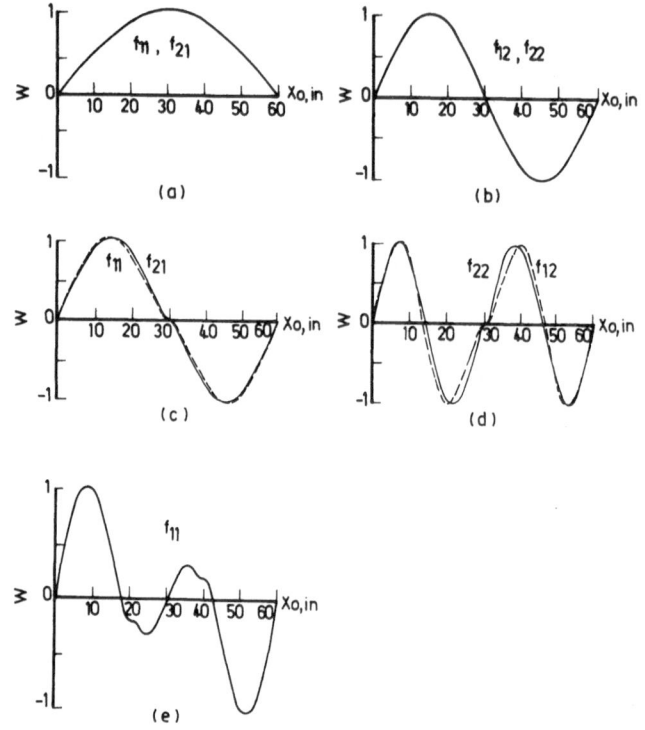

FIG. 4. Vibration mode shapes of web-stiffened foam sandwich panels

Case	Web-stiffeners
a, b	0
c, d	1
e	2

comparison of the present study is made with results obtained from the classical lamination theory for thin plates, by incorporation of no web-stiffener in the plate on purpose. Natural frequencies of the first four modes f_{11}, f_{12}, f_{21} and f_{22} are presented in Fig. 3a. Trends of curves of both groups are similar except that the frequencies obtained according to the classical theory are extremely high at lower aspect ratios and considerably low at higher aspect ratios. As a result of releasing the 'normal remains normal' hypothesis and relatively large thickness of a thick foam sandwich panel, as the plate aspect ratio changes, it is unlikely to have such high stiffness at the lower end and such low stiffness at the higher end. As to the web-stiffened foam sandwich panel structures, natural frequencies for the case of one and two stiffeners embedded in the sandwich plate are shown in Fig. 3b. As expected, significant increase in the frequencies occurs in all modes with increase in number of stiffeners because of increased stiffness in the presence of these stiffeners.

Normalized vibrational mode shapes in the transverse displacement $|w$ of the web-stiffened foam sandwich panel structures are shown in Figs 4a–4e. In Fig. 4a, which is the non-web-stiffener case, the mode shape of f_{11} coincides with that of f_{21}. So does f_{12} with f_{22} in Fig. 4b. Furthermore, these are sine curves as obtained from the classical lamination theory. Mode shapes of one and two web-stiffeners are shown in Fig. 4c–4e. A kink is noted at the location of each web-stiffener because of its higher stiffness.

ACKNOWLEDGEMENT

The authors wish to thank the National Science Council for support of this research.

REFERENCES

1. WAH, T., Vibration of stiffened plates, *Aeronautical Quarterly*, **15** (1964).
2. LONG, B. R., Vibration of eccentrically stiffened plates, *Shock Vibration Bulletin*, **38** (1968).
3. CHAO, C. C. and LEE, J. C., Vibration of eccentrically stiffened laminates, *J. Composite Materials*, **14** (1980), 233–244.
4. CHEN, Y. N., RANLET, D. and KEMPNER, J., Web-stiffened sandwich structures, *J. Applied Mechanics* (1971), 964–970.
5. CHEN, Y. N., CICERO, F. and KEMPNER, J., Refinements in the approximate analysis of web-stiffened sandwich structures, *J. Applied Mechanics* (1973), 992–996.
6. CHONG, K. P. and HARTSOCK, J. A., Flexural wrinkling in foam-filled sandwich panels, *J. Engineering Mechanics Division, ASCE*, **100** (EM1) (1974), 95–110.
7. HARTSOCK, J. A. and CHONG, K. P., Analysis of sandwich panels with formed faces, *J. Structural Division, ASCE*, **102** (ST4) (1976), 803–819.
8. JONES, R. M., *Mechanics of Composite Materials*, New York, McGraw-Hill, 1975.
9. TSAI, S. W. and HAHN, H. T., *Introduction to Composite Materials*, Westport, CT, Technomic, 1980.
10. SUN, C. T. and TAN, T. M., *Wave Propagation in a Graphite/Epoxy Laminate*, Proc. NCKU/AAS Int. Symposium on Engineering Sciences and Mechanics, Tainan, R.O.C., 1981, 1320–1337.

8

Global Transverse Shear in Laminated Composite Plates

R. Girard

*Office National d'Etudes et de Recherches Aérospatiales,
29 Avenue de la Division Leclerc, BP 72, 92322 Chatillon, France*

ABSTRACT

The theory of Reissner for homogeneous isotropic plates is generalized to composite laminates. Based on stress and displacement hypotheses, this extension is intended primarily to establish a global constitutive law including transverse shear effects. To get this law, the mixed functional of Reissner is made stationary with respect to the assumed stress field. The associated primal displacement type formulation is quite similar to the corresponding one of classical plates theories and its numerical implementation through existing finite element codes requires only the change of shear strain–stress laws. Comparison of the proposed theory with the more classical theories issued from the hypotheses of Kirchhoff or Mindlin, is performed on a test problem whose exact analytical solution is available.

INTRODUCTION

Classical laminated plate theory based on the Kirchhoff hypotheses,[1] gives satisfactory results only when applied to plates thin enough to be unaffected by transverse shear phenomena. Application of this theory is restricted not only by the small thickness of plates under consideration but also by the heterogeneity of the mechanical characteristics.

Therefore, theories including transverse shear deformation have been developed.[2] These theories are founded on the hypothesis of Hencky and Mindlin:[3,4] variation of in plane displacements is linear with respect to the

102

thickness co-ordinate, whereas transverse displacement is independent of this co-ordinate. Taking into account global transverse shear deformation, these theories are of a wider application than the classical one.[5]

But transverse shear deformation for anisotropic laminates can be introduced in a somewhat different way by attempting to generalize Reissner's theory of homogeneous plates:[6] displacements are assumed to be those of the preceding theories and in plane stresses are deduced from plane stress state constitutive laws, while an approximate expression of transverse stresses is obtained from the thickness integration of tridimensional equilibrium equations. All the equations of the new theory are then computed from the stationarity of the Reissner functional[7] with respect to the aforementioned hypotheses.

The various theories differ principally by the expression of the coefficients which connect shear resultants and transverse shear strains. In fact, if we restrict ourselves, as it will be supposed later on, to monoclinic materials and laminates whose transverse normal stress is neglected, equilibrium equations and in-plane constitutive laws of the new theory are exactly of the classical theories.

This paper deals with the generalized Reissner theory,[8] such as it has been outlined above. The proposed extension results in a primal principle which allows the direct use of finite element methods based on displacement unknowns. To assess the range of application of the new theory and compare it with classical theories, some laminated composite plate test problems are considered (these problems admit an analytical solution[9]).

I. Theoretical Formulation

Let us consider a laminated plate of thickness $2h$ and middle plane Σ defined by $z = 0$ (x and y are in-plane co-ordinates, z is thickness co-ordinate), subjected on its upper face to a uniform pressure of intensity p.

I.1. Mixed functional of Reissner

The equations governing plate equilibrium are equivalent to the determination of stresses σ_i ($i = 1, \ldots, 6$) (engineering notations) and kinematically admissible displacements u_i ($i = 1, 2, 3$) which make stationary the functional:

$$J(u, \sigma) = \int_{\Sigma \times [-h, +h]} \{\sigma_i \varepsilon_i(u) - \phi(\sigma)\} \, d\Sigma \, dz - \int_{\Sigma} p u_3 \, d\Sigma \qquad (i = 1, \ldots, 6)$$

In this expression, the strains ε_i ($i = 1, \ldots, 6$) must be considered as u-dependent through the classical linearized relationships, and the complementary elastic strain energy density $\phi(\sigma)$ is given, for a monoclinic material where the normal transverse stress effect is neglected, by:

$$\phi(\sigma) = \tfrac{1}{2} S_{ij}\sigma_i\sigma_j + \tfrac{1}{2} K_{\alpha\beta}\tau_\alpha\tau_\beta \qquad (i, j = 1, 2, 6) \ (\alpha, \beta = 1, 2)$$

The coefficients S_{ij} and $K_{\alpha\beta}$ are respectively the in-plane and transverse shear compliances (from now on transverse shear stresses are quoted τ_α). For a laminate plate, these coefficients are given functions of the z-coordinate through the mechanical properties of each layer.

I.2. Displacement and stress hypotheses

Just like classical plate theories, in-plane plate displacements are assumed to be z-linear, while the transverse displacement is supposed to be constant across the plate thickness:

$$\begin{cases} u_\alpha = u_\alpha^0 + z u_\alpha^1 \quad (\alpha = 1, 2) \\ u_3 = w \end{cases} \qquad \begin{cases} u_\alpha^p = u_\alpha^p(x, y) \quad (p = 0, 1) \\ w = w(x, y) \end{cases}$$

As for in plane stresses σ_i $(i = 1, 2, 6)$, we suppose that the constitutive law coming from the expression of the energy density ϕ has to be true, i.e.

$$\varepsilon_i = S_{ij}\sigma_j \qquad (i, j = 1, 2, 6)$$

Taking into account strains expressions implied by displacements hypotheses, it appears from the inverse form of the last equation that in-plane stresses may be expressed as:

$$\sigma_i = C_{ij}(\varepsilon_j^0 + z\varepsilon_j^1) \qquad (i, j = 1, 2, 6)$$

where components ε_j^0 and ε_j^1 are functions of x and y, while coefficients C_{ij} are z-dependent through the mechanical characteristics of the laminate (reduced stiffnesses for the plane stress state $\sigma_{zz} = 0$).

Bringing out these expressions into the tridimensional equilibrium equations,

$$\begin{cases} \sigma_{1,x} + \sigma_{6,y} + \tau_{1,z} = 0 \\ \sigma_{6,x} + \sigma_{2,y} + \tau_{2,z} = 0 \end{cases} \qquad \tau_\alpha|_{z=-h} = \tau_\alpha|_{z=+h} = 0 \qquad (\alpha = 1, 2)$$

an integration of these equations along the z-co-ordinate with regard to interlayer continuity and external faces equilibrium, shows that transverse shear stresses may be written:

$$\tau_\alpha = \sum_{h=1}^{10} R_{\alpha k}\mu_k \qquad (\alpha = 1, 2)$$

where μ_k $(k = 1, \ldots, 10)$ are unknown functions of x and y, and $R_{\alpha k}$ some known functions of z which depend only upon the mechanical properties of the laminate through the following integral coefficients:

$$a_{ij}^p = \int_{-h}^{z} \zeta^p C_{ij} \, d\zeta \qquad (p = 0, 1)$$

From now on we assume that transverse shear stresses are expressed as above.

I.3. Displacements functional

In the case of monoclinic materials, the functional of Reissner involves transverse shear stresses only through the integral:

$$J_c = \int_{\Sigma \times [-h, +h]} (\tau_\alpha \gamma_\alpha - \tfrac{1}{2} K_{\alpha\beta} \tau_\alpha \tau_\beta) \, d\Sigma \, dz \qquad (\alpha, \beta = 1, 2)$$

where γ_α $(\alpha = 1, 2)$ are the displacements dependent expressions of transverse shear strains. In our previous hypotheses, these strains do not depend upon the z-co-ordinate and the stresses τ_α are linear combinations of the μ_k functions. With respect to these latter, making J stationary yields the following algebraic linear system:

$$\left[\int_{-h}^{h} (K_{\alpha\beta} R_{\alpha k} R_{\beta l}) \, dz \right] \mu_l = \left[\int_{-h}^{h} R_{\alpha k} \, dz \right] \gamma_\alpha \qquad (k, l = 1, \ldots, 10)$$

The solution of this system makes possible the expression of the functions μ_k and hence the stress τ_α as functions of the strains γ_α (transverse shear constitutive law). Then, we obtain:

$$\int_{-h}^{h} (K_{\alpha\beta} \tau_\alpha \tau_\beta) \, dz = \int_{-h}^{h} (K_{\alpha\beta} R_{\alpha k} R_{\beta l}) \mu_k \mu_l \, dz = \int_{-h}^{h} (R_{\alpha k} \gamma_k \mu_k) \, dz = \int_{-h}^{h} (\tau_\alpha \gamma_\alpha) \, dz$$

so that the functional J_c may be equally written:

$$J_c = \frac{1}{2} \int_{\Sigma} (Q_\alpha \gamma_\alpha) \, d\Sigma \qquad \text{with} \qquad Q_\alpha = \int_{-h}^{h} \tau_\alpha \, dz$$

In the same way, in-plane stresses σ_i are connected to in-plane strains according to the in-plane constitutive law and we have:

$$\int_{-h}^{h} (\sigma_i \varepsilon_i - \tfrac{1}{2} S_{ij} \sigma_i \sigma_j) \, dz = \frac{1}{2} \int_{-h}^{h} (\sigma_i \varepsilon_i) \, dz = \tfrac{1}{2} (T_i \varepsilon_i^0 + M_i \varepsilon_i^1)$$

with

$$(T_i, M_i) = \int_{-h}^{h} (1, z)\sigma_i \, dz$$

Thus for stresses σ_i and τ_α which are computed from the previously defined constitutive laws, the mixed functional of Reissner becomes a functional of displacements alone:

$$J(u) = \frac{1}{2} \int_\Sigma (T_i \varepsilon_i^0 + M_i \varepsilon_i^1 + Q_\alpha \gamma_\alpha) \, d\Sigma - \int_\Sigma pw \, d\Sigma$$

This functional is identical to that of plate theories with transverse deformation. Therefore, the equations which govern the present theory are those of classical plates theories, except for the determination of the elastic shear modulus of the transverse shear constitutive law (relationships between shear resultants Q_α and shear strain γ_α). As far as these moduli are computed in some independent way, the implementation of the proposed theory in existing displacement type finite elements codes is straightforward.

II. Numerical Applications

In order to appreciate the usefulness of the new theory with respect to the classical theories, the solutions for a test problem of these various approximate theories are compared with the corresponding one of the elasticity theory.

II.1. Test problem

Let us consider a laminated plate composed of orthotropic layers with the orthotropy directions parallel to the plates axes. This plate is supported along the four edges so that normal displacements are allowed while tangential ones are prevented. Moreover, the upper face of the plate is subjected to a transverse sinusoidal pressure.

The exact solution of this problem has been obtained analytically by Pagano.[9] For this solution, displacements within each layer are as follows:

$$u_1 = U(z)\cos px \sin qy \qquad u_2 = V(z)\sin px \cos qy \qquad u_3 = W(z)\sin px \sin qy$$

where the functions U, V, W are the solutions of a linear first-order differential system whose general solution requires the determination of 6 constants. For the whole laminate, the solution is then brought back to the resolution of a linear system with $6N$ unknowns (N = number of layers) resulting from boundary conditions (3 on each external face) and interlayer continuity (6 conditions on each interlayer face).

For the same problem, there exists also an analytical solution within the frame of approximate theories dealing with transverse deformation (whatever the theory we consider: either the theory based on the Hencky–Mindlin hypotheses or the proposed extension of Reissner theory). This solution looks like the exact one, but this time for the laminate as a whole and with functions U, V, W such that:

$$U(z) = U_0 + zU_1 \qquad V(z) = V_0 + zV_1 \qquad W(z) = W_0$$

where the five constants U_0, U_1, V_0, V_1, W are determined so that the five equations of plate equilibrium are satisfied.

As for classical plate theory, the problem has been solved by Whitney and Leissa.[10] The analytical solution they have obtained is quite similar to the preceding one, except that the functions U, V, W are now given by:

$$U(z) = A - zpC \qquad V(z) = B - zqC \qquad W(z) = C$$

where the three constants A, B, C are computed from the three equilibrium equations of the classical plate theory (these equations are those of the shear deformation theories in which shear resultants have been eliminated).

II.2. Numerical results

All the above-mentioned solutions have been computed for the bidimensional laminates described in Pagano.[9] These laminates are constructed of quasi-transversally isotropic layers whose mechanical properties are:

$$E_L = 25 \times 10^6 \, \text{psi} \qquad E_T = 10^6 \, \text{psi} \qquad G_{LT} = 0.5 \times 10^6 \, \text{psi}$$

$$G_{TT} = 0.2 \times 10^6 \, \text{psi} \qquad v_{LT} = v_{TT} = 0.25$$

where L signifies the direction parallel to the fibers and T the transverse direction.

Four lamination types have been studied:

—a 2-layer unsymmetrical laminate with the L-directions aligned parallel to plate axes x and y respectively;

—a 3-layer laminate with the L-directions aligned parallel to x in the outer layers, while parallel to y in the central layer;

—a 9-layer laminate with the L-directions aligned parallel to x and y respectively;

—a symmetrical sandwich with face sheets constructed of the above

R. Girard

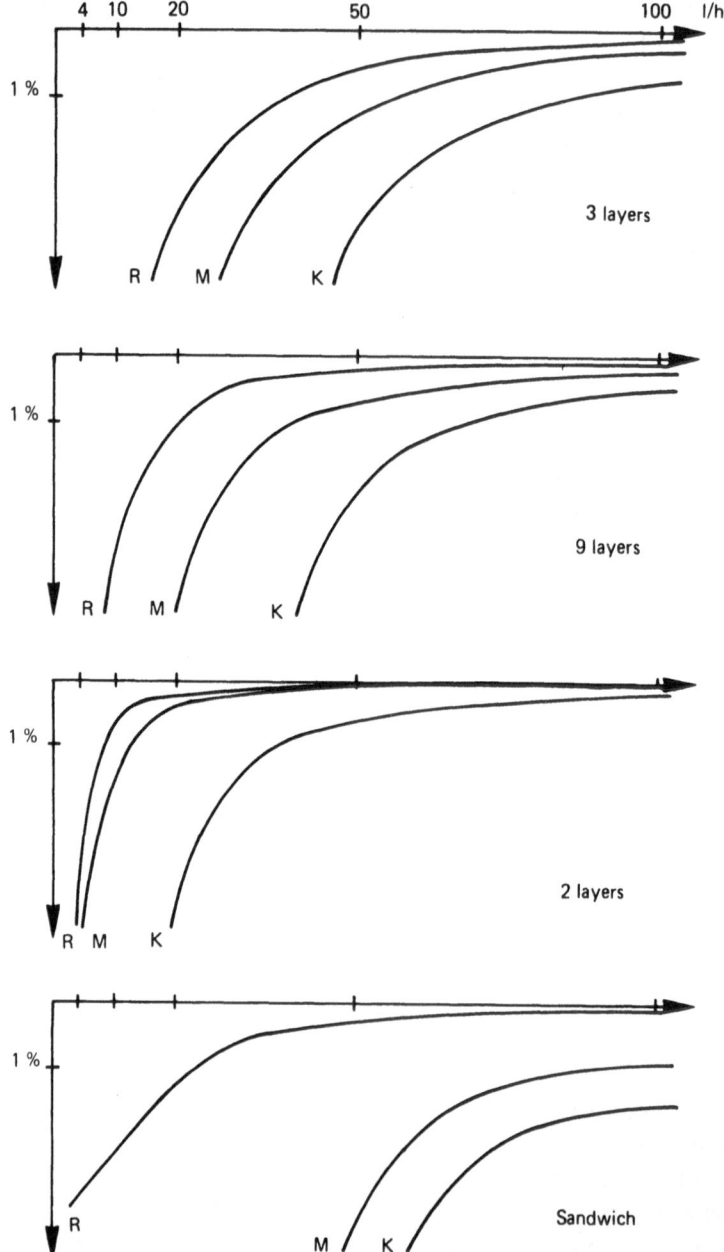

FIG. 1. Center deflection, accuracy of solutions of Kirchhoff (K), Hencky–Mindlin (M), Reissner (R).

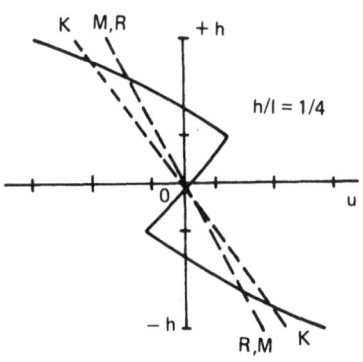

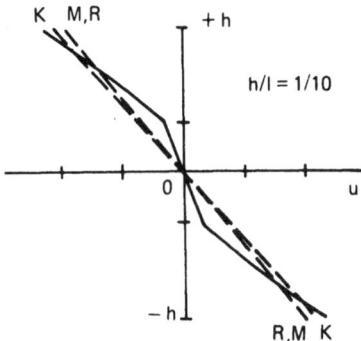

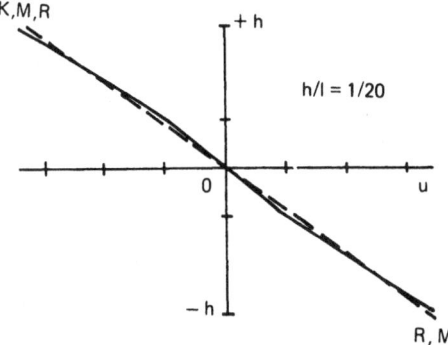

FIG. 2. Displacement u at point $x = 0$, $y = a/2$. —— Exact solution; --- approximate solutions (K, M, R).

transversally isotropic material and an orthotropic core characterized by
the following properties:

$$E_{xx} = E_{yy} = 0{\cdot}04 \times 10^6 \text{ psi} \qquad G_{xy} = 0{\cdot}016 \times 10^6 \text{ psi} \qquad v_{zx} = v_{zy} = v_{xy} = 0{\cdot}25$$
$$E_{zz} = 0{\cdot}5 \times 10^6 \text{ psi} \qquad G_{xz} = G_{yz} = 0{\cdot}06 \times 10^6 \text{ psi}$$

For the sandwich plate, the thickness of face sheets is $h/10$ (h = plate
thickness) while for other studied laminates, the layers are of equal
thickness.

For plate|center|deflection (Fig. 1), the results clearly demonstrate, in
each case we have studied, the better behavior of the extended theory of
Reissner. In order to keep in mind some knowledge about the range of each
theory, note that for thickness ratios higher than 20, the error associated
with the theory of Reissner is no more than 3 % (3-layer case) while it is 16 %
(3-layer case) and 28 % (sandwich case) for the classical plate theory. This
latter theory gives correct results (1 % in error) only for thickness ratios of
about 100, while this limit is about 50 for the theory with shear
deformation.

For in-plane displacement, differences are less apparent. Let us notice at
first (Fig. 2) that the approximate theories can account only for mean values
of displacements and rotations (displacements are assumed to vary linearly
across the plate thickness). In the 3-layer case (Fig. 2) and with a thickness
ratio equal to 20, the errors on the mean rotation are 1·1 %, 1·6 % and 2·8 %
respectively for the theories of Reissner, Hencky–Mindlin and Kirchhoff.
But differences are increasing when thickness ratios decrease: 4 %, 6 % and
11 % for a thickness ratio equal to 10.

Similar remarks can be made for the 9-layer laminate or the sandwich
plate, while in the 2-layer case the three approximate theories give
practically the same results with a very good precision (1/1000 for thickness

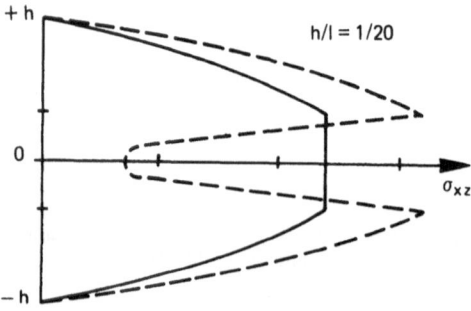

FIG. 3. Stress σ_{xz} at point $x = 0$, $y = a/2$. —— Exact solution; - - - Reissner solution.

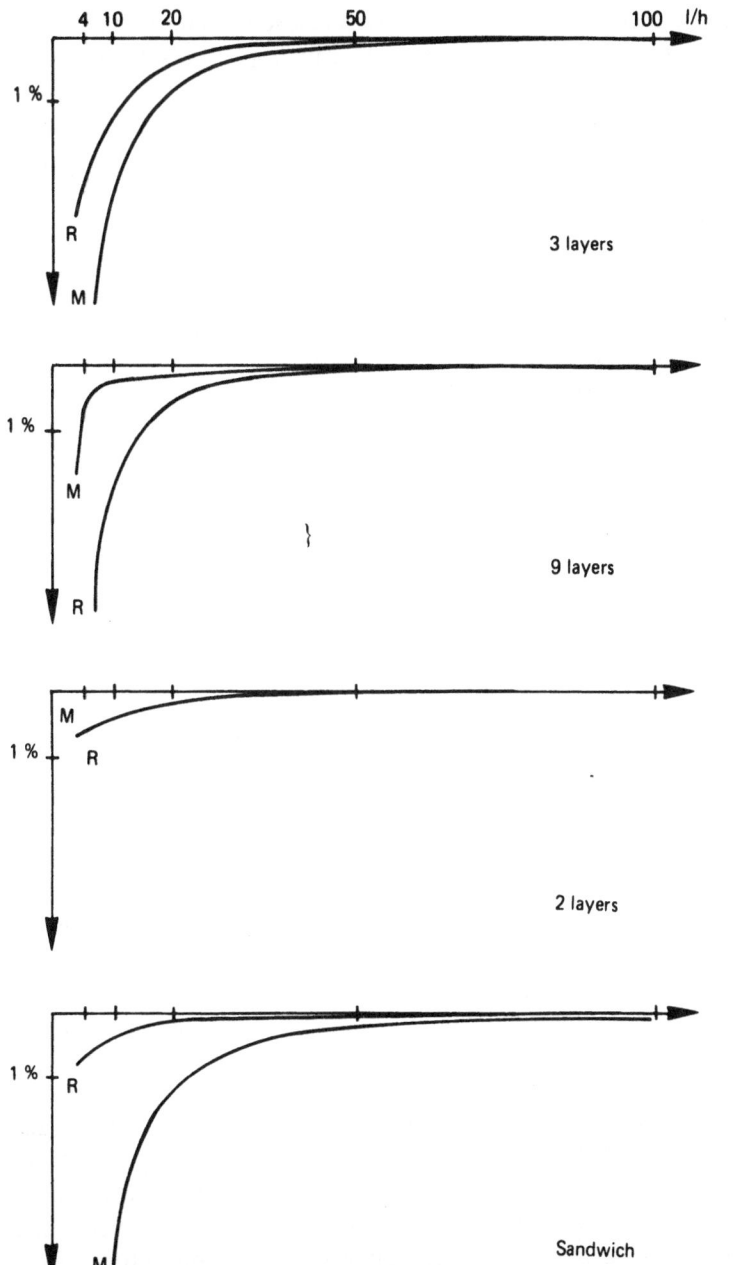

FIG. 4. Shear resultant Q_x at point $x = 0$, $y = a/2$. Accuracy of solution of Mindlin (M) and Reissner (R).

ratios higher than 20). These remarks are also valid for in-plane stress and strain values since they are computed from in-plane displacement for each theory.

As for transverse shear stresses, it appears that their description within the plate thickness according to the proposed theory, is no longer correct (*see* Fig. 3, wide oscillations around the exact curve). This is due to the fact that exact equilibrium equations are not locally satisfied but only in the mean sense. So, only a good value of mean stresses (shear resultants) can be expected.

This latter point has been confirmed through the results we have obtained (Fig. 4). One can see that the theories of Reissner and Hencky–Mindlin are equivalent as far as thickness ratios are higher than 20 (for a required precision of about 1 %). If beyond this limit the theory of Reissner is clearly better for the sandwich and the 3-layer laminate; this is not true for the 9-layer laminate. Nevertheless, it must be emphasized that, even in this latter case, the extended theory of Reissner still works with reasonable precision.

CONCLUSION

On the whole, the results we have presented show that the extended theory of Reissner, such as it has been outlined for laminated plates, brings some noticeable improvements when compared to more classical theories: in the cases we have quoted, an admissible accuracy of 2 or 3 % makes possible the application of the classical plate theory only for thickness ratios higher than 50, while the Reissner theory allows this limitation to be brought back to 20 or 10. Although these values are rather approximate, they clearly demonstrate that the extended theory of Reissner can be reasonably applied to structural problems of engineering interest. Moreover, this theory is general enough to take into account composite laminates and sandwiches as well, without having need, in the latter case, to resort to a special theory.

However, the proposed extension of Reissner theory remains a plate theory: making use of a small number of unknown displacements (and this is a practical need for engineering applications), this theory, as well as other approximate theories, ignores the transverse normal stress and cannot describe in a satisfactory way the transverse shear stresses. Nevertheless, it must be emphasized that shear resultants in plane stresses (or at least their mean values for thick plates) are in good agreement with exact values.

As a conclusion, among the three approximate theories we have experienced, it is the proposed extension of Reissner theory which seems to be actually best fitted for the analysis of laminated composite plates.

REFERENCES

1. REISSNER, E. and STAVSKY, Y., Bending and stretching of certain types of heterogeneous aeolotropic elastic plates, *J. Applied Mechanics*, **28**, 1961, 402.
2. WHITNEY, J. M. and PAGANO, N. J., Shear deformation in heterogeneous anisotropic plates, *J. Applied Mechanics*, **37**, 1970, 1031.
3. SANDER, G., *Application de la Méthode des Eléments Finis à la Flexion des Plaques*, Université de Liège, Faculté des Sciences Appliquées. Collection des Publications no. 15, 1969.
4. MINDLIN, R. D., Influence of rotatory inertia and shear on flexural motions of isotropic elastic plates, *J. Applied Mechanics*, **18**, 1951, 31.
5. WHITNEY, J. M., Stress analysis of thick laminated composite and sandwich plates, *J. Composite Materials*, **6**, 1972, 426.
6. REISSNER, E., The effect of transverse shear deformation on the bending of elastic plates, *J. Applied Mechanics*, **12**, 1945, 69.
7. REISSNER, E., On a variational theorem in elasticity, *J. Mathematics and Physics*, **29**, 1950, 90.
8. REISSNER, E., A consistent treatment of transverse shear deformation in laminated anisotropic plates, *AIAA J.*, **10**(5), 1972, 716.
9. PAGANO, N. J., Exact solution for rectangular bidirectional composites and sandwich plates, *J. Composite Materials*, **4**, 1970, 20.
10. WHITNEY, J. M. and LEISSA, A. W., Analysis of heterogeneous anisotropic plates, *J. Applied Mechanics*, **28**, 1969, 261.

9

Vibration and Elastic Stability of Polar Orthotropic Variable Thickness Circular Plates Subjected to Hydrostatic Peripheral Loading

D. G. GORMAN

Department of Mechanical Engineering,
Queen Mary College, London E1 4NS, England

ABSTRACT

The method of annular finite elements is applied to the problem of predicting the natural frequencies of free transverse vibration and critical buckling parameters of linearly varying thickness discs exhibiting polar orthotropic characteristics subjected to hydrostatic peripheral loading. By separating the flexural and geometric components of the structural stiffness, values of natural frequencies and critical buckling parameters can be obtained for a wide range of specimens by the application of a simple relationship and a few graphs. The axisymmetric and first two antisymmetric modes of vibration are considered.

1. INTRODUCTION

General analyses of plates exhibiting anisotropic characteristics have received much attention recently due to the ever increasing use of fibre reinforced plastics in a wide range of industrial applications. In the specific case of vibration analysis of circular and annular plates, Refs 1–17 detail various methods whereby the natural frequencies of free small transverse vibrations are predicted for a specific type of anisotropic plate, namely polar or orthotropic, for the case where no form of initial in-plane stressing is present. Under normal working conditions, however, these plates may well be subjected to forms of in-plane stressing resulting from centrifugal, thermal, or hydrostatic loading and consequently their natural frequencies

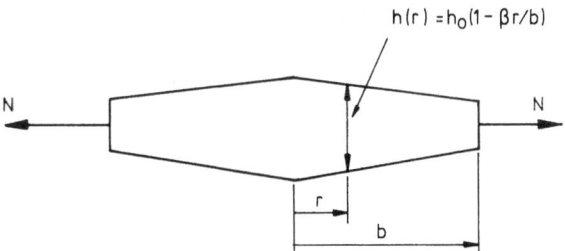

$$h(r) = h_0(1 - \beta r/b)$$

FIG. 1.   Variable thickness disc subject to hydrostatic in-plane peripheral load of N per unit circumferential length.

and, in general, dynamic characteristics, may change considerably as a consequence. When considering the effect of hydrostatic in-plane loading, useful work has been carried out by Laura et al.[18] who considered the case of polar orthotropic circular plates of linear varying axial thickness profile for the axisymmetric modes only (*see* Fig. 1). The aims of the work, described in this paper, are to continue the subject of the work of Laura et al. by considering higher modes of vibration and exam methods whereby the variations of natural frequencies for varying degrees of hydrostatic peripheral loading can be better represented. For completeness the critical buckling load values are also computed using the method of annular finite elements employed throughout this study.

2.   ANALYSIS

As shown in Figs 2a and 2b a circular disc of varying axial thickness form can be accurately modelled by a series of annular finite elements of linearly varying axial thickness form. Since for each element there are four degrees of freedom, the transverse displacement function across each element is assumed to be of the general form described by:

$$w(r, \theta, t) = [a_0 + a_1 r + a_2 r^2 + a_3 r^3]\cos m\theta \cos \omega_{mn}t \qquad (1)$$

for $\bar{R}1 < r < \bar{R}2$.

On the basis of eqn. (1), the elemental mass and flexural stiffness matrices, $[m]$ and $[k_f]$ respectively, can be derived by way of standard finite element displacement method analysis, to be finally represented as eqns (A1) and (A2) of the Appendix. In this study, however, since we are further concerned with the additional action of in-plane loading, it is required to derive for each element a geometric stiffness matrix $[k_g]$, which is then added to $[k_f]$ to form the total stiffness matrix for that element.

D. G. Gorman

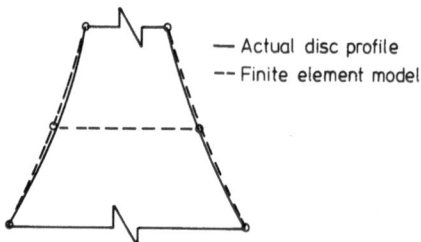

FIG. 2a. Finite element modelling of a variable thickness disc.

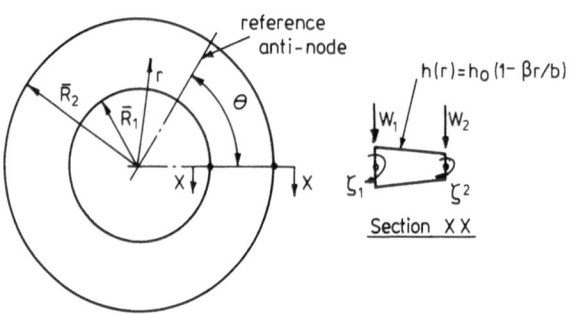

FIG. 2b. Annular finite element.

Subsequently, the natural frequencies ωT_{mn} and related displacement vector $\{\Delta_{mn}\}$ are obtained by solution of the standard eigenvalue problem, namely:

$$\{[K] - \omega T_{mn}^2[M]\}\{\Delta_{mn}\} = \{0\} \tag{2}$$

where $[K]$ and $[M]$ are the overall structural stiffness and mass matrices respectively.

2.1. Elemental Geometric Stiffness Matrix $[k_g]$

On the basis of eqn. (1) and the derivations performed by Smith,[19] the geometric stiffness matrix for each element can be written as eqn. (A3) of the Appendix, a prerequisite being that the in-plane stress distribution σ_r and σ_θ across each element be defined. In this study the in-plane stress distribution was obtained using the finite element method described in Refs 20 and 21. Note that in this study of circular plates, close representation of a circular plate can be achieved by analysing the problem of an annular plate but setting the inner peripheral radius to approximately 0·0001 of the outer peripheral radius. This also permits the use of only annular finite elements in the analysis.

2.2. Estimation of Critical Buckling Load Intensity (NC_{mn})

In this case the standard eigenvalue equation described by eqn. (2) is modified by setting $\omega T_{mn} = 0$ and expanding the $[K]$ matrix thus:

$$\{[K_F] - NC_{mn}[K_G]\}\{\Delta_{mn}\} = \{0\} \tag{3}$$

where $[K_F]$ and $[K_G]$ are the flexural and geometric structural stiffness matrices respectively and NC_{mn} = critical buckling load intensity (compressive) for the mode described by m nodal diameter and n nodal circles.

3. RESULTS

Since the convergence characteristics of the annular finite element method employed have been adequately described in Ref. 17, the following results have been arrived at by using 10 annular finite elements. The axial thickness variation of the discs analysed in this study is described by the equation:

$$h(r) = h_0(1 - \beta r/b) \tag{4}$$

3.1. Natural Frequency Variations

In presenting results for the natural frequency variations of discs subjected to varying degrees of peripheral radial load intensity, N, use is made of the relationship described by Lamb and Southwell,[22] namely:

$$\omega T_{mn}^2 = \omega F_{mn}^2 + N_{mn}^2$$

or

$$\omega T_{mn}^2 = (D/\rho h_0 b^4)\lambda F_{mn} + (N/\rho h_0 b^2)\lambda N_{mn} \tag{5}$$

It can be seen from eqn. (5) that λN_{mn} is the parameter which describes the variation of ωT_{mn} for a disc of given physical and dimensional properties, and the value of N (*see* Fig. 3). Figures 4–7 illustrate the variations of λN_{mn} and λF_{mn} against a base of $k^2 = E_\theta/E_r$, for the modes of vibration consisting of 0–2 nodal diameters only, and for the cases of clamped and simply supported at peripheral radius b for various values of β.

3.2. Critical Buckling Parameter, $-NC_{oo}b^2/D$ Variations

In all cases investigated it was found that the critical buckling mode was that mode containing zero nodal diameters or circles. Therefore by way of solution of eqn. (3) the values of non-dimensionalised critical buckling load intensity $= \lambda B = -NC_{oo}b^2/D$ are plotted against a base of k^2 for various

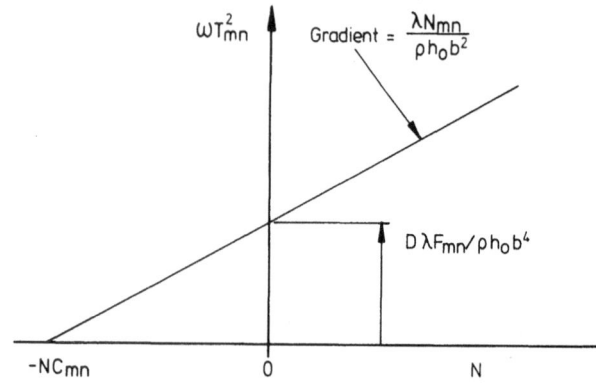

FIG. 3. Variation of ωT_{mn}^2 with hydrostatic in-plane load intensity N. $D = k^2 E_r/12(k^2 - \mu_\theta^2)$, ρ = material density.

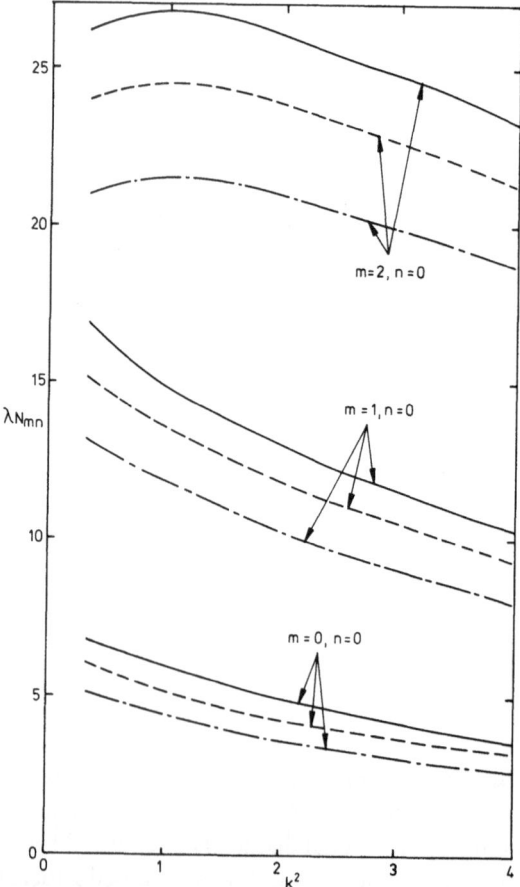

FIG. 4. Variation of λN_{mn} with k^2, disc clamped at b, —— $\beta = 0$, --- $\beta = 0\cdot2$, —·— $\beta = 0\cdot4$.

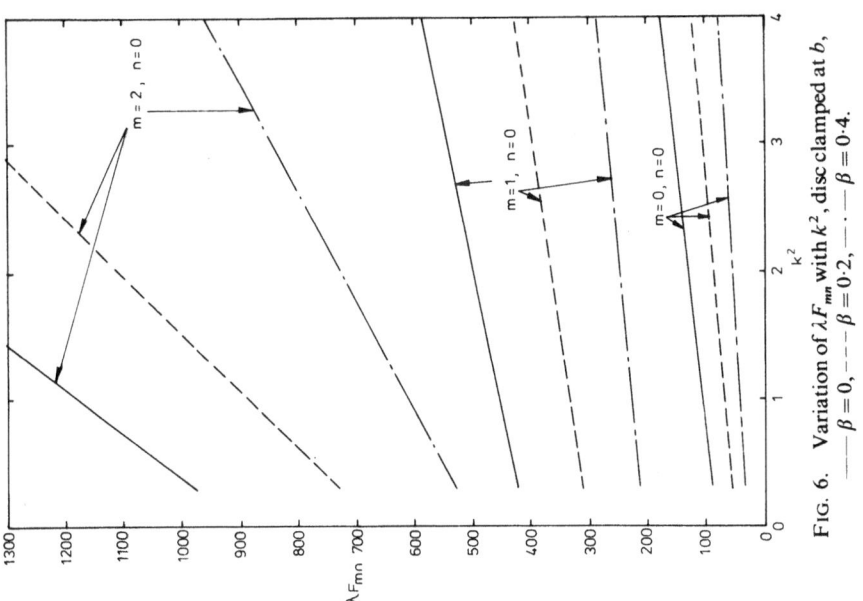

FIG. 6. Variation of λF_{mn} with k^2, disc clamped at b,
—— $\beta = 0$, – – – $\beta = 0.2$, – · – · – $\beta = 0.4$.

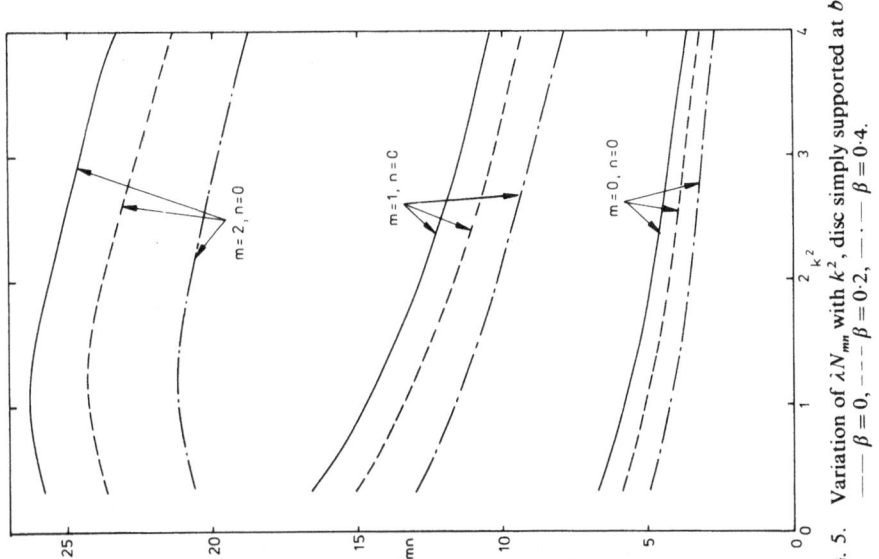

FIG. 5. Variation of λN_{mn} with k^2, disc simply supported at b,
—— $\beta = 0$, – – – $\beta = 0.2$, – · – · – $\beta = 0.4$.

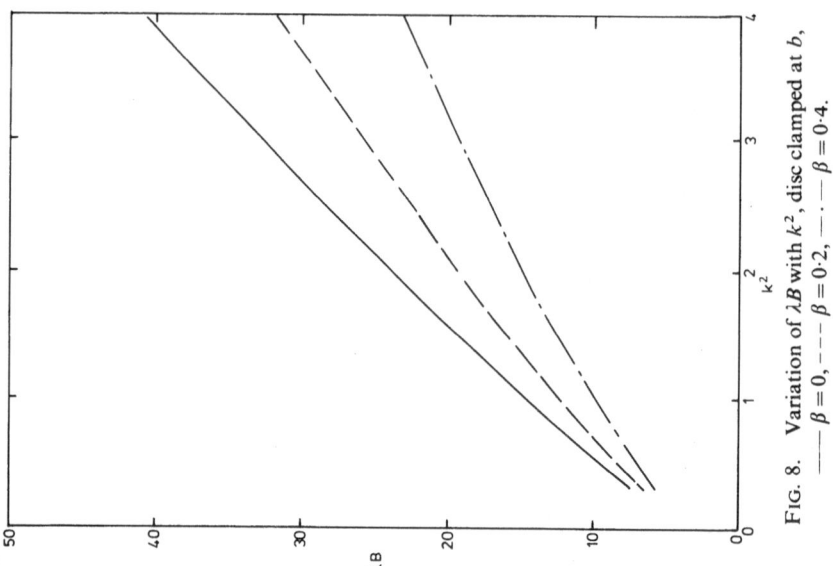

FIG. 8. Variation of λB with k^2, disc clamped at b,
———— $\beta = 0$, – – – $\beta = 0.2$, —·—·— $\beta = 0.4$.

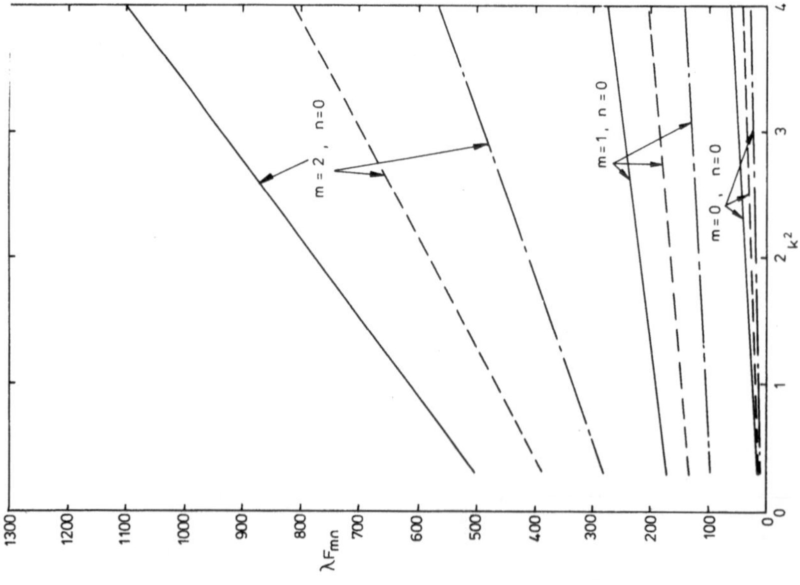

FIG. 7. Variation of λF_{mn} with k^2, disc simply supported at b,
———— $\beta = 0$, – – – $\beta = 0.2$, —·—·— $\beta = 0.4$.

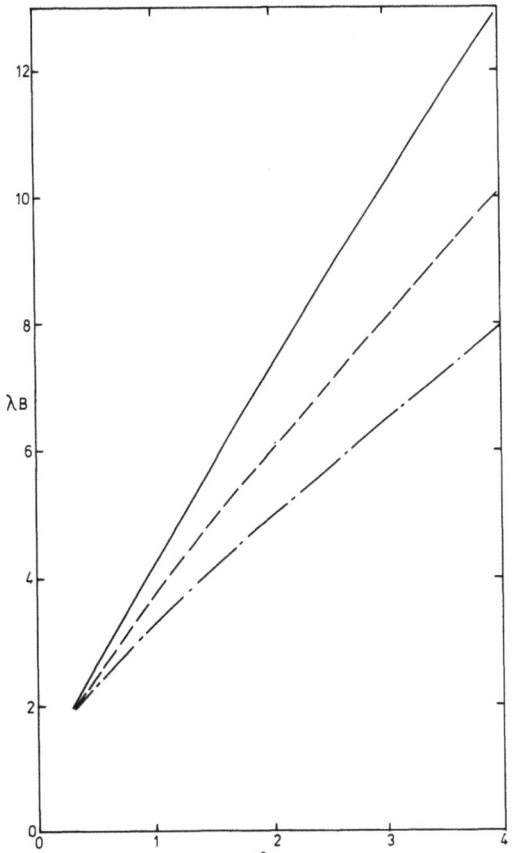

FIG. 9.   Variation of λB with k^2, disc simply supported at b, —— $\beta = 0$, – – – $\beta = 0\cdot 2$, — · — $\beta = 0\cdot 4$.

values of β in the axial thickness expression (eqn. (4)) and for the conditions of clamped and simply supported at peripheral radius b (*see* Figs 8 and 9).

DISCUSSION

The method of annular finite elements has been applied to the investigation of natural frequency variation and elastic stability of polar orthotropic discs of linearly varying axial thickness form subjected to a peripheral load N per unit length of the outer periphery. Separation of the flexural and geometric components of the total stiffness matrix resulted in the natural frequencies of transverse vibration being represented by the eqn. (5), where λF_{mn} and λN_{mn} are plotted against a loss of $(k^2 = E_\theta/E_r)$ for various values of β and modes of vibration. From the plots of λF_{mn} against k^2 (*see* Figs 6 and 7) it was observed that in all cases a linear relationship existed. This can be explained by inspection of the components constituting the [*P*] matrix for the Appendix, i.e.

$$[P](I, J) = A(I, J) + B(I, J)k^2$$

where A and B are constants.

Considering the plots of λN_{mn} against k^2 (see Figs 4 and 5), the trend was not so well defined as the geometric stiffness matrix is a function of both the tangential and radial stress distributions, σ_θ and σ_r, respectively, each of which are independently related to k^2. It can be observed, however, that the form of the plots did noticeably change for an increasing number of nodal diameters (m). This can be explained by considering the elements constituting the [*R*] matrix of the Appendix, i.e.

$$R(I, J) = f_1(r)\sigma_r + f_2(r)\sigma_\theta m^2$$

where $f_1(r)$ and $f_2(r)$ are functions of radius r. It can be seen from the above equation that as m increases, the contribution of σ_θ becomes more pronounced.

Considering the plots of the non-dimensional buckling load intensity λB against k^2 (Figs 8 and 9), it is observed that λB always increases with increasing k^2, whilst λF_{oo} increases, λN_{oo} decreases. Results obtained for λB by way of the application of the finite element method were compared to corresponding values of λB obtained by the exacting method described in Ref. 22 and are shown in Table 1, where BCB = boundary condition at b, numbers in brackets are those presented in Ref. 22, and upper values are those obtained using the finite element method.

TABLE 1

k^2 \\ BCB	Clamped	Simply supported
1·000	14·68 (14·66)	4·28 (4·28)
1·211	16·74 (16·71)	4·94 (4·94)
1·441	18·93 (18·88)	5·65 (5·64)

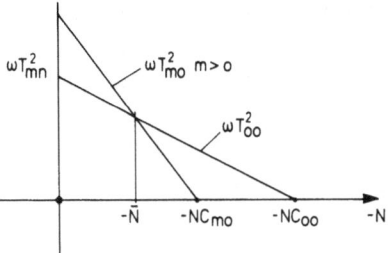

FIG. 10. Variation of ωT_{mn}^2 with hydrostatic compressive in-plane loading.

In previous reported work,[18] relating only to axisymmetric vibration and stability, it has been assumed that the critical buckling load intensity is that which reduces ωT_{oo} to zero. In this extended study which includes the antisymmetric modes, it can be noticed from Figs 4–7 that for a given value of k^2, although λF_{mn} increases with number of nodal diameters, so also does λN_{mn}. The possible effect of this can be seen by way of Fig. 10, whereby the mode shape corresponding to the lowest buckling load intensity can be deduced by finding the minimum value of $\bar{N}$ (if any) when the characteristics of all the modes are superimposed on a graph similar to Fig. 10. Alternatively, the presence of a buckling mode other than that of the first axisymmetric mode can be detected by considering the general form of eqn. (5), i.e.

$$\omega T_{oo}^2 = (D/\rho h_0 b^4)\lambda F_{oo} - (N/\rho h_0 b^2)\lambda N_{oo} \tag{6}$$

and for $m > 0$

$$\omega T_{mo}^2 = (D/\rho h_0 b^4)\lambda F_{mo} - (N/\rho h_0 b^2)\lambda N_{mo} \tag{7}$$

(note the minus sign indicating compression).

Then if a 'cross-over' occurs at $N = \bar{N}$ as in Fig. 10, i.e. $\omega T_{oo}^2 = \omega T_{mo}^2$

$$\bar{N}b^2/D = \frac{(\lambda F_{oo} - \lambda F_{mo})}{(\lambda N_{oo} - \lambda N_{mo})} \tag{8}$$

or

$$\bar{N}b^2/D = \frac{\lambda F_{oo}}{\lambda N_{oo}} \frac{(1 - \lambda F_{mo}/\lambda F_{oo})}{(1 - \lambda N_{mo}/\lambda N_{oo})} \tag{9}$$

and since $\lambda F_{oo}/\lambda N_{oo}$ can be approximated as $NC_{oo}b^2/D$ then by substituting in eqn. (9) it can be deduced that if the first critical buckling mode is not NC_{oo} then:

$$\frac{\bar{N}}{NC_{oo}} = \frac{(\lambda F_{mo}/\lambda F_{oo} - 1)}{(\lambda N_{mo}/\lambda N_{oo} - 1)} < 1 \tag{10}$$

From the range of results presented in this study it was found that the above relationship was never satisfied. In the general context however it is advisable that similar checks should be carried out as it has been previously discovered that for the case of certain forms of in-plane stressing, i.e. that due to thermal loading[24] the critical buckling mode can be that other than the mode containing zero nodal diameters.

REFERENCES

1. RAMAIAH, G. K. and VIJAYAKUMAR, K., Natural frequencies of polar orthotropic annular plates, *Journal of Sound and Vibration*, **26**(4) (1973), 517–531.

2. WOO, H. K., KIRMSER, P. G. and HUANG, C. L., Vibration of orthotropic circular plates with concentric isotropic core, *Journal of the American Institute of Aeronautics and Astronautics*, **11** (1973), 1421–1422.

3. IMER, S. and ZIMMERMANN, P., Free vibration of polar orthotropic discs, *Ing. Arch.*, **46**(3) (1973), 395–410.

4. RAO, K. S. and GANAPATHI, K., Vibration of cylindrically orthotropic circular plates, *Journal of Sound and Vibration*, **36**(3) (1974), 433–434.

5. RAMAIAH, G. K. and VIJAYAKUMAR, K., Estimation of higher natural frequencies of polar orthotropic plates, *Journal of Sound and Vibration*, **32**(2) (1974), 265–278.

6. RUBIN, C., Vibrating modes of simply supported polar orthotropic sector plates, *Journal of the Acoustical Society of America*, **58**(4) (1975), 841–845.

7. RAO, B. V. A. and RAMABHAT, B., Vibration of an orthotropic annular disc supported on the inside edge, *Journal of Sound and Vibration*, **42**(4) (1975), 510–514.

8. PRATHAP, G. and VARADAN, T. K., Axisymmetric vibrations of polar orthotropic circular plates, *Journal of the American Institute of Aeronautics and Astronautics*, **14**(11) (1976), 1639–40.

9. SONI, S. R. and AMBA-RAO, C. L., On radially symmetric vibrations of orthotropic non-uniform discs including shear deformation, *Journal of Sound and Vibration*, **42**(1) (1975), 57–63.

10. LUISONI, L. E., LAURA, P. A. A. and GROSSI, R., Antisymmetric modes of vibration of a circular plate elastically restrained against rotation and of linearly varying thickness, *Journal of Sound and Vibration*, **55**(3) (1977), 461–466.

11. GREENBERG, J. B. and STAVSKY, Y., Axisymmetric vibration of orthotropic composite circular plates, *Journal of Sound and Vibration*, **61**(4) (1978), 531–545.

12. GINESU, F., PICASSO, B. and PRIOLO, P. V., Vibration analysis of polar orthotropic annular discs, *Journal of Sound and Vibration*, **65**(1) (1979), 97–105.

13. GREENBERG, J. B. and STAUSKY, Y., Flexural vibration of certain full and annular composite orthotropic plates, *Journal of the Acoustical Society of America*, **66**(2) (1979), 501–508.

14. LAURA, P. A. A., PARDEON, G. C., LUBONI, L. E. and AVALOS, D., Transverse vibration of axisymmetric polar orthotropic circular discs elastically restrained against rotation along the edges, *Fibre Science and Technology*, 15 (1981), 65–77.

15. LAI, R. and GUPTA, U. S., Axisymmetric vibration of polar orthotropic annular plates of variable thickness, *Journal of Sound and Vibration*, 83(2) (1982), 229–240.

16. GORMAN, D. G., Natural frequencies of polar orthotropic annular plates, *Journal of Sound and Vibration*, 80(1) (1982), 145–154.

17. GORMAN, D. G., Natural frequencies of transverse vibration of polar orthotropic variable thickness annular plates, *Journal of Sound and Vibration*, 86(1) (1983), 47–60.

18. LAURA, P. A. A., AVALOS, D. R. and GALLES, E. D., Vibration and elastic stability of polar orthotropic circular plates of linearly varying thickness, *Journal of Sound and Vibration*, 82(1) (1982), 151–156.

19. SMITH, M. C., The transverse vibration of stationary and spinning discs, University of Strathclyde, Ph.D. Thesis (1974).

20. GORMAN, D. G. and HUISSOON, J. P., Analysis of thermally stressed variable thickness composite discs, in *Composite Structures*, Proc. 1st International Conference on Composite Structures, Paisley College of Technology (1981), 135–143, Applied Science Publishers.

21. GORMAN, D. G. and HUISSOON, J. P., Finite element analysis of the buckling of variable thickness discs, *Proceedings of the Canadian Society of Mechanical Engineers* (1983), in press.

22. LAMB, A. and SOUTHWELL, R. V., Vibrations of a spinning disc, *Proc. Roy. Soc. (London)* (1921), Set A, 99, 272–280.

23. PANDALAI, K. A. U. and PATEL, S. A., Buckling of orthotropic circular plates, *Journal of the Royal Aeronautical Society*, 69 (1965), 279–280.

24. KENNEDY, W. and GORMAN, D. G., Vibration analysis of variable thickness discs subjected to centrifugal and thermal stress, *Journal of Sound and Vibration*, 53(1) (1977), 83–101.

APPENDIX: ELEMENT MATRICES

For the linearly varying thickness annular finite element shown in Fig. 2(b) the mass $[m]$, flexural stiffness $[k_f]$ and geometric stiffness matrix $[k_g]$ can be written as:

$$[m] = C_1 \pi \rho [B]^T [Q][B] \tag{A1}$$

$$[k_f] = \frac{k^2 E_r}{12(k^2 - \mu_\theta^2)} [B]^T [P][B] \tag{A2}$$

$$[k_g] = C_1 \pi [B]^T \int_{\bar{R}_1}^{\bar{R}_2} h(r)[R] \, dr [B] \tag{A3}$$

where

$$[Q] = \begin{vmatrix} Q1 & Q2 & Q3 & Q4 \\ & Q3 & Q4 & Q5 \\ & & Q5 & Q6 \\ & & & Q7 \end{vmatrix}$$

$$Q_i = \int_{\bar{R}1}^{\bar{R}2} hr^i \, dr$$

and

$$[B] = \begin{bmatrix} \dfrac{\bar{R}2(\bar{R}2 - 3\bar{R}1)}{(\bar{R}2 - \bar{R}1)^3} & -\dfrac{\bar{R}1\bar{R}2^2}{(\bar{R}2 - \bar{R}1)^2} & \dfrac{\bar{R}1(3\bar{R}2 - \bar{R}1)}{(\bar{R}2 - \bar{R}1)^3} & -\dfrac{\bar{R}1^2\bar{R}2}{(\bar{R}2 - \bar{R}1)^2} \\[2ex] \dfrac{6\bar{R}1\bar{R}2}{(\bar{R}2 - \bar{R}1)^3} & \dfrac{\bar{R}2(\bar{R}2 + 2\bar{R}1)}{(\bar{R}2 - \bar{R}1)^2} & -\dfrac{6\bar{R}1\bar{R}2}{(\bar{R}2 - \bar{R}1)^3} & \dfrac{\bar{R}1(\bar{R}1 + 2\bar{R}2)}{(\bar{R}2 - \bar{R}1)^2} \\[2ex] -\dfrac{3(\bar{R}2 + \bar{R}1)}{(\bar{R}2 - \bar{R}1)^3} & -\dfrac{(\bar{R}1 + 2\bar{R}2)}{(\bar{R}2 - \bar{R}1)^2} & \dfrac{3(\bar{R}2 + \bar{R}1)}{(\bar{R}2 - \bar{R}1)^3} & -\dfrac{(\bar{R}2 + 2\bar{R}1)}{(\bar{R}2 - \bar{R}1)^2} \\[2ex] \dfrac{2}{(\bar{R}2 - \bar{R}1)^3} & \dfrac{1}{(\bar{R}2 - \bar{R}1)^2} & -\dfrac{2}{(\bar{R}2 - \bar{R}1)^3} & \dfrac{1}{(\bar{R}2 - \bar{R}1)^2} \end{bmatrix}$$

and

$$[P] = \begin{bmatrix} P11 & P12 & P13 & P14 \\ & P22 & P23 & P24 \\ & & P33 & P34 \\ & & & P44 \end{bmatrix}$$

where

$$P11 = P_1\{C_1 m^4 k^2 + 4Go\}$$

$$P12 = P_2\{C_1(m^4 - m^2)k^2\}$$

$$P13 = P_3\{C_1[(m^4 - 2m^2)k^2 - 2m^2\mu_\theta] - 4Go\}$$

$$P14 = P_4\{C_1[(m^4 - 3m^2)k^2 - 6m^2\mu_\theta] - 8Go\}$$

$$P22 = P_3 C_1(1 - m^2)^2 k^2$$

$$P23 = P_4\{(m^4 - 3m^2 + 2)k^2 - 2\mu_\theta(m^2 - 1)\}C_1$$

$$P24 = P_5\{(m^4 - 4m^2 + 3)k^2 - 6\mu_\theta(m^2 - 1)\}C_1$$

$$P33 = P_5\{C_1[(m^4 - 4m^2 + 4)k^2 - 4\mu_\theta(m^2 - 2) + 4] + 4Go\}$$

$$P34 = P_6\{C_1[(m^4 - 5m^2 + 6)k^2 - \mu_\theta(8m^2 - 18) + 12] + 8Go\}$$

$$P44 = P_7\{C_1[(m^4 - 6m^2 + 9)k^2 - 12\mu_\theta(m^2 - 3) + 36] + 16Go\}$$

where

$$Go = \frac{Gr_\theta m^2(k^2 - \mu_\theta^2)}{k^2 E_r} = 0 \cdot 35\, m^2 \text{ throughout}$$

and

$$C_1 = 2 \text{ when } m = 0$$
$$= 1 \text{ otherwise}$$

$$P_1 = \frac{1}{2}\,\bar{\alpha}^3 \frac{(\bar{R}_2^2 - \bar{R}_1^2)}{\bar{R}_1^2 \bar{R}_2^2} + \frac{3(\bar{R}_2 - \bar{R}_1)\bar{\alpha}^2\bar{\beta}}{\bar{R}_1\bar{R}_2} + 3\log_e\left(\frac{\bar{R}_2}{\bar{R}_1}\right)\bar{\alpha}\bar{\beta}^2 + (\bar{R}_2 - \bar{R}_1)\bar{\beta}^3$$

$$P_2 = \frac{(\bar{R}_2 - \bar{R}_1)}{\bar{R}_1\bar{R}_2}\,\bar{\alpha}^3 + 3\log_e\left(\frac{\bar{R}_2}{\bar{R}_1}\right)\bar{\alpha}^2\bar{\beta} + 3(\bar{R}_2 - \bar{R}_1)\bar{\alpha}\bar{\beta}^2 + \tfrac{1}{2}(\bar{R}_2^2 - \bar{R}_1^2)\bar{\beta}^3$$

$$P_3 = \log_e\left(\frac{\bar{R}_2}{\bar{R}_1}\right)\bar{\alpha}^3 + 3(\bar{R}_2 - \bar{R}_1)\bar{\alpha}^2\bar{\beta} + \tfrac{3}{2}(\bar{R}_2^2 - \bar{R}_1^2)\bar{\alpha}\bar{\beta}^2 + \tfrac{1}{3}(\bar{R}_2^3 - \bar{R}_1^3)\bar{\beta}^3$$

$$P_4 = (\bar{R}_2 - \bar{R}_1)\bar{\alpha}^3 + \tfrac{3}{2}(\bar{R}_2^2 - \bar{R}_1^2)\bar{\alpha}^2\bar{\beta} + (\bar{R}_2^3 - \bar{R}_1^3)\bar{\alpha}\bar{\beta}^2 + \tfrac{1}{4}(\bar{R}_2^4 - \bar{R}_1^4)\bar{\beta}^3$$

$$P_5 = \tfrac{1}{2}(\bar{R}_2^2 - \bar{R}_1^2)\bar{\alpha}^3 + (\bar{R}_2^3 - \bar{R}_1^3)\bar{\alpha}^2\bar{\beta} + \tfrac{3}{4}(\bar{R}_2^4 - \bar{R}_1^4)\bar{\alpha}\bar{\beta}^2 + \tfrac{1}{5}(\bar{R}_2^5 - \bar{R}_1^5)\bar{\beta}^3$$

$$P_6 = \tfrac{1}{3}(\bar{R}_2^3 - \bar{R}_1^3)\bar{\alpha}^3 + \tfrac{3}{4}(\bar{R}_2^4 - \bar{R}_1^4)\bar{\alpha}^2\bar{\beta} + \tfrac{3}{5}(\bar{R}_2^5 - \bar{R}_1^5)\bar{\alpha}\bar{\beta}^2 + \tfrac{1}{6}(\bar{R}_2^6 - \bar{R}_1^6)\bar{\beta}^3$$

$$P_7 = \tfrac{1}{4}(\bar{R}_2^4 - \bar{R}_1^4)\bar{\alpha}^3 + \tfrac{3}{5}(\bar{R}_2^5 - \bar{R}_1^5)\bar{\alpha}^2\bar{\beta} + \tfrac{1}{2}(\bar{R}_2^6 - \bar{R}_1^6)\bar{\alpha}\bar{\beta}^2 + \tfrac{1}{7}(\bar{R}_2^7 - \bar{R}_1^7)\bar{\beta}^3$$

$$[R] = \begin{bmatrix} \dfrac{m^2}{r}\sigma_\theta & m^2\sigma_\theta & m^2 r\sigma_\theta & r^2 m^2\sigma_\theta \\ \cdot & r\sigma_r + R(1,3) & 2r^2\sigma_r + R(1,4) & 3r^3\sigma_r + m^2 r^3\sigma_\theta \\ & & 4r^3\sigma_r + m^2 r^3\sigma_\theta & 6r^4\sigma_r + m^2 r^4\sigma_\theta \\ & & & 9r^5\theta_r + m^2 r^5\sigma_\theta \end{bmatrix}$$

10

A New Approach to the Nonlinear Dynamic Analysis of Composite Plates

M. Sathyamoorthy

*Department of Mechanical and Industrial Engineering,
Clarkson College of Technology, Potsdam, New York 13676, USA*

ABSTRACT

Several approximate solutions are available in the literature for plates of various geometries undergoing large amplitude flexural vibrations. In many cases the results are based on a single-mode approximation. Since closed-form solutions do not exist for many of these nonlinear problems, it is very difficult to evaluate the accuracy of any approximate solution. In this paper a new approach is used to check the accuracy and therefore improve the results in the nonlinear regime. Self-generating functions of zero, first and second order are used to investigate the effects of amplitude, geometry and material constants on the dynamic behavior of rectangular composite plates.

INTRODUCTION

Geometrically nonlinear static and dynamic analysis of composite plates of various geometries have been extensively dealt with in the literature in recent years.[1,2] Many of these studies present approximate solutions to dynamic problems based on a single-mode assumption consisting of trigonometric, polynomial or hybrid functions. In the case of plates of rectangular and square planforms, Sathyamoorthy and Pandalai,[3-6] Niyogi,[7] Prabhakara and Chia,[8] Wu and Vinson,[9] Wah,[10] and several others[2] have presented solutions based on these functions. In Refs 3, 4, 6-8, the problem is either formulated in terms of stress function, F, and lateral displacement, w, or in terms of the three median surface displacement components u^0, v^0 and w. An approach which is much simpler is based on the Berger[10] approximation and has been carefully applied in Refs 5, 9, 10.

128

This approximation simplifies the detailed calculations significantly by eliminating the in-plane displacement u^0 and v^0 from the governing nonlinear equations. The procedure in all these cases involves assuming an appropriate shape function for the lateral displacement, w, and satisfying the nonlinear governing equations approximately by Galerkin or other methods. Finite element methods[11,12] have also been used recently to solve this class of problems. In all these cases, the nature of numerical results depends very much upon the choice of shape functions. Since closed form solutions do not exist for many of these nonlinear problems, it becomes difficult to assess the accuracy of various approximate solutions. While this does not present a significant problem in the case of isotropic plates, it does seem to have an effect in orthotropic and anisotropic composite as well as laminated plates. A new method, therefore, is presented in this paper to be able to check the accuracy and thus improve the nature of the numerical results obtained to nonlinear static as well as dynamic problems.

The approach presented in this paper is based on the use of the so-called Self-generating Functions which have been successfully applied to various nonlinear dynamic beam problems.[13] Self-generating Functions of zero-order are polynomials of fourth degree for any chosen boundary conditions of the plate. Similarly first-order polynomials are of eighth degree, second-order of twelfth degree and so on. Higher order polynomials can be readily generated from the preceding lower order polynomials for all possible combinations of boundary conditions. These functions can be conveniently used in conjunction with the governing equations of von Kármán-type or with those of the Berger approximation. Nonlinear equations are presented in this paper for both cases while only equations corresponding to the Berger approximation are solved to obtain numerical results. The nonlinearities investigated here arise due to large deformation and are included in the nonlinear strain–displacement relations. The stress–strain relationship, however, is linear and the material constants of the composite plate material are with reference to an orthogonal system of axes. Numerical results are presented for clamped and simply supported square as well as rectangular composite plates. Load-deflection values are presented for nonlinear static problems while amplitude-frequency results are tabulated for dynamic problems. Effects of amplitude, geometry and material constants on the static and dynamic behaviors are discussed. Results corresponding to zero-order, first-order and second-order functions have been tabulated to show clearly the convergence. Present results are in complete agreement with existing solutions for both nonlinear vibration and bending problems.

M. Sathyamoorthy

ANALYSIS PROCEDURE

For an orthotropic rectangular plate of dimensions a and b and uniform thickness h with principal elastic axes parallel to the edges of the plate, the governing equations of von Kármán-type are

$$F_{,\zeta\zeta\zeta\zeta} + k^2 r^4 F_{,\eta\eta\eta\eta} + m^2 r^2 F_{,\zeta\zeta\eta\eta} = E_T r^2 (w^2_{,\zeta\eta} - w_{,\zeta\zeta} w_{,\eta\eta}) \tag{1}$$

$$a^4 [\rho h w_{,tt} - q(\zeta, \eta)] + D_1 L(w) = h r^2 (F_{,\eta\eta} w_{,\zeta\zeta} + F_{,\zeta\zeta} w_{,\eta\eta} - 2 F_{,\zeta\eta} w_{,\zeta\eta}) \tag{2}$$

where

$$L(w) = w_{,\zeta\zeta\zeta\zeta} + r^4 k^2 w_{,\eta\eta\eta\eta} + 2 r^2 (q^2 + 2 p^2) w_{,\zeta\zeta\eta\eta}$$

and

$$k^2 = \frac{E_T}{E_L} \qquad m^2 = (k^2 - q^4 - 2 p^2 q^2)/p^2$$

$$q^2 = v_{TL} \qquad p^2 = G_{LT}(1 - v_{LT} v_{TL})/E_L$$

$$D_1 = G_{LT} h^3/12 p^2 \qquad \frac{x}{a} = \zeta \qquad \frac{y}{b} = \eta \tag{3}$$

In eqn. (3), E_L, E_T are elastic moduli, v_{LT}, v_{TL} are Poisson's ratios and G_{LT} is the shear modulus. L and T are the longitudinal and transverse directions respectively of the principal elastic axes.

A much simpler set of equations can be easily obtained[5] by means of the Berger approximation given below.

$$\varepsilon_1 + k\varepsilon_2 = \frac{\partial^2 h^2}{12} = e \tag{4}$$

$$a^4 \left\{ C_1 e \left(\frac{1}{a^2} w_{,\zeta\zeta} + \frac{k}{b^2} w_{,\eta\eta} \right) - \rho h w_{,tt} + q(\zeta, \eta) \right\} = D_1 L(w) \tag{5}$$

where

$$\varepsilon_1 = \frac{1}{a} u^0_{,\zeta} + \frac{1}{2a^2} w^2_{,\zeta}$$

$$\varepsilon_2 = \frac{1}{b} v^0_{,\eta} + \frac{1}{2b^2} w^2_{,\eta} \quad \text{and} \quad C_1 = E_L h/(1 - v_{TL} v_{LT}) \tag{6}$$

ε_1 and ε_2 are the median surface strains written in terms of median surface displacements u^0, v^0 and w.

For any predetermined shape function w, eqns (1) and (2) or (4) and (5) must be solved in conjunction with the corresponding in-plane boundary conditions of the plate. Although eqn. (1) can be solved exactly for circular and elliptical plates, it becomes necessary to assume the stress function F in addition to w in the case of rectangular plates. It is thus possible to solve eqns (1) and (2) approximately for certain predetermined functions for F and w. It will be expected that the ultimate results will depend not only on the choice of function for w but also on F. A clever choice of functions for F compatible with a good choice for w is difficult, if not impossible. To avoid this difficulty in the present problem eqns (4) and (5) are used for which a proper choice of w alone will be sufficient. For simply supported and clamped plates, the following Self-generating Functions are applicable.

SS: $\quad w(\zeta, \eta, t) = hA_{ss}(t)(\zeta^4 - 2\zeta^3 + \zeta)(\eta^4 - 2\eta^3 + \eta)$ (7)

zero-order

$$w(\zeta, \eta, t) = hB_{ss}(t)(\zeta^8 - 4\zeta^7 + 14\zeta^5 - 28\zeta^3 + 17\zeta)$$
$$\times (\eta^8 - 4\eta^7 + 14\eta^5 - 28\eta^3 + 17\eta) \quad (8)$$

first-order

$\cdots$

CC: $\quad w(\zeta, \eta, t) = hA_{cc}(t)(\zeta^4 - 2\zeta^3 + \zeta^2)(\eta^4 - 2\eta^3 + \eta^2)$ (9)

zero-order

$$w(\zeta, \eta, t) = hB_{cc}(t)(3\zeta^8 - 12\zeta^7 + 14\zeta^6 - 14\zeta^3 + 9\zeta^2)$$
$$\times (3\eta^8 - 12\eta^7 + 14\eta^6 - 14\eta^3 + 9\eta^2) \quad (10)$$

first-order

$\cdots$

It is possible to readily generate third and higher order polynomials and therefore they are not presented here. It can be shown that $\bar{w} = w_{max}/h$ which are given below for different cases.

$$\bar{w}_{ss} = \left(\frac{25}{256}\right) A_{ss} \qquad \bar{w}_{cc} = \left(\frac{1}{256}\right) A_{cc} \qquad \text{for zero-order} \quad (11)$$

$$\bar{w}_{ss} = \left(\frac{1385}{256}\right)^2 B_{ss} \qquad \bar{w}_{cc} = \left(\frac{163}{256}\right)^2 B_{cc} \qquad \text{for first-order} \quad (12)$$

Substituting eqns (7)–(10) in eqn. (4) and assuming that the plate edges are immovable, i.e.

$$u^0 = 0 \qquad \zeta = 0, 1$$
$$v^0 = 0 \qquad \eta = 0, 1 \tag{13}$$

eqn. (4) can be integrated to obtain the value of e. Substituting this value of e in eqn. (5) and averaging the error by an integration procedure, the following time-differential equations are obtained for plates with simply supported and clamped boundaries.

$$\bar{w}_{,\tau\tau} + X\bar{w} + Y(\bar{w})^3 = Zq_0^* \tag{14}$$

In eqn. (14) $\bar{w}$ is the nondimensional amplitude, τ is the non-dimensional time given by $\tau^2 = t^2 D_1/(a^4 \rho h)$ and $q_0^* = q_0 a^4/E_L h^4$. When there is no external loading on the plate, this equation reduces to a Duffing-type equation for which an exact solution exists. This equation has been solved using the elliptic integral method to obtain the numerical results for amplitude dependent nonlinear frequencies. In the case of nonlinear static problems, $\bar{w}$ is independent of time and therefore eqn. (14) reduces to a nonlinear algebraic equation in terms of the nondimensional maximum deflection. These algebraic equations have been solved to obtain nondimensional loads for various values of $\bar{w}$. In Tables 1–3, the coefficients X, Y, Z are tabulated for isotropic (ISO) and graphite-epoxy

TABLE 1
X, Y, Z in eqn. (14) for plates with $r = 0.5$

X, Y	ISO			GRE		
	Zero	First	Second	Zero	First	Second
X_{ss}	152·36	152·20	152·20	99·46	99·32	99·32
			152·20*			99·32*
Y_{ss}	231·92	228·35	228·30	160·39	157·92	157·89
			228·30**			157·89**
Z_{ss}	17·62	17·70	17·70	19·33	19·42	19·42
X_{cc}	607·50	607·41	607·55	507·39	504·09	504·09
			638·57*			523·46*
Y_{cc}	222·91	223·08	223·13	154·16	154·28	154·31
			228·30**			157·89**
Z_{cc}	18·81	19·01	19·01	20·64	20·85	20·86

ISO: Isotropic; GRE: graphite-epoxy.
 * Results from Refs 2 and 3.
** Results from Refs 5, 10 and 14.

TABLE 2

X, Y, Z in eqn. (14) for square plates

X, Y		ISO			GRE	
	Zero	First	Second	Zero	First	Second
X_{ss}	389·97	389·64	389·64	107·04	106·90	106·90
			389·64*			106·90*
Y_{ss}	593·71	584·57	584·46	199·07	196·01	195·97
			584·45**			195·97**
Z_{ss}	17·62	17·70	17·70	19·33	19·42	19·42
X_{cc}	1 296·00	1 303·35	1 303·95	527·03	524·02	524·04
			1 385·37*			545·04*
Y_{cc}	570·65	571·09	571·21	191·34	191·49	191·53
			584·45**			195·97**
Z_{cc}	18·81	19·01	19·01	20·64	20·85	20·86

ISO: Isotropic; GRE: graphite-epoxy.
* Results from Refs 2 and 3.
** Results from Refs 5, 10 and 14.

(GRE) plates with simply supported and clamped edges using zero to second-order functions. Typical numerical results for nondimensional loads and frequency ratios are given in Tables 4 and 5 at different non-dimensional amplitudes. Wherever possible, comparisons are made with existing results in the literature.

TABLE 3

X, Y, Z in eqn. (14) for plates with $r = 1·5$

X, Y		ISO			GRE	
	Zero	First	Second	Zero	First	Second
X_{ss}	1 029·85	1 028·89	1 028·89	125·77	125·66	125·61
			1 028·88*			125·61*
Y_{ss}	1 567·76	1 543·63	1 543·34	272·82	268·62	268·57
			1 543·32**			268·57**
Z_{ss}	17·62	17·70	17·70	19·33	19·42	19·42
X_{cc}	3 703·50	3 714·67	3 716·01	591·25	588·54	588·58
			3 928·83*			613·48*
Y_{cc}	1 506·88	1 508·05	1 508·35	262·23	262·43	262·48
			1 543·32**			268·57**
Z_{cc}	18·81	19·01	19·01	20·64	20·85	20·86

ISO: Isotropic; GRE: graphite-epoxy.
* Results from Refs 2 and 3.
** Results from Refs 5, 10 and 14.

TABLE 4
Values of nondimensional load q_0^* at various amplitudes $(r = 1\cdot0)$

| $\bar{w}$ | | SS | | | CC | |
	ISO	BE	GRE	ISO	BE	GRE
0	0	0	0	0	0	0
0·5	15·134	4·845	4·013	38·052	15·368	13·709
1·0	55·034	19·502	15·593	98·641	39·662	34·303
1·5	144·463	53·782	42·308	204·301	81·805	68·671

ISO: Isotropic; BE: boron-epoxy; GRE: graphite-epoxy.

TABLE 5
Nondimensional frequencies $(\omega/\omega_0)10^4$ for clamped rectangular and square plates at various amplitudes

| $\bar{w}$ | | $r = 1\cdot0$ | | | $r = 1\cdot5$ | |
	ISO	BE	GRE	ISO	BE	GRE
0	10 000	10 000	10 000	10 000	10 000	10 000
0·5	10 404	10 395	10 338	10 375	10 472	10 411
1·0	11 538	11 507	11 297	11 432	11 785	11 564
1·5	13 200	13 141	12 733	12 996	13 664	13 249

ISO: Isotropic; BE: boron-epoxy; GRE: graphite-epoxy.

NUMERICAL RESULTS AND DISCUSSION

Self-generating Functions of zero-order and first-order are listed in this paper for plates with simply-supported and clamped boundary conditions. Although higher order functions are not listed here, they can be easily generated from the corresponding lower order polynomials. It must be pointed out here that the complexity increases as higher order polynomials are considered in the solution procedure. The results obtained from zero-order, first-order and second-order functions have been tabulated in Tables 1–3, which will give the coefficients, X, Y, Z in eqn. (14) for simply-supported and clamped immovable isotropic and composite plates. Results based on trigonometric functions (single-mode) are presented and compared with the present results obtained from the highest-order function. In the case of simply-supported plates the agreement is excellent. For clamped plates, however, the differences indicate a clear need for improving the solutions obtained by trigonometric functions. The method suggested here provides for a continuous improvement in the values of the coefficients of eqn. (14) by considering higher order functions each time. It is also observed that each coefficient converges to a steady value fairly quickly. Some numerical results for large deflection and large amplitude vibration problems are presented in Tables 3 and 4 for the sake of completeness. The nonlinearity can be clearly seen to be of the hard-spring type for both static as well as dynamic problems. Self-generating Functions could be conveniently used to obtain reliable solutions for anisotropic plates and for plates with skew geometry where single-term trigonometric approach presents several difficulties.

REFERENCES

1. SATHYAMOORTHY, M., Nonlinear vibration of plates, *Shock and Vibration Digest* (1983) (to appear).
2. CHIA, C. Y., *Nonlinear Analysis of Plates*, New York, McGraw-Hill, 1980.
3. SATHYAMOORTHY, M. and PANDALAI, K. A. V., Nonlinear flexural vibrations of orthotropic rectangular plates, *Journal of the Aeronautical Society of India*, **22** (1970), 264–266.
4. SATHYAMOORTHY, M. and PANDALAI, K. A. V., Nonlinear flexural vibrations of orthotropic skew plates, *Journal of Sound and Vibration*, **24** (1972), 115–220.
5. SATHYAMOORTHY, M. and PANDALAI, K. A. V., Large amplitude flexural vibration of simply supported skew plates, *AIAA Journal*, **11** (1973), 1279–1282.
6. SATHYAMOORTHY, M., Nonlinear vibration of rectangular plates, *ASME Journal of Applied Mechanics*, **46** (1979), 215–217.
7. NIYOGI, A. K., Nonlinear bending of rectangular orthotropic plates, *International Journal of Solids and Structures*, **9** (1973), 1133–1139.
8. PRABHAKARA, M. K. and CHIA, C. Y., Nonlinear flexural vibrations of orthotropic rectangular plates, *Journal of Sound and Vibration*, **52** (1977), 511–518.
9. WU, C. and VINSON, J. R., On the nonlinear oscillations of plates composed of composite materials, *Journal of Composite Materials*, **3** (1969), 548–561.
10. WAH, T., Large amplitude flexural vibration of rectangular plates, *International Journal of Mechanical Sciences*, **5** (1963), 425–438.
11. KANAKA RAJU, K. and VENKATESWARA RAO, G., Nonlinear vibrations of orthotropic plates by a finite element method. *Journal of Sound and Vibration*, **48** (1976), 301–303.
12. REDDY, J. N. and CHAO, W. C., Nonlinear oscillations of laminated anisotropic rectangular plates, *ASME Journal of Applied Mechanics*, **49** (1982), 396–402.
13. SATHYAMOORTHY, M., Large amplitude vibrations of moderately thick beams, *Proceedings of the International Modal Analysis Conference, Orlando, Florida*, November 1982, 136–140.
14. SATHYAMOORTHY, M. and PANDALAI, K. A. V., Nonlinear vibration of elastic skew plates exhibiting rectilinear orthotropy, *Journal of the Franklin Institute*, **296** (1973), 359–369.

11

A Combined Experimental and Numerical Technique for the Determination of the Material Properties of Laminates

J. L. WEARING and C. PATTERSON

Department of Mechanical Engineering, University of Sheffield, Mappin Street, Sheffield S1 3JD, England

ABSTRACT

A non-destructive technique for the determination of the material properties of composites, which take the form of thin laminates, is discussed. The method uses experimental natural frequencies in conjunction with the Rayleigh–Ritz method to determine initial values of constants which are related to the laminate's material properties. These initial values are then used in an iterative procedure, using the finite element method to calculate successively, improved values of the constants until converged results are obtained. Details of the procedure and the results, which were obtained, are presented in the paper.

INTRODUCTION

In a previous paper[3] the authors discussed the problems associated with static testing methods for determining the elastic constants of composite materials and proposed a non-destructive technique for the determination of the elastic constants of composite materials in the form of thin laminates of a crossply construction. The method is based on determining, experimentally, the natural frequencies and mode shapes of cantilevered rectangular plate specimens of the material under consideration. These experimentally derived results were used in conjunction with a frequency expression derived from the Rayleigh–Ritz method to determine the relevant material properties. In that initial study the assumed deflected forms of the normal modes of the plate were beam functions and the

136

inaccuracies which occurred in the calculations led to results which were assessed to be inaccurate by around 20%.

It was felt, however, that the basic technique was worthy of further investigation and in this paper further refinements of the method are discussed. Using the experimental results and the frequency expression derived from the Rayleigh–Ritz method a set of constants (*see* eqn. (17)) related to the material properties were obtained. These constants were then used in conjunction with the finite element method to obtain a set of mode shapes which were used to evaluate, numerically, the expressions for the maximum strain and kinetic energies of the plate giving an improved set of constants. These constants were used in a further round of calculations and the iterative procedure was continued until convergence was achieved. The results of the study and full details of the technique are presented in the paper.

Material Properties and Stress–Strain Relationships

The material from which the specimens were manufactured consisted of bonded layers of fibre such that the fibres in alternate layers were orientated in the same direction with the fibre directions of the intervening layers being in a perpendicular direction (i.e. the fibre formation was $0°–90°–0°$). The plates were assumed therefore, to be orthotropic and the stress–strain relationships for such a plate lying in the xy plane is given by the expression

$$\begin{Bmatrix} \sigma_x \\ \sigma_y \\ \tau_{xy} \end{Bmatrix} = \begin{bmatrix} C_{11} & C_{12} & 0 \\ C_{21} & C_{22} & 0 \\ 0 & 0 & |C_{33} \end{bmatrix} \begin{Bmatrix} \varepsilon_x \\ \varepsilon_y \\ \gamma_{xy} \end{Bmatrix} \tag{1}$$

The relationship between the constants, C_{ij}, in the above equation and the plate's material properties is given by the following equations:

$$C_{11} = \frac{E_x}{1 - v_x v_y} \tag{2}$$

$$C_{22} = \frac{E_y}{1 - v_x v_y} \tag{3}$$

$$C_{12} = \frac{v_y E_x}{1 - v_x v_y} \tag{4}$$

$$C_{21} = \frac{v_x E_y}{1 - v_x v_y} \tag{5}$$

$$C_{33} = G_{xy} \tag{6}$$

In the above equations only four of the five elastic constants are independent and as the matrix of coefficients in eqn. (1) is symmetrical, the following relationship can be obtained from eqns (4) and (5)

$$\frac{E_x}{E_y} = \frac{v_x}{v_y} \tag{7}$$

from which the value of the fifth property can be obtained.

THE RAYLEIGH–RITZ METHOD

When using the Rayleigh–Ritz method to determine the natural frequencies of rectangular plates[2,4] the maximum strain and kinetic energies of the vibrating plate are equated. In the case of an orthotropic plate, the maximum strain energy is given by the expression

$$U_{max} = \frac{1}{2} \int_0^\alpha \int_0^b \left[D_x \left(\frac{\partial^2 W}{\partial x^2} \right)^2 + D_y \left(\frac{\partial^2 W}{\partial y^2} \right)^2 \right.$$
$$\left. + 2D_1 \left(\frac{\partial^2 W}{\partial x^2} \right) \left(\frac{\partial^2 W}{\partial y^2} \right) + 4D_{xy} \left(\frac{\partial^2 W}{\partial x \, \partial y} \right)^2 \right] dy \, dx \tag{8}$$

in which the constants D_x, D_1, D_y and D_{xy} are given by the expressions

$$\left.\begin{array}{ll} D_x = \dfrac{C_{11}h^3}{12} & D_1 = \dfrac{C_{12}h^3}{12} \\[2ex] D_y = \dfrac{C_{22}h^3}{12} & D_{xy} = \dfrac{C_{33}h^3}{12} \end{array}\right\} \tag{9}$$

The maximum kinetic energy of the plate is obtained from:

$$T_{max} = \tfrac{1}{2}\rho h \omega_n^2 \int_0^\alpha \int_0^b W^2 \, dy \, dx \tag{10}$$

When eqns (8) and (10) are equated the following expression is obtained:

$$U_{max} - T_{max} = 0 \tag{11}$$

The first step in the analysis is to assume that the deflected form of the vibrating plate is a series as indicated by eqn. (12):

$$W = \sum\sum A_{mn}\phi_m(x)\theta_n(y) \tag{12}$$

Equation (12) is substituted into eqn. (11) and the constants A_{mn} adjusted to make it a minimum. This is achieved by differentiating the resulting expression with respect to each of the constants A_{mn} to give a set of equations of the type:

$$\frac{\partial U_{max}}{\partial A_{mn}} - \frac{\partial T_{max}}{\partial A_{mn}} = 0 \tag{13}$$

In the analysis, for the determination of the material properties, $\phi_m(x)$ and $\theta_n(y)$ were assumed to be beam functions having appropriate boundary conditions for the rectangular plate[2] and m and n relate to the number of node points on the deflected forms of the beam functions.

Following the relevant differentiations and integrations, when eqn. (12) is substituted into eqn. (13), a set of equations, which may be expressed in matrix form, as given by eqn. (14), is obtained:

$$([K_{mn}] - \lambda[M_{mn}])\{A_{mn}\} = 0 \tag{14}$$

The elements of the matrix $[K_{mn}]$ and $[M_{mn}]$ are obtained from eqns (8) and (10) respectively. The natural frequencies of the plate are obtained from the determinant

$$|[K_{mn}] - \lambda[M_{mn}]| = 0 \tag{15}$$

but, if only the dominant term in each of eqns (14) is used, they are reduced to single term expressions of the form

$$(K_{mn} - \lambda)A_{mn} = 0 \tag{16}$$

which gives the natural frequency having a mode shape with m nodal lines parallel to the y axis and n nodal lines parallel to the x axis of the plate.

The term K_{mn} in eqn. (16) is given by the expression

$$K_{mn} = D_x H_1^{(mn)} + 2D_1 H_2^{(mn)} + D_y H_3^{(mn)} + D_{xy} H_4^{(mn)} \tag{17}$$

and λ is obtained from

$$\lambda = \rho h \omega_n^2 \tag{18}$$

If eqns (17) and (18) are substituted into eqn. (16), the natural frequency of mode m/n is obtained from the expression:

$$\rho h \omega_n^2 = D_x H_1^{(mn)} + 2D_1 H_2^{(mn)} + D_y H_3^{(mn)} + D_{xy} H_4^{(mn)} \tag{19}$$

The integrals for the evaluation of the terms $H_i^{(mn)}$ in eqn. (19) are tabulated by Young[4] for various values of m and n.

FINITE ELEMENT ANALYSIS OF PLATES

Using the isoparametric formulation, the deflected form of an element of a plate in bending in the xy plane (Fig. 1(a)) is expressed in terms of the co-ordinates (ξ, η) (Fig. 1(b)) by the equation

$$
\begin{aligned}
w = {} & \alpha_1 + \alpha_2\xi + \alpha_3\eta + \alpha_4\xi\eta + \alpha_5\xi^2 + \alpha_6\eta^2 + \alpha_7\xi^2\eta + \alpha_8\xi\eta^2 + \alpha_9\xi^3 + \alpha_{10}\eta^3 \\
& + \alpha_{11}\xi^3\eta + \alpha_{12}\xi\eta^3 + \alpha_{13}\xi^4 + \alpha_{14}\eta^4 + \alpha_{15}\xi^5 + \alpha_{16}\eta^5 + \alpha_{17}\xi^3\eta^2 \\
& + \alpha_{18}\xi^2\eta^3 + \alpha_{19}\xi^4\eta + \alpha_{20}\xi\eta^4 + \alpha_{21}\xi^4\eta^2 + \alpha_{22}\xi^2\eta^4 + \alpha_{23}\xi\eta^5 + \alpha_{24}\xi^5\eta
\end{aligned}
$$

$$(20)$$

Equation (20) can be written in matrix form as

$$\{w\} = [B(\xi, \eta)]\{\alpha\} \tag{21}$$

The displacements at each node, i, in the (ξ, η) plane are

$$\{w_i\} = \left\{ \begin{array}{c} w \\ w_\xi \\ w_\eta \end{array} \right\} \tag{22}$$

in which

$$w_\xi = \frac{\partial w}{\partial \xi} \qquad \text{and} \qquad w_\eta = \frac{\partial w}{\partial \eta}$$

The nodal displacements for the complete element are obtained from eqn. (20) by substituting the co-ordinates ξ and η of each node to obtain a series of equations which may be written in matrix form as

$$\{w_e\} = [A]\{\alpha\} \tag{23}$$

from which

$$\{\alpha\} = [A]^{-1}\{w_e\} \tag{24}$$

Substituting eqn. (24) into eqn. (21) gives:

$$\{w\} = [B(\xi, \eta)][A]^{-1}\{w_e\} \tag{25}$$

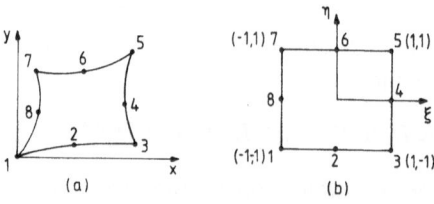

FIG. 1. Eight noded isoparametric element.

The deflections $\{w_i\}$ at each node in the (ξ, η) plane can be related to the deflections $\{\delta_i\}$ at each node in the (x, y) plane by the following transformation

$$
\left\{ \begin{array}{c} w \\ w_\xi \\ w_\eta \end{array} \right\} = \begin{bmatrix} 1 & 0 & 0 \\ 0 & \dfrac{\partial y}{\partial \xi} & -\dfrac{\partial x}{\partial \xi} \\ 0 & \dfrac{\partial y}{\partial \eta} & -\dfrac{\partial x}{\partial \eta} \end{bmatrix} \left\{ \begin{array}{c} w \\ \theta_x \\ \theta_y \end{array} \right\} \tag{26}
$$

which may be written as

$$
\{w_i\} = [T_i]\{\delta_i\} \tag{27}
$$

Hence for the whole element the relationship between the nodal deflections $\{w_e\}$ in the (ξ, η) plane and the deflections $\{\delta_e\}$ in the (x, y) plane is given by the equation:

$$
\{w_e\} = [T_e]\{\delta_e\} \tag{28}
$$

in which the matrix $[T_e]$ is a diagonal matrix made up of submatrices $[T_i]$. Substituting eqn. (28) into eqn. (25) gives:

$$
\{w\} = [B(\xi, \eta)][A]^{-1}[T_e]\{\delta_e\} \tag{29}
$$

To obtain the derivatives $\partial x/\partial \xi$, $\partial x/\partial \eta$, $\partial y/\partial \xi$ and $\partial y/\partial \eta$ in the transformation matrix $[T_e]$ the x and y coordinates of the element are expressed in terms of ξ and η by the following equations:

$$
x = M_1 + M_2\xi + M_3\eta + M_4\xi\eta + M_5\xi^2 + M_6\eta^2
$$
$$
+ M_7\xi^2\eta + M_8\xi\eta^2 \tag{30}
$$
$$
y = M_9 + M_{10}\xi + M_{11}\eta + M_{12}\xi\eta + M_{13}\xi^2 + M_{14}\eta^2
$$
$$
+ M_{15}\xi^2\eta + M_{16}\xi\eta^2 \tag{31}
$$

The x and y co-ordinates of all the nodes and their corresponding (ξ, η) co-ordinates can be inserted into eqns (30) and (31) to give the following matrix expressions from the resulting sets of equations

$$
\{x_e\} = [C]\{M\} \tag{32}
$$
$$
\{y_e\} = [C]\{M'\} \tag{33}
$$

From eqns (32) and (33):

$$
\{M\} = [C]^{-1}\{x_e\} \tag{34}
$$

and

$$\{M'\} = [C]^{-1}\{y_e\} \tag{35}$$

From eqn. (30) expressions for x and its derivatives, x_ξ, x_η, $x_{\xi\xi}$, $x_{\eta\eta}$ and $x_{\xi\eta}$, can be obtained and written in matrix form as

$$\{x_{\xi\eta}\} = [P_{\xi\eta}]\{M\} \tag{36}$$

A similar equation can be obtained for y and its derivatives from eqn. (31):

$$\{y_{\xi\eta}\} = [P_{\xi\eta}]\{M'\} \tag{37}$$

Substituting eqns (34) and (35) into eqns (36) and (37) gives the following equations

$$\{x_{\xi\eta}\} = [P_{\xi\eta}][C]^{-1}\{x_e\} \tag{38}$$

$$\{y_{\xi\eta}\} = [P_{\xi\eta}][C]^{-1}\{y_e\} \tag{39}$$

Hence the appropriate derivatives for the transformation matrices $[T_e]$ are obtained from eqns (38) and (39).

The deflection w and its derivatives, w_ξ, w_η, $w_{\xi\xi}$ and $w_{\eta\eta}$, can be obtained from eqn. (29) and written in matrix form as

$$\{S_{\xi\eta}\} = [B_{\xi\eta}][A]^{-1}[T_e]\{\delta_e\} \tag{40}$$

The derivatives of w in the (ξ, η) plane can be related to the derivatives in the (x, y) plane by the transformation

$$\{S_{\xi\eta}\} = [Q]\{S_{xy}\} \tag{41}$$

Hence

$$\{S_{xy}\} = [Q]^{-1}\{S_{\xi\eta}\} \tag{42}$$

Substituting eqn. (40) into eqn. (42) gives

$$\{S_{xy}\} = [Q]^{-1}[B_{\xi\eta}][A]^{-1}[T_e]\{\delta_e\} \tag{43}$$

PROCEDURE FOR THE DETERMINATION OF THE ELASTIC CONSTANTS

The first step in the procedure is to determine, experimentally, the natural frequencies of the rectangular plate specimens using the method described by the authors in a previous paper.[3] Four of these natural frequencies are

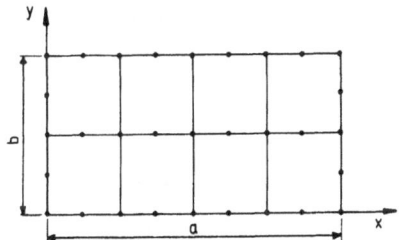

FIG. 2.   Finite element model of plate.

then used in conjunction with eqn. (19) to evaluate the constants D_x, D_y, D_1 and D_{xy}. These initial values of the constants are then used in a finite element analysis to determine the natural frequencies and mode shapes of the vibrating plate. The mode shapes are in the form of deflections at the node points on the plate (*see* Fig. 2) and can be substituted into eqn. (43) to determine the derivatives $\{S_{xy}\}$ at any point on each element.

Using values of the derivatives obtained at the Gauss points on the element for three point integration as outlined by Hinton and Owen,[1] the integrals in the energy expressions as given by eqns (8) and (10) are evaluated numerically in the (ξ, η) plane using expressions of the form

$$\int_{-1}^{+1} \int_{-1}^{+1} \left(\frac{\partial^2 W}{\partial x^2}\right)^2 |J| \, d\xi \, d\eta \tag{44}$$

The results of these numerical integrations and the natural frequencies are substituted into eqns (8) and (10) to calculate a new set of constants D_x, D_y, D_1 and D_{xy} which are used to obtain a further set of mode shapes and natural frequencies for the evaluation of the strain and kinetic energies and hence the determination of the next set of constants. This iterative procedure, which is illustrated in Fig. 3, is continued until converged values

TABLE 1
Experimental natural frequencies of cantilevered plate ($a = 0.25$ m, $b = 0.15$ m, and $h = 0.85 \times 10^{-3}$m)

n \ m	1	2	3	4
0	15·000	88·000	241·000	489·000
1	38·000	120·000	279·000	509·000
2	165·000	254·000	400·000	1 017·000
3	554·000	667·000	676·000	
4	1 080·000			

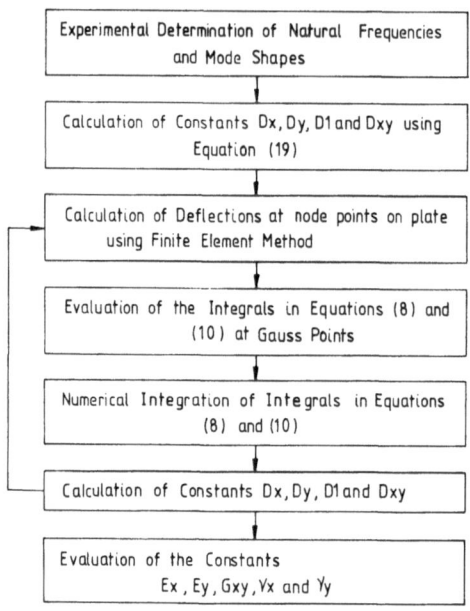

FIG. 3.   Flow diagram of iterative procedure for the evaluation of elastic constants.

TABLE 2
Variation of constants D_x, D_y, D_1 and D_{xy} with successive iterations

D	Rayleigh–Ritz analysis	Finite element iterations					
		1st	2nd	3rd	4th	5th	6th
D_x	4·7439	4·7024	4·6529	4·6326	4·6168	4·6062	4·6021
D_y	2·1242	2·1275	2·1332	2·1453	2·1507	2·1565	2·1581
D_1	0·4809	0·3923	0·3275	0·2798	0·2221	0·1970	0·2018
D_{xy}	1·6934	1·6724	1·6520	1·5993	1·5444	1·5042	1·4892

TABLE 3
Material properties of three-ply laminate

Constants	Rayleigh–Ritz analysis	Converged value from finite element analysis
E_x	$92·53 \times 10^9 \, \text{N/m}^2$	$91·15 \times 10^9 \, \text{N/m}^2$
E_y	$6·049 \times 10^9 \, \text{N/m}^2$	$4·01 \times 10^9 \, \text{N/m}^2$
G_{xy}	$10·71 \times 10^9 \, \text{N/m}^2$	$7·27 \times 10^9 \, \text{N/m}^2$
v_x	0·0371	0·0429
v_y	0·5682	0·0979

of the constants are obtained. The final values of D_x, D_y, D_1 and D_{xy} are used to evaluate the elastic constants E_x, E_y, v_x, v_y and G_{xy} of the material. Throughout the analysis the plate was subdivided into eight elements as shown in Fig. 2 using eight-noded isoparametric elements of the type shown in Fig. 1.

The values of the experimental natural frequencies of a three-ply, laminated, rectangular cantilevered plate and, which were used in conjunction with eqn. (19), are shown in Table 1. Variation of the constants D_x, D_y, D_1 and D_{xy} with successive iterations are presented in Table 2 and the final values of the elastic constants are compared with those obtained from eqn. (19), using the Rayleigh–Ritz method, in Table 3.

DISCUSSION OF RESULTS AND CONCLUSIONS

Table 2 shows the results of the convergence study for the constants D_x, D_y, D_1 and D_{xy} using results obtained from the finite element calculations to evaluate, numerically, the integrals in eqns (8) and (10). When the final iterative values were used to calculate the natural frequencies of the plates using the finite element method, it was found, that on comparing them with the experimental results, an average improvement of around 5 %, over the Rayleigh–Ritz results, was obtained. The values of the elastic constants which were obtained using these final values of the constants, D_x, D_y, D_1 and D_{xy} are shown in Table 3. On considering the improvement in the calculated natural frequencies the results of the elastic constants shown in Table 3, using the finite element results, are considered to be an improvement compared to those obtained using the Rayleigh–Ritz technique.

REFERENCES

1. HINTON, E. and OWEN, D. R. J., *Finite Element Programming*, Academic Press, London (1977).
2. WARBURTON, G. B., The vibration of rectangular plates, *Proc. Instn Mech. Engrs*, **168** (1954), 371–384.
3. WEARING, J. L. and PATTERSON, C., Vibration testing of composite materials, in: *Composite Structures*. Proc. 1st Int. Conf. on Composite Structures, Paisley College of Technology, Scotland (September 1981), 463–474, Applied Science Publishers, London.
4. YOUNG, D., Vibration of rectangular plates by the Ritz method, *ASME Trans., J. appl. Mech.*, **17** (1950), 448–453.

12

Torsion of a Composite Beam

JEAN-JACQUES BARRAU and SERGE LAROZE

Ecole Nationale Supérieure de l'Aéronautique et de l'Espace,
10 Avenue Edouard Belin, 31055 Toulouse, France

and

DANIEL GAY

Département de Mécanique INSA, Avenue de Rangueil,
31062 Toulouse Cedex, France

ABSTRACT

The problem of the torsion of a composite beam made up with several orthotropic materials has been formulated using a warping function. By using a finite difference method or a boundary integral method in the case of transversely orthotropic phases we can calculate torsion stiffness, locate the shear stresses and shear center. Experimental data have been performed for a biphase straight cantilever beam made out of aluminium and epoxy. The theoretical result for the torsion stiffness is in agreement with experimental data.

NOTATION

a, b	Co-ordinates of the shear center
$[c]$	Elastic constants
E_i	Young's modulus in x direction
GJ	Torsional stiffness
G_i	Shear modulus of phase i (isotropic transverse)
I_y^p, P_z^p	Equivalent moment of area
K	Elastic center
M_y, M_z	Components of the bending moment

Mt Twisting couple
S Area of phase i
x, y, z Cartesian co-ordinates
X, Y, Z Principal cartesian co-ordinates
τ_{xy}, τ_{xz} Shear stresses
γ_{xy}, γ_{xz} Components of the strain tensor
ϕ Torsion function
θ Angular twist

INTRODUCTION

Actual composite beams often have complex external and internal boundaries. The components are generally orthotropic with variable orientation on the cross-sectional plane. This is typically exemplified by the helicopter rotor blade.

In the twisting of such beams the problems to be solved are the following:

—The prediction of the order of magnitude of the shear stresses.
—The torsion rigidity.
—The localization of the shear center.

Solving this problem by using a three-dimensional finite element is often very expensive.

In the present investigation we are developing a method to determine the characteristic values of a cross section.

The results will be used either directly or for determining a beam element in a finite element code.

DETERMINATION OF THE SHEAR STRESSES

The beam is composed of N cylindrical phases of any cross-sectional shapes. Let D be the cross section and δD the outer boundary. The boundary of the phase ni is l_i. We suppose that the several phases are perfectly stuck and we disregard the glue. The materials are orthotropic and one axis of orthotropy is along the longitudinal axis of the beam (Fig. 1).

Let the beam be subjected to a twisting moment M^t applied at the right end. An equilibrium torque $-M^t$ acts on the left end. Let origin C of the co-ordinate system (x, y, z) be chosen at the shear center of the cross section

Jean-Jacques Barrau, Serge Laroze and Daniel Gay

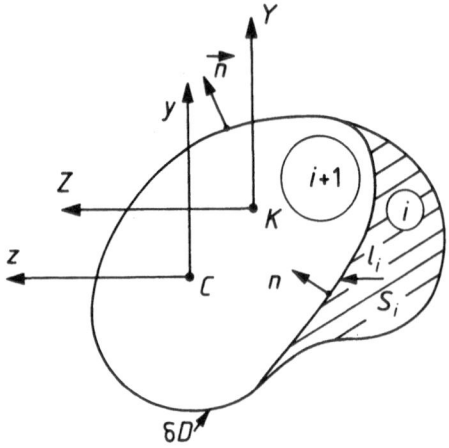

FIG. 1.

with the x axis parallel to the longitudinal of the beam and the y and z axes taken in the plane of the cross section. The orientation of the y and z axes are arbitrary. In these axes we write the strain–stress relation in the form:

$$\begin{bmatrix} \tau_{xy} \\ \tau_{xz} \end{bmatrix} = \begin{bmatrix} c_{11} & c_{12} \\ c_{21} & c_{22} \end{bmatrix} \begin{bmatrix} \gamma_{xy} \\ \gamma_{xz} \end{bmatrix} \qquad \text{or} \qquad [\tau] = [c][\gamma] \tag{1}$$

We now assume that the displacement components are:

$$U_x = \frac{d\theta}{dx} \phi(y, z) \qquad U_y = -z\theta \qquad U_z = y\theta \tag{2}$$

where θ is the twist of a section and ϕ is the torsion function. Then:

$$\begin{bmatrix} \tau_{xy} \\ \tau_{xz} \end{bmatrix} = \begin{bmatrix} c_{11} & c_{12} \\ c_{21} & c_{22} \end{bmatrix} \begin{bmatrix} \dfrac{\partial \phi}{\partial y} - z \\ \dfrac{\partial \phi}{\partial z} + y \end{bmatrix} \dfrac{d\theta}{dx} \tag{3}$$

Inserting the relation (3) in the equation of equilibrium yields:

$$\begin{bmatrix} \dfrac{\partial}{\partial y} \\ \dfrac{\partial}{\partial z} \end{bmatrix}^T \begin{bmatrix} c_{11} & c_{12} \\ c_{21} & c_{22} \end{bmatrix} \begin{bmatrix} \dfrac{\partial \phi}{\partial y} - z \\ \dfrac{\partial \phi}{\partial z} + y \end{bmatrix} = 0 \tag{4}$$

$\vec{n}$ is a unit vector normal to the outer boundary δD or to the inner boundary l_i. This vector $\vec{n}$ can be written in the component form:

$$n = (0, n_y, n_z)$$

If there are no external forces acting on the external surface δD the boundary condition gives:

$$\tau_{xy} n_y + \tau_{xz} n_z = 0$$

The boundary condition for the line l_i separating material i from material $i+1$ leads to:

$$(\tau_{xy})_i n_y + (\tau_{xz})_i n_z = (\tau_{xy})_{i+1} n_y + (\tau_{xz})_{i+1} n_z$$

Now, if we insert relation (3) in the boundary condition we must solve the following problem:

$$\begin{bmatrix} \dfrac{\partial}{\partial y} \\[2mm] \dfrac{\partial}{\partial z} \end{bmatrix} \begin{bmatrix} c_{11} & c_{12} \\ c_{21} & c_{22} \end{bmatrix} \begin{bmatrix} \dfrac{\partial \phi}{\partial y} - z \\[2mm] \dfrac{\partial \phi}{\partial z} + y \end{bmatrix} = 0$$

along l_i

$$(n)^T [c]^i \, \nabla\phi = (n)^T [c]^{i+1} \, \nabla\phi + z n_y (c_{11}^i - c_{11}^{i+1})$$
$$+ y n_z (c_{22}^{i+1} - c_{22}^i) + (z n_z - y n_y)(c_{12}^i - c_{12}^{i+1})$$

along δD

$$(n)^T [c] \nabla\phi = z n_y c_{11} + (z n_z - y n_y) c_{12} - y n_z c_{22}^i$$

$$\left. \right\} \quad (5)$$

and we have solved this problem by the finite element method. When the torsion has been determined the shear stresses have been calculated by using relation (3) and the torsional stiffness can be determined by:

$$GJ = \frac{M^t}{d\theta/dx} = \iint_D (y \tau_{xz} - z \tau_{xy}) \, ds \qquad (6)$$

If the phases are transversely isotropic, the equilibrium equation becomes $\Delta\phi = 0$ and the torsional stiffness, by using the Ostrogradsky's formula becomes:

$$GJ = \sum_{i=1}^N G_i \int_{S_i} (y^2 + z^2) \, ds - \sum_{i=1}^N G_i \int_{l_i} \phi \frac{\partial \phi}{\partial n} \, dl \qquad (7)$$

where G_i is shear modulus.

In this case the solution can be obtained very quickly by using the integral method.

DETERMINATION OF THE SHEAR CENTER

Let E_i be the Young's modulus of the phase i along the longitudinal axis of the beam. We define an elastic center K and principal directions KY and KZ in the plane of the cross section such that:

$$\sum_{i=1}^{N} \iint_{S_i} yE_i\, ds = 0$$

$$\sum_{i=1}^{N} \iint_{S_i} zE_i\, ds = 0$$

$$\sum_{i=1}^{N} \iint_{S_i} yzE_i\, ds = 0 \qquad (8)$$

In the $KXYZ$ co-ordinate system, we define the equivalent moment of area I_y^p, I_z^p by:

$$I_y^p = \frac{1}{E_0} \sum_{i=1}^{N} E_i \iint_{S_i} z^2\, ds$$

$$I_z^p = \frac{1}{E_0} \sum_{i=1}^{N} E_i \iint_{S_i} y^2\, ds \qquad (9)$$

where

$$E_0 = \frac{1}{N} \sum_{i=1}^{N} E_i$$

If M_y and M_z are the components of the bending moment M and ω_y and ω_z the components of the rotation vector of the section, we have:

$$\frac{d\omega_y}{dx} = \frac{M_y}{E_0 I_y^p} \qquad \frac{d\omega_z}{dx} = \frac{M_z}{E_0 I_z^p}$$

The shear stresses due to the twisting moment have been determined in the $Cxyz$ axes (C is the shear center).

Let a and b be the unknown co-ordinates of the shear center C in the $KXYZ$ co-ordinate system.

The components of the displacement vector in $KXYZ$ become:

$$U_x = \frac{d\theta}{dx} \phi(y, z)$$

$$U_y = -(z - b)\theta$$

$$U_z = -(y - a)\theta \tag{10}$$

Put $\psi = \phi - az + by$. From eqn. (3) and using relation (10) we get:

$$\begin{bmatrix} \tau_{xy} \\ \tau_{xz} \end{bmatrix} = \frac{d\theta}{dx} \begin{bmatrix} c_{11} & c_{12} \\ c_{21} & c_{22} \end{bmatrix} \begin{bmatrix} \dfrac{\partial \psi}{\partial y} - z \\ \dfrac{\partial \psi}{\partial z} - y \end{bmatrix}$$

Then we can easily verify that ψ is given by problem (5) in which ϕ has merely turned into ψ.

Now, for the part of the beam between section S_1 and left section S_0, we apply the following loadings (Fig. 2).

First loading. A concentrated force that lies in the plane of the cross-section acts at the shear center C.

Second loading. A twisting moment M' is applied to the free end of the beam.

We calculate the work produced by the first force owing to displacement due to the twisting moment.

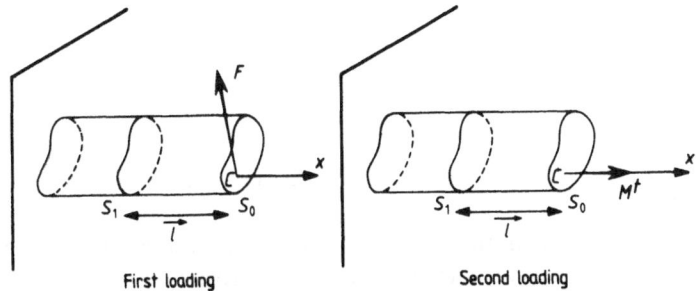

First loading   Second loading

FIG. 2.

On S_0: The point of application of the force is the shear center so the work of F is zero.

On S_1: a force $-F$ and a bending moment $\vec{M} = -\vec{F} \wedge \vec{l}$ act on the section S_1. Force F does not work but the bending moment works because of the warping of S_1.

The work done by $M(M_y, M_z)$ is given by:

$$W_{12} = \sum_{i=1}^{N} \iint_{S_i} \frac{d\theta}{dx} (\psi + az - by) \frac{E_i}{E_0} \left(\frac{M_y}{I_y^p} z - \frac{M_z}{I_z^p} y \right) ds$$

We are now going to calculate the work done by the twisting moment owing to displacement resulting from force F. This work can be neglected because the cross sections S_0 and S_1 do not rotate around the Cx axis due to the application of the force F.

$$W_{21} = 0$$

and since

$$W_{21} = W_{12}$$

$$\sum_{i=1}^{N} \iint_{S_i} \frac{d\theta}{dx} (\psi + az - by) \frac{E_i}{E_0} \left(\frac{M_y}{I_y^p} z - \frac{M_z}{I_z^p} y \right) ds = 0$$

This relation must be verified whatever the values of M_y and M_z so the co-ordinates of the shear center are:

$$a = -\frac{\sum\limits_{i=1}^{N} \iint_{S_i} E_i \psi z \, ds}{E_0 I_y^p}$$

$$b = +\frac{\sum\limits_{i=1}^{N} \iint_{S_i} E_i \psi y \, ds}{E_0 I_z^p}$$

EXPERIMENTAL DATA

We have performed the experiment using a straight beam made of aluminium and epoxy as shown in Fig. 3. The cross section of each material is rectangular.

FIG. 3.

The cantilever beam is subject to a pure twisting moment M^t at the free end. In order to eliminate the difficulty near the cantilever section we measure the rotation θ_0, θ_1 of the final section S_0 and of an intermediate cross section S_1; the distance between S_0 and S_1 is L.

Then the torsion stiffness is easily obtained as:

$$GJ = \frac{M^t}{(\theta_0 - \theta_1)} \, L$$

Experimental data and the corresponding theoretical values are presented in Table 1.

TABLE 1

	Theoretical	Experimental
GJ_{MKSA}	637	685

CONCLUSION

As can be seen, the prediction of the torsional stiffness is in fairly good agreement with experimental data. This first result shows that this theory is adequate. We are now improving the numerical resolution of the entire problem by using the finite element method and we are carrying out an experiment with a helicopter rotor blade.

REFERENCES

GAY, D., *Shear Deformation of Composite Beam*, I.N.S.A., Toulouse, France, 1980.

BARRAU, J. J., *Strength of Composite Materials*, E.N.S.A.E. Edition, Toulouse, France, 1982.

13

Elasto-Plastic Analysis of Fibrous Composite Shells Using 'Semiloof' Finite Elements

STEVAN MAKSIMOVIĆ

Vazduhoplovnotehnički Institut, Niška bb, 11132 Žarkovo-Beograd, Yugoslavia

ABSTRACT

A finite element formulation is presented for conducting nonlinear analysis of fibrous composite shells structures. The Semiloof elements, which have proved to be one of the most efficient families of finite elements available for linear elastic thin shell analysis, have been extended for elasto-plastic fibrous composite shell analysis. Inelastic material behaviour is modelled with flow theory of plasticity adopting the von Mises yield criterion.

The geometry, constitutive equations and stiffness relations of the shell element represent the displacement method of analysis. Solution of the nonlinear equilibrium equations is obtained with a Newton–Raphson type iteration technique.

This paper also considers static fracture of notched composite shells based on damage zones.

Some examples are furnished to demonstrate the versatility and accuracy of presented efficient finite elements in modelling structure of aircraft applications.

INTRODUCTION

The continuing search for lighter, stronger, more economical structures, particularly in the aircraft and missile industry, has led to the investigation of various composite materials as a possible applicable type of construction. Many of these materials are anisotropic, multi-layered and have nonlinear physical properties.

155

The purpose of this investigation is to study the elasto-plastic and large deflection behaviour of anisotropic, laminated composite plates and shells under static loads. Various studies on the elastic behaviour of anisotropic materials have been undertaken by different researchers.[1-3] The large deformation analysis of laminated composite shells can be found in Refs 4 and 5. Theories of anisotropic yield functions have been studied by a number of investigators. Hill[6] generalized the Huber–von Mises criterion to include anisotropic parameters.

A finite element analysis technique for an arbitrarily laminated anisotropic shell is described. Application of the composite structures exhibiting elasto-plastic deformation has been very limited. In this paper the use of the Semiloof finite element[7,8] is extended to nonlinear situations in which elasto-plastic or large deflection effects fibrous composite shells are present.

The extensive introduction of fibre reinforced composite materials into aerospace structures has led to attempts at predicting their behaviour when cracked. Because these structures are generally rectilinear anisotropic in nature and possess complicated boundary conditions, closed form analysis of their fractured state is not often possible, being largely only amenable to numerical technique. The finite element method is the obvious choice because of its ability to handle in a unified manner any complex geometry, general loading, and in particular, laminated anisotropic properties of multilayered composite materials.

THE YIELD CRITERION

The yield criterion determines the stress level at which plastic deformation begins. No essential difference occurs with derivation of yield loci for anisotropic shells in comparison with the isotropic case.

The anisotropic yield function, f, given by Hill[6] can be written in the tensorial form

$$f = \sigma_0^2 = \tfrac{1}{2}\Lambda_{ijkl}S^{ij}S^{kl} \tag{1}$$

where σ_0 is equivalent yield stress, Λ_{ijkl} the anisotropic parameters, and S^{ij} is the deviatoric stress tensor. For the case of a shell, f assumes the form

$$f = \tfrac{3}{2}\sigma^{\rho\alpha}\sigma^{\gamma\beta}g_{\rho\beta}g_{\gamma\alpha} - \tfrac{1}{2}(\sigma^{\gamma\beta}g_{\alpha\beta})^2 \tag{2}$$

Here $g_{\alpha\beta}$ denotes the covariant metric tensor of the layer. The influence of shear forces on the yielding of the shell wall is neglected.

ELASTO–PLASTIC CONSTITUTIVE RELATION

If the elastic strains are not negligible in comparison to the plastic strains, the total strain increment at a point on the body is given by the relation

$$d\varepsilon_{ij} = d\varepsilon_{ij}^E + d\varepsilon_{ij}^P \tag{3}$$

For linearly elastic anisotropic material the generalized Hooke's law can be written in the form

$$d\sigma_{ij} = D_{ijkl}d\varepsilon_{kl}^E = D_{ijkl}(d\varepsilon_{kl} - d\varepsilon_{kl}^P) \tag{4}$$

where D_{ijkl} are the anisotropic elasticity moduli. The incremental plastic strain tensor is related by an associated flow rule

$$d\varepsilon_{ij}^P = d\lambda \frac{\partial f}{\partial \sigma_{ij}} \tag{5}$$

in which $d\lambda$ is as yet undetermined. Consistent with eqn. (5), the projection of a stress increment $d\sigma_{ij}$ onto the yield surface normal is proportional to the projection of the corresponding increment of plastic strain, $d\varepsilon_{ij}^P$. Therefore

$$d\sigma_{ij} \frac{\partial f}{\partial \sigma_{ij}} = H d\varepsilon_{ij}^P \frac{\partial f}{\partial \sigma_{ij}} \tag{6}$$

where H is a strain-hardening parameter determined from uniaxial stress–strain data. By multiplying eqn. (4) by $\partial f/\partial \sigma_{ij}$, using eqn. (6) to substitute for $d\sigma_{ij}.\partial f/\partial \sigma_{ij}$, and eliminating $d\varepsilon_{ij}^P$ by means of eqn. (5) the final result is

$$d\lambda = \frac{D_{ijkl}d\varepsilon_{kl} \dfrac{\partial f}{\partial \sigma_{ij}}}{\left(D_{ijkl} \dfrac{\partial f}{\partial \sigma_{kl}} + H \dfrac{\partial f}{\partial \sigma_{ij}} \right) \dfrac{\partial f}{\partial \sigma_{ij}}} \tag{7}$$

Equations (5) and (7) are next used to eliminate the plastic strains, $d\varepsilon_{kl}^P$ in eqn. (4). Equations (3)–(7) lead to the following elasto-plastic constitutive equations

$$d\sigma_{ij}^{EP} = C_{ijkl}d\varepsilon_{kl} \tag{8}$$

in which

$$C_{ijkl} = D_{ijkl} - D_{ijmn}A_{mnkl} \tag{8.1}$$

$$A_{mnkl} = \cfrac{\cfrac{\partial f}{\partial \sigma_{mn}} \cfrac{\partial f}{\partial \sigma_{ij}} D_{ijkl}}{D_{ijkl} \cfrac{\partial f}{\partial \sigma_{ij}} \cfrac{\partial f}{\partial \sigma_{kl}} - \left\{ \cfrac{\partial f}{\partial \varepsilon_{ij}^P} + \cfrac{\partial f}{\partial H} \cfrac{\partial H}{\partial \varepsilon_{ij}^P} \right\} \cfrac{\partial f}{\partial \sigma_{ij}}} \tag{8.2}$$

and

$$d\varepsilon_{mn}^P = A_{mnkl} d\varepsilon_{kl} \tag{8.3}$$

To determine the anisotropic parameters Λ_{ijkl} in eqn. (1) we require two tensile tests, one shear test and another tensile axis of anisotropy following the plan in Ref. 9.

THE SEMILOOF SHELL ELEMENT

The Semiloof element is the result of an evolution process dating from the isoparametric elements.[10] Complete details of this element may be found elsewhere.[11] The quadrilateral Semiloof shell element is first formulated with 45 degrees of freedom (Fig. 1). Values of local displacements and their derivatives can be obtained at any point of the element in the form

$$a = N_i a_i \tag{9}$$

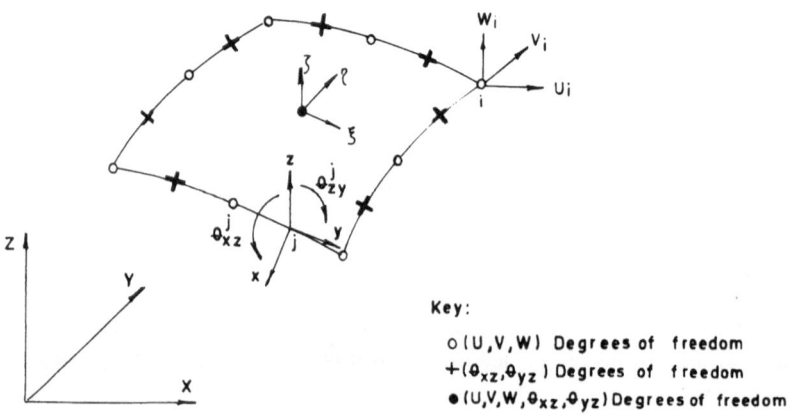

Key:
o (U,V,W) Degrees of freedom
+ (θ_{xz},θ_{yz}) Degrees of freedom
● (U,V,W,θ_{xz},θ_{yz}) Degrees of freedom

FIG. 1. Nodal configuration of the Semiloof element.

Here N_i is shape function, a_i is the element nodal freedom and

$$a = \left\{ u, v, w, \frac{\partial u}{\partial x}, \frac{\partial u}{\partial y}, \frac{\partial v}{\partial x}, \frac{\partial v}{\partial y}, \frac{\partial u}{\partial z}, \frac{\partial v}{\partial z}, \frac{\partial^2 u}{\partial x \, \partial z}, \frac{\partial^2 u}{\partial y \, \partial z}, \frac{\partial^2 v}{\partial x \, \partial z}, \frac{\partial^2 v}{\partial y \, \partial z} \right\} \qquad (9.1)$$

Of these, 13 are condensed to 32 by the constraints on the shear behaviour of the element. The eleven constraint equations can be written in matrix form[11]

$$|K_A \quad K_B|\{u_A \quad u_B\} = 0 \qquad (10)$$

where u_A represents the 32 degrees of freedom to be retained and u_B those to be eliminated. We can write eqn. (9) as

$$a = [S_A \quad S_B]\{u_A \quad u_B\} \qquad (11)$$

On the use of constraints (10), eqn. (9) becomes

$$u = N_S u_A \qquad (12)$$

$$N_S = [S_A - S_B K_B^{-1} K_A] \qquad (12.1)$$

where $N_S (13 \times 32)$ is constrained shape function array used in this analysis.

FINITE ELEMENT EQUATIONS

For nonlinear analysis which takes account of both material and geometrical nonlinearities, an incremental solution scheme is required. We can choose either a total or updated Lagrangian formulation. Here the total Lagrangian formulation is derived from the virtual work equation assuming small strains and conservative loading. The virtual work statement can be written as

$$\int_\Omega [\delta \varepsilon_n]^T \sigma_n \, d\Omega - \int_\Omega [\delta u_n]^T b_n - \int_{\Gamma_t} [\delta u_n]^T t_n \, d\Gamma = 0 \qquad (13)$$

where σ_n are the Piola–Kirchhoff stresses, δu_n is the vector of virtual displacements, $\delta \varepsilon_n$ is the vector of associated virtual strains, b_n is the vector of applied body forces and t_n is the vector of surface tractions.

The components of the vector of Green's strains ε_n can be written as

$$\varepsilon_n = \sum_{i=1}^{m} ([B_{Li}]_n + \tfrac{1}{2}[B_{NLi}]_n)[u_i]_n \qquad (14)$$

where B_L and B_{NL} are linear and nonlinear strain–displacement matrices respectively. The virtual strains can be expressed as

$$\delta\varepsilon_n = \sum [B]_n [\delta u]_n \tag{15}$$

where

$$[B]_n = [B_L]_n + [B_{NL}]_n \tag{15a}$$

The virtual work eqn. (13) may be discretized for an element in the usual manner.[12] Since virtual displacements du are arbitrary, the nonlinear equilibrium equations for an element, or indeed the total assembled structure, become

$$\Psi(u) = \int_\Omega B^T \sigma_n \, d\Omega - R \tag{16}$$

where Ψ once again represents the sum of internal and external generalized forces. The solution of eqn. (10) will have to be approached iteratively. If the Newton–Raphson is to be adopted we have to find a relation between du and $d\Psi$. Thus taking appropriate variations of eqn. (16) we have

$$d\Psi = \int_\Omega d[B_L]^T \sigma d\Omega + K_T du \tag{17}$$

where

$$K_T = K_L + K_{NL}$$

$$K_L = \int_\Omega B_L^T C_{ijkl} B_L \, d\Omega$$

$$K_{NL} = \int_\Omega (B_L^T C_{ijkl} B_{NL} + B_{NL}^T C_{ijkl} B_{NL} + B_{NL}^T C_{ijkl} B_L) \, d\Omega$$

Here K_T is the tangent stiffness matrix, K_L is the large deflection stiffness matrix and K_{NL} is the geometric stiffness matrix.

FRACTURE ANALYSIS OF COMPOSITES

In the case of fracture mechanics of multilayered composite shells we can use 'singular elements' to describe the crack tip singularities. Barsoum[13] has demonstrated that by collapsing one side of an 8-node element to one point and moving the nodes to a quarter-distance from the crack tip the elements contain both $1/\sqrt{r}$ and $1/r$ singularities. These are used to model a small but finite region of the crack tip while the rest of the domain under consideration has to be modelled using 'regular' curved shell elements.

The performance of Semiloof quadrilateral singular elements are similar to the degenerate solid thick shell elements used in Refs. 14 and 15.

NUMERICAL EXAMPLES

The foregoing theory and procedures were applied to a simply supported laminated plate subjected to a uniformly distributed lateral load. The plate was composed of four layers of fibre reinforced laminae of equal thicknesses.

TABLE 1

Material properties of boron–epoxy composites

Modulus of elasticity, GPa	$E_L = 219{\cdot}8$
	$E_T = 21{\cdot}4$
Shear modulus, GPa	$G_{LT} = 6{\cdot}9$
Poisson's ratio	$v_{LT} = 0{\cdot}208$
Plastic modulus, GPa	$E_{LP} = 2{\cdot}41$
	$E_{TP} = 0{\cdot}04$
Plastic shear modulus, GPa	$G_{LT}^{P} = 0{\cdot}008$
Yield stresses, GPa	$(\sigma_y)_L = 1{\cdot}1$
	$(\sigma_y)_T = 0{\cdot}06$
	$(\tau_y)_{LT} = 0{\cdot}02$
σ_0, GPa	$5{\cdot}374$
n	$1{\cdot}33$

The material properties presented in Table 1 are for boron–epoxy composites.[16] Here subscript L refers to the direction of the fibres and subscript T to the transverse direction.

Results have been obtained by using Semiloof shell elements and a 4×4 grid in the plate quarter for ratio of the plate $t/a = 0{\cdot}0375$.

The results of the elastic–plastic plate are shown in Figs 2 and 3. The results obtained are compared with the classical solution.[16] Good agreement is generally evident for both displacements and stresses.

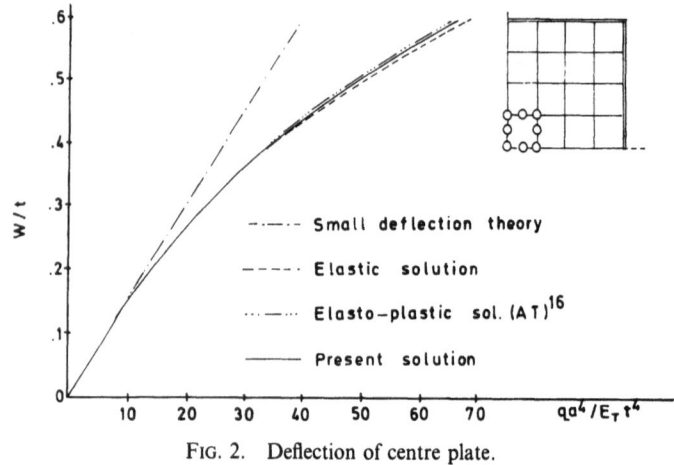

FIG. 2. Deflection of centre plate.

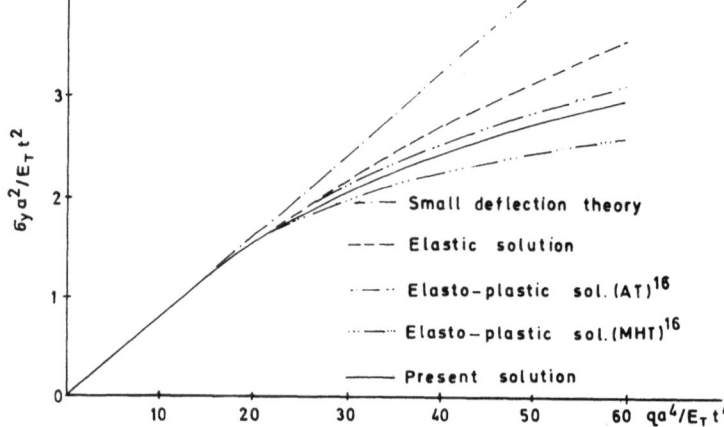

FIG. 3. Variation of the extreme fibre stress σ_y with load at the centre of the plate.

CONCLUDING REMARKS

The Semiloof elements, which proved to be one of the most efficient families of finite elements available for linear elastic, isotropic thin shell analysis, have now been extended for large-deflection and elasto-plastic fibrous composite shell analysis.

The numerical problems presented give results which compare well with solutions from other sources. Good agreement is generally evident for both displacements and stresses.

REFERENCES

1. Pryor, C. W. and Baker, R. M., A finite element analysis including transverse shear effects for application to laminated plates, *AIAAJ*, 9 (1971), 912–917.
2. Panda, S. C. and Natarajan, R., Finite element analysis of laminated composite plates, *Int. J. Numer. Meth. Engng*, 14 (1967), 69–79.
3. Maksimović, S., Finite element analysis of laminated composite plates and shells, *Proc. 15th Yugoslav Congr. Mech.*, Kupari, June (1981).
4. Noor, A. K. and Hartley, S. J., Nonlinear shell analysis via mixed isoparametric elements, *Comput. Structures*, 7 (1977), 615–626.
5. Baher-El-Din, Y. A., Dvorak, G. J. and Utku, S., *Comput. Structures*, 13 (1981), 321–330.
6. Hill, R., *The Mathematical Theory of Plasticity*, Clarendon Press, Oxford (1950).
7. Iron, B. M., The Semiloof shell element, in *Finite Elements for Thin Shells and Curved Members* (Edited by Gallagher, R. H. and Ashwel, D. G.), Ch. 11, Wiley, New York (1976).
8. Martins, R. A. F., Structural instability and natural vibration analysis of thin arbitrary shells by use of the Semiloof element, *Int. J. Numer. Meth. Engng*, 11 (1977), 481–498.
9. Whang, B., Elastoplastic orthotropic plates and shells, *Proc. of the Symp. on application of finite element methods in civil engineering*, Vanderbilt University (1969).
10. Ahmad, S., Irons, B. M. and Zienkiewicz, O. C., Analysis of thick and thin shell structures by curved elements, *Int. J. Numer. Meth. Engng*, 2 (1970), 419–451.
11. Martins, R. A. F. and Owen, D. R. J., Elastoplastic and geometrically nonlinear thin shell analysis; the Semiloof element, *Comput. Structures*, 13 (1981), 505–513.
12. Zienkiewicz, O. C., *The Finite Element Method in Engineering Science*, McGraw-Hill, London (1977).
13. Barsoum, R. S., On the use of isoparametric finite elements in linear fracture mechanics, *Int. J. Numer. Meth. Engng*, 10 (1976), 25–37.
14. Barsoum, R. S., Loomis, R. W. and Stewart, B. D. S., Analysis of through cracks in cylindrical shells by quarter-point elements, *Int. J. Fracture*, 15 (1979), 259–280.
15. Maksimović, S., Finite elements in elasto-plastic fracture mechanics, *The 4th Seminar Numer. Meth. Tech.*, Stubličke Toplice (1982).
16. Gorji, M. M., Plastic constitutive relations for unidirectionally reinforced composites: Application to large deflection of plates, *Int. J. Solids Structures*, 16 (1980), 149–160.

14

Failure Mechanisms and Strength Reduction in Composite Laminates with Cut-outs—A 3-D Finite Element Numerical Autopsy

A. DE ROUVRAY, E. HAUG and J. DUBOIS

Engineering System International SA,
20 rue Saarinen, SILIC 270, 94578 Rungis-Cedex, France

ABSTRACT

Organic composite laminates exhibit a 'size effect' when perforated with cut-outs, such as holes or notches: the strength of components with a larger cut-out is smaller than predicted by the effective strength criterion (i.e. constant average stress across the ligament). A Finite Element simulation model is presented to predict the failure strength of composite laminates, for arbitrary cut-out shapes and sizes, ply stacking sequences, and mechanical loading conditions. The model performance is demonstrated on the prediction of the failure loads of carbon–epoxy tension specimens with a central circular cut-out, for two hole sizes and two stacking sequences $|0°/45°/90°/-45°|_s$ and $|90°/-45°/0°/45°|_s$. Comparison with experimental results shows good qualitative and quantitative agreement.

The model is used to investigate the sequence of failure events inside the laminate: initiation and propagation of matrix shear damage and crazing in each ply, with the load transfer mechanisms inside the ply and towards adjacent plies; progressive fiber fracture and delamination between plies near the free edges.

The numerical model is further exploited to identify the physical (material and geometric) parameters responsible for the cut-out strength reduction.

1. THE STRENGTH REDUCTION OF COMPOSITE LAMINATES WITH CUT-OUTS

1.1. Size Effect and Scale Effect

Organic composite laminates (carbon–epoxy or carbon–carbon) exhibit a 'size effect' when perforated with cut-outs, such as holes or notches, i.e.

the strength of the component with a larger size cut-out is smaller than predicted by the mere reduction of the uncut section (or 'ligament'). A typical strength reduction curve for tension specimens with a central circular hole is shown in Fig. 1 (Whitney and Kim).[1] The Effective Strength Reduction Factor, $R = \sigma_N^R / \sigma_0^R$, is the ratio of the average stress at failure across the ligament or 'effective strength' for the notched specimen to the average stress at failure for the intact specimen. R is a measure of the size effect. $R = 1$, the 'effective strength criterion' often used for ductile materials, implies no notch sensitivity: the material flows perfectly to redistribute all stress concentrations. Conversely a brittle material will exhibit an R value close to $1/K$, where K is the notch stress concentration shape factor ($K = 3$ for circular cut-outs in homogeneous isotropic materials), and sharp notches or cracks will reduce R to very small values. The lower bound of R for brittle materials cannot be evaluated by stress concentration factor considerations only, but necessitates a 'fracture mechanics' approach to account for the 'characteristic length' across which the stress concentration needs to spread to initiate cracking. This gradient or 'scale effect' explains the reduced fracture level of larger scale specimens; it also accounts for the transition from ductile to brittle behavior for larger specimens (Wilkins *et al.*[2]): thin specimens plastify across their entire thickness before the crack tip stress concentration has spread over a distance large enough to initiate crack advance. To account for both 'size' and 'scale' effect in composite laminates with cut-outs, several authors have introduced a characteristic length in their failure criterion (Nuismer[3,4]; Byron Pipes *et al.*[5]; Garbo and Ogonowski[6]).

1.2. Stress Concentrations and Strength Reduction in Laminates

Typical organic composite laminates consist of a ductile matrix reinforced in large proportions (e.g. 40–60 %) by long brittle fibers.

Stress concentrations in laminates with cut-outs are of many types. At the macroscopic scale of an individual ply, four levels need to be considered: (Haug *et al.*[7]; de Rouvray *et al.*[8]):

—the 0° layer concentrates the average specimen axial stress with a ratio of typically 5 to 50 respectively to the ±45° and 90° plies (Fig. 2);

—within each orthotropic ply, the cut-out raises the stresses by a factor of 2–4 (circular hole) to no limit (sharp notches or cracks) (Fig. 2);

—next to the free edges, where the fibers are cut, stress equilibrium between adjacent anisotropic layers raises the stresses also by an order of magnitude or more (Haug *et al.*[7]; Engrand[9]);

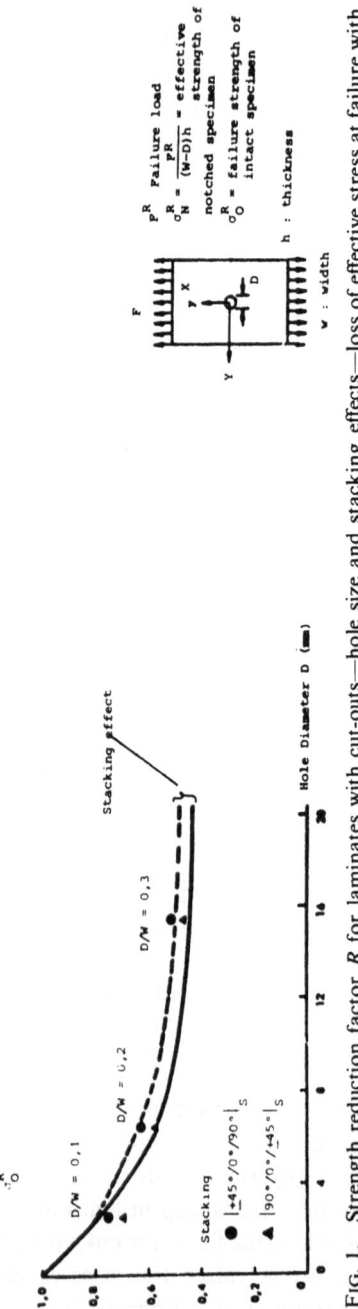

FIG. 1. Strength reduction factor R for laminates with cut-outs—hole size and stacking effects—loss of effective stress at failure with increasing hole size (after Whitney and Kim[1]).

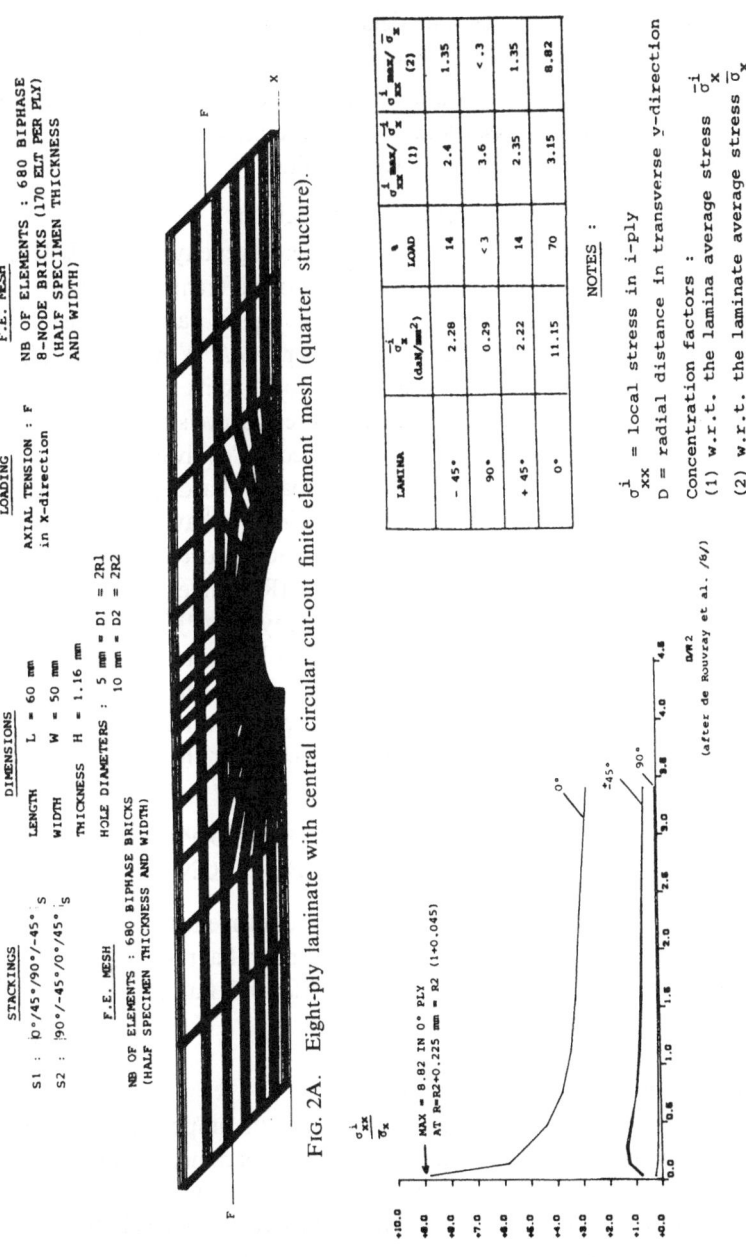

FIG. 2A.   Eight-ply laminate with central circular cut-out finite element mesh (quarter structure).

NOTES :

σ_{xx}^i = local stress in i-ply

D = radial distance in transverse y-direction

Concentration factors :
(1) w.r.t. the lamina average stress $\dfrac{\sigma_x^i}{\overline{\sigma}_x^i}$

(2) w.r.t. the laminate average stress $\dfrac{\overline{\sigma}_x^i}{\overline{\sigma}_x}$

LAMINA	$\overline{\sigma}_x^i$ (daN/mm²)	LOAD	σ_{xx}^i max/ $\overline{\sigma}_x^i$ (1)	σ_{xx}^i max/ $\overline{\sigma}_x$ (2)
- 45°	2.28	14	2.4	1.35
90°	0.29	< 3	3.6	< .3
+ 45°	2.22	14	2.35	1.35
0°	11.15	70	3.15	8.82

FIG. 2B.   Laminate with circular cut-out ($D_2 = 10$ mm) 'global' axial stress concentrations in each ply with respect to the average laminate stress.

—long before ultimate load, the above stress concentrations have created local cracks, fiber ruptures and delaminations, which in turn act as additional stress concentrators (Beaumont[16]).

The resulting laminate compounded stress concentration factor may be very high indeed at critical locations. For a material with low ductility, the reduction factor R should turn out very small.

1.3. Heterogeneity and 'Equivalent Limited Ductility'

Composite laminates exhibit two levels of heterogeneity: at the ply level between fibers and matrix, and between plies of different orientation.

When failure begins in one ply, it follows a typical pattern to reach a 'characteristic damage state' (CDS: Reifsnider *et al.*[10]). Matrix crazing initiated in the 90° plies transfers the 90° ply loads to its fibers and to the adjacent more resistant angle and 0° plies. These two levels of load transfer, for laminate strength purposes, may be viewed as 'equivalent' to plastic yielding with strain hardening. In fact, the apparent hardening coefficient would be so high that the observed global laminate behavior still appears nearly linear, although irreversible damage indeed occurs long before ultimate load, as indicated by experiments (Reifsnider *et al.*[10]) and numerical simulations (Haug *et al.*[7]; de Rouvray *et al.*[8]).

This 'equivalent limited ductility' governs the laminate R-value. If the redistributions between matrix, fibers and plies operate efficiently, R will be closer to 1 (e.g.: ca. 0·6 or 0·7 for the tension plate with a circular hole). With no redistributions R could drop to values lower than 0·3.

2. A NUMERICAL SIMULATION MODEL TO PREDICT THE CUT-OUT STRENGTH REDUCTION

The numerical technique used here is the finite element method, with a specialized version of the ESI general nonlinear Lagrangian program PAM-NL developed for incremental collapse and failure analyses of complex nonlinear structures (Haug *et al.*[11]).

2.1. Modelling for Physical Simulation

The numerical model must incorporate all important physical phenomena with the minimum number of physical parameters (for clarity), plus a few more (for safety), including their interaction.

As described above the strength reduction effect will result from the combination of several stress concentration mechanisms with the relative ability of the individual plies and the laminate to redistribute them, as they locally and gradually damage the composite materials.

2.2. Identification of the Relevant Physical Parameters

2.2.1. Stress concentration parameters—'biphase stiffness' model

(a) *Heterogeneous fiber–matrix lamina model.* In each ply (or stack), the fibers and matrix are individually modelled and the lamina stiffness characteristics are obtained by a superposition procedure, with interaction (Fig. 3(A)). The fibers are represented by an orthotropic thick membrane (smeared membrane elements) and the matrix by an orthotropic continuum (brick elements) over one ply thickness but with a reduced density and directional stiffness to account for the voids where the fibers fit. Both elements are condensed into one 8-node brick with 8 integration points.

Fiber–matrix interaction is implemented by strain compatibility, but allows imperfect bonding through the material law. An external load applied to the ply tends to cause equal strains in fibers and matrix, and so concentrates the stresses in the stiffer fibers (until either one breaks).

Ply interaction in the laminate is established by imposing displacements continuity between layers. Similarly the stiffer plies for a given load direction (i.e. the $0°$ plies) concentrate the loading and the stresses.

(b) *Geometric concentrations.* Stress concentrations due to the cut-out shape are accounted for by a locally finer mesh. For a circular hole, this gradient is mild, extending over about one hole diameter, and the meshing can be reasonably coarse (Fig. 2(A)).

The free edge stress concentrations are much steeper, extending over about one laminate thickness. They can be introduced with a very fine meshing (e.g. several elements over one ply thickness width bordering the free edge). Such a meshing, although feasible, is impractical, and expensive, and requires special techniques to follow the free edge effects when the delamination and fibers fracture propagate inward (remeshing; extrapolation).

It was decided in a first model to ignore the free edge effect, and to examine its influence subsequently in a finer model.

Similarly sub-ply or microscopic effects at the fiber level were also ignored, assuming they could be accounted for by their global influence on macroscopic material parameters.

2.2.2. Stress redistribution and hardening parameters—'biphase material' model

Within the precision of the stiffness model retained, damage initiation, cumulation and stress redistributions have to be realistically modelled to capture the equivalent laminate hardening effect. This is achieved with the 'biphase material' model (Fig. 3(B)).

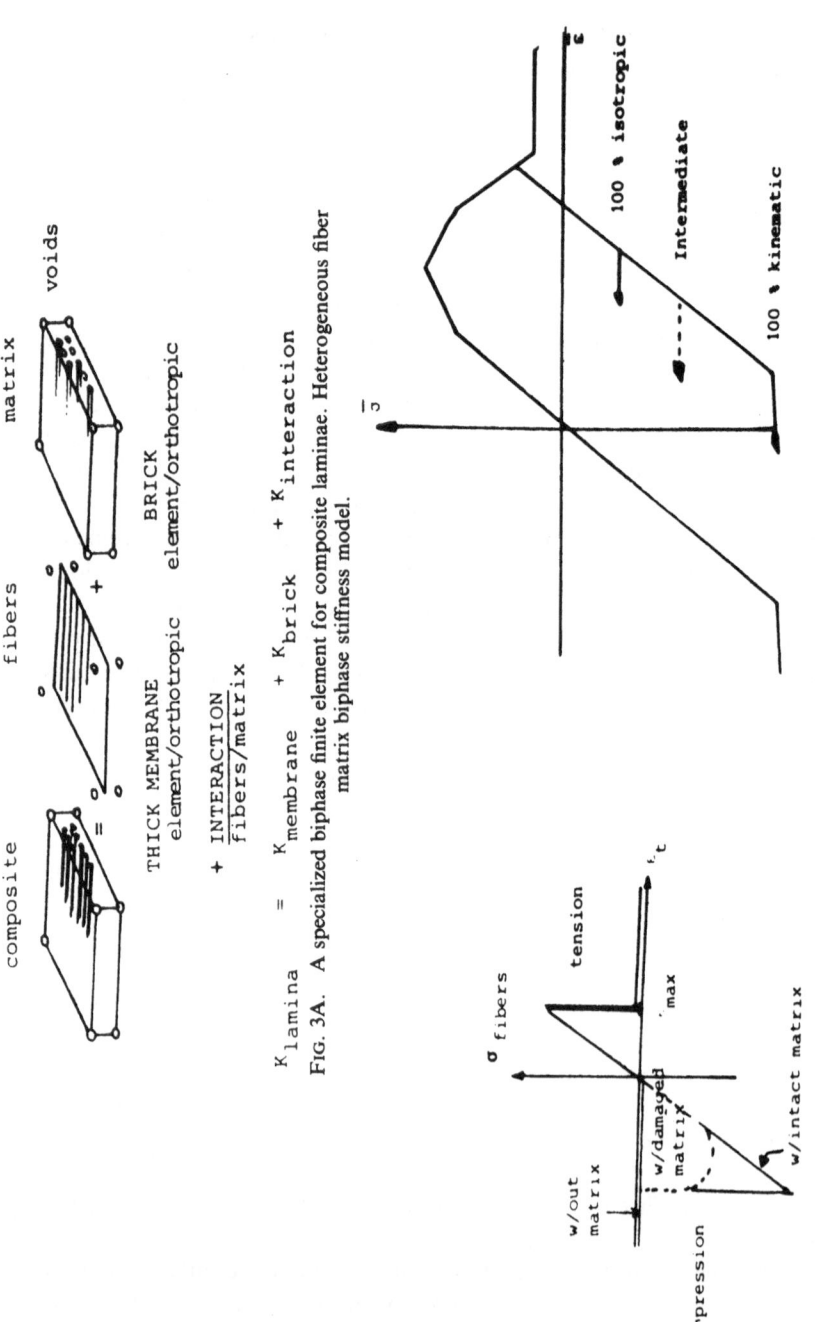

composite   fibers   matrix   voids

THICK MEMBRANE   BRICK
element/orthotropic   element/orthotropic

+ INTERACTION
  fibers/matrix

K_{lamina} = $K_{membrane}$ + K_{brick} + $K_{interaction}$

FIG. 3A.   A specialized biphase finite element for composite laminae. Heterogeneous fiber
matrix biphase stiffness model.

B.2. MATRIX PHASE, IN THE EQUIVALENT OCTAHEDRAL
      STRESS-STRAIN SPACE (ANIPLA MODEL)

100 % isotropic

Intermediate

100 % kinematic

B.1. FIBER PHASE, IN THE FIBER DIRECTION,
      INCLUDING MATRIX INTERACTION

FIG. 3B.   Fiber–matrix biphase material model.

The fibers constitutive law describes the fibers immersed in the matrix (Fig. 3(B1)). It is orthotropic elastic–brittle in tension, and elastic–plastic with strain softening in compression. The softening rate is controlled by the amount of matrix damage. A severely crazed matrix will confine the fibers poorly and they will contribute low properties in compression.

The matrix constitutive law (Fig. 3(B2)), is initially orthotropic under low strains to account for the fiber voids. It is assumed elastic–perfectly plastic (generalized von Mises) in compression. In tension, it is elastic–plastic–brittle with directional cracking. Matrix crack opening strains cumulate to cause a damage parameter to increase and reduce the global matrix strength through a combined isotropic–kinematic hardening orthotropic plasticity model (ANIPLA model, Dubois and Curnier[12]; Dubois and Habiballah[13]).

Delamination is modelled globally by allowing excessive shear and tension stresses between plies to vanish through matrix plastification. This tends to smear the effect of delamination across one ply thickness.

Note that such material models, although complex to derive, are not expensive to run and their input is made simple by an automatic preprocessor.

2.3. Loading to Failure

The loading is applied by displacement increments to the specimen ends, until maximum load is reached ('failure load'), and continued beyond to observe the beginning of the specimen total collapse ('post peak' behavior).

At each load increment, the current (elastic or damaged) material properties are used to compute the incremental strains and the solution is iterated upon until stress equilibrium is reached and all the material laws are satisfied for the local strain state.

2.4. Model Results

Loading the specimen to failure requires 10–20 load increments (or Newton iterations). Typical results are now presented.

3. NUMERICAL SIMULATION AND RESULTS

3.1. Specimens and Material Properties

3.1.1. Experimental data and test results

A 50 mm wide, 60 mm long, 1·16 mm thick, symmetric eight-ply tension specimen with a central circular cut-out is loaded to total failure. Two hole sizes ($D_1 = 5$ mm and $D_2 = 10$ mm diameter) small with respect to the

TABLE 1
Experimental results vs several theoretical models

Notched laminate			A. Experimental results (after Verchery et al. (Ref. 14))				B. Theoretical 'effective' strength values: σ_N^T and $R^T = \sigma_N^T/\sigma_0$					
Config.	Stacking	Hole diam.	Material load F_R^E (daN)	Effective strength σ_N^E (daN/mm²)	Material strength σ_0 (daN/mm²)	$R^E = \sigma_N^E/\sigma_0$	Effective strength criterion $\sigma_N^{T1} = \sigma_0$	R^{T1}	Fracture mechanics criterion (1) σ_N^{T2}	R^{T2}	Present F.E. model w/continuum criterion (2) σ_N^{T3}	R^{T3}
1·0	S1 $[0/45/90/-45]_s$ H1 = 1·16 mm	DO (2·5 mm)	1 790	32·5	42·2	0·77	42·2	1·0	34·8	0·82		
3·1	S3 $[0_2/45_2/90_2/-45_2]_s$ H3 = 2H1	D1 (5 mm)	3 405	32·6	41·5	0·78	41·5	1·0	(31·0)	(0·75)		
1·1	S1 $[0/45/90/-45]_s$ H1	D1	1 710	32·8	42·2	0·78	42·2	1·0	31·6	0·75	32·8	0·78
2·1	S2 $[90/-45/0/+45]_s$ H2 = H1	D1	1 555	29·8	42·4	0·70	42·4	1·0	(31·6)	(0·75)	30·1	0·71
1·2	S1	D2 (10 mm)	1 355	29·2	42·2	0·69	42·2	1·0	29·4	0·69	27·1	0·64
2·2	S2	D2	—	—	42·4	—	42·4	1·0	(29·4)	(0·69)	25·5	0·60
REMARKS			The 'Effective Strength' Reduction K seems identical for Configurations 1.0, 3.1 and 1.1 (no apparent SCALE effect; Asymptotic SIZE effect)				Predicts NO Effective Strength Reduction		Predicts SIZE + SCALE effects		Predicts only a SIZE effect	

$$\sigma_N^x/\sigma_0 = \frac{2(1-z)}{d - z_2 - z_4} \qquad z = \frac{R}{R + a_0} \qquad \sigma_N^T = \sigma_N^x \times \frac{\text{Intact Section}}{\text{Notched Section}}$$

Notes: (1) Whitney–Nuismer average stress criterion (Ref. 3):

with a_0 = characteristic length = 4 mm (best fit for Stacking S1 and hole sizes D1 and D2 (Verchery et al. (Ref. 14)).

(2) Present Finite Element Simulation Model, with Failure Loads calibrated on configuration 1.1 (S1/D1).

TABLE 2

Experimental lamina properties (after Ref. 14) and biphase material model/input parameters

Material parameters	E_{11} (daN/mm²)	$E_{22}=E_{33}$ (daN/mm²)	$G_{12}=G_{13}$ (daN/mm²)	G_{23} (daN/mm²)	$v_{12}=v_{13}$	v_{23}	σ_{11}^+ (daN/mm²)	σ_{11}^- (daN/mm²)	σ_{22}^+ (daN/mm²)	σ_{22}^- (daN/mm²)	S_{12} (daN/mm²)	ε_f (%)
'Biphase' orthotropic brick (=)	14 125 (14)	945 (14)	455 (14)	340 (14)	0·35 (14)	0·39 (14)	156·7 (14)	127·7 (15)	5·65 (14)	11·2 (15)	9·1 (15)	
Matrix w/o fibers	320 (17)	945	455	340	0·35	0·39	2 (17)	2 (17)	5·65	11·2	9·1	0·14 (16)
Fibers alone + interaction (+)	22 000 (17), (18)	—	—	—	—	—	245 (17), (18)	200	—	—	—	

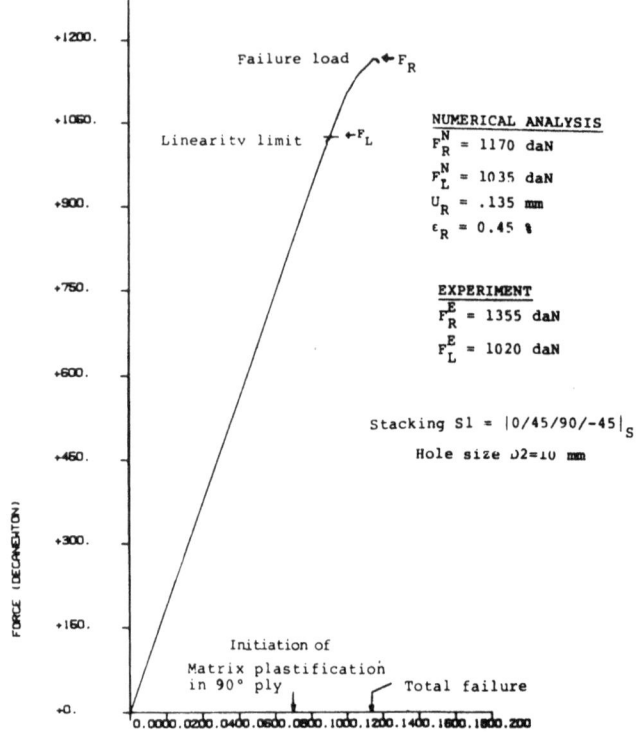

FIG. 4. Numerical load versus displacement curve up to specimen total failure.

specimen width, and two stacking sequences $S_1 = |0°/+45°/90°/-45°|_s$ and $S_2 = |90°/-45°/0°/+45°|_s$ are analyzed. The experimental results obtained at ENSTA (Verchery et al.[14]), are shown in Table 1A (average of four specimens per result point), together with the unidirectional elastic, plastic and failure lamina properties (Table 2).

3.1.2. Numerical data

The material input data for the biphase model are also shown in Table 2, completed from test results from typical published data.[15-18] The following numerical results are obtained with the reasonably coarse mesh shown earlier (Fig. 2(A)).

3.2. Stacking S_1—Hole Size $D_2 = 10$ mm (Configuration S_1/D_2)

3.2.1. Elastic Analysis

A first load increment (i.e. imposed end displacement U_1) is applied to

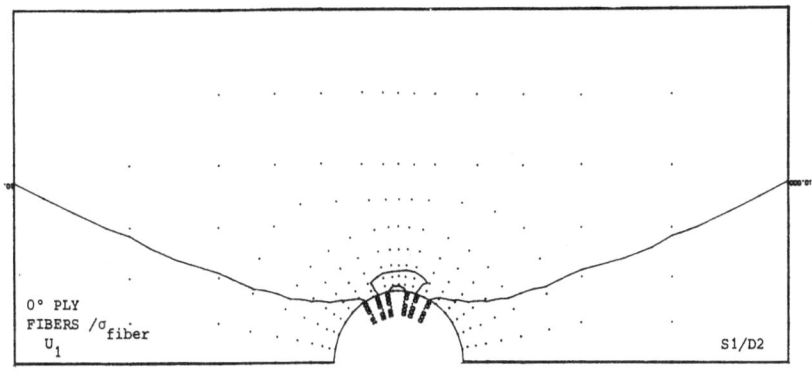

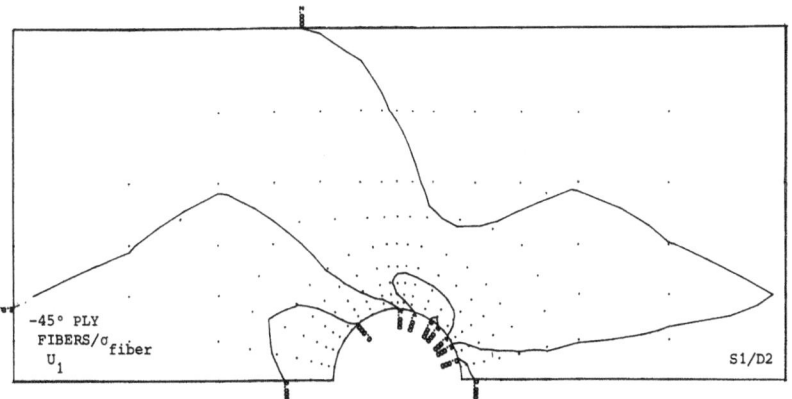

FIG. 5. Stacking S_1: $|0°/45°/90°/-45°|_s$. Hole size $D_2 = 10$ mm. Laminate elastic analysis; fiber iso-σ_{fiber} contours.

the specimen: $F_1 = 231$ daN, $U_1 = 0.025$ mm, which creates in the laminate an average axial stress $\sigma_x^1 = 3.98$ daN/mm² and strain $\varepsilon_x^1 = 0.08\%$, corresponding to about 20 % of the failure load ($F_1/F_R = 0.2$, $\varepsilon_1/\varepsilon_R = 0.18$).

For each ply, the lamina stresses are derived as the volume weighted average of the fiber and matrix stresses computed individually by the heterogeneous biphase lamina model (Fig. 3(A)).

Typical elastic isostress contours in each ply are shown in Fig. 5, for the fibers, and Fig. 6(A, B) for the matrix. Figure 2(B) sums up the results. The 0° plies carry 70 % of the load, the 45° plies 14 % each, the 90° plies less than 3 %. The stress concentration factors for σ_{xx} in each ply i, with respect to the ply average stress $\bar{\sigma}_x^i$, and with respect to the laminate average stress σ_x, are also shown. The largest laminate factor is obtained for the 0° ply, where it

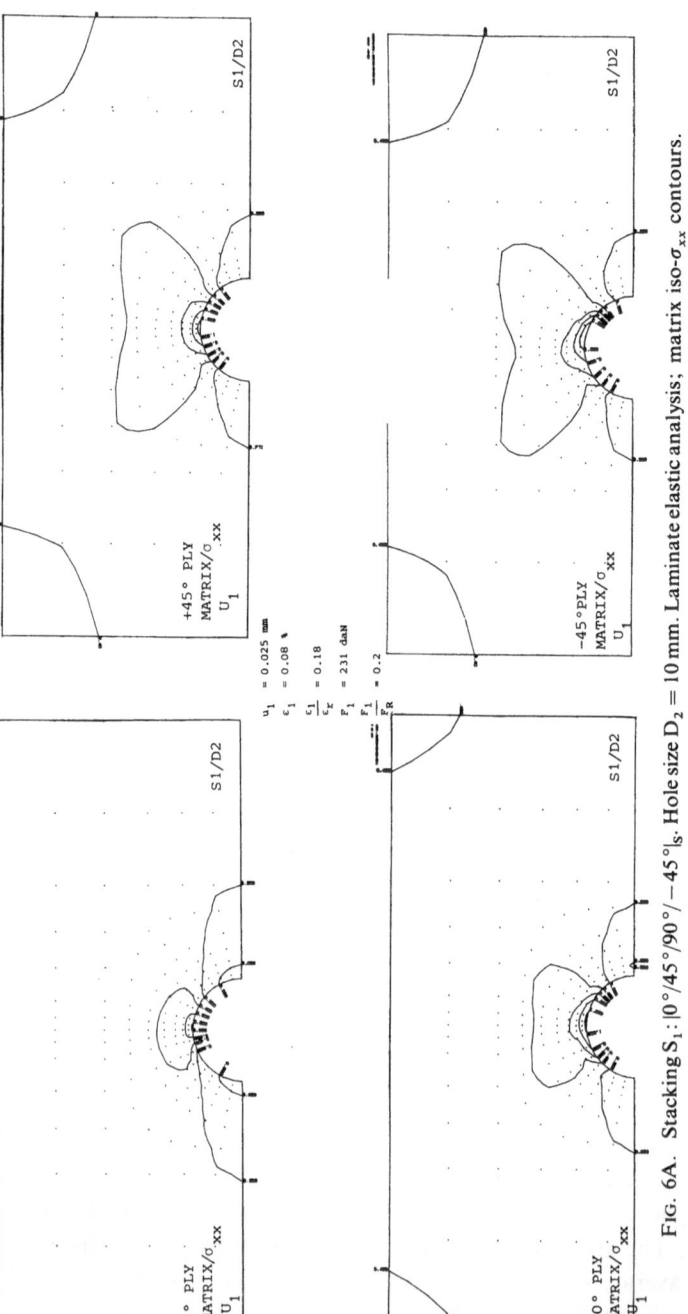

FIG. 6A.   Stacking S_1 : $|0°/45°/90°/-45°|_s$. Hole size $D_2 = 10$ mm. Laminate elastic analysis;   matrix iso-σ_{xx} contours.

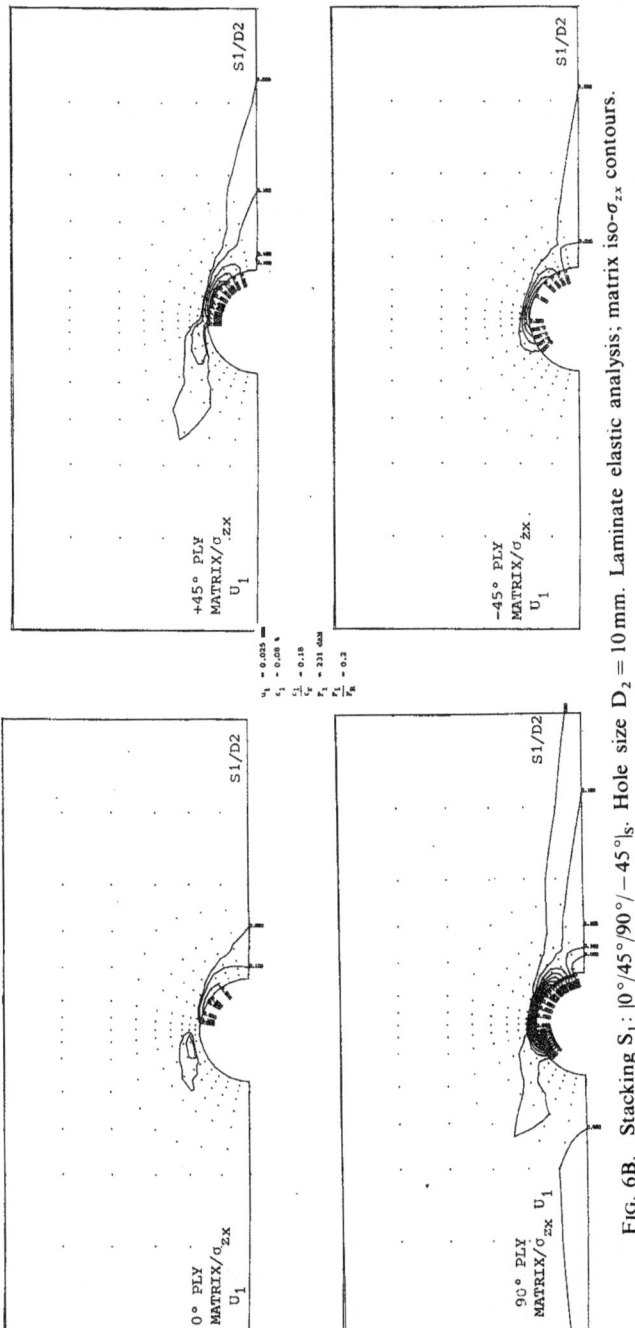

FIG. 6B. Stacking S_1: $|0°/45°/90°/-45°|_s$. Hole size $D_2 = 10$ mm. Laminate elastic analysis; matrix iso-σ_{zx} contours.

reaches 8·8 calculated at a distance of $0·045R = 0·225$ mm from the hole free edge.

Note: (a) Figure 5, for the fiber stresses: in the 0° ply the highest concentrations on the 90° axis; in the −45° plies the highest stresses near the hole in the zones of least cut fibers (+45°), exhibiting a disymmetry in the specimen longitudinal direction; the 90° fibers hardly stressed at all (stress contours not shown). (b) Figure 6(A, B) for the matrix stresses: the most stressed is the 90° ply, with high normal σ_{xx} and shear σ_{zx} stresses near the 90° axis.

3.2.2. Failure analysis

Additional incremental end displacements $U_2, U_3, \ldots, U_{12}$, are now applied until specimen maximum load capacity ('failure load') and a little beyond, U_{13}, is reached. The numerical specimen load–displacement curve is plotted in Fig. 4. *Damage* (i.e. plasticity) begins at ca. 55% of failure load. Fracture sensitive zones in each ply are clearly apparent on the matrix equivalent plastic deformation contours, $|\bar{\varepsilon}_p|$ (Fig. 7); e.g. matrix crazing should begin in the 90° ply, on the transverse axis.

With increasing load, the damage increases and spreads. As predicted, *matrix crazing* begins in the 90° ply at ca. 70% of failure load, followed in the ±45° and the 0° plies at ca. 85% of failure load (Fig. 8). *Fiber fracture* begins in the 0° ply and +45° ply and extends radially ca. 0·2 mm at 75% of failure load (Fig. 9), which marks the departure from global linearity on the load–displacement curve (Fig. 4).

Specimen ultimate load is reached when fiber breakage in the 0° ply has extended about one radius width (5 mm), matrix plastification about two radii, and matrix crazing and delamination about one radius (Figs 9, 10).

Using the elastic moduli and unidirectional tensile strength values measured by ENSTA, and completing with typical literature values for the fibers and missing matrix material strength parameters (*see* Table 2), the numerical failure load obtained is:

$$F_{12}^{N} = 1170 \, \text{daN}$$

which is 13·6% smaller than the experimental value $F_{12}^{E} = 1355 \, \text{daN}$.

3.2.3. Calibration

To account for the material strength input uncertainty, and allow a meaningful comparison of the variants analyzed hereunder, the numerical failure loads are scaled to the experimental failure load F_{11}^{E} for the

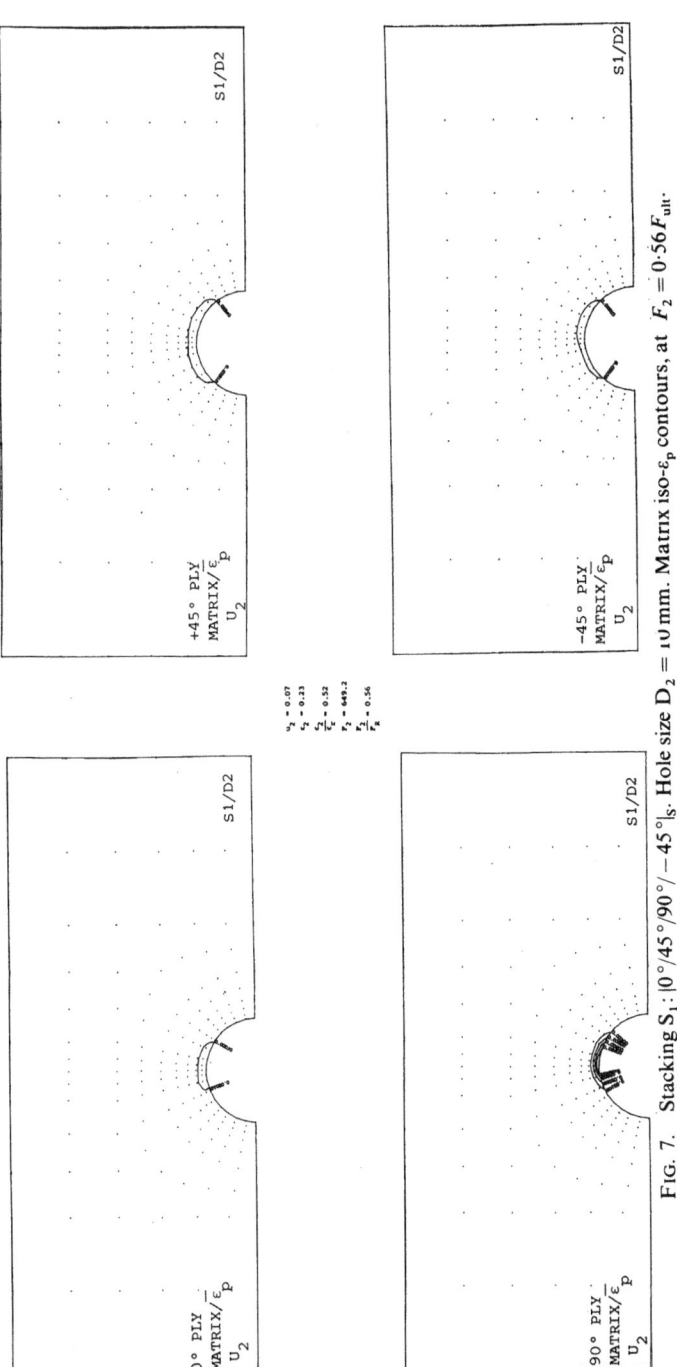

FIG. 7. Stacking S_1: $|0°/45°/90°/-45°|_s$. Hole size $D_2 = 10$ mm. Matrix iso-ε_p contours, at $F_2 = 0.56 F_{ult}$.

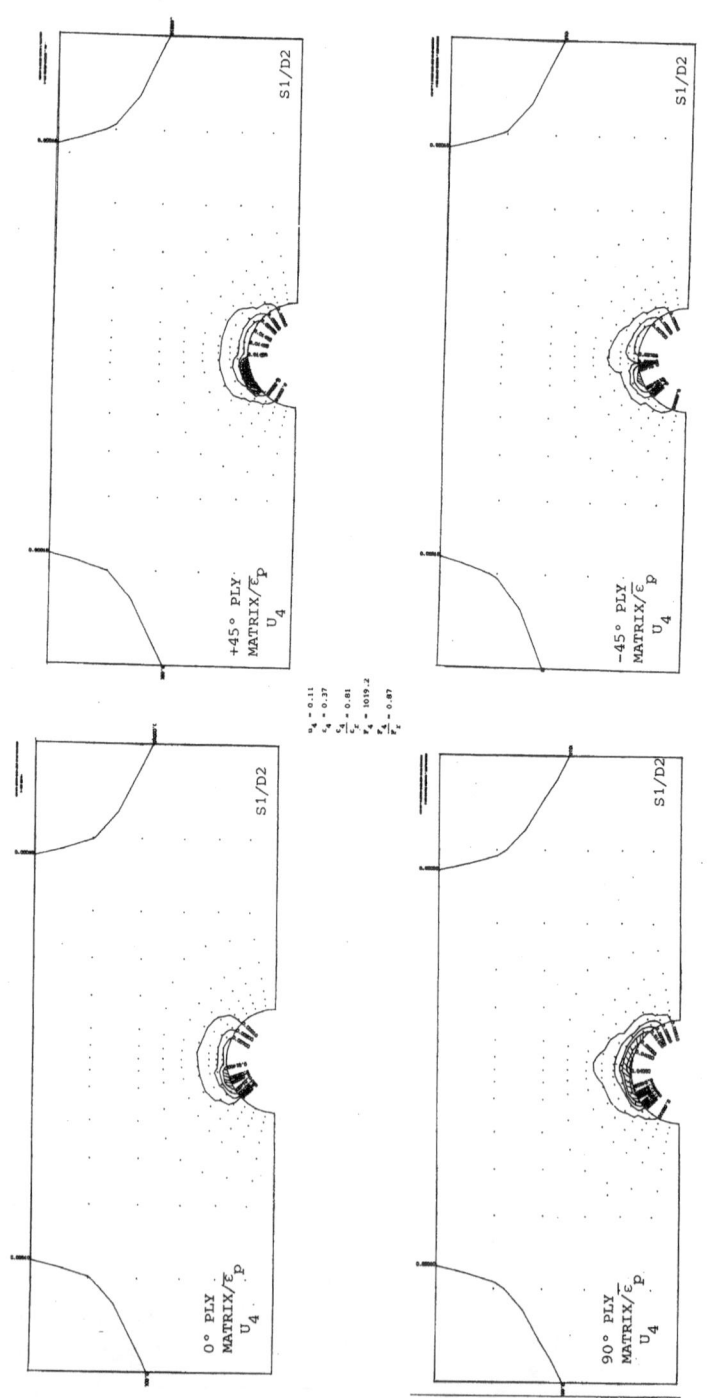

$u_4 = 0.11$
$u_4 = 0.37$
$u_4 = 0.81$
$\frac{u_c}{u_c} = 1019.2$
$\frac{F_4}{F_R} = 0.87$

▨ crazed matrix

Fig. 8. Stacking $S_1 = |0°/45°/90°/-45°|_s$. Hole size $D_2 = 10$ mm. Matrix iso-$\bar{\varepsilon}_p$ contours, at $F_4 = 0.87 F_R$.

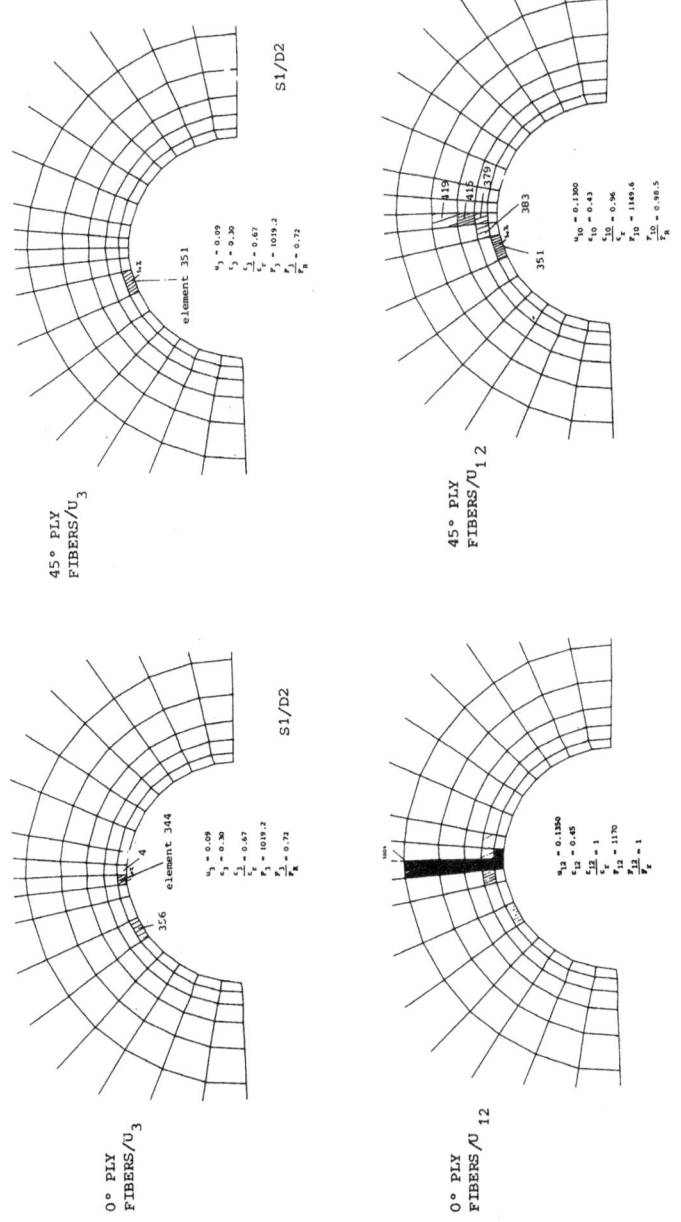

FIG. 9.   Stacking $S_1 = |0°/45°/90°/-45°|_s$. Hole size $D_2 = 10$ mm. Fiber fracture initiation and propagation in 0° and 45° plies.

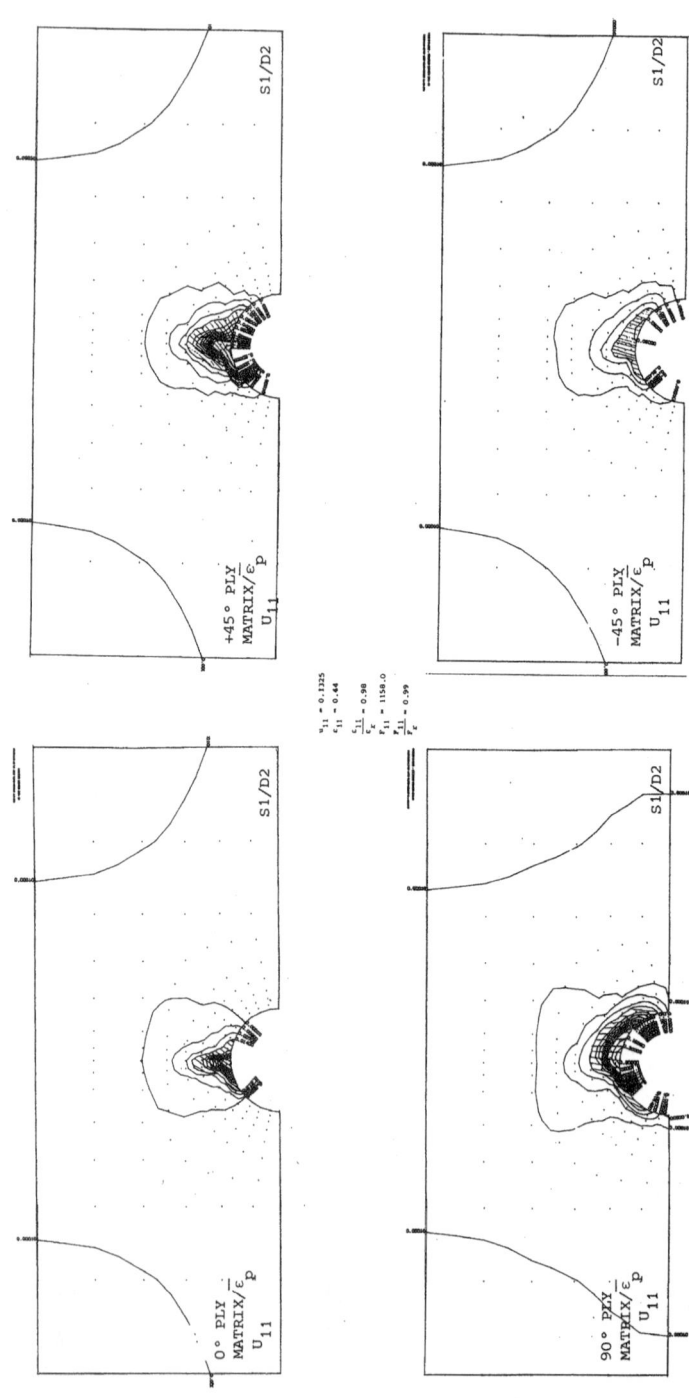

□ crazed matrix

FIG. 10. Stacking $S_1 = |0°/45°/90°/-45°|_s$. Hole size $D_2 = 10$ mm. Matrix iso-ε_p contours, at $F_{11} = 0.99 F_R$.

$S_1/D_1 = 5$ mm specimen. With this correction, the estimated numerical failure load for the S_1/D_2 specimen becomes:

$$F_{12}^{N*} = 1258 \text{ daN}$$

now 7% smaller than the experimental value F_{12}^E and within the experimental scatter (9·6%).

3.3. Stacking S_1—Hole Size $D_1 = 5$ mm (Configuration S_1/D_1)

Qualitatively the results of the numerical failure analysis of the specimen with reduced hole size $D_1 = 5$ mm are similar to those for $D_2 = 10$ mm described above.

Quantitatively, due to the constant specimen scale (e.g. thickness), the gradients near the hole differ, and the D_1 specimen turns out significantly stronger. The predicted numerical failure load for the S_1/D_1 specimen is:

$$F_{11}^N = 1590 \text{ daN}$$

which after calibration as above, yields exactly

$$F_{11}^{N*} = F_{11}^E = 1710 \text{ daN}$$

The experimental value, F_{11}^E (9·5% scatter) is 26% higher than F_{12}^E (9·6% scatter) to be compared with 36% for the numerical model.

3.4. Stacking S_2—Hole Sizes D_1 and D_2 (Configurations S_1/D_1 and S_1/D_2)

Changing the stacking sequence from S_1 to S_2, to place the $0°$ ply between the two $\pm 45°$ layers instead of above, modifies the damage mechanisms significantly. As a result of their combined shearing effect on the $0°$ ply, matrix damage in the $0°$ ply initiates earlier (Fig. 11 vs 7, 8) transferring an extra loading to the $0°$ fibers which break faster, to cause an ultimate load ca. 10% lower.

The calibrated numerical failure loads obtained are:

$$F_{21}^{N*} = 1570 \text{ daN} \qquad F_{22}^{N*} = 1183 \text{ daN}$$

The smaller hole size D_1 has a similar effect (33% strength increase with respect to D_2) on the specimen with the S_2 stacking. The corresponding experimental values are:

$$F_{21}^E = 1555 \text{ daN} \qquad F_{22}^E = \text{N.A.}$$

These results are summarized in Table 1B.

Note that F_{21}^E differs from the calibrated numerical value F_{21}^{N*} by only 1%.

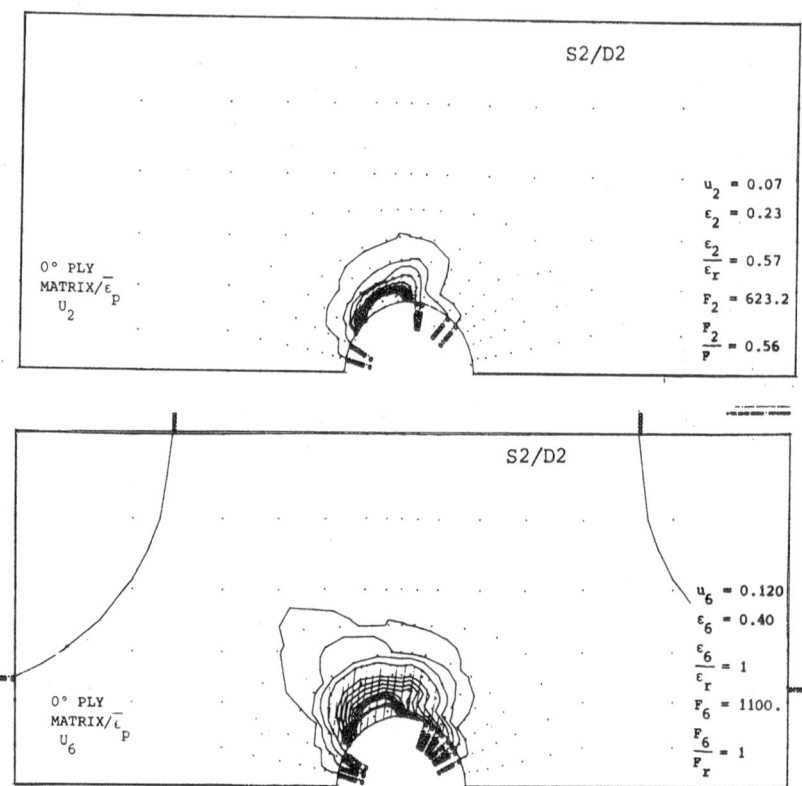

FIG. 11. Stacking $S_2 = |90°/-45°/0°/45°|_s$. Hole size $D_2 = 10$ mm. Matrix iso-$\bar{\varepsilon}_p$ contours, in 0° ply at $F_2 = 0.56F_R$; $F_6 = 1.0F_R$.

3.5. Free Edge Effect—Stacking S_1—Hole Size D_1

To check how the hole free edge high stress concentrations evolve as the loading increases, one specimen (S_1/D_1) was reanalyzed to failure with a very fine mesh close to the hole (Fig. 12).

Around the 5 mm hole is placed a crown of 5 equal radial width elements (0·0625 mm), providing 2·3 elements on a radial distance of one-ply thickness (0·145 mm). The mesh then grows coarser, to reach a 0·33 mm width at one hole radius distance. The maximum computed global stress concentration factor in the 0° ply is now 12·2, at $r = 0.031\,25$ mm ($0.0125R_1$) from the hole edge.

At $r = 0.0937$ mm ($0.0375R_1$) this factor decreases to 8·8, which is identical to the value obtained at $r = 0.045R_2$ with the standard coarse mesh.

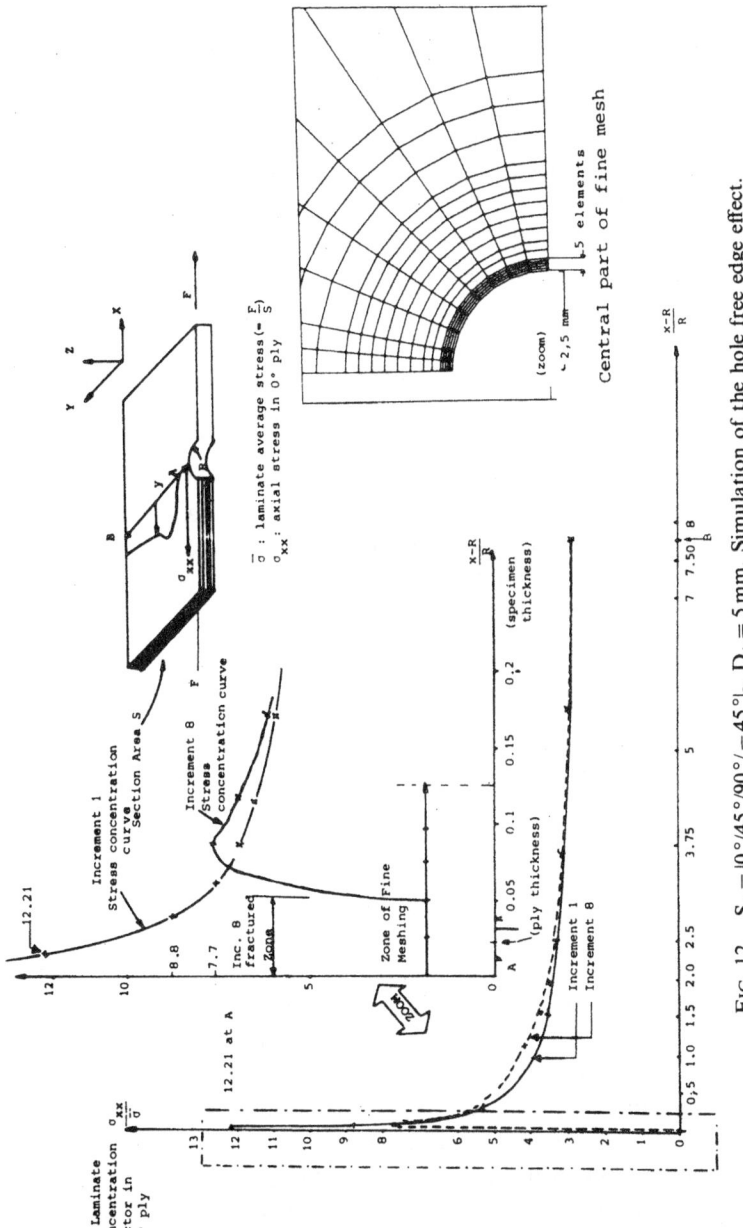

FIG. 12. $S_1 = |0°/45°/90°/-45°|_S$; $D_1 = 5$ mm. Simulation of the hole free edge effect.

Increasing the load, the 0° fibers break at the hole border, and the peak stress recedes inside the laminate. When the 0° fibers are broken on a radial distance of over one ply thickness (load increment 8), the peak stress has receded at ca. $0.218\,\text{mm} = 0.0875R_1$ from the edge, and the maximum concentration factor decreases to 7·7 (Fig. 12). Such a concentration is now mild enough to be captured by the standard coarse mesh. As expected, loading to failure from then on yields identical results with the fine and standard meshes.

In other words, the hole free edge stress concentrations blunt out, as the 0° fibers break to release the load and redistribute it inside the laminate.

4.   SENSITIVE PARAMETERS OF THE CUT-OUT, EFFECTIVE STRENGTH REDUCTION

Cut-outs reduce the laminate effective strength. Since this effect is predicted by a numerical model which does *not* include a 'characteristic length' or material gradient dependent failure criterion, it is interesting to investigate what causes the observed hole size effect, by briefly analyzing the failure of the same notched specimens, but with a single ply.

4.1. Material Effects

Several material types are studied:

A homogeneous reference material: isotropic elasto–ductile (von Mises or Tresca perfect plasticity); an isotropic matrix material (elasto–plastic with limited ductility; an istropic quasi-brittle material similar to the matrix–fibre composite); an isotropic–brittle material with the fiber properties; a heterogeneous unidirectional ply.

As before, the numerical analysis is carried to total failure, and the R factor is computed (Table 3). The axial normal stress profile σ_{xx} at failure across the ligament (y-direction) is presented in Fig. 13.

The strength reduction value R measures the loss of area at failure under the curve, σ_{xx} vs y (Fig. 14). For a perfectly ductile material, there is none (Tresca) or a small gain (von Mises). For increasingly brittle materials, the strength loss due to the hole damaged zone is only partially compensated by the increased stress in the ligament intact fraction. This stress redistribution is less pronounced when brittleness increases. Typically, at failure, for *this* type of notched specimen and tensile loading, the width of the damaged (or 'process') zone, is about one hole radius, whatever the material type.

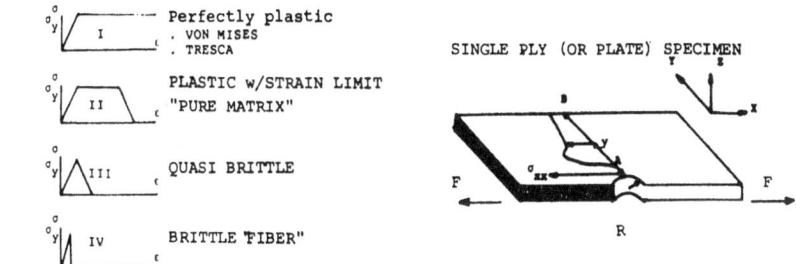

UNIAXIAL TENSION MATERIAL STRESS-STRAIN LAWS

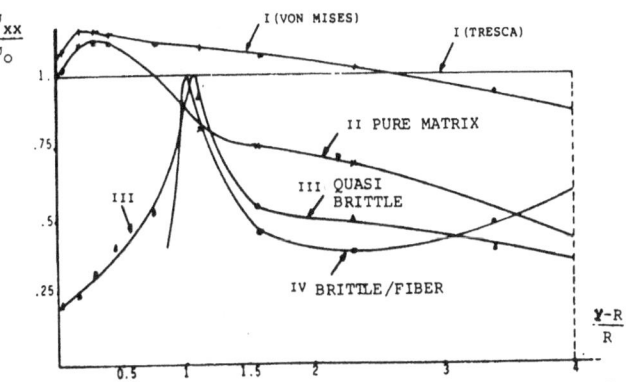

FIG. 13. Axial stress distribution at failure across the ligament.

TABLE 3
Hole size effect—notched specimen, D = 10 mm

	TYPE	HOLE SIZE EFFECT	MATERIAL LAW	MATERIAL TYPE	$\sigma_N \div \sigma_O = R$
HOMOGENEOUS MEDIUM (plate model)	Elasto perfectly plastic (TRESCA)	NO		matrix law w/no strain limit	1.0
	Elasto perfectly plastic (v. Mises)	NO		idem	1.07
	Elastic-plastic brittle	YES		pure matrix	0.73
	Elastic quasi-brittle	YES strong		intermediate	0.51
	Elastic-brittle	YES max		fiber	<0.4
HETEROGENOUS MEDIUM single or multiple lamina model.	Single ply 0°	YES max	biphase model	fiber + matrix composite	0.4
	Laminate	YES strong	biphase model	idem	0.6 to 0.7

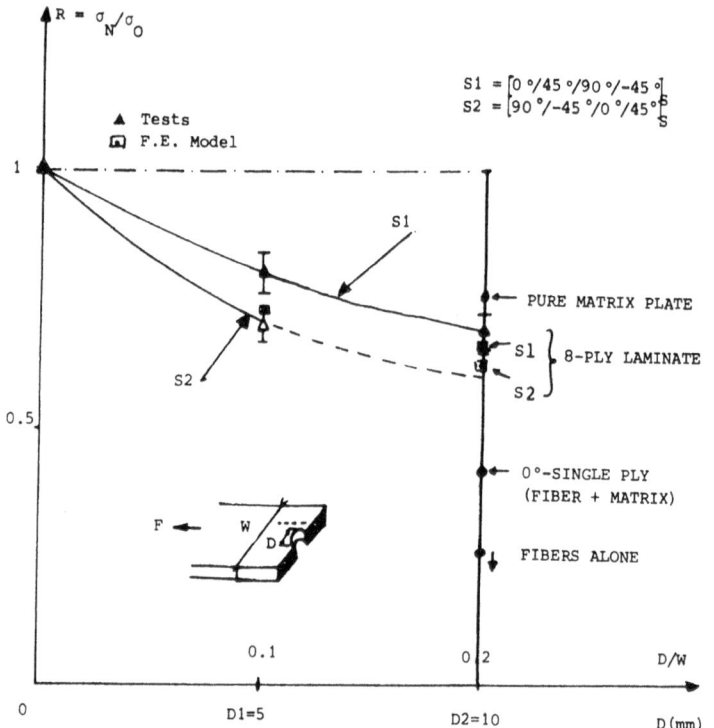

FIG. 14. Notched plate W/hole diameter $D_2 = 10$ mm. Effect of material and ply stacking on effective strength reduction.

Very brittle materials, such as fibers, exhibit a maximum strength reduction effect, since the redistributions away from the notch are nearly impossible ($R < 0.4$). The addition of a ductile matrix, permitting some redistribution, limits the reduction effect ($R = 0.4$). Ply superposition increases the redistribution capabilities, and the resulting R is higher ($R = 0.6$–0.7), but remains sensitive to the stacking sequence which influences interply redistributions limited by early delaminations (as noted in the experiments, the weakest specimens exhibit the most delamination[14]).

4.2. Geometric Effects

The hole 'size' effect and the specimen 'scale' effect can be compared on the notched single ply specimens.

If a new 'scale' configuration is analyzed, i.e. with a double ply thickness and a double hole size, it yields trivially an identical numerical failure load, since the model used here does not include a material gradient effect.

If a new 'size' configuration is analyzed at the same scale (i.e. plate thickness), with half a hole diameter, the redistribution mechanisms are qualitatively similar, but the affected damage zone at failure may be different and so will be the relative loss of area under the σ_{xx} vs y curve (by an amount dependent on the material ductility).

4.3. Failure Criteria for Composite Laminates

Table 3 summarizes the Failure Loads experimentally measured and predicted by a widely used 'Fracture Mechanics' criterion (Whitney–Nuismer[3]), and the present numerical model without material gradient effects. The size effect is similarly predicted by both models for the range studied; however the scale effect is not. Additional *ad hoc* experimental results would be welcome to clarify this matter further.

5. CONCLUDING REMARKS

Numerical simulation strives to identify and incorporate the appropriate physics in the modelling and analysis process. This approach applied to the damage and failure analysis of notched composite laminates suggests that for the configurations and loadings analyzed:

—internal damage initiates and propagates early while the component behavior remains globally linear, exhibiting an 'equivalent' limited material ductility;
—free edge effects around cut-outs blunt out, and bear little influence on the failure load;
—the effective strength reduction due to cut-outs of several sizes and for different stacking sequences can be explained by stress concentration shape factors and material brittleness effects in heterogeneous continua;
—fracture mechanics criteria introduce a material characteristic length, and add a scale effect to the size effect, which may not be present when the stress concentration shape factor is relatively mild, and the cut-out dimensions large with respect to the ply thickness.

ACKNOWLEDGEMENTS

This work was supported by the French Ministry of Defense, DRET/SDR-G8 Mr Marcoin/Mme Corvino, with the collaboration of ENSTA, Mr

Verchery, for the experiments, and of Avions Marcel Dassault (AMD-BA, Suresnes), MM. Petiau and Chaumette, for industry feedback.

At ESI, MM. J. F. Levy and M. Habiballah, performed most numerical analyses.

REFERENCES

1. WHITNEY, J. M. and KIM, K. Y., Effect of stacking sequence on the notched strength of laminated composites, *Composite Materials: Testing & Design* (4th Conf.), *ASTM STP 617*, 1977, pp. 229–242.

2. WILKINS, M. L., STREIT, R. D. and REAUGH, J. E., Cumulative-strain-damage, model of ductile fracture: simulation and prediction of engineering fracture tests, *UCRL-53058 LLNL,* Oct. 1980.

3. NUISMER, R. J. and WHITNEY, J., Uniaxial failure of composite laminates containing stress concentrations, *Fracture Mechanics of Composites, ASTM STP 593*, 1975, pp. 117–142.

4. NUISMER, R. J. and LABOR, J. D., Applications of the average stress failure criterion Part II—Compression, *J. Comp. Mat.*, **13,**49, Jan. 1979.

5. BYRON PIPES, R., WETHEHOLD, R. C. and GILLESPIE, J. W., Jr, Notched Strength of Composite Material, *J. Comp. Mat.*, **13,**148, April 1979.

6. GARBO, S. P. and OGONOWSKI, J. M., Strength predictions of composite laminates with unloaded fastener holes, *AIAA Journal*, **18**, no. 5, May 1980.

7. HAUG, E., LEVY, J. F. and DE ROUVRAY, A., Mécanismes et critères de rupture des mat. composites, *ESI/RT/ED 77-192*, Vol. I/II, DRET/76-462, April 1979.

8. DE ROUVRAY, A., HABIBALLAH, M., DUBOIS, J. and HAUG, E., Mécanismes et critères de rupture des mat. comp., effets spécifiques: d'echelle, d'empilement, de bord libre, *ESI-RT/ED 80-298*, DRET 80-245, June 1982.

9. ENGRAND, D., Calcul des contraintes de bords libres dans les plaques composites symétriques avec ou sans trou—Comparaison avec l'expérience, *C.R. JNC3 AMAC*—Paris, Editions Pluralis, Sept. 1982.

10. REIFSNIDER, K. L., HENNEKE, E. G. and STINCHCOMB, W. W., Defect–property relationships in composite materials, Virginia Polytechnic Institute & State Univ.—*AFML-TR-76-81*, Part IV, Final Report, June 1979.

11. HAUG, E., LOCCI, J. M. and ARNAUDEAU, F. C., PAM-NL: a general finite element program for the nonlinear thermomechanical analysis of structures, *E.S.I., Trans. 5th S.M.I.R.T. Intl. Conf.*, Berlin, W.G., Aug. 1979.

12. DUBOIS, J. and CURNIER, A., Développement de modules de plasticité adaptés aux composites tridimensionnels, *E.S.I., R. T./ED 77-212*, April 1980.

13. DUBOIS, J. and HABIBALLAH, M., Anipla material model, *E.S.I., T. R.*, 1/82.

14. VERCHERY, G., DIERNAZ, C. and VONG, T. S., Essai de rupture en traction de plaques composites trouées, *R. T. ENSTA, 1978*, Marché DRET 76-462/77-493.

15. NARAYANASWANI, R., Evaluation of the tensor polynomial and Hoffman strength theories for composite materials, *J. Comp. Mat.*, **11**, 366, Oct. 1977.

16. BEAUMONT, P. W. R., Micromechanisms of fracture of fibrous composites in monotonic loading, *Mech. Comp. Review*, AFML, Nov. 1979.
17. ISHIKAWA, T., KOYAMA, K. and KOBAYACHY, S., Elastic moduli of carbon-epoxy composites and carbon fibers, *J. Comp. Mat.*, 332, July 1977.
18. ISHIKAWA, T., Thermal expansion coefficients of unidirectional composites. *J. Comp. Mat.*, **12**, 153, April 1978.

15

Structural Applications for Pultruded Profiles

TREVOR F. STARR

*Technolex—Consultants in FRP Technology, Middle Path,
Crewkerne, Somerset TA18 8BG, England*

ABSTRACT

*Analysis of fibre reinforced composite material working processes shows
pultrusion gaining an ever-increasing share of the market. After a review
which also outlines the process, the materials and specifications employed,
the reasons for that situation are answered through description of selected
and recent, pultruded profile application developments. These confirm the
excellent, reproducible quality and structural properties of pultrusions,
which further enhance those of for example, low weight, high strength and
corrosion resistance for which composite materials are now widely
recognized.*

The sole purpose of this discourse is to further confirm that fibre reinforced
polymeric composites have 'come of age'. The excellent track record,
increasing use and acceptance of pultruded profiles for structural
application provides ideal examples. But the ability of 'pultrusions' to be
tailored to meet precise mechanical, physical property or economic
demands which increasingly challenge traditional design concepts and
materials is no longer unusual. With the better availability of improved and
more sophisticated reinforcements and matrices, that ability is now true for
all composites. The pultrusion process is but one alternative, well-
established manufacturing technique whether labour- or capital-intensive,
by which composites are processed to offer a continuing, exciting challenge
for this decade and beyond. Application diversity is equally as wide, from

building and construction, through corrosion engineering to marine and aerospace engineering.

But what is pultrusion? It is simply the continuous length 'pulled-extrusion' of constant cross-section fibre reinforced profiles. As shown by Figs 1 and 2, size-range and profile complexity is considerable with current section thickness limits of typically 1·75–40 mm. Pultruded profiles are

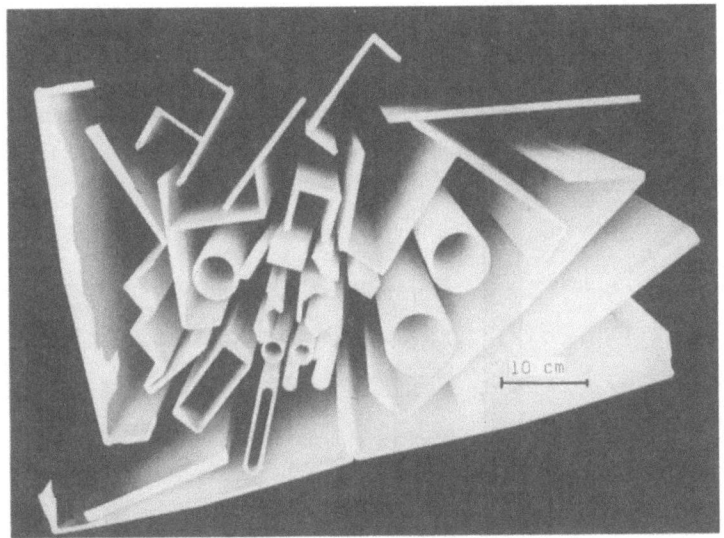

FIG. 1.

therefore analogous to the extrusion of steel, aluminium or thermoplastic sections, or even the mechanical profiling of timber or the casting of steel reinforced concrete beams.

The pultrusion process was introduced some 30 years ago to answer the electrical industry's need for one-stage manufacture of close-dimensioned insulating components such as commutator slot-wedges and transformer spacer-bars. Now there are more than 80 companies world-wide well able to meet a wider and ever-increasingly severe, customer demand. In 1981 the USA output approached 30 000 tonnes at a value of £6¼ million,[1] but that was still only 3% of their total market for composites. Many authorities,[2,3] have stated that pultrusion is and will remain, the fastest growing sector of the composite materials industry. Two other facts are offered in support. Although the world trade recession meant there was

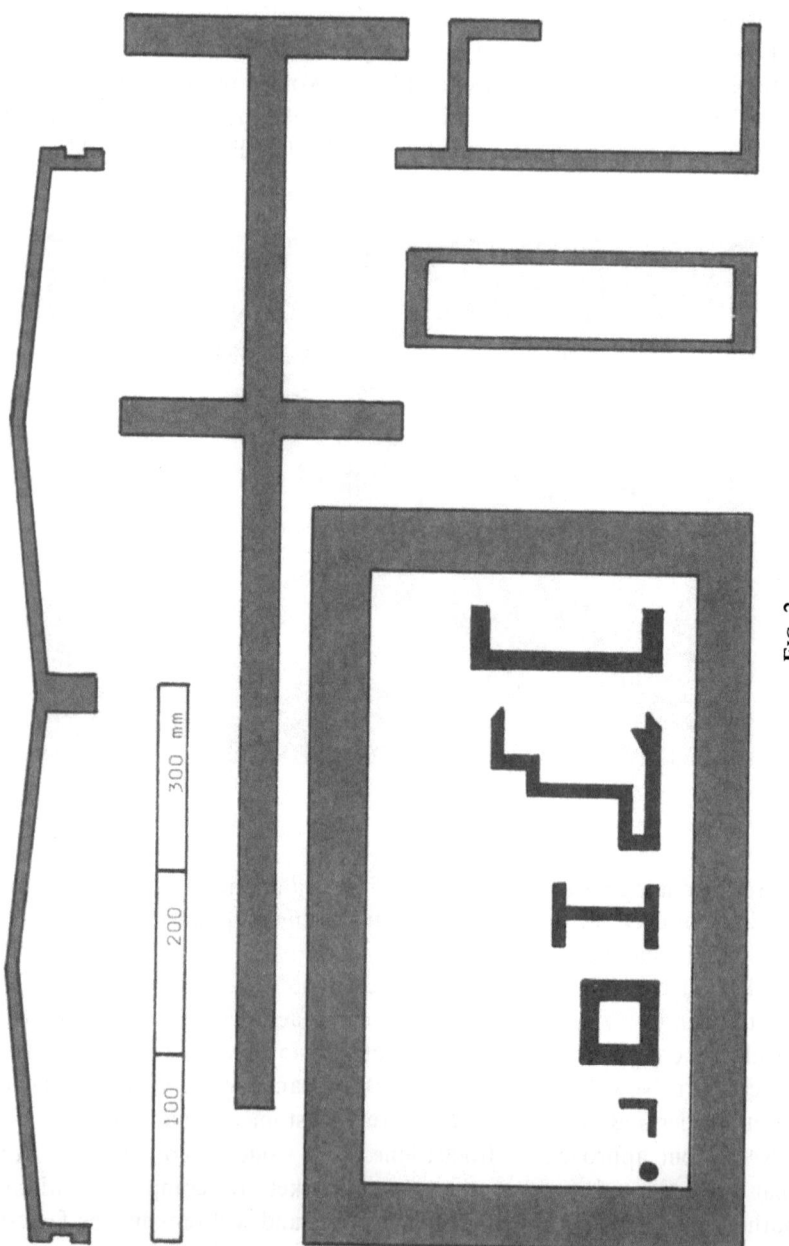

FIG. 2.

little or no growth in 1980 and 1981, over the last 12 years the pultrusion industry has shown an average annual growth rate of 17 %. But perhaps a better indicator was provided by the Awards at the 1983 SPI Reinforced Plastics/Composite Conference in Houston, Texas. The majority went to pultruded profiles.

Even so too many design engineers and potential users answer the suggestion to employ pultrusions with a blank stare! The process demands a better understanding, appreciation and application acceptance. Faster escalating metal costs—particularly for aluminium—have now begun to cross-over those for comparable load-bearing pultruded sections. No longer is the consumer required to pay an excessive premium for the accepted composite advantages such as light weight and corrosion resistance. Improvements in section complexity, pulling-speed, die design and life, or other process changes such as more accurate reinforcement location, remain continuous. Many are frequently related to advances in the materials employed, with exciting possibilities on the horizon. Indeed it is a feature of pultrusion that considerable process development is more wide open than for most alternative composite material manufacturing techniques.[3]

THE PROCESS

Compared to those alternatives pultrusion is basically a simple process, a pulling of the required reinforcement impregnated with an initially viscous liquid matrix, through a heated metal die. Here the pre-formed composite is both consolidated to the desired shape, and the matrix brought to a cured or polymerized condition. The finished composite up to a present maximum of some $520\,cm^2$ therefore exits from the die as a solid but possibly hollow shape or profile, of excellent surface finish and to dimensions matching the die geometry. With all the sophistication built into the machine and die, the process can be semi or fully automated and thus operated by a low, unskilled labour force. They may be required to only ensure continuity of material supply. Skilled maintenance and comprehensive technical support is however essential. Since the process is continuous, costs, given a high volume requirement, are low relative to other methods of forming and moulding composite material components.

Whilst some companies pultrude very simple shapes on 'Rolls-Royce' equipment costing upwards of £300 000, others offer much greater complexity from *ad-hoc* in-house designed and manufactured machines.

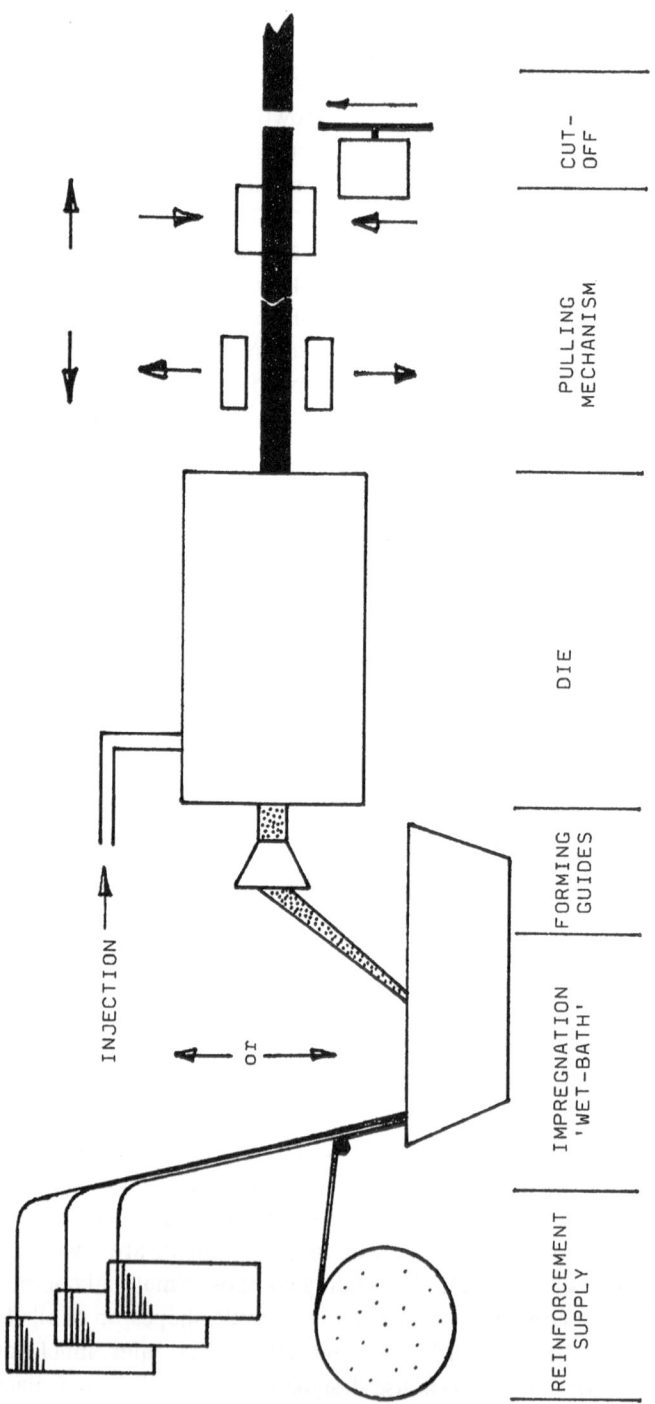

FIG. 3. The pultrusion process—schematic.

There is every shade in between. Although machines vary, each must comprise at least six parts:

(1) Reinforcement supply.
(2) Pre-forming, direction guides.
(3) Matrix impregnation.
(4) Heated metal die—commonly by strap-on and/or cartridge electrical elements, with close temperature control.
(5) Pulling mechanism.
(6) Cut-off table.

Process variations are clearly possible relative to the nature of the operation, the type of reinforcement and matrix, and the type and size of profile being manufactured. For example as shown by Fig. 3, the reinforcement can be impregnated by passing it through an open trough or 'wet-out bath', or by injecting the chosen matrix directly into the die just beyond the reinforcement entry point. Again die-heat can be supplemented in the former technique by a radio-frequency cure stage after some pre-consolidation prior to die entry. This variation is more often employed for thicker sections, or where faster pulling-speed commensurate with the profile shape is possible. Polymerization control is critical to the process and temperature control is usually 'zoned' and now solid-state pro-grammed to hold the exotherm at an optimum position as the profile moves, at typically 0·25–0·75 m per minute, through the die.

The purpose of the pre-forming guides is obvious—to gather the reinforcement grades together and ensure at the same time their accurate location to the tapered die-mouth. Die design for the 'wet-bath' and injection techniques differs, with the form varying along its length—typically 0·45–1·50 m—down to the exit point where the dimensions and profile shape are 'matched'.

Tool steel is the usual die material, chrome-plated and highly polished to reduce the coefficient of friction between the stationary and moving surfaces. This extends die-life and allows effective, economic re-furbishment. Front-end supported and balanced mandrels, or core-rods, are clearly necessary for hollow profiles such as tubes. Their alignment difficulty can be considered one of the few disadvantages of the process; the reinforcement can 'wedge' the mandrel out of position.

Multiple die operation, positioned across the machine-bed—and sometimes stacked—is frequently practised, particularly where the cross-sectional area of each profile does not typically exceed 30 cm². The usual number is 2–6 dies, and as long as their optimum pulling-speeds match, their geometry can differ as can their length.

A number of alternative pulling arrangements are employed ranging from simple powered rollers or belt-type pullers to tandem caterpillar-type tractors fitted with reciprocating pneumatic product-clamps. The driving force may be mechanical or hydraulic with the latter possessing the advantage of easier control, programmed to 'follow' the profile. This permits compensating speed changes as the pulling-force alters and possibly an alarm indication under a situation of rising and undesirable die-friction. Cut-off arrangements are usually diamond-tipped 'flying saws' fitted with dust extraction if they are not water-cooled.

THE MATERIALS

In terms of reinforcement, glass fibre currently continues to predominate, and this situation can be expected to apply for perhaps many years. However, textiles such as terylene can be readily introduced or, as in the case of carbon, employed exclusively to achieve certain property advantages. Kevlar, polypropylene, steel and aluminium fibres and wire are also suggesting many exciting possibilities for the future. Natural fibres such as cotton and sisal may feature with time and where mechanical properties do not dominate.

Glass fibre feedstock ranges from continuous, spun or hooped roving, through mats to virtually every type of woven format available with tapes and specially prepared 'mixed' reinforcements assuming greater importance.[4]

Unsaturated polyester resins, often formulated specifically for pultrusion, predominate for the continuous matrix. However, bisphenols, vinyl esters, phenolics and epoxy resins are increasing in popularity when for example, optimum corrosion or fire resistance is required. Certain thermoplastics, such as polyethersulphone, are also beginning to show distinct possibilities. Many offer enhanced mechanical or physical properties—and the ability to post-form the profile to geometries which cannot be built into the die.

Such material advances although requiring equipment and process modification, will only widen the application potential which can be realized. Some developments in turn will obviously be translated back to the present standard techniques. It is for example a truism that for all their success, few if any pultruders really understand what is happening to the materials within the die, which is the key to the process!

THE PROPERTIES

It is the excellent longitudinal tensile, and flexural strength properties for which pultrusions are especially noted, which allows their consideration as true structural elements. The continuous longitudinal reinforcement at high loading—typically 45–75 % by wt—is of course essential, for without it profiles could not be pulled through and from the die.

Table 1 provides a property summary, but should be read with care. Figures are for typical specifications only, the reinforcement, matrix alterations readily possible as noted, and the type of profile, clearly have pronounced effects. Whenever their use is contemplated close reference should be maintained with the potential supplier from conception and through all design stages. In addition the 'standards'—and therefore costs—from competitive suppliers may also vary between very wide limits.

Pultruded profiles which are 'through' pigmented, often contain as well as a die release agent, small additions (perhaps 10 % by wt maximum) of inert mineral filler. Although they retain all the basic property advantages of fibre reinforced composites, like hot-press moulded SMC/DMC compounds there is no true gel-coat. However, a resin-rich outer layer can be achieved by selective techniques. Enhanced surface protection and finish can be provided by post-coating, powder application.

TABLE 1

Typical profile properties

			Solids	Sections
Tensile	—strength	N/mm^2	800–900	500–650
	—modulus	N/mm^2	38 000–42 000	18 000–30 000
Flexural	—strength	N/mm^2	1 000–1 100	400–700
	—modulus	N/mm^2	36 000–38 000	18 000–24 000
Compressive	—strength	N/mm^2	380–450	125–280
	—modulus	N/mm^2	16 500–17 500	11 000–12 000
Shear	—strength	N/mm^2	35–40	30–40
Barcol			45–50	
Dielectric		VPM	200	
Coef. thermal expansion		in/in/°C	$5 \cdot 2 \times 10^{-6}$	
Density		g/cm^3	1·6–1·9	

THE COSTS

Table 2 attempts to compare the cost performance of pultruded profiles with traditional structural materials; only concrete has an overall better score. However, that summary suffers from—and for the same reason—the

TABLE 2
Cost performance

	Structural mild steel	Stainless steel	Aluminium alloy	Timber	Concrete
Cost per kilogram	0·3	1·6	1·4	0·1	<0·1
Cost to density	1·2	7·2	2·2	<0·1	<0·1
Strength	2·1	5·6	7·1	1·8	0·9
Stiffness—tensile	0·2	1·2	1·0	0·4	0·2
—bending	0·6	3·9	1·6	2·2	0·8

Pultruded profiles taken as unity. Thus values lower than 1.0 indicate pultrusions are more expensive, values higher than 1·0 indicate pultrusions are cheaper.

disadvantages of Table 1. In other words the supplier must be carefully selected; there is little point in paying for property gains not required. The analysis of 1982 American and European price structures suggests for the latter, and a reasonably simple profile at £1·10–£1·40 per kilogram ex-works, that there is around 15% disadvantage.

But any comparison must also consider the lower delivery, erection/installation and long-term maintenance costs which apply to all composites. Even a price differential of 400% is acceptable if the cheaper component is worthless after simple damage which the pultruded profile can resist many times over.[1]

THE APPLICATIONS

It should by now be clear that pultrusions have a commonality to the structural elements, angles, girders, beams, hollow sections and pipes. They can be sawn, drilled, threaded or otherwise post-machined, bolted or riveted together. Whilst welding is not possible where thermosetting polyester/epoxy resins have been employed—but possibly at a later date for those using selected thermoplastic matrices—adhesive bonding is practical.

Space makes it impossible to provide a definitive application list, but generally the classifications are, building and construction, engineering, processing and corrosion, transport, leisure, electrical and others. As the examples to be quoted will hopefully indicate, the application limits are only those of the Designer's imagination and his better understanding—and education—of the process and the many advantageous properties of pultrusions. In this context the Design Manuals[5] now published by several pultruders provide to the Structural Engineer that data which he justifiably demands, and has become accustomed to, when dealing with concrete, timber, steel or aluminium and its alloys. That in itself, confirms the maturity now reached by this sector of the composite materials industry, as noted at the start of this discourse. Only the preparation of specific pultruded product standards remains to complete the picture.

Any component of regular, straight section, required in volume quantity warrants examination, and is the reason why the larger pultrusion companies now stock an ever-increasing range of standard profiles matching in dimensions those of traditional material supplies. To that must be added the rapidly growing range of custom profiles many of which will, with time, become standards for differing application. Finally, in these examples, the word 'structural' has been taken to assume profiles subjected to any form of load condition—in effect perhaps 95 % of all profiles currently manufactured.

1. Housing

For many, structures mean housing, cabins or containment enclosures. Three have been selected which merit mention. The first, to shortly be constructed in upstate New York for a sensitive computer research facility, employs pultrusions for the entire two-storey structure—walls, roof and floors. The system comprises a standard hollow tubular panel (Fig. 4) with over 24 different connectors to allow the high degree of design flexibility required. They uniquely interlock to ensure a sturdy, mechanically jointed seam. Foam insulation will then be pumped into the panel's core to withstand the sub-zero winters and blistering summers of that location. Pultrusions provided the best total package in respect of initial fabrication, erection, finishing and insulation costs. In addition, the 40 % glass-reinforced fire retardant resin specification offered in competition with timber, better life-cycle maintenance/corrosion costs for a non-conducting structure in an electromagnetic environment.

That type of system can of course equally be employed for smaller structures to contain for example, electrical switch-gear, or as temporary and permanent site cabins. The longest section shown in Fig. 2 is another

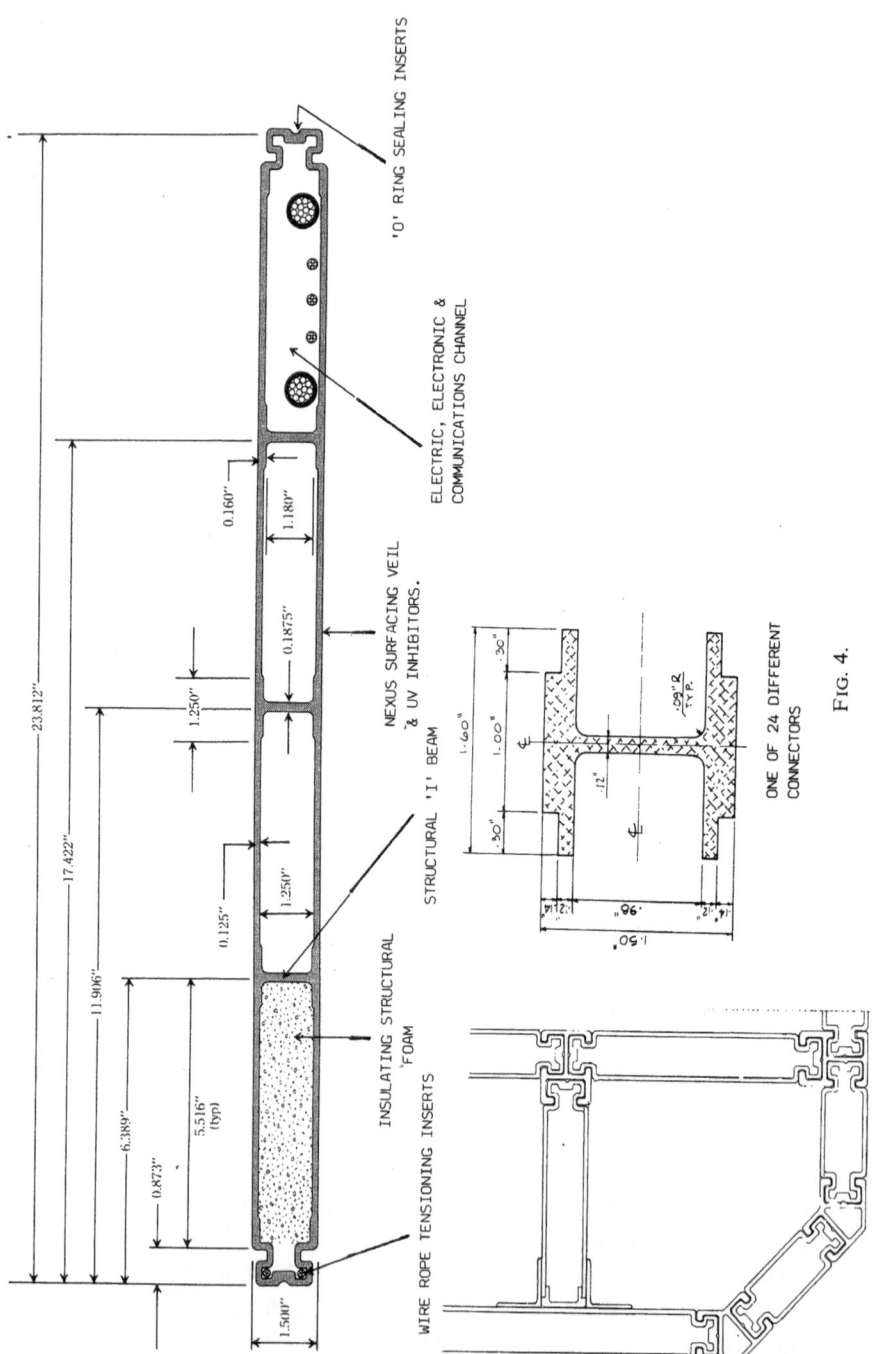

'O' RING SEALING INSERTS

ELECTRIC, ELECTRONIC &
COMMUNICATIONS CHANNEL

NEXUS SURFACING VEIL
& UV INHIBITORS.

STRUCTURAL 'I' BEAM

INSULATING STRUCTURAL
'FOAM'

WIRE ROPE TENSIONING INSERTS

23.812"
17.422"
11.906"
6.389"
5.516" (typ)
0.873"
1.500"
0.160"
1.180"
0.1875"
1.250"
1.250"
0.125"

ONE OF 24 DIFFERENT
CONNECTORS

FIG. 4.

manufactured for that purpose, but here the jointing across suitable sealants is by conventional nuts and bolts with internal galvanized steel fixtures. Like the above building, one indirect advantage of this form of construction—where high voltage equipment is enclosed—is the high dielectric and load-bearing strength of pultrusions. The need for separate insulators is therefore markedly reduced, but when required they can again be standard angle and channel pultrusions. Such structures are also highly resistant to vandalism, an important factor resulting from the good impact resistance of composites generally.

Many of the properties of pultrusions are now being realized for the third example—low and medium cost housing. Due to the considerable sales potential which exists, the development of a total composite material house[6] has been the aim of numerous companies and entrepreneurs since the late 1960s. Whilst monocoque design was initially favoured owing to the ability to fabricate large one-piece mouldings, their futuristic appearance met with considerable customer resistance. A simpler, more easily shipped and erected, smaller pre-fabricated module form of design is now seen as a more practical way forward. Here pultrusions are beginning to show their worth, to provide a basic framework or skeleton which permits completion with a wider variety of wall/roof panels and treatments. Thus the design flexibility is improved and the results are a more attractive, acceptable building at economic cost.[7]

2. Construction and Building

Four examples from, again, many, have this time been chosen. The first, shown in Fig. 5, is a roof support beam fabricated from several standard profiles. The bolted truss is currently manufactured in lengths up to 12 m, and allows a design loading of 900 kg per linear metre. Although developed

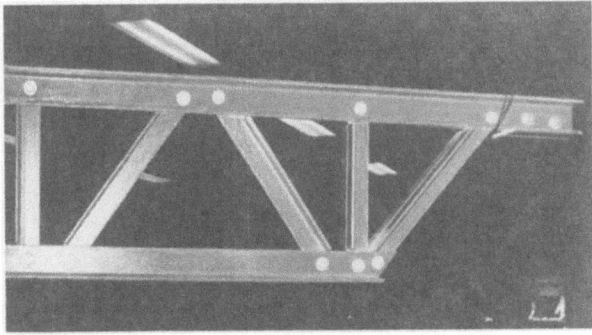

FIG. 5.

FIG. 6.

FIG. 7.

for corrosive environments the light-weight and high strength of the assembly rates investigation for many additional load-bearing applications.

Standard profiles are also exclusively employed in the sunscreen illustrated in Fig. 6, erected on a South African building. This carefully designed assembly, angled to permit the passage of winter rays of the sun but not those of the summer, is also secured to the building with prefabricated pultruded gussets. Again the corrosion resistant, light-weight properties score heavily, but are combined in this example with a material which also does not readily deteriorate under the action of UV radiation.

All the properties noted in this section are of importance in the example shown in Fig. 7. Conventional un-manned lighthouse towers require constant maintenance due to the joint action of salt-spray, wind, rain and sunlight. This 10-m high pultruded structure employed by the Brazilian Navy has very successfully answered a 4-year replacement cycle problem of the former metal towers. In addition, the new design is so lightweight that it can be located on rocky seacoasts in a single helicopter operation. Electricity pylons, telegraph poles, lamp standards[8] and even crane-jibs, are not too divorced for inclusion alongside, as similar successful applications.

Whilst pultruded pipework can in certain instances compete with filament-wound and centrifugal-cast composite pipe, the final example (Fig. 8) in this section concerns a 'People's Pipe' being developed as a walkway for American Airlines. Here a two-part 'U' section corner pultrusion forms the junction, and provides the primary structure to each length of a tube section whose walls, roof and floor use a laminate-foam sandwich construction. The whole assembly is adhesive bonded, and on completion will closely resemble the predominantly hand-laminated walkways extensively employed at Terminal 2, Heathrow Airport, London.[9]

3. Corrosion

Additional applications for the previously mentioned roof beam are not too dissimilar to a rapidly growing use of pultrusions in the construction of gratings, support-platforms and walkways subjected to corrosive environment. The end section of such an assembly and comprising also a short staircase, is illustrated in Fig. 9. Its form, and reason for that acceptance should by now be clear but Fig. 10 is included to clarify the grating assembly which competes with other one-piece moulded or cast forms of construction.

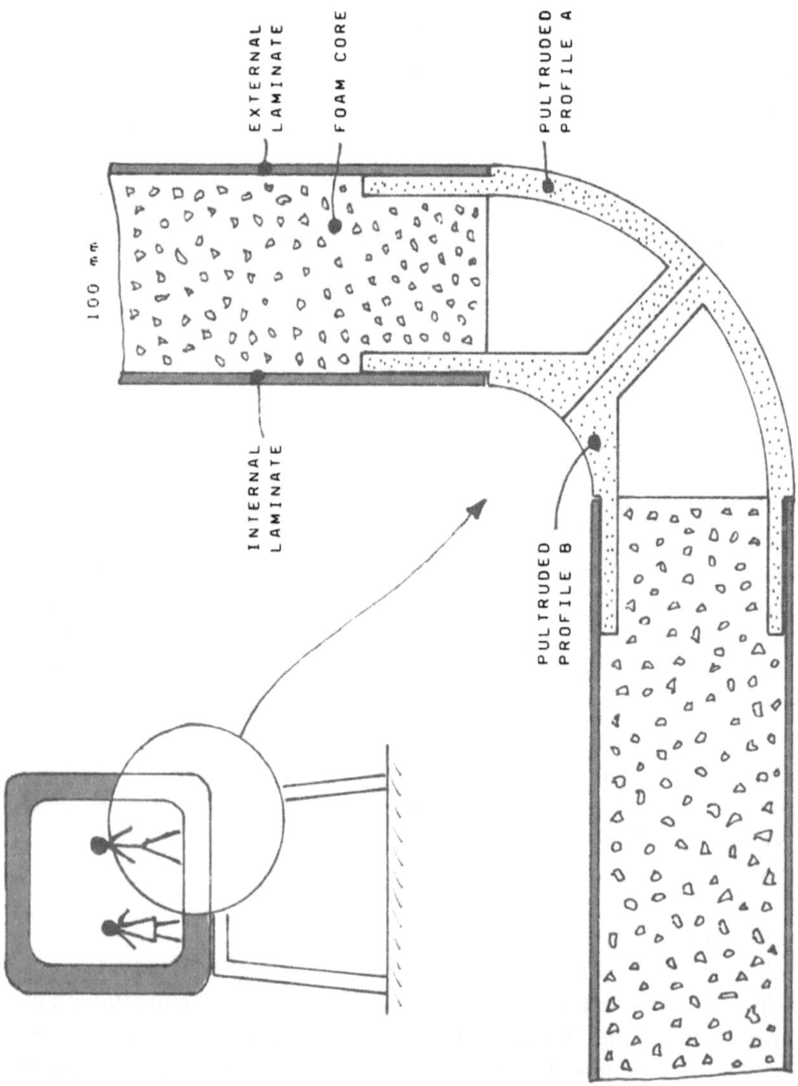

EXTERNAL LAMINATE

FOAM CORE

PULTRUDED PROFILE A

100 mm

INTERNAL LAMINATE

PULTRUDED PROFILE B

FIG. 8.

FIG. 9.

This type of structural section can also be effectively employed in the manufacture of tooling jigs and fixtures,[8] and non-corrodable 'open' instrument tube and electric cable support trays.[10] There are several well developed systems on the market, plus directly pultruded cable trays. In addition and not unlike the assemblies described, there are now many domestic and commercial pultruded ladders available[11] which, as well as corrosion, offers electrical resistance.

Perhaps the heaviest pultrusion at the present time is the 10-m beam shown in Fig. 11. At an overall height of 600 mm and incorporating a 165 × 230 mm 'I' section, it weighs 100 kg per linear metre. These beams support demister units in a flue gas desulphurization air scrubber at a New

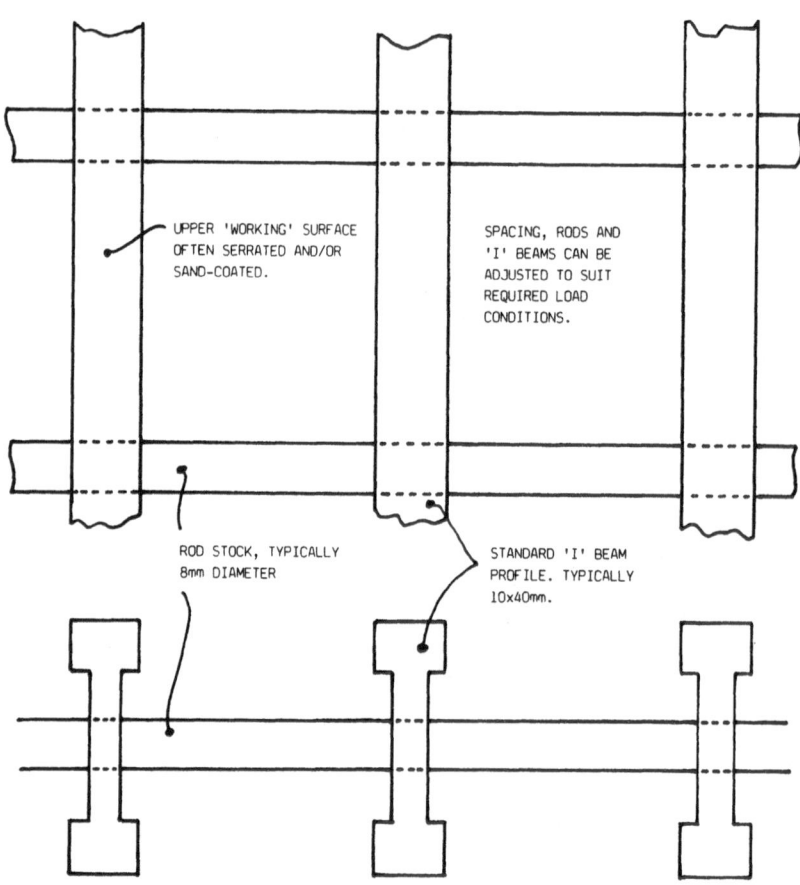

FIG. 10.   Assembled grating, either adhesive bonded or with spacers (not shown) supported by pultruded channels/angles.

Mexico coal-fired power station. Centre-line deflection (under an unknown load), had to be limited to 18 mm. Because the beam had to effectively resist the corrosive effects of the station's emissions as well as the wide swings of climate at that location, vinyl-ester was the chosen matrix. Reinforcement comprised mat and woven rovings overlaid with a synthetic nexus veil.

4. Transport

Whilst many custom and standard profiles are employed to enhance the structural integrity of vehicle bodies—particularly refrigerated container lorries—and for alternative load accepting members such as wind

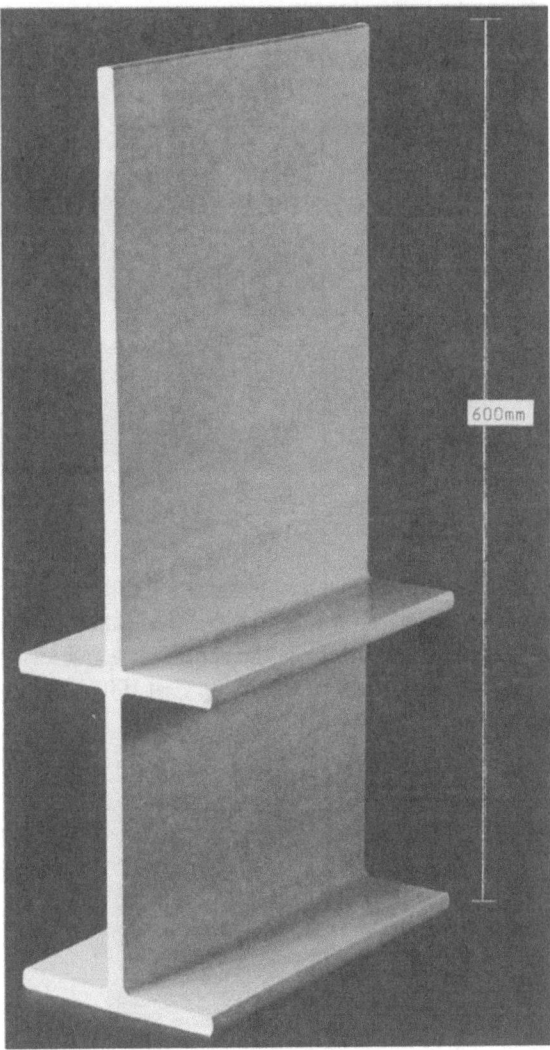

FIG. 11.

deflectors,[12] other transport applications are being secured. These cover rail, marine and aerospace, with the latter one area where for corrosion and weight reasons, the replacement of aluminium and its alloys is increasing. Figure 12, which illustrates a recently developed windshield support post for a prototype composite helicopter, also provides a necessary carbon-fibre example. This 1·5-m long trapezoidal section at a wall thickness of just

Trevor F. Starr

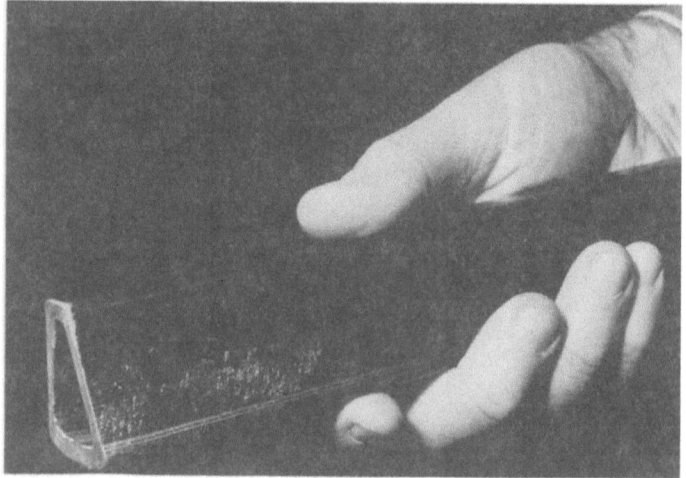

FIG. 12.

over 3 mm, runs from the nose-wheel to the roof and has been designed to withstand a simulated crash-landing where that wheel strikes first. The matrix is vinyl-ester, which provides the required degree of fire resistance and temperature cycling from freezing to 70 °C at 100 % RH to be achieved without degradation.[13]

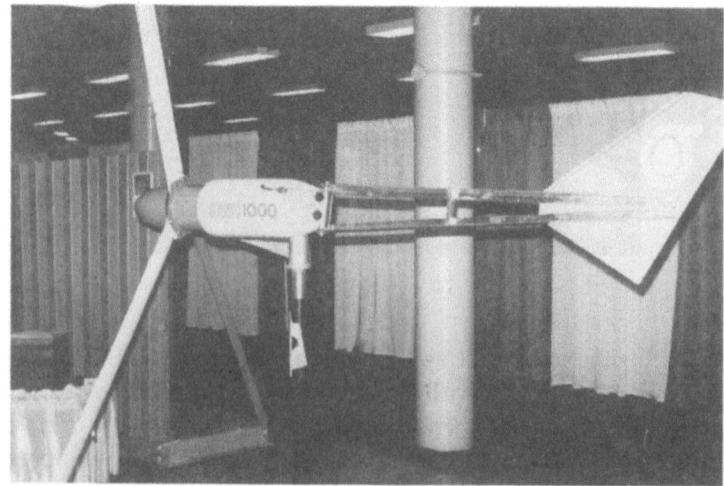

FIG. 13.

5. Other

Finally here are two examples—one which utilizes the full flexural characteristics of a pultrusion, the other their adaptability.

Energy generation from the wind is assuming growing importance and here again pultruded profiles are—and for helicopter blades—beginning to replace aluminium alloys for reasons of economy and performance. The generator shown in Fig. 13 employs three pultruded 1·2-m long blades offering variable pitch without moving parts. Pitch weights near the tip cause aerodynamic and centrifugal forces to act together to twist the blade to the optimum angle of incidence for each wind speed.[2] Compared with aluminium the blades have shown superior bearing stress resistance,

FIG. 14.

torsional modulus and reduced fatigue, to an extent which allows essentially unlimited life.

Figure 14 illustrates a demonstration section of an overhead monorail system where a 5:1 weight reduction has been achieved by extensively employing pultrusions. No lubrication is required and the system is therefore ideal for food processors and other applications where lubricant contamination is a problem. The light-weight makes installation easy and, being electrically non-conductive, safer and less hazardous in use.

CONCLUSIONS

Given an attempt to provide a really up-to-date picture might suggest a limited choice for the examples presented. That the reverse was true—with indeed a selection difficulty—once again confirms the health of the pultrusion industry. Whilst no specific United Kingdom examples were discussed, that market too is active. Additional and larger machines are being considered and although the maximum profile size remains lower than in, say, America, technology is otherwise equivalent. By 1990 there will be additional composite material manufacturing techniques, which although perhaps classified under other headings, will have based their development on the pultrusion process known today.

REFERENCES

1. Martin, J. and Sumerak, J., A review of the market for pultruded applications—and factors affecting its growth. *RP/C Institute Conf. Proc.*, **38th**, Soc. Plast. Ind. (1983), 6F, 1–5.
2. Bergey, M. and Anderson, R., Design and development of a pultruded FRP laminate to replace aluminium shape in a high stress high fatigue application. *RP/C Institute Conf. Proc.*, **38th**, Soc. Plast. Ind. (1983), 6D, 1–4.
3. Beck, D. E., New processes and prospects in pultrusion. *RP/C Institute Conf. Proc.*, **38th**, Soc. Plast. Ind. (1983), 6B, 1–4.
4. Florentine, R. A., Low-cost high performance composites—magnaweave reinforcements for pultrusion. *RP/C Institute Conf. Proc.*, **37th**, Soc. Plast. Ind. (1982), 10A, 1–3.
5. *Design Guide*, Creative Pultrusions, Inc. (1982) *et al.*
6. Starr, T. F., RP & housing: an examination. *RP/C Institute Conf. Proc.*, **38th**, Soc. Plast. Ind. (1983), 22C, 1–7.
7. For sale—GRP house & plant to produce it. *Reinforced Plastics*. MacDonald Publications of London Ltd, **26** (Feb. 1982), 58.

8. SPENCER, R. A. P., Pultrusion of glass reinforced polyester—a new approach in the U.K. *Reinf. Plast. Conf.* Brit. Plast. Fed. (1980), 135–141.

9. *Polyester Resins for Building,* Scott Bader Company Ltd (1978).

10. Mounting praise for RP cable trays. *RP/C Institute Conf. Product Showcase,* **38th,** Soc. Plast. Ind. (1983), 33.

11. Pultruded ladder, *European Plastics News,* IPC Ind. Press Ltd, **8** (Sept. 1981), 111.

12. PAUL, J. C. and SIMMONS, C., Pultrusion & RIM combined with computer-aided design in the development of an optimum add-on truck aerodynamic drag reducing device. *RP/C Institute Conf. Proc.,* **37th,** Soc. Plast. Ind. (1982), 10C, 1–11.

13. Copter post survives crash test with flying colours. *RP/C Institute Conf. Product Showcase,* **38th,** Soc. Plast. Ind. (1983), 7.

16

Large GRP Butterfly Valves

M. H. Bryan-Brown and D. M. Walker

Central Electricity Generating Board, South Western Region, Bedminster Down, Bristol BS13 8AN, England

ABSTRACT

At present the lives achieved by large cast-iron valves in the cooling water circuits of seaside power stations are often limited by corrosion. In order to investigate the potential of GRP as an alternative material to cast iron, two 60-in diameter butterfly valves were manufactured by Boving for installation at Fawley Power Station in 1980. Instrumented pressure tests to 4 bar pressure were performed on both valves and a material test programme has been carried out; it is found that operational strains and deflections should be satisfactory for a 30-year life. There will be an inspection of one valve in 1983.

1. INTRODUCTION

The lives of cast iron components in the cooling water (CW) circuits of seaside stations are often significantly reduced by a number of factors which include graphitic corrosion, localized corrosion due to mixed metals, pitting of surfaces, seizure of sealing surfaces during outages and brittle fracture.

The condenser inlet and outlet valves at Fawley and Pembroke Power Stations have been particularly affected by the above factors. Therefore, the CEGB chose to investigate the manufacture and installation of prototype 60-in diameter butterfly valves in GRP material, since there was a potential for significant long term savings.

Since the project was of a development nature, it was agreed that single tender action was acceptable. Boving and Co. Ltd were the manufacturers of the Fawley cast iron valves; they were interested in a GRP valve, and so, after a satisfactory tender, became responsible for the detailed design and manufacture of the valve.

2. FEASIBILITY STUDY

The first investigations were made in 1978, by the SW Region of the CEGB, into the feasibility of large GRP butterfly valve manufacture. These investigations resulted in an approach being made to Boving and Co. Ltd in October 1978 with a view to carrying out a feasibility study into the design of a 60-in diameter valve for Fawley Power Station.

Boving produced a preliminary design in April 1979. The main features of the design are shown on Fig. 1 and are as follows.

2.1. Disc
The seal is made of nitrile rubber and is based on the geometry for the current cast iron design. The GRP disc thickness of 190 mm ($7\frac{1}{2}$ in) is a

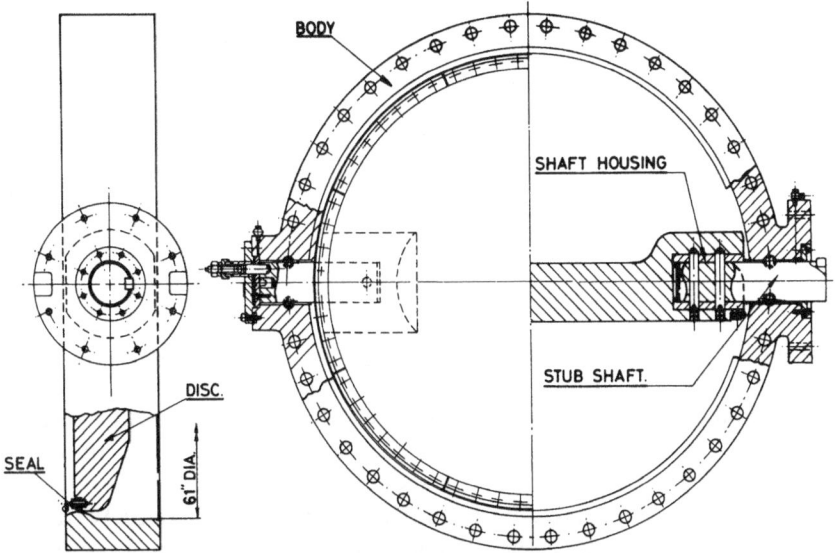

FIG. 1. GRP butterfly valve components.

compromise between that required for satisfactory disc deflection in the valve-closed position and that required for low hydraulic head loss; CEGB's requirement is for a nominal flow velocity of only 0·6 m/s (2 ft/s), whereas Boving are often involved with other high velocity installations where a thicker disc would be unacceptable. The disc is an all woven roving GRP construction, except for the seal clamping ring and bolts, the stub shafts and their housings, all of which are constructed from stainless steel. A woven roving and isophthalic resin system was chosen, being of a relatively high modulus, of proven durability and a reasonably economic form of composite material.

2.2. Body

The body is of a 'wafer' design, being clamped by long through-bolts between the adjacent pipe flanges. This design avoids the conventional flanges in the body, which could be subject to unknown system loads, and also provides a sufficient material thickness to limit the body's bending deflections due to pressure. The construction is, basically, of alternating layers of woven rovings and chopped strand mat; this type of construction has a higher bond strength than the all woven roving design, but a lower modulus. The resin was isophthalic.

The surface mating with the rubber seal on the disc is the moulded gel-coat, which should therefore avoid the corrosion or seizing mechanism which occurs in the cast iron valve. The flange mating with the actuator gearbox must be capable of withstanding the motor stall torque of 27 000 Nm (20 000 lb ft) and the bending moments due to the gearbox and actuator dead weight. Lifting lugs were omitted from the design as it was considered that the relatively low weight of the valve would enable it to be installed by a fork lift truck, rather than requiring a crane.

3. FINAL DESIGN

In June 1979, Boving submitted a quotation for the supply of 2 valves. The cost of the moulds for the body and the disc were amortized over 10 valves, and on this basis the valve unit cost was considered to be competitive with standard grey cast iron valves, and cheaper than valves in corrosion-resisting materials or supplied with special coatings.

Following recommendations from the CEGB, Boving selected W. & J. Tod Ltd of Weymouth as the sub-contractor for the manufacture of the GRP valve sections. An order was placed for the manufacture of the moulds for the body and the disc in June 1979, due to the 6 months' lead time required for these items.

Discussions took place in the period June to November 1979 between CEGB and Boving and a number of modifications to the preliminary design were finally agreed in November 1979, as follows.

(a) The stainless steel hexagonal inserts in the disc were set into the disc, and finally bored out (after the GRP moulding) to accept circular stub shafts. Connections between the hexagonal inserts and the stub shafts were by 2 radial pins at the drive end and 1 pin at the non-drive end (*see* Fig. 1).

(b) On the body, the drive flange thickness was increased to 76 mm (3 in), to cater for the motor stall torque.

(c) Stub shafts and hexagonal inserts to be martensitic stainless steel, BS 970, 431-S29.

(d) Seal ring to be 18/8/3 stainless steel, BS 970, 316-S16.

(e) A locking facility was included so that a valve could be locked in the closed or open positions as desired.

The cost of these modifications added 15 % to the valve unit cost, and an order for the two valves was placed by the CEGB in November 1979.

4. MANUFACTURE

The manufacture of the GRP sections of the first and second valves was completed in May 1980 and July 1980 respectively. Some problems arose on both valves due to the slight distortion of the body and the disc during moulding; this distortion only became evident after detailed and careful measurement by Boving's engineers. When manufacturing sections up to 190 mm ($7\frac{1}{2}$ in) thick, the exothermic reaction that occurs during moulding can cause significant thermal stress to be set up; in thick sections it is much more difficult for the heat to be dissipated than in thin sections. Polyester resin always shrinks during moulding and the presence of thermal effects could aggravate the shrinkage and cause the part to distort. On the first valve, the following distortions were measured:

(a) Body elliptical by 3·8 mm (0·150 in), larger at 90° to the shaft centreline.

(b) Body shaft bearing housings distorted by up to 0·4 mm (0·017 in) and tapered.

FIG. 2.   The completed valve.

(c)   Disc seal spigot slightly irregular by up to 0·75 mm (0·030 in) on diameter.

(d)   Disc distorted so that at 90° to the shaft centreline the seal seating was 2·5 mm (0·100 in) low with the flat face uppermost.

The only corrective action required by Tod's was to build up the shaft bores, item (b) above, for subsequent remachining to correct tolerances. In order to obtain satisfactory clearance between the body and the disc at 90° to the shaft centreline, due predominantly to item (a) above, the blade spigot was built up with PVC tape to a maximum thickness of 1·5 mm (0·060 in).

The distortions in the second valve manufactured were significantly less than in the first valve. The body elliptical distortion, item (a) above, was reduced to 2·3 mm (0·090 in), and the other points required no corrective action.

The drilling of the $36 \times 1\frac{1}{2}$ in diameter holes through the body proved to be a very lengthy manufacturing process. The drill had to be withdrawn every $\frac{1}{4}$ in and the hole blown out before proceeding, and the drill life before regrinding was short.

The manufacture and assembly in all other respects proceeded satisfactorily, and the final fabrication is seen in Fig. 2.

5. MATERIAL TESTS

The valve geometry had been designed to material properties previously found to be typical for the two laminate types, based on the chosen resin (Scott Bader 625) and glass cloths.

The valve body was constructed with alternating layers of woven roving and chopped strand mat. For this material, a modulus of $10\,000\,\text{N/mm}^2$ and a strength of $145\,\text{N/mm}^2$ were assumed (*see* Table 1). The valve blade on the other hand had been constructed from all woven roving, because of its higher modulus, which was expected to be $14\,500\,\text{N/mm}^2$. Its anticipated strength was $215\,\text{N/mm}^2$.

TABLE 1
Material test results
(a) *Valve body material*
(Average results of eight separate test boards from both bodies)

		Design assumption	BS 4994 level	Measured average
Glass, % by weight		37	—	40
Tensile modulus	1	10 000	—	10 570
Tensile strength	1	145	—	159
Unit modulus	2	18 200	14 700	17 110
Unit strength	2	267	258	255
Shear strength	1	7	7	9·4

Units: 1. N/mm^2
2. N/mm/kg/m^2 glass

(b) *Valve blade material*
(Average results of eight boards, four from each blade)

		Design assumption	BS 4994 level	Measured average
Glass, % by weight		48	—	48
Tensile modulus	1	14 500	—	13 450
Tensile strength	1	215	—	214
Unit modulus	2	19 000	16 200	18 110
Unit strength	2	280	300	287
Shear strength	1	7	7	12·3

Units: 1. N/mm^2
2. N/mm/kg/m^2 glass

Test laminates were laid alongside the main components during the course of their construction. A selection of these were cut into specimens for strength, modulus, shear and glass content tests. The results of these tests are summarized in Table 1.

It can be seen from the tabulated results that the measured moduli were reasonably close to those assumed for design, as were the strengths. The higher values for the all woven roving construction (the blade material) were confirmed but the expected sacrifice in shear strength was not apparent.

The BS 4994 levels were those that were originally put forward as the minimum acceptable, but are currently being revised. The unit strength level for WR will be lowered to 250 N/mm/kg/m^2 glass.

Some other tests were carried out to confirm that the in-plane swelling of the woven roving laminates in water was not significant. Jamming of the blade might have otherwise been a possibility. The mean increase in length in water at 40 °C for one year was found to be only 0·26 %, and had virtually stabilized at this level.

6. PRESSURE TESTING

Machining, final assembly and pressure testing of the two valves manufactured to date was carried out at Brit Engineering Ltd, Bridport, in October and November 1980 respectively.

The opportunity was taken of instrumenting both of the valves prior to the pressure tests and the instrumentation fitted was agreed between the CEGB and Boving. Figure 2 shows the first valve assembly at the time of its pressure test.

6.1. Details of Instrumentation

On the first valve to be tested in October 1980, forty strain gauge channels were used for the valve disc. Full details of their locations are found in Ref. 1. The positioning of the gauges was intended to measure the following:

(a) Maximum strains on the disc flat surfaces.
(b) Check on theoretical strains for a homogeneous material.
(c) Interlaminar shear strain.
(d) Strains in stress concentrations.
(e) Symmetry between the strains on the two sides of the disc.

The instrumentation fitted to the second valve to be tested in November 1980 was intended to complement the results obtained from the first test, rather than simply to repeat the first test. Again, the details of the strain gauge locations are found in Ref. 1; 32 channels were used on the disc and in addition, two strain gauges were fitted to the outside of the body. The positioning of the strain gauges was intended to measure the following:

(a) More complete data in stress concentration regions.
(b) Check on strains on disc 'wings'.
(c) Interlaminar shear strains at different locations.
(d) Principal stresses at 3 locations where only partial strain data was obtained from the first test.
(e) Maximum strains on the valve body.

Disc deflections were measured using a beam, supported on 2 fulcrums, and 4 clock gauges.

7. RESULTS

Inspection of the measured strains and deflections showed that the results varied linearly with pressure. Also, strains and deflections at a given pressure in the reverse-sealing direction tests were equal and opposite to those in the normal-sealing direction tests. Hence all the calculations of stresses and comparisons with theoretical levels have been based on a pressure of 4·1 bar (60 lb/in^2) from the normal sealing direction, which was the maximum reached in both the first and second valve tests.

7.1. Disc Stresses

A full presentation of the strain gauge results is given elsewhere.[1] Stresses at all points on the valve discs measured have been calculated from the strain gauge results using the conventional equations,[2] and assuming the material to be homogeneous; these are presented in Table 2 for the first valve test and in Table 3 for the second valve test. The Appendix presents the theoretical stress analysis of a valve disc, assuming constant thickness and homogeneous material properties; also presented in Tables 2 and 3 are the theoretical stresses obtained from this analysis, and the comparisons between the measured and the theoretical stresses, where applicable. Also the measured and theoretical stresses are presented graphically in Figs 3–6.

It was found that, on areas of the disc away from stress concentrations, there was agreement between the measured and theoretical stresses to

TABLE 2
First valve test disc stresses
Pressure = 60 lb/in²
Young's modulus = 1·95 × 10⁶ lb/in²

Location	θ° Measured from shaft centre line	Stress direction	Measured stress (lb/in²)	Calculated stress (lb/in²)	Ratio calculated/measured stress
Centre (pressure face)	—	Max princ	+462	+585	1·27
		Min princ	−2 547	−2 840	1·12
		Max shear	1 504	1 713	1·14
Centre (non-pressure face)	—	Max princ	+2 742	+2 840	1·04
		Min princ	−563	−585	1·04
		Max shear	1 653	1 713	1·04
Disc perimeter (non-pressure face)	0	Tangential	−1 353	−2 950	2·18
	22	Tangential	−263	−500	1·90
	45	Tangential	+1 047	+1 550	1·48
	66	Tangential	+1 425	+1 650	1·16
	90	Tangential	+1 977	+2 300	1·16
18⅝ in Radius (non-pressure face)	0	Tangential	−1 332	−1 671	1·25
		Radial	+2 438	+2 089	0·86
		Max shear	1 885		
	22	Tangential	−291	−320	1·10
		Radial	+938	+1 380	1·47
	45	Tangential	+1 214	+1 380	1·14
		Radial	−15	+390	
	66	Tangential	+1 914	+2 050	1·07
		Radial	−240	0	—
	90	Tangential	+2 216	+2 450	1·11
		Radial	−364	−167	0·46

14 in Radius (pressure face)	90	Tangential	−2 572	−2 607	1·01
(non-pressure face)	90	Tangential	+2 685	+2 607	0·97
Shaft housing fillet radius	90	Radial	−5 470		
Shaft housing	90	Tangential	+1 034		
		Radial	−159		
Shaft housing fillet radius	17	Max princ	+1 577		
		Min princ	−2 901		
		Max shear	2 239		
		Tangential	+416	+724	1·74
		Radial	−1 740	−1 743	1·00
17 in Radius (non-pressure face)	45	Max princ	+461		
		Min princ	−1 740		
		Max shear	1 178		
		Tangential	−992	−1 448	1·46
		Radial	−132	−251	1·90
Disc edge	12	Interlaminar shear	610		
Disc wing 27 in radius	90	Tangential	−985	−2 340	2·38

TABLE 3
Second valve test disc stresses
Pressure $= 60\,\text{lb/in}^2$
Young's modulus $= 1.95 \times 10^6\,\text{lb/in}^2$

Location	$\theta°$ Measured from shaft centre line	Stress direction	Measured stress (lb/in^2)	Calculated stress (lb/in^2)	Ratio calculated/measured stress
Centre (non-pressure face)	—	Max princ	$+2\,783$	$+2\,840$	1·02
		Min princ	-595	-585	0·98
		Max shear	$1\,689$	$1\,713$	1·01
$18\frac{5}{8}$ in Radius (non-pressure face)	22	Tangential	-396	-320	0·82
		Radial	$+872$	$+1\,380$	1·58
		Max shear	$1\,571$		
	45	Tangential	$+1\,103$	$+1\,380$	1·25
		Radial	$+129$	$+390$	3·02
		Max shear	$1\,348$		
	66	Tangential	$+1\,872$	$+2\,025$	1·10
		Radial	-191	-20	
		Max shear	$1\,285$		
Disc perimeter (non-pressure face)	0	Tangential	$-2\,679$	$-2\,950$	1·10
		Radial	-22	$+624$	
		Max shear	$1\,329$		

Disc perimeter (pressure face)	12	Tangential	+7 370	+2 300	0·31
		Radial	+158		
		Max shear	3 609		
	0	Tangential	+1 668	+3 090	1·85
		Radial	−193	−420	2·18
		Max shear	930		
Shaft housing fillet radius	0	Tangential	+125	+1 392	
		Radial	−6 746	−2 256	
		Max shear	3 436		
	17	Tangential	+2 238	+613	
		Radial	−2 270	−1 727	
		Max shear	2 639		
Disc wing 27 in radius	90	Tangential	−833	−2 340	
		Radial	+181	+28	
		Max shear	507		
Disc edge	12	Interlaminar shear	550		
		Interlaminar shear	450		

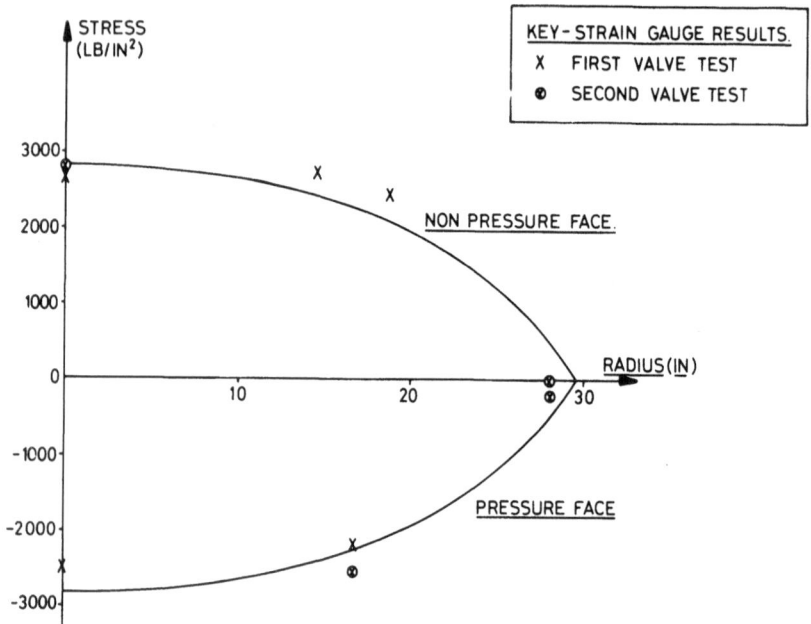

FIG. 3.   Radial stress, $\theta = 0$.

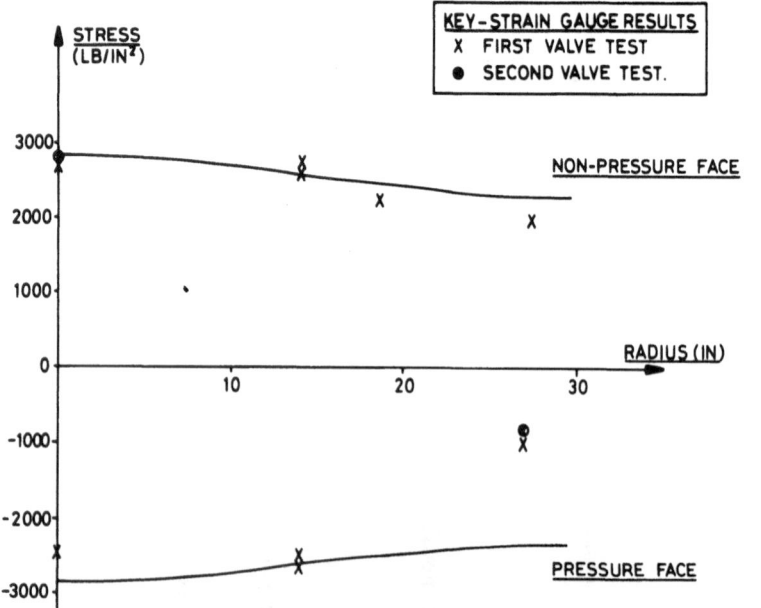

FIG. 4.   Tangential stresses, $\theta = 90°$.

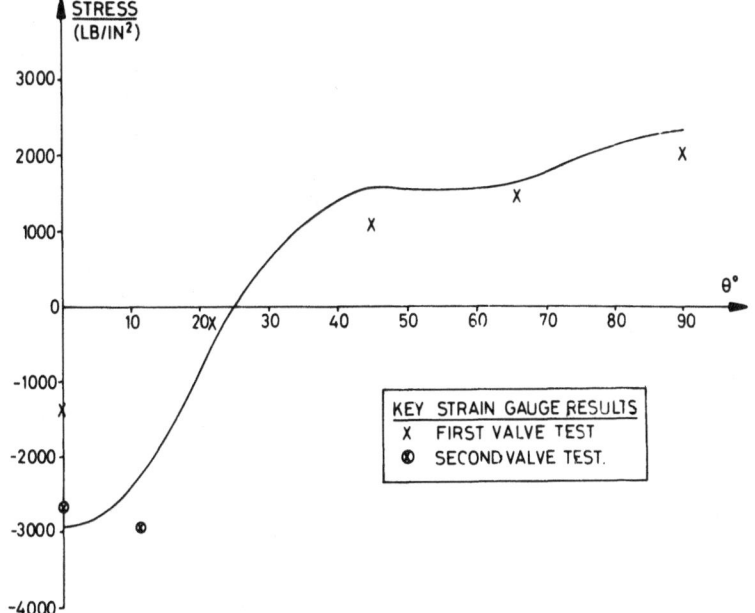

FIG. 5.   Tangential stress at rim $r/a = 0.93$ (non-pressure face).

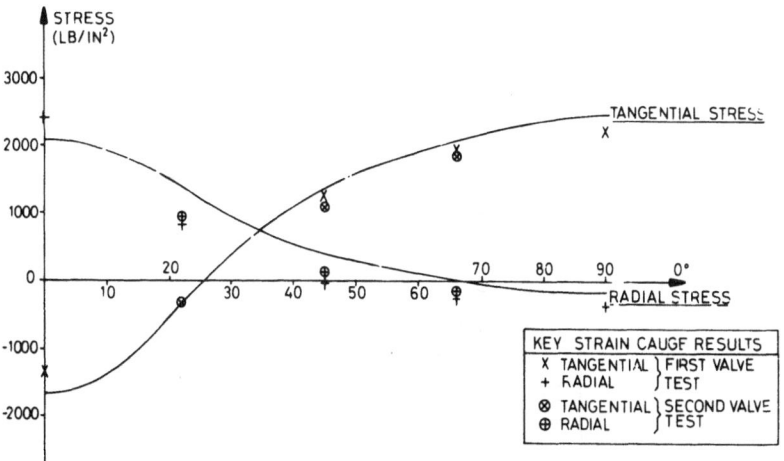

FIG. 6.   Stresses at $18\frac{5}{8}$ in radius, $r/a = 0.63$.

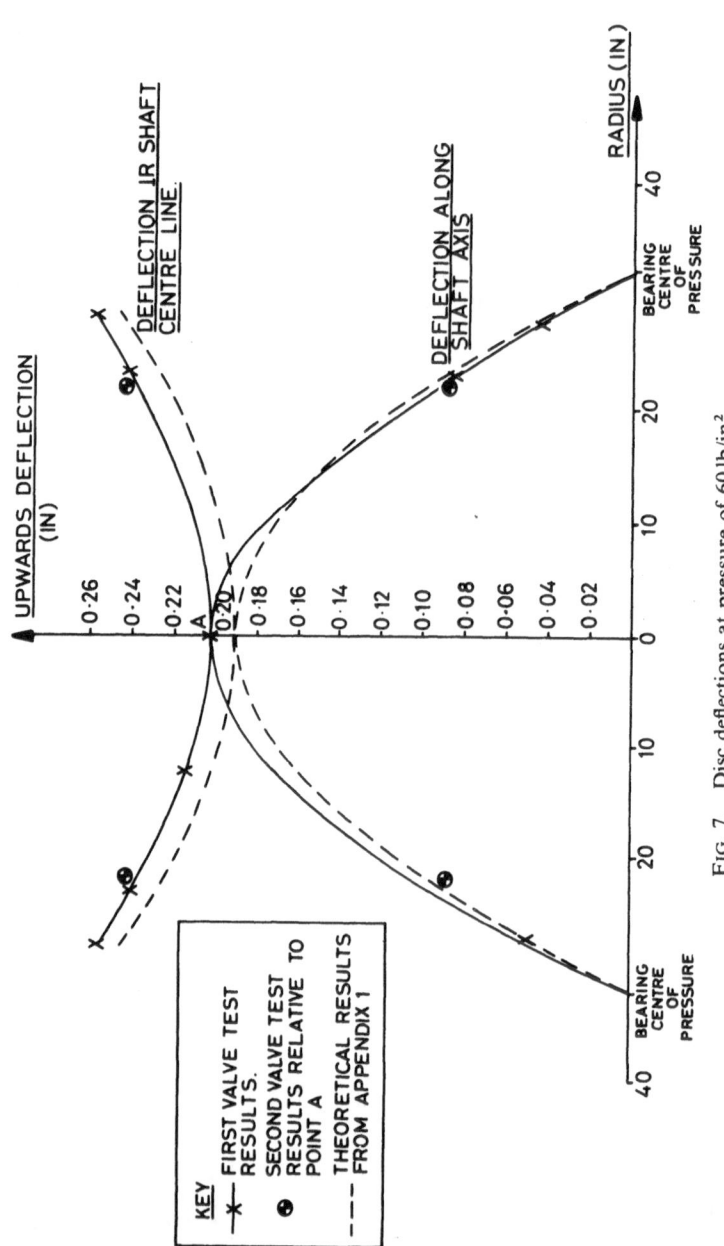

FIG. 7. Disc deflections at pressure of 60 lb/in^2.

within $\pm 14\%$. Also, there was agreement between the measured and theoretical disc deflections to within $\pm 5\%$. Thus, even though the theoretical analysis is based on homogeneous material properties, and a constant thickness disc, it may be used with confidence at the design stage for thickness calculations.

The fillet radius at the disc rim adjacent to the stub shaft housing produced the highest measured strain, $+0\cdot38\%$, equivalent to $+0\cdot19\%$ at the maximum operating pressure of 2 bar (30 psi). It is seen that this is less than the BS 4994 criterion of $0\cdot2\%$ strain. Away from stress concentrations the maximum operational stress occurred at the disc centre and was $9\cdot5$ MN/m^2 (1370 lb/in^2), giving a Reserve Factor of 22 on the material tensile strength.

The strains produced in the disc due to the maximum torque applied through the gearbox were found to be small, less than 100×10^{-6}, so that there should be no operational problems due to this cause. Tables 2 and 3 present the interlaminar shear stresses, calculated from the strain gauge results by assuming Young's Modulus/Shear Modulus = 8. Comparison of the theoretical and measured shear strains has shown this value to be appropriate for the all woven roving construction of the disc.

7.2. Disc Deflections

The measured valve disc deflections for the first valve test are presented graphically in Fig. 7 for the pressure loading from the normal-sealing direction. The disc deflection curve was obtained from the clock gauge readings up to a 711 mm (28 in) radius, and the curve up to the bearing centres of pressure was obtained by extrapolation. The limited results of the disc deflections obtained for the second valve are given as spot points in Fig. 7 for comparison with the first valve deflections.

The Appendix presents a theoretical analysis for the disc deflections, assuming homogeneous material properties and constant thickness; the results of this analysis are shown on Fig. 7 for comparison with the measured deflections.

7.3. Valve Body Deflections and Stresses

The valve body diametral deflections were measured during the first valve test and the results are presented in Table 4; it is seen that in a direction perpendicular to the shaft centreline, the diametral growth of 2 mm (0·078 in) was over $2\frac{1}{2}$ times the growth of 0·75 mm (0·030 in) along the shaft axis.

TABLE 4
Measured diametral growths of valve body

Pressure (psi)	Shaft centreline		Along shaft centreline	
	Dia (in)	Growth (in)	Dia (in)	Growth (in)
0	60·037	—	59·950	—
20	60·054	0·017	59·960	0·010
30	60·068	0·031		
40	60·085	0·048	59·970	0·020
48	60·095	0·058		
60	60·115	0·078	59·980	0·030

8. DISCUSSION

Experience to date suggests that the design concept employed is satisfactory for GRP butterfly valves up to a diameter of 60 in. The distortions produced during manufacture were capable of rectification; however, it should be noted that W.&J. Tod is a firm well used to the manufacture of GRP products to high standards. It is not known how many manufacturers in the country are capable of achieving the required degree of accuracy. Also, Boving and Co. Ltd have a major interest in developing corrosion resistant valves, cheaper than the current valve designs, because of their worldwide interests, particularly in the Middle East. Moreover, Boving do not have their own manufacturing facilities, but sub-contract manufacture of their valves, and are therefore in a position to sub-contract work to a GRP manufacturer of proven ability. Other valve manufacturers with their own plant are less likely to promote the sales of GRP valves which they themselves would not be in a position to manufacture. In order to obtain truly competitive quotes from different manufacturers, it would be necessary to tighten up the specification, in order to reflect the required tolerances and other factors discussed in this report. A significant cost-saving design modification that could be investigated for the future would be the use of preformed holes in the body, instead of drilling the solid body, as carried out to date; this would probably require some development to find the optimum manufacturing techniques.

The pressure tests on the two valves showed that the valve structure should be satisfactory for extended service under all normal operating conditions. To date, the first installed valve has performed satisfactorily for 3 years, but a more extended period of operation is required before definite conclusions can be drawn. This valve will be inspected in Summer 1983, when the economic argument for the GRP valve should become clearer.

Tests on the GRP materials used in both blades and bodies showed that the materials were close to having the properties assumed in the design. The modulus of the blade material (all woven roving) was required to be high so that deflections at the sealing edge would remain small (without having an unduly thick profile with consequential hydraulic drag). The measured blade modulus was $13\,450\,N/mm^2$ (2×10^6 psi). An all woven roving construction is said to have a lower inter-laminar shear strength, but in fact the measurements made showed the blade construction to have a higher shear strength than the valve body. Interlaminar strength is, nevertheless, probably the most vulnerable property of the blade construction and design calculations suggest a minimum safety margin in this mode.

Future developments of the 60-in diameter GRP valve are clearly towards the investigation of large diameter valves and of their simplified construction; future CEGB policy indicates a predominance of $1\cdot6\,m$ and $2\cdot2\,m$ diameter valves. Using the same construction method as for the 60-in valve would require disc and body thicknesses scaled up linearly; however, it is clear from the experience to date that such thicknesses might result in unacceptable distortion during manufacture. Smaller valves can be made more economically from thermoplastic materials,[3] or by using different design concepts, e.g. diaphragm valves. The feasibility of using hot pressed GRP valves up to 300 mm diameter is also being investigated by the CEGB. Other manufacturing techniques, such as the use of a pre-formed core for the centre of the disc, might be required to overcome this distortion problem. An alternative blade design comprising two flat plates joined by an open-webbed central section is one way of increasing stiffness without increasing the hydraulic drag. Although this would reduce the thickness of the GRP sections (and the consequent exothermic distortion) it would be difficult to manufacture. The joining of the plates and webs would be a very vulnerable feature.

The prices given below are based on Boving's figures for June 1981, and refer to the price per 60-in diameter valve on an 'apples to apples' basis:

	£
Grey cast iron	10 000
Coated grey cast iron	12 000
GRP	14 000
Corrosion resistant (Niresist or bronze)	20 000

Hence, on this comparison, the GRP valve would seem to be up to 16 % more expensive than its obvious competitor, the coated cast iron valve. It is notable that of the £14 000 GRP valve price, only £4200 (30 %) is attributable to the cost of the GRP; hence the scope for cutting the price by alteration to the GRP components would seem to be limited. On the other hand, there would seem to be some scope for cost cutting in the other manufacturing operations, in particular the hole drilling mentioned above.

Currently, cast-iron valves on seaside stations often require refurbishing every 2–4 years during the Unit major overhauls. Inspection of the GRP valve on Unit 1 in 1983 should indicate whether any significant refurbishing of the GRP valve is necessary after 3 years' operation; the valve has been located in the outlet position which has the more arduous duty due to turbulent flow with the outlet valve throttled, and the Unit has been 2-shifting in general, involving a significant degree of pressure cycling. If the GRP valve is seen to have a potential for providing much longer lives than the current cast iron valves, then the economic assessment of reduced maintenance costs versus higher first cost of the valve can be made. At the present, however, evidence of the condition of GRP after 4 years' service at Fawley was seen in Summer 1981 during the internal inspection of an inlet GRP condenser waterbox; it was found that the internal surface was in excellent condition and that the gel coat thickness had not been significantly thinned over the 4-year period. However, turbulence and consequent material erosion is likely to be more severe in the valve than in the waterbox, but can be readily recovered by a local repair. During the first 2 years in service, each valve has carried out approximately 450 operational cycles.

9. CONCLUSIONS

Successful design, fabrication, test and installation of two 60-in diameter GRP condenser butterfly valves have been achieved, and service experience after 3 years has been satisfactory.

At 1981 prices, the GRP valve is estimated to be only 16 % more expensive than a comparable coated cast iron valve, and 30 % cheaper than a metallic corrosion resistant valve.

The long-term maintenance benefits of the GRP valve will be reviewed in 1983 but it is anticipated that these will outweigh the higher initial installation cost.

There are future opportunities for established valve and GRP manufacturers to exploit the use of the GRP for large diameter butterfly valves.

There is the need for high quality fabrications to produce a satisfactory valve assembly.

ACKNOWLEDGEMENTS

Thanks are due to Boving and Co. Ltd, with whose agreement this paper is published. Also the help of the staff at Fawley Power Station and other colleagues in the CEGB is gratefully acknowledged. We also acknowledge the permission to publish this paper, given by Mr D. A. Pask, the Director-General of the South Western Region, CEGB.

REFERENCES

1. BRYAN-BROWN, M. H., *Structural Analysis of 60 inch Diameter Valves in GRP Material*, CEGB Report SSD/SW/81/N29, July 1981.
2. BRUEL and KJAER, Technical Publication, October 1975.
3. SULLY, S., *Non-metallic Valves for Auxiliary Cooling Systems*, CEGB Report SSD/SW/82/N164, 1982.
4. TIMOSHENKO, S. and WOINOWSKY-KRIEGER, S., *Theory of Plates and Shells*, McGraw-Hill, 1959.

APPENDIX: STRESS ANALYSIS OF VALVE DISC WITH HOMOGENEOUS MATERIAL PROPERTIES

The fundamental equations for the stress analysis of a constant thickness disc, supported at two diametrically opposite points on the perimeter, under constant pressure loading, are given in Ref. 4. These equations have been used in Ref. 1 to give the theoretical bending moments and deflections for the butterfly-valve disc. Homogeneous material properties are assumed, and polar coordinates are used. The relationships derived are as follows:

$$\frac{M_r}{pa^2} = \frac{(1-\rho^2)}{(3+v)}\left[\frac{(3+v)^2}{16} + \cos 2\theta + \frac{3-v}{4}\cdot\cos 4\theta\cdot\rho^2 + \frac{2-v}{3}\cdot\cos 6\theta\rho^4\right]$$

$$\frac{M_t}{pa^2} = \frac{1}{12(3+v)}\ [\tfrac{3}{4}(3+v)[(3+v)-\rho^2(1+3v)] - 12\cos 2\theta(1+v\rho^2)$$

$$- 3\cos 4\theta\rho^2[(3-v)-(1-3v)\rho^2][-4\cos 6\theta\rho^2[(2-v)-(1-2v)\rho^2]]$$

where M_r and M_t = radial and tangential bending moments per unit length, respectively;

p = water pressure on disc;

v = Poisson's ratio;

a = disc radius;

r, θ = polar co-ordinates;

$\rho = r/a$.

These expressions have been evaluated for $v = 0.18$, for varying values of ρ and θ, and the results are shown, for comparison with the strain gauge results, in Figs 3–6.

The disc transverse deflection, w, is given by the expression:

$$\frac{2w \cdot D(3 + v)}{pa^4} = (2 \log_e 2 - 1) + \frac{1 + v}{1 - v}\left(2 \log_e 2 - \frac{\pi^2}{12}\right)$$

$$- \rho^2 \cos 2\theta \left(\frac{1}{2} + \frac{1}{2}\frac{1 + v}{1 - v} - \frac{\rho^2}{6}\right)$$

$$- \rho^4 \cos 4\theta \left(\frac{1}{12} + \frac{1}{24}\frac{1 + v}{1 - v} - \frac{\rho^2}{20}\right)$$

$$- \rho^6 \cos 6\theta \left(\frac{1}{30} + \frac{1}{90}\frac{1 + v}{1 - v} - \frac{\rho^2}{42}\right)$$

where

$$D = \text{disc flexural rigidity} = \frac{Et^3}{12(1 - v^2)}$$

This expression has been evaluated for the 60-in diameter disc, and the results are presented in Fig. 7. The following data were used:

$E = 1.95 \times 10^6 \, \text{lb/in}^2$;

$v = 0.18$;

$t = 7.5 \, \text{in}$;

$p = 60 \, \text{lb/in}^2$;

$a = 29.5 \, \text{in}$.

17

Thermal Control of Tubular Composite Structures in Space Environment

ROBERT D. KARAM

Fairchild Space Company,
Germantown, Maryland 20874-1181, USA

ABSTRACT

Thermal control of spacecraft tubular composites is discussed. The equations used to calculate orbital temperatures are presented with a description of the design techniques which limit temperature excursions and associated distortions. Laminate fiber orientation is related to heat transfer characteristics, and it is shown that orientations selected to yield high axial strength and least thermal deformations will generally lead to excessive fin effect heat losses. A perforation procedure is proposed to eliminate in-plane deflections in a space environment independently of laminate construction. The lamination sequence may then be optimized to meet strength and heat loss requirements.

INTRODUCTION

The need for very stable spacecraft support structures is becoming more evident as stricter accuracy requirements are introduced for pointing sensors, telescopes, and antennas. Current specifications for communications satellites cite dimensional alignments of less than one arc second, and even tighter control can be projected for the near future.

A prevailing technique for limiting thermal distortions is to provide massive mounting platforms in which potential orbital variations in temperature are dampened by thermostatically controlled heaters. The heater duty cycle is adjusted automatically to compensate for changes in

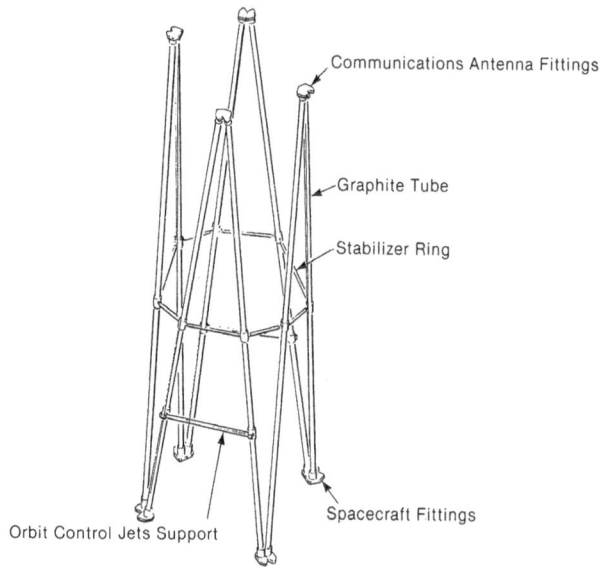

Communications Antenna Fittings

Graphite Tube

Stabilizer Ring

Spacecraft Fittings

Orbit Control Jets Support

FIG. 1.   ATS-6 truss assembly.

the environmental flux inputs. The advantage in this approach lies in its engineering simplicity. The disadvantages include heavy structure and performance unpredictability associated with transient response and the existence of a 'dead band' in the neighborhood of the thermostat set point. Considerable enhancement of the situation may be realized by resorting to structural materials which exhibit low coefficients of thermal expansion. Composites have been used on a number of spacecraft with satisfactory results. ATS-6 (Ref. 1) employed a truss structure consisting of eight graphite reinforced plastic tubes, each 4·32 m long and 6·35 cm in diameter, to support a 9·14 m diameter parabolic reflector (Figs 1 and 2). The spacecraft was NASA's final satellite in the Applications Technology Satellite program and was designed and built by Fairchild Space and Electronics Company, Germantown, Maryland.

ATS-6 laminate orientation was optimized to obtain a high axial modulus while maintaining low hoop and longitudinal coefficients of thermal expansion.[2] Since conduction heat transfer along the graphite fibers was high, the ply arrangement presented a potentially serious radiating fin leakage from the spacecraft electronics compartment. The problem was alleviated by wrapping multilayer insulation around the tubes. However, the resulting increase in the effective diameter of the truss system led to increases in signal blockage between communications

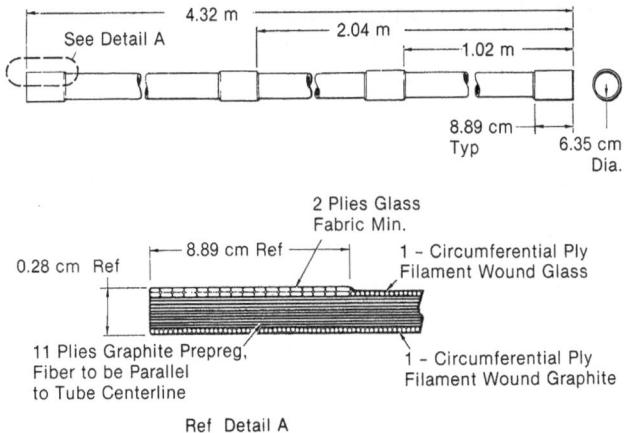

FIG. 2.   ATS-6 graphite composite tube.

elements on the main body of the spacecraft and the reflector. This application suggested a necessity for determining a laminate construction that offers an optimum balance among the thermal and mechanical parameters which define the design.

In this paper the methods of spacecraft temperature control are applied to tubular composites with anisotropy of the material included in the formulation and solution of the energy equations. Studies are conducted in relating fiber orientation to strength, heat transfer, and thermal distortion requirements. It is found that heat losses and thermal distortions can be reduced significantly by resorting to low absorptance and low emittance surface coatings. In addition, a structure perforation technique can be devised to eliminate thermal distortions in a solar environment independently of ply orientation. The concept is based on the knowledge that symmetric temperature profiles (and, therefore, negligible thermal moments) can be obtained by proper selection of surface treatment and fraction of surface perforated. Strength and heat conduction limitations are then considered as being nearly independent of the resulting coefficients of expansion.

ORBITAL TEMPERATURE CONTROL

General Considerations

Extended structures in spacecraft are generally exposed to both ultraviolet heating (such as solar and reflected solar flux) and infrared heating (from planetary and adjoining sources). In most cases, flux impingement on a tube is asymmetric along the direction of the source and causes thermal gradients as well as changes in the level of temperature. For conventional materials, moderate changes in the average temperature of an unrestrained tubular structure lead to uniform extension and contraction without significant distortion in shape. Hence the element will remain approximately straight until thermal moments are created due to circumferential and longitudinal variations in temperature. In the case of a composite body, the dimensional thermal response of the matrix material is considerably higher than that of the fibers, and therefore in unidirectional composites distortion in the transverse direction far exceeds that in the longitudinal direction thus hindering free deformation. For this reason the thermal design of spacecraft composite structures must include the necessary provisions to limit the temperature variations within the body as well as orbital deviations from the average temperature at which final ground assembly and acceptance are completed.

Average Temperature

Temperature calculations for an orbiting structure usually begin by considering steady state conditions in 'hot case' and 'cold case' environments. A definition for hot case may be the position in orbit when there is maximum impingement of solar, albedo, and earth radiation. In this situation the average temperature of a cylinder can be defined from the equation

$$\sigma T_{\text{avg}}^4(\text{max}) = \frac{1}{2\pi} \int_0^{2\pi} \left\{ \frac{\alpha}{\varepsilon} \left[S(\theta) + A(\theta) \right] + E(\theta) \right\} d\theta \qquad (1)$$

where $\sigma =$ Stefan–Boltzmann constant ($5 \cdot 668 \times 10^{-8}$ W/m^2 K^4); $T_{\text{avg}} =$ average temperature (K); $\alpha =$ solar absorptance; $\varepsilon =$ infrared emittance; $S =$ solar flux (W/m^2); $A =$ albedo (W/m^2); $E =$ earth flux (W/m^2) and $\theta =$ circumferential position on the cylinder.

A possible cold case may occur when the cylinder is shaded by earth and exposed to earth flux only. Then

$$\sigma T_{\text{avg}}^4(\text{min}) = \frac{1}{2\pi} \int_0^{2\pi} E(\theta) d\theta \qquad (2)$$

Hence for a pre-determined orbit in which the impinging fluxes are known, a desired range of maximum average temperature can be achieved if the exterior surface of the cylinder is treated to yield an absorptance to emittance ratio in accordance with eqn. (1). However, as may be seen from

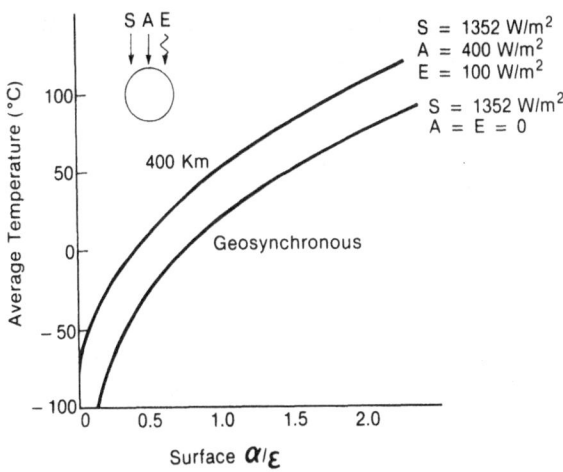

FIG. 3. Hot case average temperature of orbiting cylinder.

eqn. (2), the minimum temperature could depend only on the intensity of the source and an arbitrarily assigned value of temperature will not in general be obtained without resorting tolspecialized heating techniques. One relatively simple and inexpensive design features a heater wire wrapped over a low emittance structure. The low emissivity results in moderate heater power consumption.

Figure 3 shows the relationship between surface properties and temperature for two hot case orbits. As an example, the average temperature in sunlight at geosynchronous altitude (35 800 km) is about 22 °C when a surface coating having α/ε of one is used. The decrease from this value during equinox, when there is nearly zero flux input, can be controlled and limited by utilizing heater power as given in Fig. 4. If both the

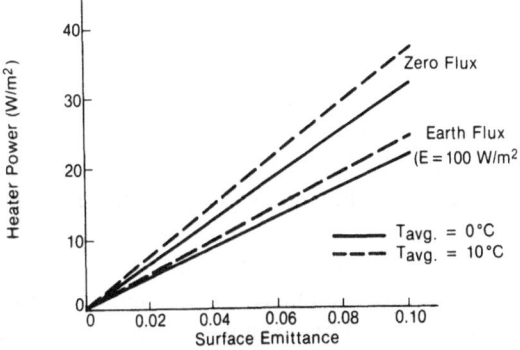

FIG. 4. Cold case heater power requirements.

absorptance and emittance can be selected to be 0·10, it can be shown that in geosynchronous orbit the temperature remains between 0 °C and 65 °C throughout the orbit if 31·5 W were made available. This range in temperature is within the values experienced during ground testing of composite structures.[2] Hence the complex and unpredictable deformation associated with changes in average temperatures can be nearly eliminated by adequate thermal design.

Temperature Gradients

Circumferential temperature variations are caused by the directional nature of fluxes in space. For long ($D/L \ll 1$) thin tubular structures of the type commonly used in spacecraft, radial conduction and longitudinal temperature changes may be neglected and the steady state energy equation is written[3]

$$\frac{d^2 T}{d\theta^2} + \frac{\varepsilon_o r^2}{k_\theta t} \left[\frac{\alpha_o}{\varepsilon_o} \{S(\theta) + A(\theta)\} + E(\theta) \right.$$

$$\left. - \left(1 + \frac{\varepsilon_i}{\varepsilon_o}\right) \sigma T^4 + \frac{\varepsilon_i}{4\varepsilon_o} \int_\theta^{\theta + 2\pi} \sigma T^4 \sin\left(\frac{\theta' - \theta}{2}\right) d\theta \right] = 0 \qquad (3)$$

with periodicity conditions

$$T(\theta) = T(\theta + 2\pi) \qquad (4)$$

$$\left.\frac{dT}{d\theta}\right|_\theta = \left.\frac{dT}{d\theta}\right|_{\theta + 2\pi} \qquad (5)$$

Here r is radius, k_θ is circumferential thermal conductivity, t is cylinder wall thickness, and the subscripts o and i refer to outer and inner surface, respectively.

Equation (3) with boundary conditions (4) and (5) is non-linear and a solution is not known which can be expressed in integrals of single variables. Approximate and numerical solutions can be readily obtained[4,5] for known functions $S(\theta)$, $A(\theta)$, and $E(\theta)$. Figures 5 and 6 were generated using the approximation techniques discussed in Ref. 5. The figures illustrate the dependence of maximum temperature difference across a cylinder on controllable parameters such as surface properties and circumferential conductance. It is significant to note the decrease in gradients with increasing conductance and internal emittance. Also, comparison between Fig. 5 and Fig. 6 shows that temperature differentials increase with increasing imbalance in the energy distribution. These

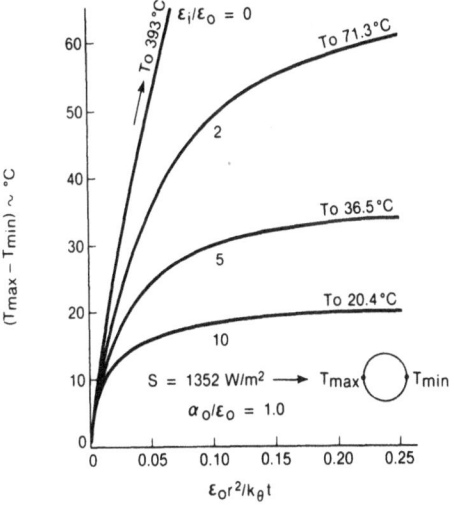

FIG. 5. Maximum gradient in sun ($\alpha_o/\varepsilon_o = 1\cdot 0$).

observations can be generalized and apply in an environment where infrared as well as ultraviolet heating exists.

Thermal Distortion

Distortion associated with circumferential temperature differentials is calculated by translating the temperature profiles into thermal bending moments and solving the differential equation of the ensuing deflection

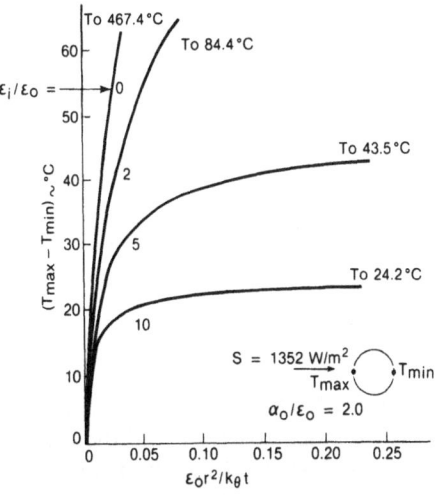

FIG. 6. Maximum gradient in sun ($\alpha_o/\varepsilon_o = 2\cdot 0$).

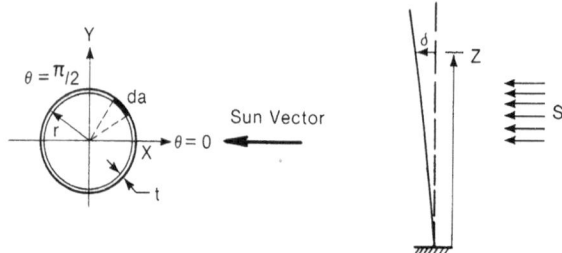

FIG. 7. Solar deflection of thin-walled cylindrical structure.

curve.[6] The problem of a cantilevered cylindrical shell in a solar field has been treated in great detail by a number of investigators (see, for example, Eby and Karam[7]). The thermal moments about two orthogonal axes x and y are given by

$$M_x = \int_a e_l E_l T(x, y) y \, da \qquad (6)$$

$$M_y = -\int_a e_l E_l T(x, y) x \, da \qquad (7)$$

where e_l is longitudinal coefficient of expansion and E_l is longitudinal modulus of elasticity. da is an elemental area as shown in Fig. 7. If the x-axis is aligned with the solar vector, then symmetry of the temperature profile about the x-axis will cancel out the contribution of M_x and the deflection will depend only on the variation of the profile between $\theta = 0$ and $\theta = \pi$. The deflection equation can be written

$$\delta(z) = K \frac{e_l[T(0) - T(\pi)]}{4r} z^2 \qquad (8)$$

where $\delta(z)$ is deflection at distance z and K is a dimensionless factor indicating the extent of deviation of the temperature profile between $\theta = 0$ and $\theta = \pi$ from a linear distribution. In most applications the distribution can be represented as the sum of two or three trigonometric terms[8] and the value of K remains nearly 1·0.

Fin Effect Heat Loss

The treatment presented above does not include the effects of the interface between the extended structure and the platform on which it is mounted. Usually the influence is localized and does not contribute significantly to the structure thermal performance. However, it is almost always desirable that heat leakage across the mounting interface be minimized. This is particularly the case when the structure is attached to the main body of a spacecraft which encloses temperature-sensitive components. Since strength requirements often dominate the design of the attachment system, heat losses must be controlled by the thermal design of the protruding structure.

Heat loss is equivalent to the energy conducted at the base of the structure. For a cylindrical shell

$$Q = -k_l l \left(\frac{\partial T_c}{\partial z} \right)_{z=0} \tag{9}$$

where Q is heat loss and T_c is average circumferential temperature at distance z from the base. For long fins, T_c eventually becomes dependent solely on the external environment and $\partial T_c / \partial z$ tends to zero as z becomes large. Under these conditions, a first integral of the energy equation gives[9]

$$Q = 2\pi r [2k_l l \{\tfrac{1}{5}\varepsilon_o (\sigma T_0^5 - \sigma T_\infty^5) - \bar{U}(T_0 - T_\infty)\}]^{1/2} \tag{10}$$

$\bar{U}$ being the integrated absorbed flux (W/m²), T_0 the temperature of the base (usually spacecraft temperature) and T_∞ is the limit of T_c as z increases indefinitely; that is

$$T_\infty = \left(\frac{\bar{U}}{\sigma \varepsilon_o} \right)^{1/4} \tag{11}$$

Heat leak calculations are commonly estimated conservatively by assuming no flux impingement ($\bar{U} = 0$) and full view to deep space ($T_\infty = 0$ K). Hence

$$Q_{max} = 2\pi r [\tfrac{2}{5}\varepsilon_o k_l l \sigma T_0^5]^{1/2} \tag{12}$$

Thus heat losses can be contained by resorting to low emittance finishes and low longitudinal conductance material.

EFFECT OF FIBER ORIENTATION

Composite structures are fabricated by a lamination process in which the plies are oriented in a known manner. The transport properties can then be determined in terms of the coefficients of a single ply. The longitudinal and circumferential heat conductivities and coefficients of expansion are related

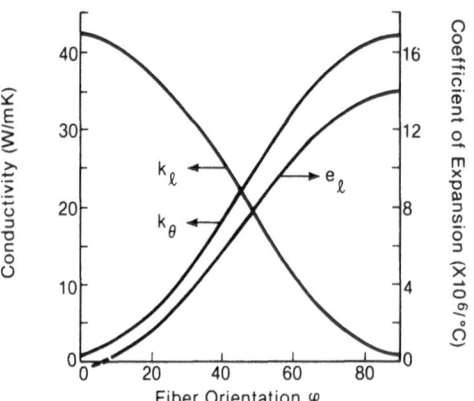

FIG. 8.   Material properties variation with laminate orientation (ATS-6 GFRP 2002M).

to the corresponding single ply properties by the following transformation equations:[10]

$$k_l = (\cos^2 \varphi)k_x + (\sin^2 \varphi)k_y$$
$$k_{l\theta} = (\sin \varphi \cos \varphi)(k_x - k_y) \qquad (13)$$
$$k_\theta = (\sin^2 \varphi)k_x + (\cos^2 \varphi)k_y$$

and

$$e_l = (\cos^2 \varphi)e_x + (\sin^2 \varphi)e_y$$
$$e_{l\theta} = (\sin \varphi \cos \varphi)(e_x - e_y) \qquad (14)$$
$$e_\theta = (\sin^2 \varphi)e_x + (\cos^2 \varphi)e_y$$

The subscripts x and y refer to direction along the principal axes and φ is lamina axes rotation. The manner by which k_l, k_θ and e_l vary with axes rotation is illustrated in Fig. 8, which represents the graphite fiber reinforced plastic materials used for ATS-6 reflector truss assembly. The construction of the truss consisted of the ply arrangement shown in Fig. 2. It is noted that a lay-up intended to decrease the longitudinal conductivity and increase circumferential conductance leads to a radical increase in the longitudinal coefficient of thermal expansion. The combined effect of $e_l[T(0) - T(\pi)]$, which is a measure for deflection, is given in Fig. 9. For the most part, the advantages of decreased circumferential gradients and longitudinal conductivity are upset by increases in e_l. Figure 9 also shows the reduction in maximum heat loss as a function of fiber orientation. The

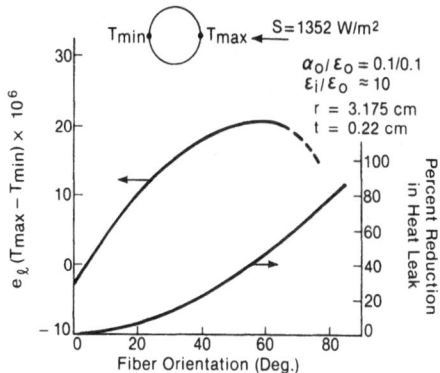

FIG. 9.   Deflection parameters versus laminate orientation (ATS-6 GFRP 2002M).

data were obtained by using Fig. 8 and Fig. 5 as applied to ATS-6 GFRP structure.

PERFORATED COMPOSITE TUBES

Perforation of tubular spacecraft structures is often used to minimize the heating flux imbalance which causes thermal distortions.[11] The technique is well established for conventional materials and, with some modifications, can be used on composites without seriously compromising structural integrity. During the last phases of composite fabrication when the laminate is being vacuum-freed of entrapped air and heated in an autoclave, the epoxy softens as the temperature increases and it becomes possible to easily penetrate the resin with pointed pegs or pins. The fibers, which offer high resistance to puncture, will remain intact during this process with only localized misalignment. The tools can be made similar to those used by Fairchild Space Company for perforating extendible metallic booms.[12] A pre-determined pattern of pins mounted on a moving platform punctures the laminate and is not removed until shortly before final curing is completed. At this time the epoxy is still soft but flow has nearly ceased. Adhesion to the epoxy can be eliminated by coating the platform with a thin layer of Teflon. The cure procedure would follow a history similar to that shown in Fig. 10.

The thermal analysis for a perforated tube can be simplified considerably by neglecting circumferential conduction. The procedure is valid since conduction can only enhance the results by dampening the effect of tolerances on material properties. Referring to Fig. 11, a heat balance on

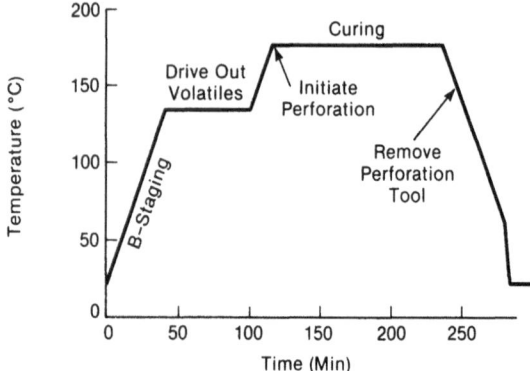

FIG. 10.   Perforation procedure during curing (typical graphite/epoxy composite).

two opposing points 1 and 2 under the condition of zero temperature differences gives

$$\alpha_o(S_1 + A_1) + \varepsilon_o E_1 + p[\alpha_i(S_2 + A_2) + \varepsilon_i E_2]$$
$$= \alpha_o(S_2 + A_2) + \varepsilon_o E_2 + p[\alpha_i(S_1 + A_1) + \varepsilon_i E_1] \qquad (15)$$

where p is percentage of surface perforated. Hence, for arbitrary distribution of solar, albedo, and infrared flux, complete temperature symmetry and negligible thermal moments will be achieved if

$$p = \frac{\alpha_o}{\alpha_i} = \frac{\varepsilon_o}{\varepsilon_i} \qquad (16)$$

In order to maintain strength as near as possible to the non-perforated configuration, α_o and ε_o must be selected as low as attainable while α_i and ε_i as high as possible. Silverized finishes on the exterior can be patterned with

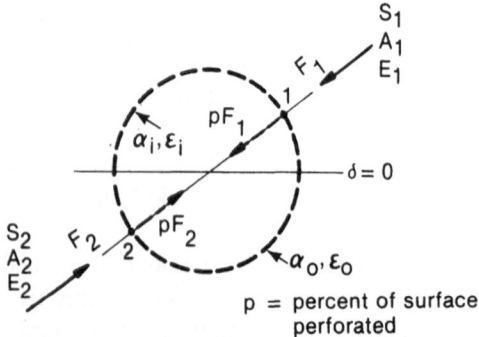

FIG. 11.   Perforated tube in space environment.

less than 5 % striping of white paint to yield $\alpha_o = \varepsilon_o = 0\cdot1$. Properties of the graphite/epoxy ply used on ATS-6 were measured at $\alpha_i \approx \varepsilon_i \approx 0\cdot9$. Hence, the ideal percentage perforation for ATS-6 GFRP structure would be $0\cdot1/0\cdot9 = 11\%$.

With distortions nearly eliminated by perforation, some freedom becomes available to change laminate orientation for the purpose of reducing longitudinal conductance and consequently heat leakage. It is seen from Fig. 9 that almost 50 % reduction in heat loss can be attained when the fibers are oriented 60 degrees off the cylinder longitudinal axis. The reduction in strength for ATS-6 tubes was 75 % at 45° orientation[2] but the values remained within spacecraft specifications. However, any design which deviates from maximum strength capabilities requires parallel structural and thermal analyses to determine the optimum selection for laminate orientation.

SUMMARY AND CONCLUSIONS

Thermal design of spacecraft tubular composite structures can be assessed using conventional analytical techniques. The hardware for temperature control is available and is qualified for space environment application. Temperature variations in orbit are dampened by treating the exterior surface with low absorptance and low emittance coatings (with $\alpha_o/\varepsilon_o = 1\cdot0$) and by ensuring maximum radiation exchange on the interior of the cylinder. Fin effect heat leaks can be reduced by orienting the laminate fibers off the principal axis which usually yields maximum strength. This procedure, however, can result in increasing coefficient of expansion which consequently leads to excessive deflections. A perforation technique can be devised to resolve the problem of distortion by equalizing the bending moment induced by the temperature distribution. Heat leakage can then be reduced in accordance with a laminate alignment and fraction of surface perforated which meet the structural requirements.

REFERENCES

1. WALES, R. O. (Ed.), *ATS-6 Final Engineering Performance Report*, Vol. I, NASA Reference Publication 1080, Nov. 1981.
2. *Final Design Assessment of ATS Reflector Support Truss*, prepared for Fairchild Industries under Contract SC71-6, Hercules Inc., Magna, Utah, Sept. 1971.

3. FRANK, I. and GRAY, E. I., Temperature distribution in long cylindrical shells, *ASME Trans.*, *J. Heat Transfer*, May 1962, 190–191.

4. KARAM, R. D., Optimum solution of linearized radiation equations, in: *Numerical Methods in Thermal Problems*, Lewis, R. W. and Morgan, K. (eds), Swansea, UK, Redwood Burn Limited, 1979, pp. 90–98.

5. KARAM, R. D. and EBY, R. J., Linearized solution of conducting-radiating fins, *AIAA Journal*, May 1978, 536–538.

6. GATEWOOD, B. E., *Thermal Stresses*, New York, McGraw-Hill, 1957, pp. 1–20.

7. EBY, R. J. and KARAM, R. D., Solar deflection of thin-walled cylindrical, extendable structures, *JSR*, **7** (1970), 577–581.

8. EDWARDS, D. K., Temperature distribution around a hollow cylinder heated in space, *TRW IOC*, dated April 18, 1977.

9. CARSLAW, H. S. and JAEGER, J. C., *Conduction of Heat in Solids*, 2nd edn, Oxford University Press, London, 1959, pp. 154–156.

10. TSAI, S. W. and HAHN, H. THOMAS, *Introduction to Composite Materials*, Technomic Publishing Co., Westport, Conn., 1980, pp. 329–333.

11. *DASS Antenna Element Study*, prepared for TRW by Fairchild Space and Electronics Co. under Subcontract No. M43695DS2S, report no. FSEC 78-1, May 1982.

12. *Deployment Mechanisms Development*, Fairchild Space Company report no. FSEC QPA 82-067 (proprietary), March 1982.

18

Simultaneous Elastic and Photoelastic Calibration of Birefringent Orthotropic Model Materials

R. Prabhakaran

*Department of Mechanical Engineering and Mechanics,
Old Dominion University, Norfolk, Virginia 23508, USA*

ABSTRACT

Transmission photoelastic analysis of composite models has been shown in recent years to be a viable stress analysis method for composite structures. However, difficulties in the fabrication of the model materials require the development of efficient mechanical and optical calibration procedures. In this paper, the methods for measuring elastic and photoelastic constants of composites are briefly reviewed. Then a new method, utilizing a relatively small calibration specimen, is described. In this method, electrical resistance strain gauges are mounted at selected points of a half-plane model and photoelastic measurements and strain readings are simultaneously taken. The data are processed by a least-squares procedure to yield the elastic and photoelastic constants. Comparison of results obtained by the proposed method with the results from conventional tests shows good agreement.

INTRODUCTION

Photo-orthotropic-elasticity or the application of transmission photo-elastic techniques by birefringent orthotropic model materials, is now a feasible method for the stress analysis of composite structures. The developments in this area have been reviewed by the author.[1] Difficulties in the fabrication of transparent composites require the model materials to be conserved through the mechanical and optical calibration processes. The model materials have to be calibrated mechanically and optically to establish the four independent in-plane elastic constants and the three fundamental photoelastic constants. A straightforward procedure for both elastic and photoelastic calibration makes use of tensile coupons. But these tensile coupons, especially since they include off-axis specimens, can deplete a relatively small sheet of the model material. Compared to isotropic model materials, there is a greater need to calibrate each sheet of a composite birefringent model material as variations in fibre volume fraction, thickness, etc., can influence the properties very significantly.

The present investigation resolves the calibration difficulties by making use of a relatively small calibration specimen to simulate an orthotropic half-plane subjected to a concentrated edge load. The theoretical stress distribution is known in closed-form. Electrical resistance strain gauges, mounted at selected points on the model, provide the data for obtaining the elastic constants. Photoelastic measurements taken from one-half of the model provide the data for obtaining the photoelastic constants. The details of the proposed method are described in the following sections, after a brief review of existing methods of elastic and photoelastic calibration.

MEASUREMENT OF ORTHOTROPIC ELASTIC CONSTANTS

The measurement of E_L, E_T, v_{LT} and v_{TL} is straightforward, using two tensile specimens with the specimen axes oriented along the major and minor symmetry axes, L, and T, and two strain gauges on each specimen. The determination of the in-plane shear modulus is difficult. A number of test specimens have been proposed by researchers, without any agreement regarding the best. Torsion of a thin-walled circular cylinder yields an accurate shear modulus but the test specimens are difficult and expensive to fabricate. The picture-frame specimen requires the attachment of reinforcing boundary members which result in stress concentrations. Byron[2] did not observe a state of pure shear in a photoelastic study of the picture-frame specimen. The rail shear test was originally proposed for isotropic materials and later adapted for composites; analysis[3] has shown that the actual stresses obtained in the specimen may be different from those assumed by dividing the applied load by the area. The cross-beam shear test, in which a cruciform shaped specimen is subjected to reversed flexural loading, has been shown[4] to result in a stress state which deviates considerably from predictions based upon an elementary bending analysis. The slotted-tension coupon is a tensile specimen which is subjected to a transverse stress and the stress-diffusion effects in the transverse direction are reduced to a minimum by introducing two axial slots. This specimen has been shown[5] to be a valid test specimen for determining the in-plane shear properties. The ten-degree off-axis specimen has been proposed[6] as a possible standard for measuring the in-plane shear properties of composites. A circular disc specimen, with cutouts positioned anti-symmetrically and loaded in diametral compression, has also been suggested[7] for shear characterization. Many of the specimens proposed for shear characterization were intended to give the elastic as well as the fracture behaviours.

In this paper, a single specimen is proposed for the determination of all the in-plane elastic constants. The proposed specimen simulates an orthotropic half-plane subjected to a concentrated edge-load, for which closed-form theoretical solutions are available. Considering a load P applied perpendicular to the edge of an orthotropic half-plane, Green[8] has given the stress components as:

$$\sigma_r = \frac{2P \sin \theta}{\pi r t [(\alpha_1)^{1/2} - (\alpha_2)^{1/2}]}$$

$$\times \left[\frac{\alpha_2 - 1}{\alpha_2 + 1 + (\alpha_2 - 1)\cos 2\theta} - \frac{\alpha_1 - 1}{\alpha_1 + 1 + (\alpha_1 - 1)\cos 2\theta} \right] \quad (1)$$

$$\sigma_\theta = \tau_{r\theta} = 0 \quad (2)$$

where t is the half-plane thickness, r is the distance from the point of application of the load, θ is the angle from the loaded edge and α_1 and α_2 are given by

$$\alpha_1 \alpha_2 = \frac{E_T}{E_L} \quad (3)$$

$$\alpha_1 \alpha_2 = \frac{E_T}{G_{LT}} - 2\nu_{TL} \quad (4)$$

The radial strain, ε_r, at any point is given by

$$\varepsilon_r = \frac{\sigma_r}{E_r} \quad (5)$$

where

$$E_r = \left[\frac{1}{E_L} \cos^4 \theta + \left(\frac{1}{G_{LT}} - \frac{2\nu_{LT}}{E_L} \right) \sin^2 \theta \cos^2 \theta + \frac{1}{E_T} \sin^4 \theta \right]^{-1} \quad (6)$$

Combining eqns (5) and (6),

$$\varepsilon_r = \frac{\sigma_r}{E_T} \left[\alpha_1 \alpha_2 \cos^4 \theta + (\alpha_1 + \alpha_2) \sin^2 \theta \cos^2 \theta + \sin^4 \theta\right] \tag{7}$$

Rewriting eqn. (7),

$$g_k(\alpha_1, \alpha_2, E_T) = 0 \tag{8}$$

where $k = 1, 2, 3, \ldots, M$ ($M \geq 3$). Expanding eqn. (8) in a Taylor's series,

$$(g_k)_{i+1} = (g_k)_i + \left(\frac{\partial g_k}{\partial \alpha_1}\right)_i \Delta\alpha_1 + \left(\frac{\partial g_k}{\partial \alpha_2}\right)_i \Delta\alpha_2 + \left(\frac{\partial g_k}{\partial E_T}\right)_i \Delta E_T \tag{9}$$

where i refers to the iteration step number i and $\Delta\alpha_1$, $\Delta\alpha_2$, ΔE_T are the corrections to the previous estimates of the corresponding parameters. Since the desired result is $(g_k)_{i+1} = 0$,

$$\left(\frac{\partial g_k}{\partial \alpha_1}\right)_i \Delta\alpha_1 + \left(\frac{\partial g_k}{\partial \alpha_2}\right)_i \Delta\alpha_2 + \left(\frac{\partial g_k}{\partial E_T}\right)_i \Delta E_T = -(g_k)_i \tag{10}$$

Rewriting eqn. (10) in matrix notation,

$$[g] = [b][\Delta E] \tag{11}$$

where

$$[g] = \begin{bmatrix} -g_1 \\ -g_2 \\ \vdots \\ -g_M \end{bmatrix} \tag{12}$$

$$[b] = \begin{bmatrix} \dfrac{\partial g_1}{\partial \alpha_1} & \dfrac{\partial g_1}{\partial \alpha_2} & \dfrac{\partial g_1}{\partial E_T} \\ \vdots & \vdots & \vdots \\ \dfrac{\partial g_M}{\partial \alpha_1} & \dfrac{\partial g_M}{\partial \alpha_2} & \dfrac{\partial g_M}{\partial E_T} \end{bmatrix} \tag{13}$$

$$[\Delta E] = \begin{bmatrix} \Delta\alpha_1 \\ \Delta\alpha_2 \\ \Delta E_T \end{bmatrix} \tag{14}$$

Solving for $[\Delta E]$ in eqn. (11),

$$[\Delta E] = [d]^{-1}[b]^T[g] \tag{15}$$

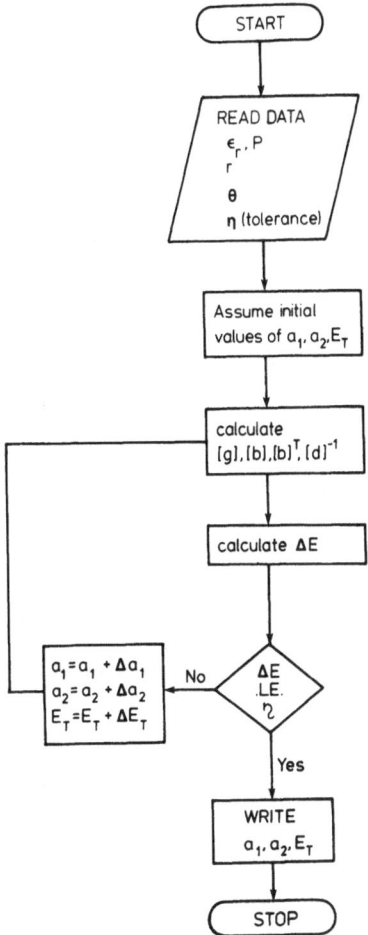

FIG. 1. Flow-chart for orthotropic elastic constants.

where

$$[d] = [b]^T [b] \tag{16}$$

In implementing the above procedure, initial values of α_1, α_2, and E_T are assumed and the matrices $[g]$ and $[b]$ are computed. Then the error vector $[\Delta E]$ is calculated and the estimates of α_1, α_2, ΔE_T are revised. The steps are repeated until the convergence is satisfactory. The flow-chart used in the computation is shown in Fig. 1.

A unidirectionally reinforced E-glass-polyester specimen, 75 mm × 50 mm × 5 mm in size, was used to verify the proposed calibration method.

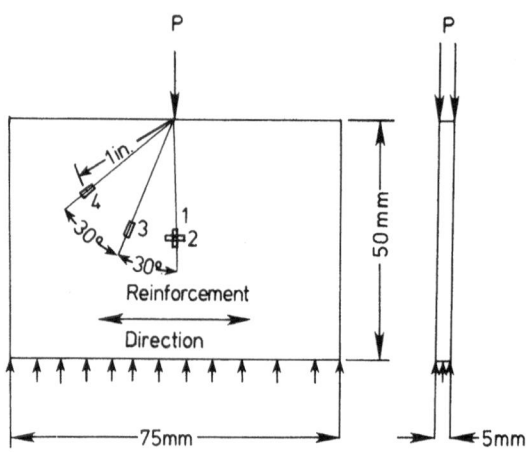

FIG. 2. Orthotropic half-plane specimen with strain gauges.

A schematic diagram of the half-plane specimen, showing the strain gauges, is given in Fig. 2. The specimen was supported along the bottom edge and was loaded by a compressive load on the top edge. An error analysis indicated that the errors due to strain gradients could be reduced to less than 5 % for gauge lengths of up to 10 mm by locating the gauges at a radial distance (from the point of loading) equal to or greater than 25 mm. The gauges, with a gauge length of about 2 mm, were mounted radially at three points and a transverse gauge was mounted at the point directly below the loading point. The strains were measured at load increments of 200 N up to 2000 N. The strains were then plotted as a function of the load and best fitting lines were drawn. The strain values corresponding to 2000 N were used as input data for the flow-chart shown in Fig. 1. The strain readings from gauges 1 and 2 were directly used to compute v_{TL}. In the computation of α_1, α_2, and E_T by the least-squares procedure, convergence

TABLE 1
Comparison of orthotropic elastic constants

Property	Half-plane specimen	Tensile specimen
E_L (GPa)	30·1	28·8
E_T (GPa)	9·8	9·4
G_{LT} (GPa)	3·3	3·2
v_{LT}	0·28	0·30

was rapid and was independent of the initial estimates. The elastic constants E_L and G_{LT} were then computed from eqns (3) and (4). Maxwell's reciprocal relationship yields v_{LT}. The results obtained from this test are shown in Table 1, along with the elastic constants measured previously with conventional tensile specimens. The agreement is seen to be good.

MEASUREMENT OF ORTHOTROPIC PHOTOELASTIC CONSTANTS

Several methods have been proposed and utilized for the measurement of the three fundamental orthotropic photoelastic constants, f_L, f_T and f_{LT}. Beam,[9] compression[10] and tension[11] calibration specimens have been used for the photoelastic characterization of orthotropic model materials. The specimens have been cut along orientations of $0°$, $45°$ and $90°$ to the major reinforcement direction to determine f_L, $f_{\pi/4}$ and f_T respectively. The shear-material fringe value, f_{LT} is then computed as

$$f_{LT} = \left[\frac{1}{f_{\pi/4}^2} - \frac{1}{4}\left(\frac{1}{f_L} - \frac{1}{f_T}\right)^2 \right]^{1/2} \tag{17}$$

The circular disc specimen under diametral compression is commonly used for photoelastic calibration of isotropic model materials. It has not been used in the case of orthotropic model materials because a closed-form analytical stress solution is not available. To overcome this difficulty, a strain-gauge rosette has been used[12] at the centre of an orthotropic circular disc. It has been shown that this method yields f_L and f_T but not f_{LT}.

Bugakov and Grakh[12] expressed the three basic stress-fringe values in terms of two stress-optic coefficients and the elastic constants. These relations are:

$$\frac{1}{f_L} = C_\sigma + \frac{1 + v_{LT}}{E_L} C_e \tag{18}$$

$$\frac{1}{f_T} = C_\sigma + \frac{1 + v_{TL}}{E_T} C_e \tag{19}$$

$$\frac{1}{f_{LT}} = C_\sigma + \frac{1}{2G_{LT}} C_e \tag{20}$$

It has been shown[13] that the 'characterization plot', in which the reciprocal

R. Prabhakaran

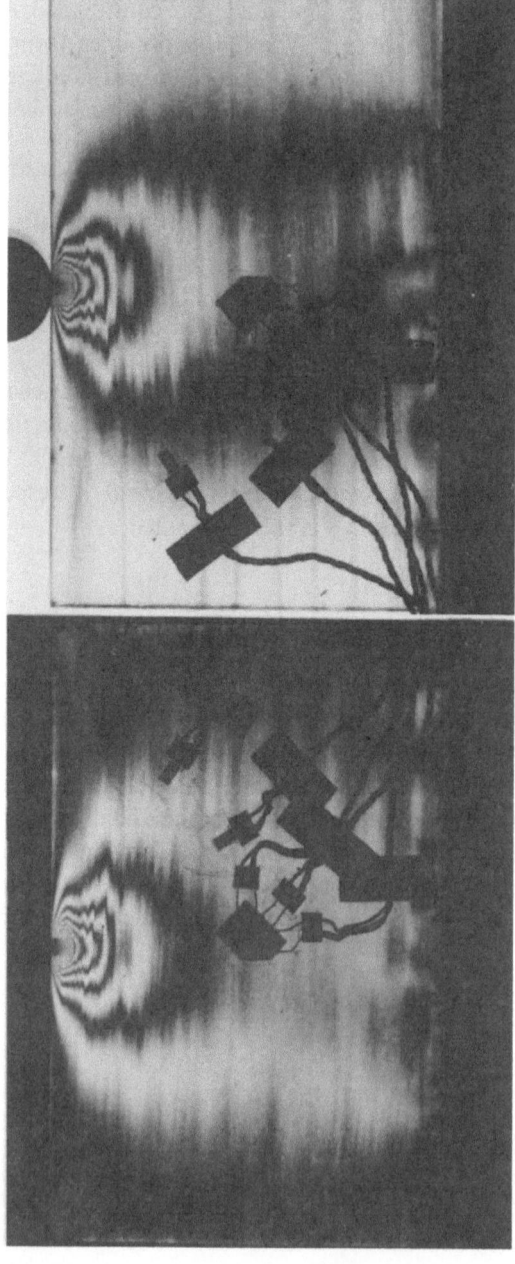

FIG. 3. Dark-field and light-field isochromatic fringe patterns.

of the stress-fringe value is shown as a function of the coefficient of C_e, is a straight line for a given material system, irrespective of the fibre volume fraction. Knowing the elastic constants of the composite model material, the values of f_L, f_T and f_{LT} can be determined from the characterization plot, if the matrix material and a composite specimen of any convenient fibre orientation are photoelastically calibrated. However, the matrix material may not always be available with the composite model material.

In the present investigation, all the three stress-fringe values are obtained from the same orthotropic half-plane model described earlier in connection

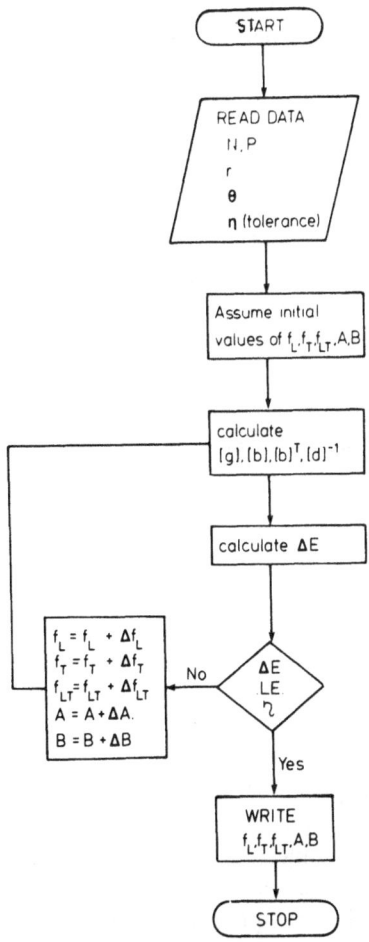

FIG. 4.   Flow-chart for orthotropic photoelastic constants.

TABLE 2
Comparison of orthotropic photoelastic constants

Property	Half-plane specimen	Tensile specimen
f_L (kPa$^{m/fringe}$)	153·1	156·7
f_T (kPa$^{m/fringe}$)	74·9	78·4
f_{LT} (kPa$^{m/fringe}$)	72·6	70·2

with the elastic calibration. The photoelastic isochromatic fringe patterns were recorded at each load. Typical dark-field and light-field patterns at 1800 N, recorded with monochromatic sodium light, are shown in Fig. 3. The isochromatic fringe order per unit thickness, N, can be expressed as

$$N = \sigma_r \left[\left(\frac{\cos^2 \theta}{f_T} - \frac{\sin^2 \theta}{f_L} \right)^2 + \frac{\sin^2 2\theta}{f_{LT}^2} \right]^{1/2} + Ar + B \qquad (21)$$

where σ_r is given by eqn. (1), and the term $Ar + B$ accounts for the residual birefringence, due to the matrix shrinkage during the fabrication. The least-squares method outlined earlier for the elastic constants is employed in determining f_L, f_T, f_{LT}, A and B. The flow-chart used in the computation is shown in Fig. 4. The elastic constants already determined in the first part of this investigation were utilized in the computation of stresses. The residual birefringence was uniform over the model ($A = 0$) and the photoelastic constants measured for the half-plane specimen are compared with the values obtained earlier for conventional tensile specimens in Table 2. The agreement is seen to be good.

CONCLUSIONS

The measurement of the shear modulus and the shear stress-fringe value pose difficulties in the case of birefringent composite models. These values have been so far measured with calibration specimens different from the ones used for the other properties. In the case of photoelastic calibration of birefringent orthotropic model materials, there is a need to minimize the amount of material used in the calibration. To address these problems, a half-plane specimen subjected to a compressive edge-load has been proposed for simultaneous elastic and photoelastic calibration. It has been demonstrated that all the in-plane elastic and photoelastic properties can be obtained from this specimen, the size of which can be relatively small.

ACKNOWLEDGEMENTS

This work was supported by a research grant from the National Science Foundation, Grant No. CME 8012956 and a cooperative agreement, No. NCCI-26, with the NASA-LaRC. The author would like to thank Dr Clifford J. Astill of NSF and Dr Paul A. Coper of NASA-LaRC for their support and encouragement.

REFERENCES

1. PRABHAKARAN, R., Photo-orthotropic-elasticity: a new technique for stress analysis of composites, *Optical Engineering*, **21** (1982).
2. BYRON, E. L., Photoelastic evaluation of the panel shear test for plywood, *Symposium on Shear and Torsion Testing,*'ASTM STP289 (1961).
3. WHITNEY, J. M., STANSBARGER, D. L. and HOWELL, H. B., Analysis of the rail shear test—applications and limitations, *J. Composite Materials*, **24** (1971).
4. BERGNER, H. W., DONIS, J. G. and HERAKOVICH, C. T., *Analysis of Shear Test Methods for Composite Laminates*, Virginia Polytechnic Institute Report No. E-77-14 (1977).
5. DUGGAN, M. F., An experimental evaluation of the slotted-tension shear test for composite materials, *Experimental Mechanics*, **20** (1980).
6. CHAMIS, C. C. and SINCLAIR, J. H., Ten-degree off-axis test for shear properties in fiber composites, *Experimental Mechanics*, **17** (1977).
7. ARCAN, M., HASKIN, Z. and VOLOSHIN, A., A method to produce uniform plane-stress states with applications to fiber-reinforced materials, *Experimental Mechanics*, **18** (1978).
8. GREEN, A. E., Stress systems in aelotropic plates II, *Proc. Roy. Soc.* Series A, **173** (1939).
9. HAYASHI, T., Photoelastic method of experimentation for stress analysis in orthotropic structures, *Proceedings of the Fourth International Symposium on Space Technology and Science*, Tokyo (1962).
10. PIH, H. and KNIGHT, C. E., Photoelastic analysis of anisotropic fiber reinforced composites, *J. Composite Materials*, **3** (1969).
11. DALLY, J. W. and PRABHAKARAN, R., Photo-orthotropic-elasticity, *Experimental Mechanics*, **11** (1971).
12. BUGAKOV, I. I. and GRAKH, I. I., *Investigation of the Photoelastic Method for Anisotropic Bodies*, Leningradskii Universitet, Vestnik, Matematika, Mekhanika, Astronomiia, Vol. 102 (1968).
13. PRABHAKARAN, R., Photoelastic calibration of orthotropic model materials, *Optica Acta*, **27** (1980).

19

The Effect of Moisture Absorption on Composite Laminates

Chuan S. Wang and Guan C. Chang

Chung-Shan Institute of Science and Technology,
P.O. Box 1-2-5, Lung-Tan, Taiwan

ABSTRACT

The effect of moisture on composite materials was examined experimentally by means of flexural and interlaminar shear tests of a typical glass–epoxy system. CIBA Araldite 507 epoxy resin and HY956 hardener were used as a matrix material, cross-ply laminates were cut into [±45] and [0/90] specimens.

Test results of laminated specimens are compared with unfilled epoxy specimens in most conditions. It was found that an approximate method can be used to calculate the degradation of composite laminates under moisture attack, when the degradation of resin matrix is known.

1. INTRODUCTION

It is well known that the exposure of composite structures in a wet environment may cause water sorption. The absorbed water in polymeric composites not only produces changes of chemical and physical nature but also causes degradation of mechanical properties of materials. The degradation of strength and stiffness is the result of the weakening of the three composite phases, the fiber, the matrix, and the interface.[1,2] The degree of degradation depends on the moisture content in the composite and varies with the structural components and the quality of the composite. Sometimes the degradation is so remarkable that it should not be ignored in the application or design of composite structures, when they are to be exposed in a humid environment for a long period.

There have been a lot of approaches in this field of composites with different functions. In this study, an advanced examination of moisture effects has been done through the three phases of composite material. It has been tried to combine and simplify these functions to get an approximated method for calculating the strength and stiffness retention of composites after water sorption. Here, a generally used glass–fabric reinforced epoxy laminate was chosen as a basic material and a low strength carbon fabric laminate was also used for checking of equations.

2. EXPERIMENT

2.1. Specimens

A room temperature cured epoxy resin system was chosen as a matrix material because of its low coefficient of shrinkage and very low thermal stress which might be induced during curing. Here, we chose CIBA Geigy's Araldite 507 and Hardener HY956 (Ref. 3) as a matrix material. The reinforcements generally used are plain weave cloth (ASAHi MS252) and woven roving ($580\,g/m^2$) all with chrome finish. The basic GRP laminates consisted of 2 layers of fabric cloth outside and 4 layers of woven roving inside (C1/4WR/C1); we also tested some specimens with different layers of woven roving. All the specimens were made by a hand lay-up method and pressed to 3 mm thick with match moulds. After room temperature curing, it was post-cured at 80 °C for 8 h in an oven to assure specimens were fully cured and dried.

2.2. Procedures

The moisture absorption of GRP laminate in natural weather conditions was recorded first for a reference (Fig. 1), then, unfilled epoxy specimen was tested for moisture effects on cured resin. A three-point flexural test was chosen for testing, because in most plate applications failure due to bending appears more critical.[4,5] A span-to-depth ratio of 16 was set throughout the whole experiment.

Moisture was introduced into the specimen by immersing it in boiled water until a specified weight was gained. The specimens were exposed in air for 30 min before weighing to ensure that the surface moisture had evaporated.

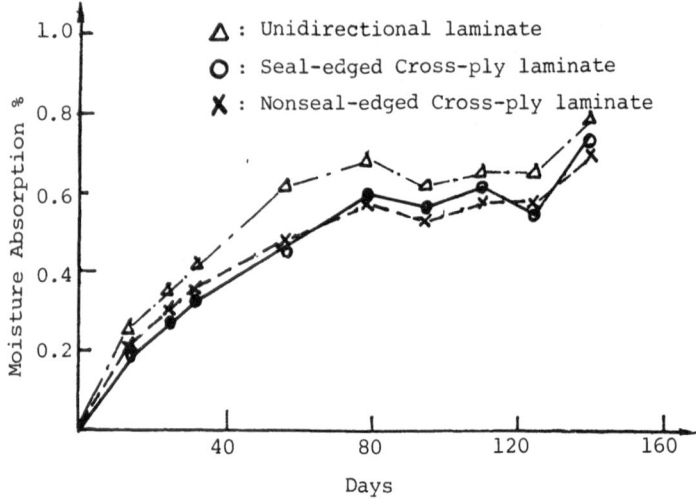

FIG. 1.   Effect of natural exposure on the moisture absorption of GRP laminate.

3.   RESULTS AND DISCUSSION

3.1.   Effect of Moisture on Matrix Material

It was observed that the strength and modulus reduction of epoxy resin matrix occurred after moisture sorption (Fig. 2). The strength retention decreased to 58 % when moisture content went up to 3·5 % (equilibrium condition). It was concluded that the degradation of mechanical properties was due to (a) moisture-induced plasticization of material, and (b) moisture-induced material damage.[6,7] The plasticization of material could be recovered by drying it, but the material damage is irreversible. Henceforth, the recovery of the strength of the epoxy material couldn't be one hundred per cent after it was strongly attacked by moisture.

Table 1 shows that the failure mode of cured epoxy varies with the moisture content. It tends to change from a brittle failure to a ductile failure when the moisture content of the specimen is over 2·5 %. It is reasonable that moisture effect on resin matrix may have a profound influence on the process of degradation of GRP.

3.2.   Effect of Moisture on Interface

It has long been recognized that glass–resin interface exerts a strong influence on the shear properties of composite materials. It also affects the ability of matrix material to transfer stress among the glass fibers. It has

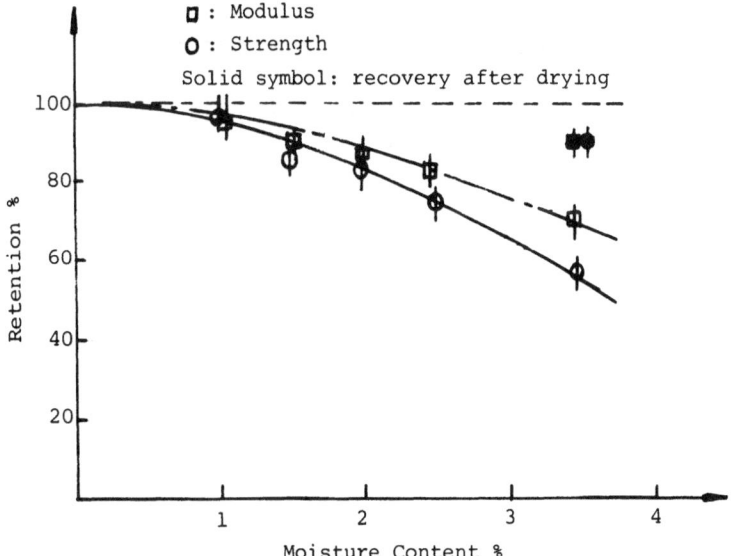

□ : Modulus

O : Strength

Solid symbol: recovery after drying

FIG. 2. Effect of moisture on the flexural properties of an unfilled epoxy specimen.

been proved that the bonding strength between resin and fibers could be decreased by moisture.[8,9] Sometimes, it may cause an irreversible effect on the composite structures. Here, a generally used interlaminar shear test was used to observe the moisture effect on composite interfacial regions. It is shown in Fig. 3 that the strength decreases when the moisture content increases and the recovery through drying seems not to be effective on these specimens. The effect of moisture absorption on the interface can also be seen in Fig. 4, which shows the fracture surface of a unidirectional

TABLE 1

Effect of moisture absorption on cured epoxy

Moisture content (%)	Average flexural strength (kg/mm^2)	Average flexural modulus (kg/mm^2)	Failure mode
0	15·7	405	Brittle
1	15·1	381	Brittle
2	13·0	366	Mixture of brittle and ductile
2·46	11·5	350	Ductile (not broken)
3·45	9·1	273	Ductile (not broken)
3·5 drying 0 ⟶	13·9	374	Brittle

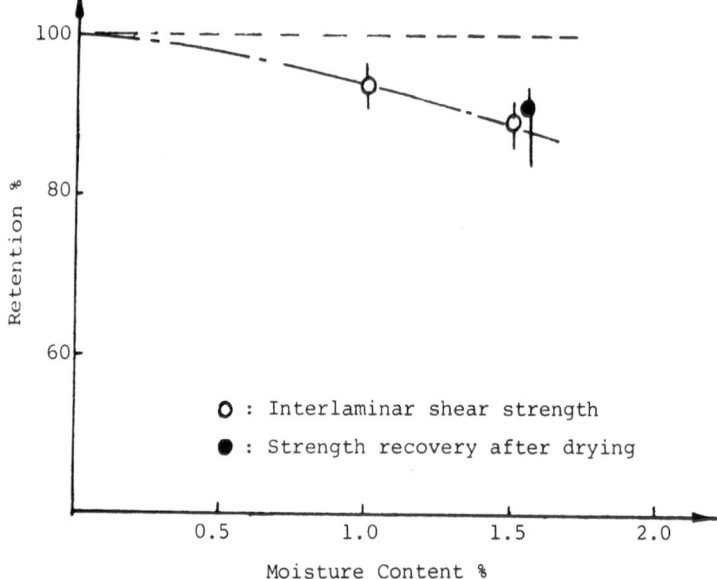

FIG. 3.   Effect of moisture on the interlaminar shear strength of GRP laminate.

composite after loading. A weak adhesion between glass fibers and epoxy matrix is found in the photomicrograph of the specimen after moisture attack. Therefore, there is no doubt that the moisture effect on interface will extend to the other site of GRP structures.

3.3. Effect of Moisture on [±45] Composite Laminate

According to the mechanical behavior of a composite, it is known that

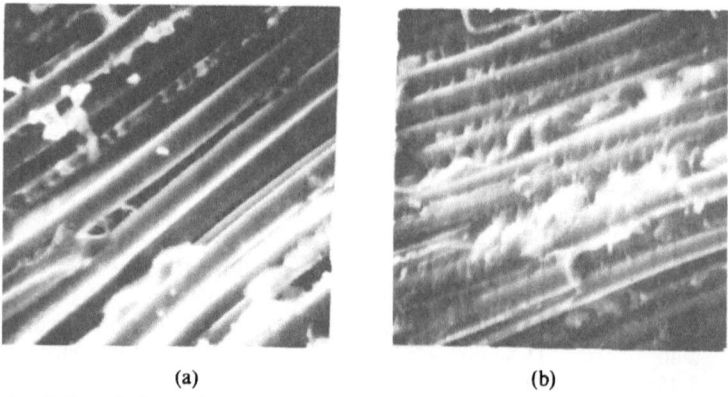

(a)   (b)

FIG. 4.   Enlarged photomicrograph of the fracture surface of a UD laminate (650 ×): (a) after moisture attack; (b) without moisture attack.

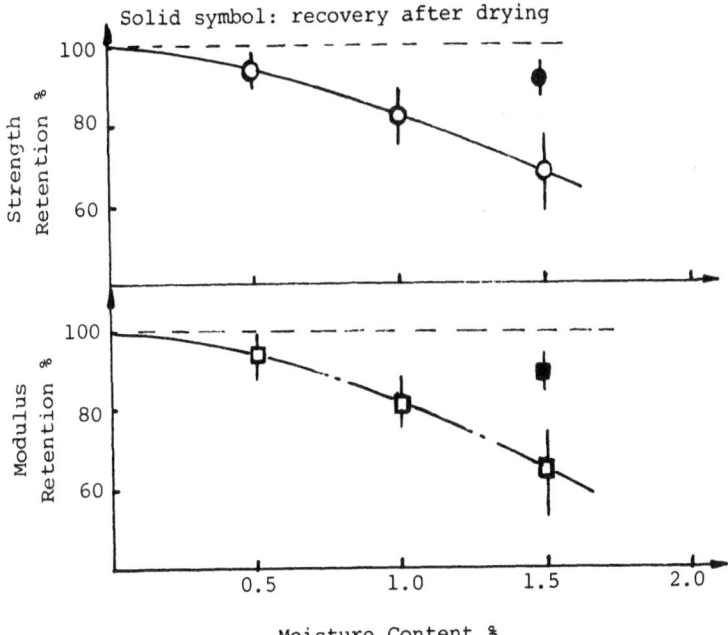

FIG. 5. Effect of moisture on the [±45] GRP specimen (C1/4WR/C1).

FIG. 6. Failed GRP specimen: (a) dry; (b) 1·0% M.C.; (c) 1·5% M.C.

the strength of $[\pm 45]$ laminate is less dominated by fibers but depends on the shear properties of resin and interface.[10,11] Thus, the loss of stiffness of matrix material and bonding strength of interface may cause a reduction of stiffness of $[\pm 45]$ laminate. And, for this reason, the moisture effect on a composite should be observed in the testing results of $[\pm 45]$ laminate. It is found that there is an apparent degradation in both strength and modulus in a high moisture content level (1·5 %) (Fig. 5).

Figure 6 shows the failed specimens under load. It is obvious that a larger crazing area due to debonding of interface can be observed in the specimen with 1·5 % moisture content.

4. AN APPROXIMATE METHOD FOR CALCULATION OF STRENGTH AND MODULUS RETENTION OF [0/90] LAMINATES

4.1. Modulus Retention

It is undoubted that degradation in resin matrix properties would result in a parallel degradation in composite properties. For unidirectional laminate, the simple rule of mixture (eqn. (1)) has been used to calculate the composite modulus from given fiber and resin matrix moduli:[12]

$$E_c = V_f E_f + V_r E_r \tag{1}$$

where the subscripts c, f and r refer to the composite, fiber and resin respectively, V is the volume fraction, and E is elastic modulus. It is obvious that if E_r decreases E_c decreases accordingly, as moisture may reduce the interfacial bonding strength of composite and cause a loss of efficiency of stress transfer between fibers. Fried and others concluded in their studies that glass fiber may be attacked by moisture on long term exposure in a wet environment.[1,13-15] It is more apparent for Kevlar composite.[5] Hence, eqn. (1) has to be modified as below involving the interface and reinforcement considerations for calculation of E_c after moisture absorption. That is

$$E_c' = V_f E_f + V_r E_r' - \alpha E_f V_f W_m \tag{2}$$

where α is an empirical constant for a fixed type of reinforcement and coupling agent; E_c', E_r' are the modulus of composite and resin matrix after moisture absorption, W_m is moisture content by weight in composite. E_r'

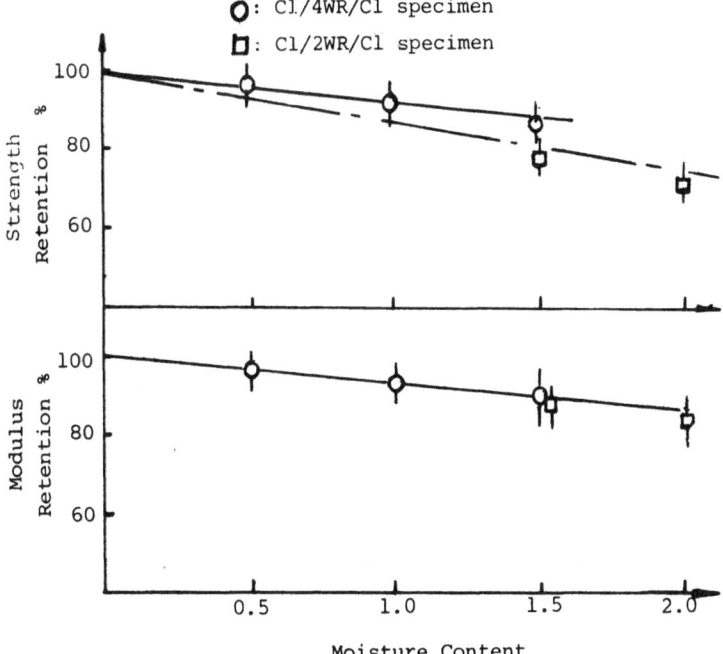

FIG. 7. Comparison of calculated and experimental results of [0/90] GRP laminates.

can be obtained from the test results referred to in Fig. 2. For a [0/90] laminate, we may approximate eqns (2) and (3)

$$E'_c = \frac{V_f}{2} E_f + V_r E'_r - \alpha E_f V_f W_m \tag{3}$$

Figures 7 and 8 show the calculated line and experimental results of modulus retention relative to the moisture content for GRP and CRP laminates respectively. It is seen that this equation correlates very close to the experimental results for both glass and carbon composites, when $\alpha = 2$ and 3 for glass and carbon respectively.

4.2. Strength Retention

Following the general engineering equation for strength of uni-directional composites[12] and the observation of moisture effect on composite materials and its components it is found that an approximate

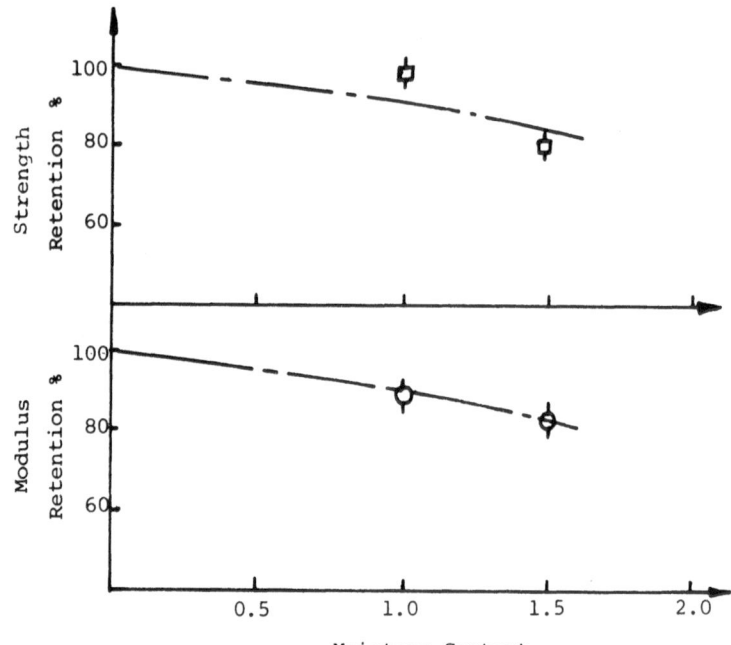

FIG. 8. Comparison of calculated and experimental results of [0/90] CRP laminate (Carborundum GSGC 2).

eqn. (4) may also be used as an engineering calculation for the strength of [0/90] composite laminate after absorbed moisture,

$$\sigma_{uc} = \sigma_{uf} \left[V_f/2 + V_r\beta \frac{E'_r}{E_f} \right] - \alpha\sigma_{uf}V_f W_m \qquad (4)$$

where β is an empirical constant for a fixed type of reinforcement and coupling agent, and σ_{uc} and σ_{uf} are strengths of composite and fiber respectively. The results of calculation and testing of GRP and CRP laminates are shown in Figs 7 and 8.

5. CONCLUSION

Based on the experiments and analysis described in this paper, the following conclusions can be formed:

(1) The mechanical properties of cured epoxy resin are strongly

affected by moisture. The degradation of this material is quite serious when the moisture content is high. The softening of resin matrix will extend to the whole composite structure, resulting in a decrease of strength and stiffness of material, especially on the [±45] laminate.

(2) The interfacial region of composite material can also be affected by moisture, which causes a loss of efficiency of stress transfer between fibers that leads to a further decrease in the flexural strength and modulus of the composite.

(3) A simple approximate method can be used to calculate the strength and modulus degradation of unidirectional and [0/90] specimens.

Other complicated factors which may influence the water absorption in composite laminates can be ignored in these calculations. Obviously, these calculations are based only on known degradation of matrix material. It was found that these calculated results correlated with the experimental data very well.

REFERENCES

1. FRIED, N., Degradation of composite materials: the effect of water on glass-reinforced plastics, *Mechanics of Composite Materials*, in: Proc. Fifth Symp. Naval Structural Mechanics, Philadelphia, Pa. (1967), 813–837.
2. ISHAI, O., Environmental effects on deformation, strength and degradation of unidirectional glass–fiber reinforced plastics, *Polym. Engng Sci.*, **15** (1975), 486–499.
3. LEE, H. and NEVILLE, K., *Handbook of Epoxy Resins*, New York, McGraw-Hill (1967).
4. WHITNEY, J. M. and HUSMAN, G. E., Use of the flexure test for determining environment behavior of fibrous composites, *Experimental Mechanics*, May (1978), 185–191.
5. ALLRED, R. E., The effect of temperature and moisture content on the flexural response of Kevlar/epoxy laminates, *J. Composite Materials*, **15** (1981), 100–132.
6. WANG, C. S. and WANG, A. S. D., Creep behavior of glass epoxy composite laminates under hygrothermal condition, *Advances in Composite Materials*, Proceedings of ICCM 3, Paris, Pergamon Press (1980), pp. 569–583.
7. BROWNING, C. E., The mechanisms of elevated temperature property losses in high performance structural epoxy resin matrix materials after exposures to high humidity environments, *Polym. Engng Sci.*, **18** (1978), 16–24.
8. STERMAN, S. and MARSELEN, J. G., Bonding organic polymers to glass by silane coupling agents, in: *Fundamental Aspects of Fiber Reinforced Plastic Composites*, John Wiley, New York (1968).

9. FIELD, S. Y. and ASHBEE, K. H. G., Weathering of fiber reinforced plastics. Progress of debonding detected in model systems by using fibers as light pipes, *Polym. Engng Sci.*, **12** (1972), 30–33.

10. ISHAI, O. and LAVENGOOD, R. E., The mechanical performance of cross-plied composites, *Polym. Engng Sci.*, **11** (1971).

11. JONES, B. H., Predicting the stiffness and strength of filamentary composites for design application, *Plastics & Polymers*, April (1968), 119–125.

12. TSAI, S. W. and HAHN, H. T., *Introduction to Composite Materials*, Westport, Technomic (1980), pp. 379–425.

13. COHEN, Y. B., MERON, M. and ISHAI, O., Nondestructive evaluation of hygrothermal effects on fiber-reinforced plastic laminates, *J. Testing Evaluation*, **7** (1979), 291–296.

14. BROWNING, C. E., HUSMAN, G. E. and WHITNEY, J. M., Moisture effects in epoxy matrix composites, *Composite Materials: Testing and Design, ASTM STP 617* (1977), pp. 481–496.

15. BROWNING, C. E. and HARTNESS, J. T., Effects of moisture on the properties of high-performance structural resins and composites, *Composite Materials: Testing and Design, ASTM STP 546* (1974), pp. 284–302.

20

The Nonlinear Viscoelastic Response of Resin Matrix Composites

C. C. HIEL and H. F. BRINSON

Department of Engineering Science and Mechanics,
Virginia Polytechnic Institute and State University,
Blacksburg, Virginia 24060, USA

and

A. H. CARDON

Department of Continuum Mechanics, Free University of Brussels,
Pleinlaan 2, B-1050 Brussels, Belgium

ABSTRACT

The current paper describes the utilization of a thermodynamic based analytical nonlinear viscoelastic approach to represent lamina properties. Test data to verify the analysis for both transverse and shear properties of a T300/934 composite are presented. Master curves as a function of stress level and temperature are generated. Favorable comparisons between the traditional graphical and the current analytical approaches are shown.

INTRODUCTION

Fiber reinforced plastics (FRP) are light, strong and can be tailored for a particular structural application. That is, they can be designed with the strong and stiff fibers in the directions of highest stress while the relatively weak and compliant polymeric matrix operates under low stress levels. However, the matrix is an important structural component as it serves to transfer the applied stress from fiber to fiber and from ply to ply. For this reason, the matrix, and hence matrix dominated properties, plays an

271

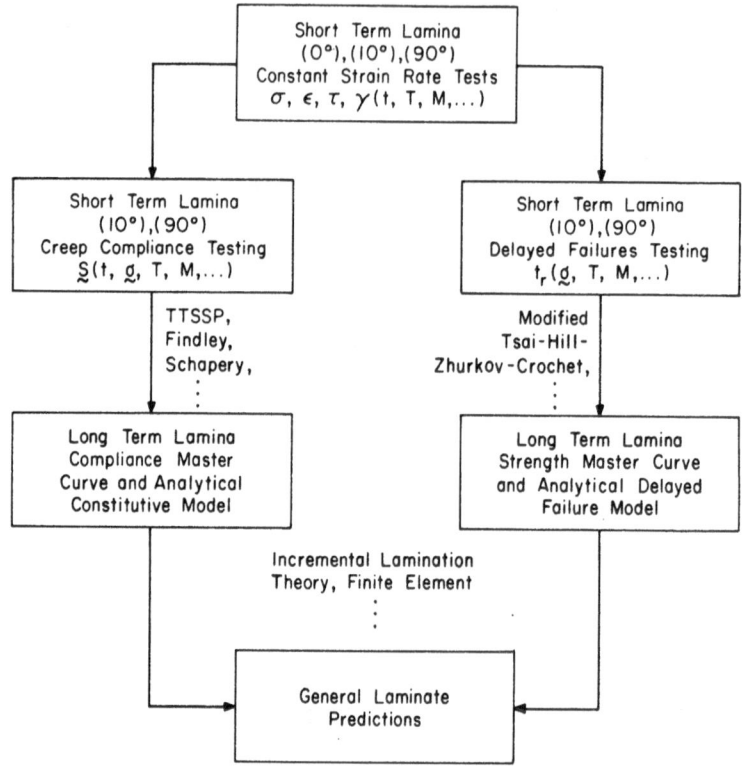

FIG. 1. Accelerated characterization method for laminated composite materials.

important role in structural design and/or behavior of a composite material and cannot be ignored.

A cooperative research program between NASA-Ames and Virginia Tech has been underway for a number of years to try to understand the matrix dominated viscoelastic properties of composite laminates. The objective of this program has been to develop an accelerated strength and stiffness characterization procedure for FRP materials in general and graphite/epoxy laminates in particular. This accelerated characterization procedure is shown schematically in Fig. 1. The method is an evolution of the various studies discussed below.

The accelerated characterization procedure which was developed for polymer based composite laminates several years ago by Brinson et al.[1] and Yeow[2] was initially based upon the well-known time–temperature–superposition principle (TTSP) for polymers and the widely used lamination theory for composite materials. More recently, however, both a time–stress–superposition principle (TSSP) and a time–temperature–stress–superposition principle (TTSSP) have been utilized.[3,4] Fundamentally, the idea is that certain environmental conditions such as temperature and stress serve to accelerate the deformation processes associated with viscoelastic properties.

The application of the TTSP, TSSP and TTSSP in our earlier work was graphical requiring a large data base creating a tedious and time consuming approach.

Subsequently, Dillard[5] and Brinson and Dillard[6] utilized a Findley[7] type nonlinear creep power law as a means of analytically representing the long term data generated by Griffith[3] with the TSSP approach. They then developed a numerical nonlinear viscoelastic lamination theory computational model which included this compliance creep power law and predictions of general laminate creep response were compared to measurements.[5,8,9]

The nonlinear creep power law referred to above is semi-empirical with several severe limitations.[5] The fact that the determination of the exponent of the power law may require a week long creep test is an example of one such limitation.[10]

For the above reasons, the present study represents the application of a thermodynamic based nonlinear viscoelastic model due to Schapery[11] and Lou and Schapery[12] to represent the lamina properties of a T300/934 graphite/epoxy composite. Specifically, long term stress dependent master curves were to be predicted from short term data as a substitute for the graphical TSSP. One great value of the present model is that it has the potential of accurately predicting more general loading conditions than creep and can be used to model unloading as well.

ANALYTICAL CONSIDERATIONS

A uniaxial stress, σ_x, acting at an angle, θ, to the fibers for a unidirectional tensile coupon gives rise to a state of plane stress as shown in Fig. 2. A uniaxial compliance model to accommodate biaxial creep loading for such an orthotropic tensile bar in a state of plane stress was given[5,6] to be,

$$
\begin{bmatrix} \varepsilon_1(t) \\ \varepsilon_2(t) \\ \gamma_{12}(t) \end{bmatrix} = \begin{bmatrix} S_{11} & S_{12} & 0 \\ S_{12} & S_{22}(t, \sigma_x) & 0 \\ 0 & 0 & S_{66}(t, \sigma_x) \end{bmatrix} \begin{bmatrix} \sigma_1 \\ \sigma_2 \\ \tau_{12} \end{bmatrix} \tag{1}
$$

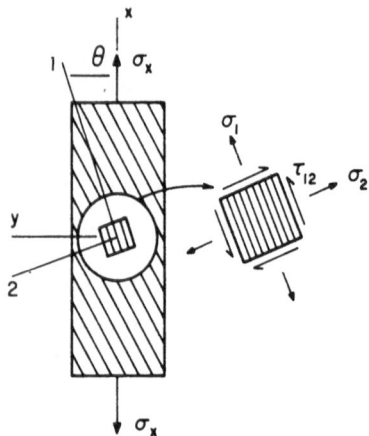

FIG. 2.   Off-axis test geometry and internal stress state.

Here ε_1, ε_2, γ_{12} and σ_1, σ_2, τ_{12} represent the strains and stresses in the material coordinates respectively. S_{11} is the compliance in the fiber direction and is assumed to be time independent while $S_{22}(t)$ and $S_{66}(t)$ are time dependent. The latter may be determined from tensile tests when θ is 90° and 10° respectively. S_{12} is found from the same test as S_{11}, i.e. $\theta = 0°$, and is assumed to be time independent.

The Schapery nonlinear viscoelastic model for isotropic materials may be written as,

$$\varepsilon = g_0 S_0 \sigma_x + g_1 \int \tilde{S}(\psi - \psi') \frac{d(g_2 \sigma_x)}{d\tau} \, d\tau \qquad (2)$$

where

$$\psi(t) = \int_0^t \frac{dt'}{a_\sigma[\sigma_x(t')]} \quad \text{and} \quad \psi'(\tau) = \int_0^\tau \frac{dt'}{a_\sigma[\sigma_x(t')]}$$

In eqn. (2) S_0 is the initial linear creep compliance, $\tilde{S}$ is the transient linear creep compliance, ψ is a reduced time parameter and g_0, g_1, g_2 and a_σ are four stress dependent nonlinearizing parameters. The quantity a_σ represents a stress induced shifting of the time scale which is similar to the temperature dependent shift function a_T used in the TTSP. As a result, eqn. (2) is a mathematical statement of the TSSP. The four parameters g_0, g_1, g_2 and a_σ may be temperature dependent as well. For such events, eqn. (2) is then a statement of the TTSSP.

Equation (2) can be used to model the time dependent compliances of

eqn. (1) and assuming a creep power law for creep in the linear range, the following representation is obtained,

$$S_{22}(t) = g_0^{(T)} S_{22}^e + \frac{g_1^{(T)} g_2^{(T)}}{a_\sigma^n} C_{22} t^n$$

$$S_{66}(t) = g_0^{(S)} S_{66}^e + \frac{g_1^{(S)} g_2^{(S)}}{a_\tau^n} C_{66} t^n$$

(3)

where the superscripts (T) and (S) refer to transverse and shear properties respectively.

These equations were first derived by Schapery[11] and he indicated that g_0, g_1 and g_2 nonlinearize the Gibbs free energy G while a_σ and a_τ nonlinearize entropy production as well as Gibbs free energy. Details on these thermodynamic aspects are also given in Ref. 10.

Equations (3) above, resulting from the inclusion of a creep power law to represent linear response, are similar to the earlier mentioned Findley model.[5,10,11] However, these equations, as developed by Schapery, are very different in that an explicit shifting of the time scale is now included. Further, once the nonlinear parameters are determined for tensile creep using eqns (3), they can be utilized in the more general eqn. (2) to model a wide variety of loadings. The Schapery method is therefore a true analytical representation of the TSSP where long term results can be predicted from short term tests using stress as an accelerating parameter. The Findley approach, on the other hand, does not have the same capability.

EXPERIMENTAL DETAILS

The laminate used in this investigation consisted of sixteen unidirectional plies of T300 graphite fibers of about 60 % volume in a matrix of Fiberite-934. The latter was a TGDDM epoxy with a 4'4' diaminodiphenyl sulfane curing agent and a boron trifluoride catalyst.

Tensile specimens were cut from a unidirectional plate at the desired load angles (0°, 10°, 90°) by means of a diamond cutting wheel. Only the 10-deg off-axis specimens were reinforced with adhesively bonded, tapered aluminum tabs.

Subsequently the specimens were post cured at 350 °F ± 10 °F (176 °C ± 5·5 °C) for 4 h ± 15 min, followed by a slow, controlled cooldown at a rate of 5 °F/h (2·75 °C/h).

After postcuring, the specimens were stored in a desiccator at ≈ 15 %

relative humidity. They were instrumented with back to back strain gauges prior to testing.

Specimens were tested in a creep frame at various temperatures from ambient to 335 °F (168 °C). The temperature was held constant for each test within ±2 °C.

Both creep and creep recovery tests were performed as required by the Schapery method. Creep tests were generally of the order of one hour while recovery was sufficient to allow all transients to cease.

RESULTS AND DISCUSSION

Only linear viscoelastic behavior was found for the transverse or 90° oriented properties at any of the temperature levels investigated. Thus, $g_0^{(T)} = g_1^{(T)} = g_2^{(T)} = a_\sigma = 1$. For shear, however, a significant nonlinear creep behavior was found. Figure 3 shows the compliance $S_{66}(t)$ at eight different stress levels for a temperature of 246 °F (119 °C). Similar curves were also obtained at 300 °F (148 °C), 320 °F (160 °C), and 335 °F (168 °C).

Utilizing the second of eqns (3) for the shear component in our laminate, it can be seen that the Schapery approach is ideal for an analytical interpretation of time–stress equivalence, that is, the equation includes a horizontal time scale shift, a_t, as well as the vertical shift $g_1^{(S)} \cdot g_2^{(S)}$ which is also often required.

Figure 4 represents a schematic of the shifting procedures used to produce stress dependent master curves via a graphical procedure and the

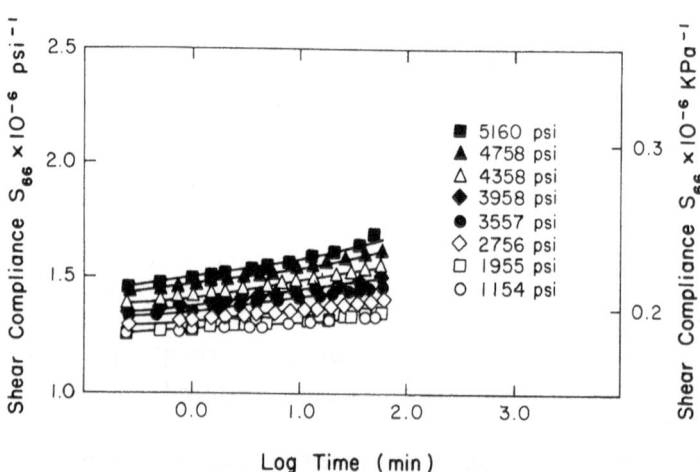

FIG. 3. Shear compliance versus time and stress level at 246 °F (119 °C).

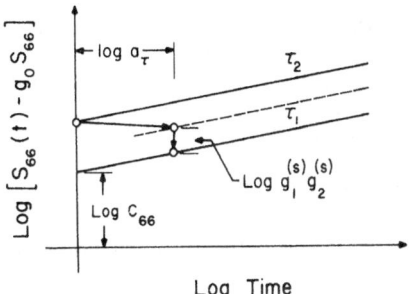

FIG. 4. Shifting procedure for master curve.

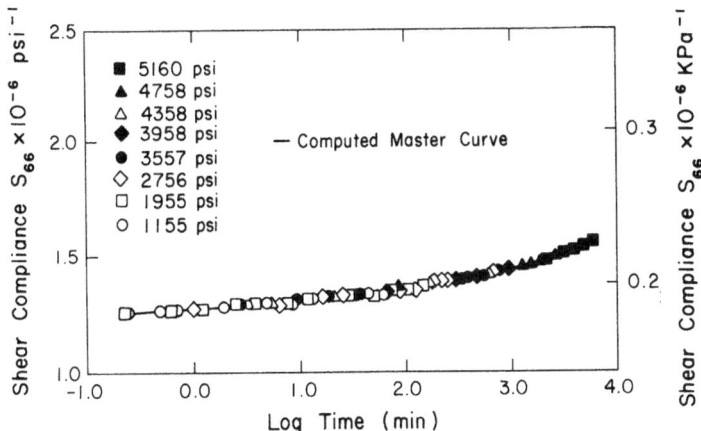

FIG. 5. Shifted $S_{66}(t)$ master curve for 246°F (119°C).

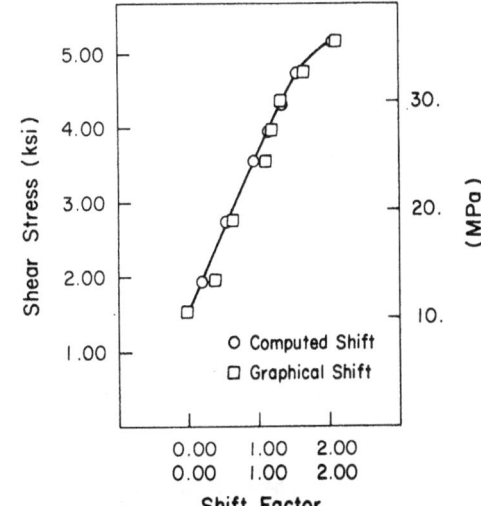

FIG. 6. Comparison of graphical and analytical horizontal shift factor for S_{66} at 246°F (119°C).

corresponding shifting functions from eqn. (3). The data of Fig. 3 were shifted horizontally and vertically as described by Fig. 4 to produce the master curve shown in Fig. 5.

The Schapery procedure as described earlier and in more detail in references 10–12 was also used to obtain the master curve shown by the solid line in Fig. 5. A comparison of the graphical and analytical horizontal shift factors is shown in Fig. 6.

The vertical shifting needed to produce the master curve of Fig. 5 was very small. The reason for this can be seen by examination of the $g_1^{(S)}$ and $g_2^{(S)}$

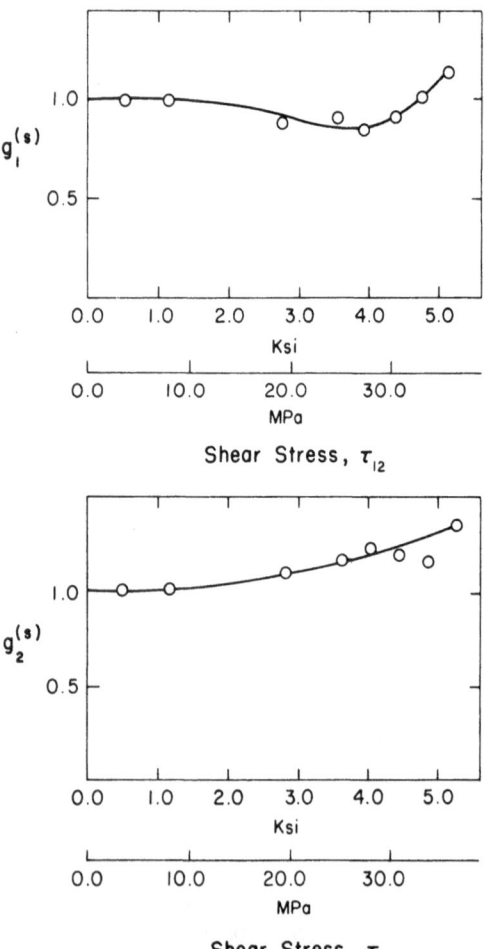

FIG. 7.   Nonlinear functions $g_1^{(S)}$ and $g_2^{(S)}$ as a function of shear stress, τ_{12}.

results presented in Fig. 7. These nonlinear functions of stress level as determined from data are observed to vary in such a way that their product is nearly unity. Hence the need for only a small vertical shift. This conclusion is also true at higher temperatures. The resulting master curve is clearly the transient component of the linear viscoelastic compliance. It can be expected, however, that this conclusion will lead to an underprediction

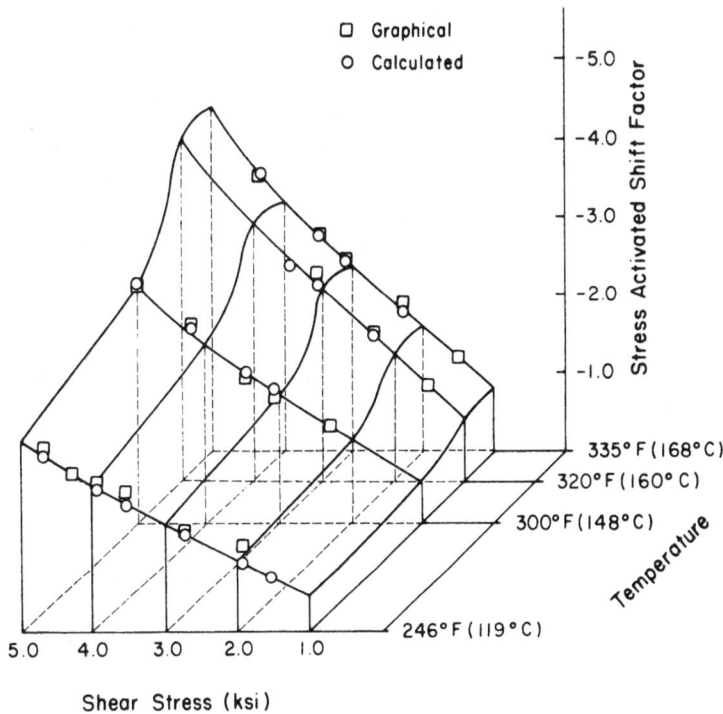

FIG. 8.   Stress activated shift (a_τ) as a function of temperature (T) and shear stress (τ_{12}).

of very long term compliance data. The reason is, as was shown in Hiel,[10] that the power law in its simple form given by eqn. (3) is only adequate to describe a limited part of the viscoelastic spectrum.

Figure 8 shows a three-dimensional plot of the horizontal shift factor, a_τ, as a function of both temperature and stress level. Thus we have the essential ingredients necessary to utilize the Schapery analysis of eqn. (3) in a TSSP sense to produce lamina compliance surfaces similar to those suggested in Brinson and Dillard[6] for any arbitrary reference temperature within the range of our data.

SUMMARY AND CONCLUSIONS

A nonlinear constitutive model developed by Schapery from thermo-dynamic considerations was shown to be a valid analytical substitute for the graphical time–stress–superposition principle (TSSP) used in previous efforts. The power law approximation of linear compliance should be replaced with a form capable of better approximating the complete retardation spectrum. Extrapolations to long time ($>10^4$ min) would then be more accurate and conservative underpredictions would be avoided.

Transverse properties were found to be linearly viscoelastic while the shear properties were shown to be nonlinearly viscoelastic for the time scale of our test, i.e. one hour. This is likely due to interaction effects between the fiber and the matrix and/or due to the higher deformations found in the shear mode.

Because the model discussed herein can be used for more general loadings, including unloading, the method needs to be incorporated into our numerical lamination procedure[9] or perhaps into a more general finite element code for general laminates. These efforts are underway.

We have shown that the methods discussed herein have the potential of predicting lamina compliance surfaces as a function of stress level and temperature. This needs to also be incorporated into our general laminate analysis. With these improvements, the prediction of detailed viscoelastic stress distributions in a general laminated composite structural component will be possible utilizing short term measurements. Local damage and/or failures as a function of time could therefore be anticipated and avoided.

ACKNOWLEDGEMENTS

The financial support provided for this work under NASA Cooperative Agreement NCC 2-71 from the Materials and Test Engineering Branch of NASA-Ames Research Center, Moffett Field, CA, is gratefully acknowledged. Further, sincere appreciation is extended to the grant monitor, Dr H. G. Nelson for his encouragement and support.

REFERENCES

1. BRINSON, H. F., MORRIS, D. H. and YEOW, Y. T., A new experimental method for the accelerated characterization of composite materials, *Sixth International Conference on Experimental Stress Analysis*, Munich, September 18–22, 1978.

2. YEOW, Y. T., The time–temperature behavior of graphite epoxy laminates, *Ph.D. Dissertation*, VPI & SU, Blacksburg, VA, 1978.
3. GRIFFITH, W. I., The accelerated characterization of viscoelastic composite materials, *Ph.D. Dissertation*, VPI & SU, Blacksburg, VA, 1980; also VPI Report VPI-E-80-15, with D. H. Morris and H. F. Brinson.
4. BRINSON, H. F., MORRIS, D. H., GRIFFITH, W. I. and DILLARD, D. A., The viscoelastic response of a graphite/epoxy laminate, *Proceedings of International Conference on Composite Structures*, Paisley, Scotland, September, Applied Science Publishers, 1981.
5. DILLARD, D. A., Creep and creep rupture of laminated graphite/epoxy composites, *Ph.D. Dissertation*, VPI & SU, Blacksburg, VA, 1981; also VPI Report VPI-E-81-3, with D. H. Morris and H. F. Brinson.
6. BRINSON, H. F. and DILLARD, D. A., The prediction of long term viscoelastic properties of fiber reinforced plastics, *Progress in Science and Engineering of Composites* (T. Hayashi, K. Kawata and S. Umekawa, eds), ICCM-IV, Tokyo, 1982.
7. FINDLEY, W. N. and PETERSON, D. B., Prediction of long-time creep with ten-year creep data on four plastic laminates, *ASTM Proc.*, **58**, 1958.
8. DILLARD, D. A., MORRIS, D. H. and BRINSON, H. F., Predicting viscoelastic response and delayed failures in general laminated composites, *ASTM Sixth Conference on Composite Materials: Testing and Design*, Phoenix, AZ, May, 1981.
9. DILLARD, D. A. and BRINSON, H. F., A numerical procedure for predicting creep and delayed failures in laminated composites, *ASTM-STP, Williamsburg Symposium on Long-Term Behavior of Composites*, March, 1982.
10. HIEL, CLEMENT, The nonlinear viscoelastic response of resin matrix composites, *Doctoral Thesis*, Vrije Universiteit Brussel, Jan. 1983. (Also VPI-E-83-6 with A. H. Cardon and H. F. Brinson.)
11. SCHAPERY, R. A., On the characterization of non-linear viscoelastic materials, *Polym. Eng. Sci.*, **9**, No. 4, 1969.
12. LOU, Y. C. and SCHAPERY, R. A., Viscoelastic characterization of a nonlinear fiber-reinforced plastic, *J. Composite Materials*, **5**, 1971.

21

Rotational Strength and Optimal Design of a Hybrid Filament-Wound Disc

MASUJI UEMURA, HISASHI IYAMA and YOSHIKO FUKUNAGA

*Institute of Interdisciplinary Research, Faculty of Engineering,
University of Tokyo, 4-6-1, Komaba, Meguro-ku, Tokyo 153, Japan*

ABSTRACT

*In order to overcome the cracking failure due to radial tensile stress in a
filament-wound composite disc and to increase the rotational velocity, the
hybrid disc for storage of energy is treated here. It is wound by both glass-
and carbon-fibers and has a cold-fitted metallic disc in the center.*

(1) *First, the analytical expressions for estimating the residual curing
stresses by which the cracking and buckling failures take place are
proposed and verified by experiments with a good agreement.*

(2) *Based on the analytical combined stresses and taking account of the
various kinds of failure modes, the limiting rotational velocities can
be predicted with parameters of inner to outer radius ratio and of
hybrid ratio in a filament-wound disc and are verified by spinning
tests.*

(3) *As a result, the optimal disc configuration and hybrid ratio required
to attain the maximum peripheral velocity are obtained.*

1. INTRODUCTION

Much attention has been paid to the high-performance fiber reinforced
plastic flywheels for storing kinetic energy. The circumferentially filament-
wound (FW) disc is the fundamental rotary disc. However, cracks are apt
to occur along fibers owing to the radial tensile stress due to the thermal
curing process and the rotation, because the tensile strength transverse to

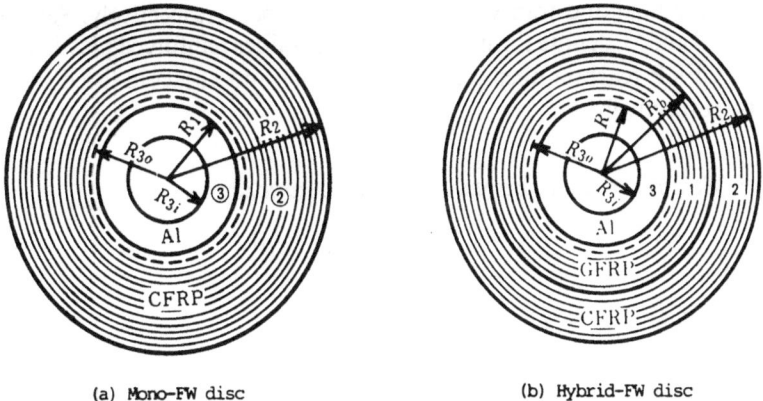

(a) Mono-FW disc   (b) Hybrid-FW disc

FIG. 1. Circumferentially filament-wound discs.

fibers is very low. Hence one of the authors has previously discussed the optimal design of the discs laminated with laminae wound by various kinds of winding patterns.[1,2] As another technique to reduce the radial tensile stress, the hybrid disc in which a metal disc is inserted by cold-fitting as shown in Fig. 1 is considered and the optimal design is discussed.

2. ANALYSIS OF VARIOUS STRESSES IN HYBRID FW DISCS

A hybrid disc wound by glass-fibers on the inner side and carbon-fibers on the outer side where a metal disc is inserted at the center by the cold-fitting technique is considered, to overcome the cracking failure due to the tensile stress normal to fibers. The Mono-FW disc wound by carbon-fibers only is also considered for convenience of comparison. The analytical methods for stresses due to (1) thermal curing process, (2) cold-fitting of metal disc and (3) rotation are briefly described below.

2.1. Analysis of Various Stress Components
2.1.1. Stress (σ)–strain (ε) relations

$$
\left.\begin{array}{l}
\sigma_{ri} = E_{ri}[\varepsilon_{ri} + v_{\theta i}\varepsilon_{\theta i} - (\alpha_{ri} + v_{\theta i}\alpha_{\theta i})\,\Delta T]/(1 - v_{ri}v_{\theta i}) \\
\sigma_{\theta i} = E_{\theta i}[\varepsilon_{\theta i} + v_{ri}\varepsilon_{ri} - (\alpha_{\theta i} + v_{ri}\alpha_{ri})\,\Delta T]/(1 - v_{ri}v_{\theta i})
\end{array}\right\}
\tag{1}
$$

where E is Young's modulus, v is Poisson's ratio, α is thermal expansion coefficient, ΔT is temperature drop after hardening during cure. Suffixes 'r', 'θ' denote the radial and circumferential directions and 'L', 'T' denote the longitudinal and transverse directions to fibers, respectively. Suffixes $i = $ '1', '2' and '3' denote the inner GFRP (glass-fiber reinforced plastic), the outer CFRP (carbon-fiber reinforced plastic) and the cold-fitted metal parts, respectively.

2.1.2. Strain (ε)–radial displacement (U) relations

$$\varepsilon_{ri} = dU_i/dr \qquad \varepsilon_{\theta i} = U_i/r \tag{2}$$

2.1.3. Equilibrium equations during rotation

$$d(\sigma_{ri} r)/dr - \sigma_{\theta i} + \rho\omega^2 r^2 = 0 \tag{3}$$

Substitution of eqns (1) and (2) into eqn. (3) gives the fundamental differential equation for U_i as follows.

$$r^2 \frac{d^2 U_i}{dr^2} + r \frac{dU_i}{dr} - \eta_i^2 U_i$$

$$= [(v_{\theta i} - \eta_i^2)\alpha_{\theta i} + (1 - v_{ri}\eta_i^2)\alpha_{ri}] \Delta T . r - \frac{\rho_i \omega^2 r^3 (1 - v_{ri} v_{\theta i})}{E_{ri}} \tag{4}$$

where $\rho_i = \gamma_i/g$, ρ_i is specific mass, g is gravity constant, γ_i is specific weight, ω is angular velocity, $\eta_i = E_{\theta i}/E_{ri} = v_{\theta i}/v_{ri}$

2.1.4. Radial displacement U_i and stress components σ_{ri}, $\sigma_{\theta i}$

$$\left.\begin{aligned}
U_i &= A_i r^{\eta_i} + B_i r^{-\eta_i} + \xi_i r + \rho_i \omega^2 (1 - v_{ri} v_{\theta i}) r^3 / E_{ri}(\eta_i^2 - 9) \\
\sigma_{ri} &= E_{ri}[A_i(\eta_i + v_{\theta i})r^{\eta_i - 1} - B_i(\eta_i - v_{\theta i})r^{-(\eta_i + 1)} + \mu_{ri}]/(1 - v_{ri} v_{\theta i}) \\
&\quad + \{(3 + v_{\theta i})\rho\omega^2 r^2/(\eta_i^2 - 9)\} \\
\sigma_{\theta i} &= E_{\theta i}[A_i(1 + \eta_i v_{ri})r^{\eta_i - 1} + B_i(1 - \eta_i v_{ri})r^{-(\eta_i + 1)} + \mu_{\theta i}]/(1 - v_{ri} v_{\theta i}) \\
&\quad + \{(1 + v_{ri})\rho\omega^2 r^2 E_{\theta i}/E_{ri}(\eta_i^2 - 9)\}
\end{aligned}\right\} \tag{5}$$

$$\mu_{ri} = \frac{(\alpha_{ri} - \alpha_{\theta i})(\eta_i^2 - v_{\theta i}^2) \Delta T}{1 - \eta_i^2}$$

$$\mu_{\theta i} = \frac{(\alpha_{ri} - \alpha_{\theta i})(1 - v_{ri} v_{\theta i}) \Delta T}{1 - \eta_i^2}$$

$$\xi_i = [(v_{\theta i} - \eta_i^2)\alpha_{\theta i} + (1 - v_{ri}\eta_i^2)\alpha_{ri}] \Delta T/(1 - \eta_i^2)$$

2.2. Thermal Residual Stress During Cure

The thermal residual stresses can be obtained by setting $\omega = 0$ in eqns (5) and by determining the four unknown constants A_i, B_i ($i = 1, 2$) under the following boundary conditions.

$$r = R_1; \; \sigma_{r1} = 0 \qquad r = R_2; \; \sigma_{r2} = 0 \qquad r = R_b; \; \sigma_{r1} = \sigma_{r2} \qquad U_1 = U_2 \tag{6}$$

where R_1 and R_2 denote the inner and outer radius of the FW disc, respectively and R_b is the boundary radius between GFRP and CFRP parts.

2.3. Stresses Due to Rotation and Cold-fitting of Metal Disc

The stresses due to rotation and cold-fitting of metal disc can be analysed simultaneously by setting $\xi_i = \mu_{ri} = \mu_{\theta i} = 0$ in eqns (5) and by determining the six unknown constants A_i, B_i ($i = 1, 2, 3$) under the following boundary conditions:

$$\left. \begin{array}{c} r = R_{3i}; \; \sigma_{r3} = 0 \qquad r = R_1; \; \sigma_{r3} = \sigma_{r1} \qquad (U_1)_{r=R_1} - (U_3)_{r=R_{3o}} = \delta \\[2mm] r = R_b; \; \sigma_{r1} = \sigma_{r2} \qquad U_1 = U_2 \qquad r = R_2; \; \sigma_{r2} = 0 \end{array} \right\} \tag{7}$$

where R_{3i} and R_{3o} denote the inner and outer radius of cold-fitted metal disc at room temperature. For the Al-alloy disc, $E_{r3} = E_{\theta3} = E_a$, $v_{r3} = v_{\theta3} = v_a$, $\eta_3 = 1$.

3. DISCUSSION OF THERMAL RESIDUAL STRESS DURING CURE

In the analysis described in Section 2.2, the thermal expansion coefficient α is assumed as constant for simplicity of analysis. The problem here is how to estimate a reasonable temperature change during cure.

3.1. Numerical Example

The material constants such as elastic constants and thermal expansion coefficients of unidirectional composites used in numerical examples are shown in Table 1. These values are those obtained by the formulas proposed by one of the authors and verified by experiments.[3,4] As an example, the stress distributions of $\sigma_\theta(\sigma_L)$ and $\sigma_r(\sigma_T)$ in the case of $\lambda = 0.44$ and $\Delta T = -100\,°C$ are shown in Fig. 2(a) and (b), respectively, with a parameter of ζ_b, where λ denotes the inner to outer radius ratio $= R_1/R_2$, $\zeta = r/R_2$ and $\zeta_b = R_b/R_2$.

TABLE 1
Material constants values used for numerical calculations

Material	Elastic constant (kg/mm^2)	Poisson's ratio	Thermal expansion coeff. (1/°C)	Density (g/cm^3)	Failure strength (kg/mm^2)
GFRP ($V_f = 60\%$)	$E_L = 4\,580$ $E_T = 1\,424$	$v_L = 0.257$ $v_T = 0.080$	$\alpha_L = 7.2 \cdot 10^{-6}$ $\alpha_T = 31 \cdot 10^{-6}$	2.07	$F_L = 120$ $F_T = 2.5\text{--}3.5$
CFRP ($V_f = 60\%$)	$E_L = 13\,940$ $E_T = 833$	$v_L = 0.316$ $v_T = 0.019$	$\alpha_L = -0.037 \cdot 10^{-6}$ $\alpha_T = 37 \cdot 10^{-6}$	1.5	$F_L = 150$ $F_T = 5.0$
Al-alloy	$E_a = 7\,400$	$v_a = 0.3$	—	2.8	$F_t = 60$

It can be seen from Fig. 2(a) that the compressive hoop stress occurs near the outer edge and may result in thermal buckling. It can be also seen from Fig. 2(b) that the radial tensile stress takes place in the intermediate region and may result in cracking because of the low value of tensile strength normal to fibers, F_T.

3.2. Experiments for Thermal Residual Stress

3.2.1. Disc specimens

The disc configurations are shown in Fig. 1 and R_1 is kept constant at 50 mm. The boundary between CFRP and GFRP is positioned at the middle of the FW disc, that is, $R_b = (R_1 + R_2)/2$. The constitutive materials used in the specimens are as follows:

(1)   Glass-fibers—glass roving ER 2310.
(2)   Carbon-fibers—Torayca T-300B, 6000 filaments.
(3)   Epoxy resin—Epikote 828 (100 parts), hardener NMA (80 parts), additive TDMP (1 part).

A rubber type resin is used only for the NM-280-C-1 specimen.
The following two series of specimens are used.

Series I

The dimensions of specimens and the experimental results are summarised in Table 1.

Curing condition; $T_{max} = 170$ °C; room temperature $T_r = 25$ °C—(4 h)—90 °C/(2 h)—(1.5 h)—170 °C/(4 h)—gradual   cooling.   Outer   radius $R_2 =$ about 75, 112.5, 140, 150 mm, thickness $h = 5.7$ mm.

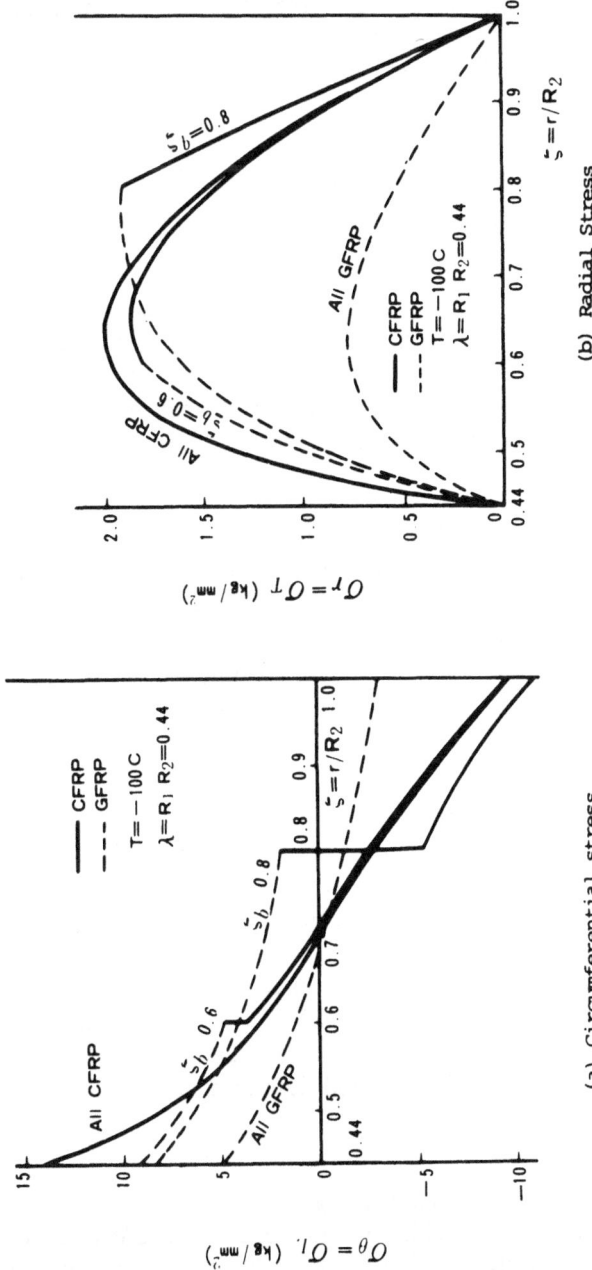

(a) Circumferential stress

(b) Radial Stress

FIG. 2. Distributions of residual curing stresses in hybrid-FW disc.

Series II

Curing condition; $T_{max} = 120$, 145, 170 °C; room temperature, $T_r = 25$ °C—(10 h)—90 °C—(10 h)—T_{max}—gradual cooling. Outer radius $R_2 = 112 \cdot 5$ mm, thickness $h = 7$ mm.

3.2.2. Comparison with analytical results

The residual internal stresses were measured by the following two methods. The strain gauges were placed at several points in both radial and circumferential directions.

(1) Method to cut away the inner edge successively—the residual stresses can be calculated from the circumferential strains on the outer edge.

(2) Method to split disc specimens into small pieces—the residual stresses can be calculated from the two-dimensional strains relieved by splitting.

The residual stress distributions thus measured in Mono-FW and Hybrid-FW discs in Series-I are compared, in Fig. 3, with the analytical ones obtained by setting $\Delta T = -125$ °C. The analytical values seem to agree with the measured ones.

As for the ΔT value necessary for calculating the residual stresses, the following expression is recommended for use.

$$-\Delta T = T_{max} - T_r - 20 \, °C \tag{8}$$

In the Series-I specimens, $T_{max} = 170$ °C and $T_r = 25$ °C and hence $\Delta T = -125$ °C is used for estimation. Ohnishi and Tanaka[5] proposed previously that the temperature at which the hardening began was about 150 °C. This agreed well with our proposal because $T_{max} - 20 \, °C = 150 \, °C$ as found from the experiments.

In Series-II specimens, too, the measured stresses in the specimens cured at $T_{max} = 120$, 145 and 170 °C agreed well with the calculated stresses derived by assuming $\Delta T = -75$, -100 and -125 °C, respectively. Accordingly, the estimation method described above was verified to be reasonable.

3.2.3. Buckling due to residual curing stress

The specimens which buckled or not after curing in FW disc are marked by O or X, respectively, in Table 2. The buckling occurs due to the residual

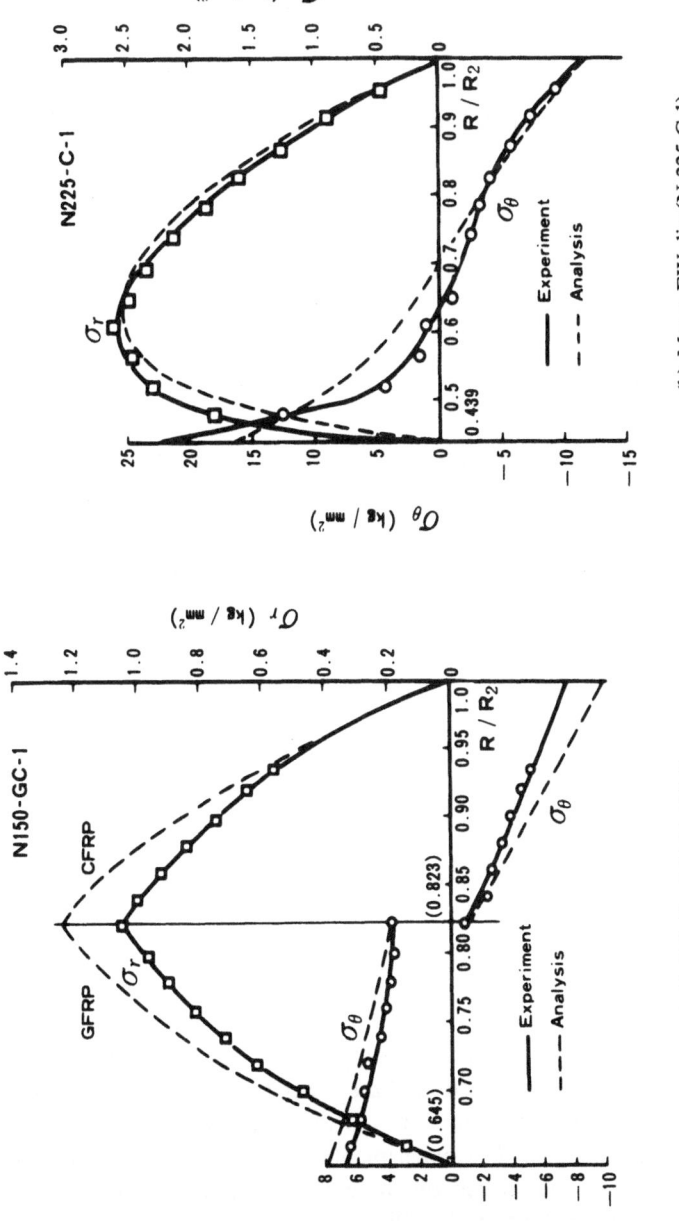

(a) Hybrid-FW disc(N-150-GC-1)

(b) Mono-FW disc(N-225-C-1)

FIG. 3. Comparison of analytical and measured residual stress distributions.

Masuji Uemura, Hisashi Iyama and Yoshiko Fukunaga

hoop compressive stress near the outer edge and the critical temperature change ΔT_{cr} can be obtained analytically based on the energy method.

(1)   Bending strain energy due to buckling:

$$U = \frac{1}{2} \int_0^{2\pi} \int_{R_1}^{R_2} \left[D_r \left(\frac{\partial^2 w}{\partial r^2} \right)^2 + 2D_r \nu_\theta \frac{\partial^2 w}{\partial r^2} \left(\frac{\partial^2 w}{\partial \theta^2} \frac{1}{r} + \frac{1}{r} \frac{\partial w}{\partial r} \right) \right.$$
$$\left. + D_\theta \left(\frac{1}{r^2} \frac{\partial^2 w}{\partial \theta^2} + \frac{1}{r} \frac{\partial w}{\partial r} \right)^2 + 2D_{r\theta} \left\{ \frac{\partial}{\partial r} \left(\frac{1}{r} \frac{\partial w}{\partial \theta} \right) \right\}^2 \right] r \, dr \, d\theta \qquad (9)$$

where w is buckling deflection, h is thickness.

$$D_r = E_r h^3 / 12(1 - \nu_r \nu_\theta) \qquad D_\theta = E_\theta h^3 / 12(1 - \nu_r \nu_\theta) \qquad D_{r\theta} = G_{r\theta} h^3 / 6$$

(2)   Work done by the residual membrane stress in buckling

$$W = \frac{1}{2} \int_0^{2\pi} \int_{R_1}^{R_2} \left[\sigma_r \left(\frac{\partial w}{\partial r} \right)^2 + \sigma_\theta \frac{1}{r^2} \left(\frac{\partial w}{\partial \theta} \right)^2 \right] h \cdot r \, dr \, d\theta \qquad (10)$$

(3)   Buckling deflection is assumed as

$$w = \sum_{n=2}^{\infty} W_n(r) \sin n\theta \qquad W_n(r) = \sum_{m=2}^{\infty} C_{mn} r^m \qquad (11)$$

(4)   Mechanical boundary conditions;

$$r = R_1, R_2 \qquad M_r = 0, Q_r + \partial M_{r\theta}/r\partial\theta = 0 \qquad (12)$$

The number of undetermined constants C_{mn} is reduced by using eqns (12). Based on the stationary principle of total potential energy $\Pi = U - W$, the minimisation of Π with respect to the independent constants C_{mn}

$$\partial(U - W)/\partial C_{mn} = 0 \qquad (13)$$

gives the critical temperature change ΔT_{cr} as eigenvalue. ΔT_{cr} is proportional to h^2. The variation of ΔT_{cr} with the outer diameter, $2R_2$, for the case of $h = 5$ mm and 7 mm are plotted by the solid and dotted lines, respectively, in Fig. 4. The experimental results on the occurrence of thermal buckling are plotted on these curves. As the boundary critical temperature on whether the buckling occurred or not, $\Delta T = -125\,°C$ is found to be reasonable as predicted by eqn. (8).

TABLE 2
Specimens used in experiment (I) and results

	Outer diameter $(2R_2)$ (mm) Nominal	Real	Thickness (mm)	Specimen No.	Resin*	Buckling Yes or no	Calculated ΔT_{cr} (°C)	Cracking in experiments Yes or no	Radial coordinate (mm)	Analytical max. radial stress Radial coordinate (mm)	$\sigma_{r,max}$ (kg/mm²)
Mono-FW	225	228	5	N-255-C-1	N	O	−123·3	X		72	2·53
Disc	280	287	7	N-280-C-1	N	X	−144·3	O	66	79	3·07
(CF only)	280	281	7	NM-280-C-1	S	X		X***	(65)**		
	300	308	5	N-300-C-1	N	O	−63·4	X		82	3·20
Hybrid	150	155	5	N-150-GC-1	N	X		X		64	1·24
FW	225	232	5	N-225-GC-1	N	O		O	81	82	2·68
Disc	280	286	7	N-280-GC-1	N	X		O	63, 69	97	3·28
(GF + CF)	280	289	7	N-280-GC-2	N	X		O	74	97	3·30
	300	306	5	N-300-GC-1	N	O		O	100	101	3·44
	300	305	5	N-300-GC-2	N	O		O	89	101	3·43

* N: normal epoxy resin; S: soft epoxy resin of rubber type.
** Crack occurrence during cutting.
*** Stress release due to large deformation buckling.

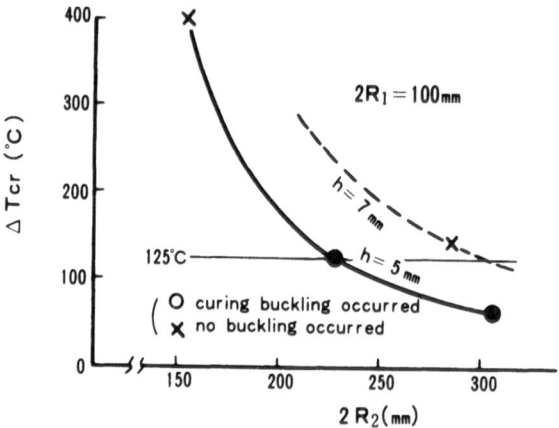

FIG. 4.   Variations of critical temperature change T_{cr} for buckling with outer diameter and
experimental results in Mono-FW disc.

3.2.4. *Cracking due to residual curing stress*

The specimens which exhibited cracks during cure are marked by O in
Table 2, together with the analytical values of $\sigma_{r,max}$ and their radial
coordinates obtained by setting $\Delta T = -125\,°C$. The variations of $\sigma_{r,max}$
with the outer diameter, $2R_2$, in Mono-FW and Hybrid-FW discs are
shown by the dotted and solid lines, respectively, in Fig. 5. The
experimental results for occurrence of cracking are plotted on these curves.
The boundary for $\sigma_{r,max}$ on whether the cracking occurs or not, seems to be
$2.5\,kg/mm^2$ which corresponds to the F_T value for unidirectional GFRP.

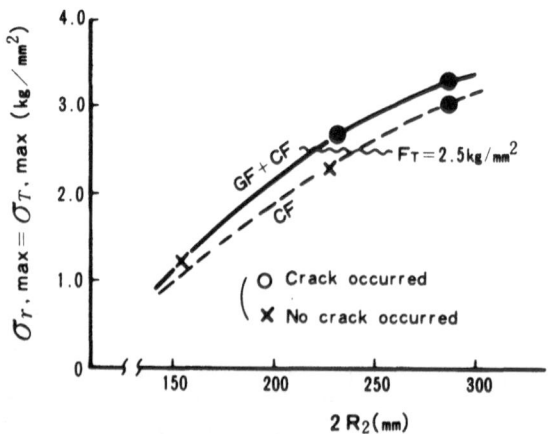

FIG. 5.   Variations of maximum radial curing stress and occurrence of cracking failure with
outer diameter of FW disc.

4.   COMBINED STRESSES IN HYBRID-FW DISC WITH COLD-FITTED METAL DISC

The combined stresses due to thermal cure, cold-fitting and rotation are analysed with use of normalised parameters of $\lambda = R_1/R_2$, $\zeta_b = R_b/R_2$ and $\bar{\delta} = \delta/R_2$ etc. The material constants shown in Table 1 and $\Delta T = -125\,°\mathrm{C}$ are used for numerical calculations.

4.1. Stress Due to Cold-fitting

An Al-alloy disc is inserted inside the FW disc after cooling to low temperature. As an example, the stresses due to cold-fitting are shown in Fig. 6 with a parameter of ζ_b for the case of $\lambda = 0.44$, $\bar{\delta} = 0.001\,33$. The

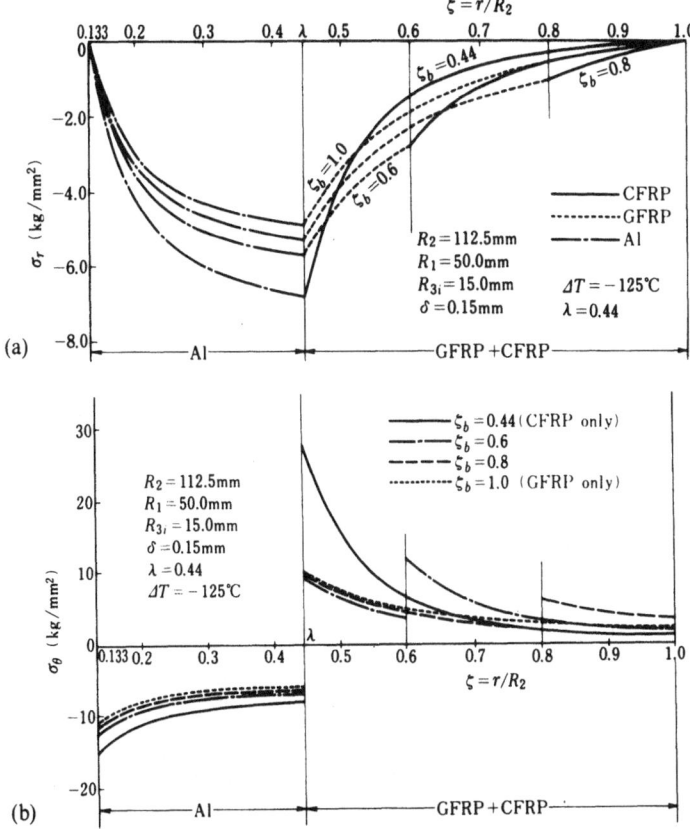

FIG. 6.   Distributions of stresses due to cold-fitting of metal disc. (a) Radial stress. (b) Hoop stress.

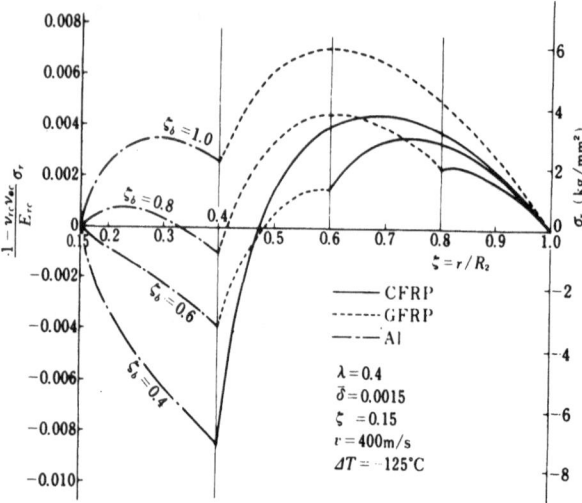

FIG. 7. Distributions of combined radial stresses under $v = 400\,\text{m/s}$ with a parameter of boundary radius in hybrid disc.

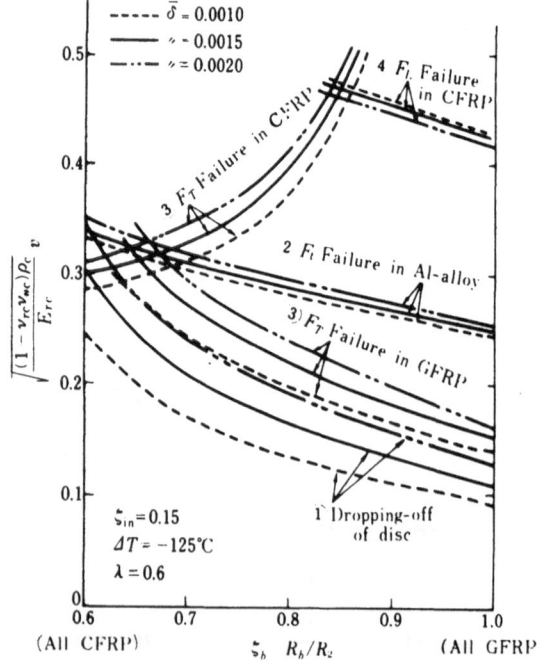

FIG. 8. Variations of limiting peripheral velocities based on four kinds of failure modes with boundary radius ζ_b.

compressive radial stress σ_r and the tensile hoop stress σ_θ occurring in the FW disc become effective in overcoming the cracking and the buckling. Needless to say, this effect is remarkable with an increase in δ.

4.2. Combined Stresses in Hybrid-FW Disc

The combined radial stresses of σ_r for the same case as in Fig. 6 and under $v = 400$ m/s are shown in Fig. 7 with a parameter of ζ_b. The combined stress is decreased by cold-fitting; however, $\sigma_{r,max}$ decrease in CFRP, but increase in GFRP. The radial stress, σ_r, at the inner edge, that is, at $\zeta = \lambda$ becomes positive with increasing ζ_b, which results in the dropping-off of the FW disc from the central disc with driving shaft. Hence it shows a need to optimise the hybridisation for ζ_b.

4.3. Maximum Rotational Limit of Hybrid-FW Disc

The following four failure modes are considered to limit the high-speed rotation.

(1) Dropping-off of FW disc when $\sigma_r = 0$ at inner edge (at $\zeta = \lambda$).
(2) Failure due to σ_r or σ_θ in Al-alloy disc.
(3) F_T-failure due to σ_r in GFRP or CFRP FW disc.
(4) F_L- failure due to σ_θ in GFRP or CFRP FW disc.

The limiting peripheral velocity v_1 due to the four failure modes

$$v_1 = \omega_1 R_2 = (2\pi N_1)R_2 \tag{14}$$

where ω_1 and N_1 are the limiting angular velocity and number of rotations, respectively, and are shown with a parameter of $\bar{\delta}$ for the case of $\lambda = 0.6$ in Fig. 8. As $\bar{\delta}$ is increased, the limiting velocities for most failure modes increase except for mode (4) F_L-failure in CFRP.

5. SPINNING TESTS OF A HYBRID DISC AND COMPARISON WITH ANALYTICAL RESULTS

5.1. Specimen Configurations and Composition

The dimensions and constitutions of disc specimens as shown in Fig. 1 are listed in Table 3, together with the analytical and experimental results on the critical limiting peripheral velocities.

5.1.1. Specimen configurations

FW disc; inner radius $R_1 = 50.1$ mm, outer radius $R_2 = 112.5$ mm, three values of boundary radius $R_b = 61.9$, 81.0, 96.8 mm

TABLE 3

Disc specimens and comparison of analytical and experimental critical peripheral speeds v_1

Kind of specimens	Boundary of GF and CF parts $\zeta_b = R_b/R_2$	Disc thickness t (mm)	Critical peripheral speeds v_1 (m/s)			Failure modes
			Analytical		Experimental Critical	
			Due to drop-off of FW disc	Due to F_T failure		
Mono-FW (CFRP only)	0·44	5	899	558**	553	X
		7			506	X
Hybrid-FW	0·55 ($V_f = 55\%$)	7	568	600**	583	X
		7		713*	577	O
	0·72	5	425	763**	434	O
		7		471*	424	O
	0·86	7	358		365	O
		7		370*	383	O
Mono-FW (GFRP only)	1·00	7	300	320*	324	O
		7			331	O

* F_T failure in GFRP.
** F_T failure in CFRP.
Failure mode X: F_T failure; O: dropping-off of FW disc.

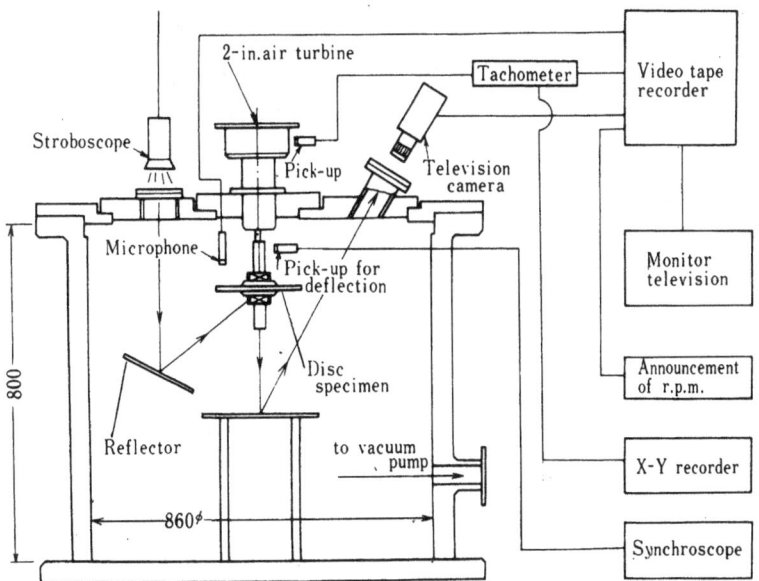

FIG. 9. Arrangement of apparatus for spinning test.

Al-alloy disc; inner radius $R_{3i} = 15 \cdot 0$ mm, outer radius $R_{3o} = 50 \cdot 25$ mm
Radial cold-fitting length $\delta = R_{3o} - R_1 = 0 \cdot 15$ mm
Thickness $h = 5$ mm, 7 mm.

5.1.2. Specimen composition and curing condition

The specimen composition and the curing condition are almost the same as those described in Section 3.2.1 for Series-I specimens. The fiber volume fraction is nearly 60%, but 55% for the specimens with $\zeta_b = 0 \cdot 55$.

5.2. Experimental Apparatus for Spinning and Instrumentation

The schematic arrangement of apparatus for spinning test is shown in Fig. 9. The driving is done by air turbine. The dynamic balance is checked before rotation by a dynamic balancing machine. The cracking pattern can be observed as a stationary figure by using a stroboscope synchronised with rotation. A television camera and a microphone are used to record the failure behaviour and the cracking sound, respectively. The vibration of the driving shaft is recorded by a synchroscope.

5.3. Experimental Results and Comparison with Analytical Results

The limiting peripheral velocity v_1 and the two observed failure modes,

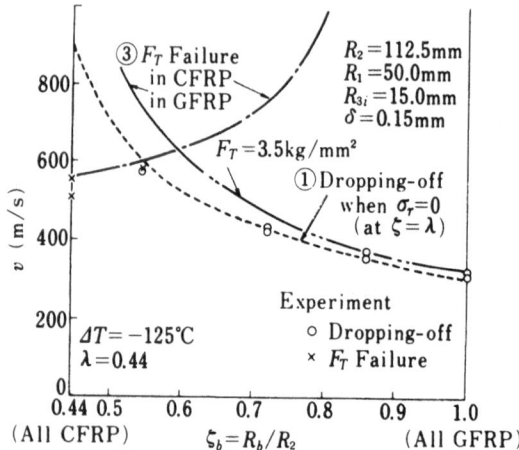

FIG. 10.   Comparison of analytical limiting peripheral velocities with experimental results in cold-fitted hybrid discs.

that is, the dropping-off of the FW disc and the F_T-failure in CFRP are shown with the marks O and X, respectively, in Table 3 and in Fig. 10. The analytical values for v_1 in Table 3 are obtained by using the material characteristic values shown in Table 1. Three points can be noted.

(1)   The limiting peripheral velocity v_1 obtained in experiments are found to agree well with the analytical results.

(2)   In the range of $\zeta_b = 0.44$–0.53, the cracks occurred along fibers in CFRP due to the high radial stress σ_r, resulting in the instantaneous scattering of the FW disc or in the dropping-off of the outer ring disc. While, in the range of $\zeta_b = 0.53$–1.0, the FW discs dropped off from the central Al-alloy disc as predicted by analysis.

(3)   The maximum limiting peripheral velocity for the case of $\lambda = 0.44$ was achieved in the specimen with $\zeta_b = 0.55$ as predicted by analysis.

6.   OPTIMAL DESIGN OF FILAMENT-WOUND DISC

Based on the experimental evidence described above, the optimal constitution of an FW disc is discussed.

The limiting peripheral velocities v_1 due to the four kinds of failure modes are given depending on the normalised configuration, the cold-fitting length and the characteristic strength as shown in Table 1. The

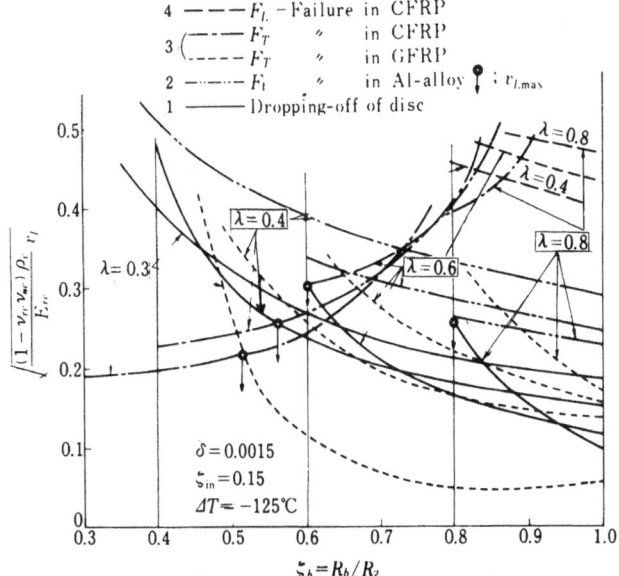

FIG. 11.   Variations of limiting peripheral velocities based on four kinds of failure modes with boundary radius ζ_b.

variation of the four kinds of v_1 with ζ_b for the case of $\zeta_{in} = R_{3i}/R_2 = 0.15$, $\bar{\delta} = \delta/R_2 = 0.0015$, $\Delta T = -125\,°C$ are shown in Fig. 11 with a parameter of λ.

6.1.1.   *Optimal hybrid ratio, $\zeta_{b,op}$*

The optimal hybrid ratio, $\zeta_{b,op}$, at which the maximum velocity $v_{1,max}$ is achieved can be determined from the maximum of the lower curves at each value of λ. It is found from Fig. 11 that the failure modes at $v_{1,max}$ are (3) F_T-failures in CFRP and GFRP at $\zeta_b = 0.51$ in the case of $\lambda = 0.3$; (3) F_T-failure in CFRP and (1) dropping-off of the FW disc at $\zeta_b = 0.56$ and $\zeta_b = 0.6$ in the case of $\lambda = 0.4$ and 0.6, respectively and so on.

6.1.2.   *Optimal inner to outer radius ratio, λ_{op}*

The optimal boundary radius $\zeta_{b,max}$ and the maximum rotational velocity $v_{1,max}$ thus determined are shown against λ in Fig. 12. It can be found from Fig. 12 that the maximum rotational velocity $v_{1,max}$ can reach $\sqrt{E_{rc}/(1 - v_{rc}v_{\theta c})\rho_c} \cdot 30 \doteqdot 700\,m/s$ at the inner to outer radius ratio $\lambda = 0.6$ with $\zeta_b = 0.6$. This means that the Mono-FW CFRP is best.

When an FW disc is used for a flywheel to store kinetic energy, the

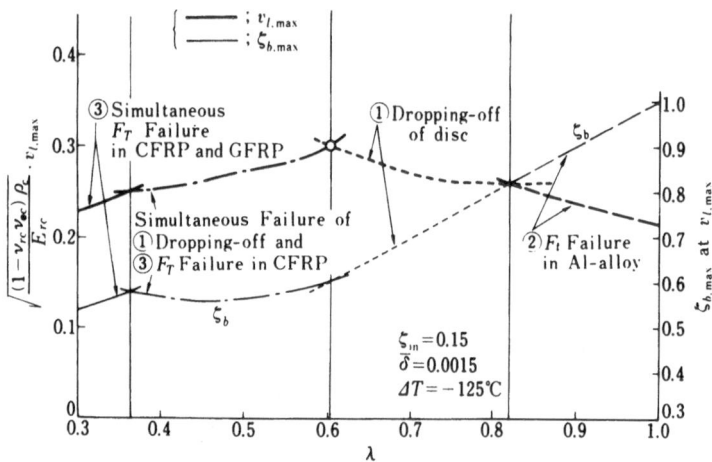

FIG. 12.   Variations of maximum limiting peripheral velocities with inner-to-outer radius
ratio λ.

following energy density, that is, energy-to-weight ratio e_w is chosen as the
objective function in place of the rotational velocity.

$$e_w = \frac{E}{W} = \frac{v^2}{4g}\left[\frac{\rho_c(1 - \zeta_b^4) + \rho_G(\zeta_b^4 - \lambda^4) + \rho_{A1}(\lambda^4 - \zeta_{in}^4)}{\rho_c(1 - \zeta_b^2) + \rho_G(\zeta_b^2 - \lambda^2) + \rho_{A1}(\lambda^2 - \zeta_{in}^2)}\right] \quad (15)$$

where E is rotational kinetic energy, W is disc weight.

Even if the energy density e_w is evaluated in place of $v_{1,max}$ discussed
above, the resulting conclusion is not so changed, because the correction
term in [ ] in eqn. (15) is not so effective. If the dropping-off of the FW disc
is prevented by any other technique when $\sigma_r = 0$ at the inner edge of FW
disc, then the rotational velocity will increase further except for the disc
with small λ.

7.   CONCLUSIONS

The hybrid filament-wound composite disc with a cold-fitted metal disc is
discussed with a view to attaining high performance and the analytical
formulae and methods for optimal design have been presented in the
present paper.

(1)   The residual curing stress can be estimated reasonably by assuming
the temperature change during cure process as $-\Delta T = T_{max} -
20\,°C - T_r$ where T_{max} is the maximum cure temperature and T_r is

the room temperature. The analytical curing stress distributions were verified well by experiments.

(2)   The experimental evidence that the buckling and cracking failures occurred after the cure process were explained with good agreement with the above analytical predictions.

(3)   The two techniques of hybrid winding and cold-fitting were found to be effective in increasing the rotational velocity in FW discs.

(4)   Taking account of several kinds of failure modes based on the combined stresses, the limiting rotational velocities can be predicted with parameters of inner-to-outer radius ratio and of hybrid ratio in FW disc and were verified by spinning tests.

(5)   The optimal values for disc configuration and hybrid ratio for attaining the maximum peripheral velocity or rotational energy density were determined.

(6)   The analytical expressions proposed by the senior author to estimate the material properties such as elastic constants and thermal expansion coefficients and the temperature change during cure were found to be reasonable.

REFERENCES

1. BYON, O., UEMURA, M. and ENDO, M., Optimization of laminated flywheels reinforced by carbon-fiber, *Trans. Japan Soc. Mech. Eng.*, **45** (1979), 505–515.
2. BYON, O. and UEMURA, M., Optimal design of fiber composite flywheels reinforced besides circumferentially, *Proc. 23rd National SAMPE Conference* (Anaheim) (1975), 728–739.
3. UEMURA, M., IYAMA, H. and YAMAGUCHI, Y., Elastic constants of carbon-fiber reinforced plastic materials, *J. Soc. Mater. Sci.*, **24** (1975), 156–163.
4. UEMURA, M., IYAMA, H. and YAMAGUCHI, Y., Thermal expansion coefficients and residual stresses in filament-wound CFRP materials, *J. Japan Soc. Aero. Space Sci.*, **26** (1978), 471–478.
5. OHNISHI, H. and TANAKA, T., Buckling analysis due to residual stress in CFRP disc, *Proc. 1st Symp. Composite Materials, Japan* (1975), 62–65.

22

Simplified Design Procedures for Composite Plates Under Flexural Loading

A. F. Johnson and G. D. Sims

Division of Materials Applications, National Physical Laboratory, Teddington, Middlesex TW11 0LW, England

ABSTRACT

The paper describes simplified procedures for the design analysis of rectangular plates loaded in flexure and compares predicted results with test data on a range of fibre reinforced plastics plates. Design formulae and design charts are given for determining maximum deflections and bending moments in rectangular orthotropic elastic plates under transverse pressure or point loads. The design methods are based on computed solutions to the orthotropic plate equations whose results are presented as simplified design charts appropriate to composite applications and depending mainly on a single stiffness ratio parameter. Design formulae, derived from approximate variational solutions to the orthotropic plate, are shown to be in good agreement with the computed solutions especially for materials with low and moderate stiffness ratios. Flexure tests are reported on a range of reinforced plastics plates with mat, woven, unidirectional or cross-ply reinforcement and subjected to centre point transverse loads. Measured deflections are compared with predicted values and the validity of the design procedures is assessed.

1. INTRODUCTION

Current work on designing with reinforced plastics is aimed mainly at aerospace applications and is based on detailed computer stress analysis for highly anisotropic materials. Such an approach is not always suitable when materials with glass or aramid fibre reinforcement are used in applications such as storage tanks, boats, cladding panels, motor vehicles, etc. In these medium technology industries, in the absence of suitable design methods, stress analysis is frequently based on conventional isotropic design formulae and the effects of material anisotropy and inhomogeneity are neglected. There is thus a need for design procedures, particularly for the high volume glass reinforced plastics, which extend isotropic formulae to take account of material anisotropy. The paper describes such simplified design procedures suitable for the design analysis of rectangular composite plates loaded in flexure, and compares predicted results with test data on a range of fibre reinforced plates containing mat, woven, unidirectional and cross ply reinforcement.

The design methods described in Section 2 are based on a detailed finite difference analysis of the small deflection orthotropic elastic plate equations.[1] Numerical results computed using a general orthotropic plate analysis program[2] are presented as simplified nondimensional design charts for a plate stiffness parameter α and stress parameters β_1, β_2, for rectangular plates under pressure and centre point loads. It is found that for many composite materials of practical interest such as glass, carbon and Kevlar fibre reinforced plastics these design parameters often depend only on a single stiffness ratio parameter. Together with elementary design formulae, the design charts enable maximum deflections, bending moments and stresses in orthotropic plates under flexural loading to be calculated. The design methods are thus an extension to anisotropic materials of the traditional methods used for isotropic plates[3,4] based on design formulae and tables of design constants.

The design charts for the stiffness and stress parameters are supplemented in Section 3 by approximate design formulae for these parameters in terms of plate geometry and flexural rigidities. The formulae are derived from approximate variational solutions to the orthotropic plate equations. Comparison with the computed solutions shows good agreement for materials with low and moderate anisotropy ratios. Use of the approximate formulae enables the design method to be extended to a wider range of plate aspect ratios and material stiffness ratios than those shown in the design charts.

Section 4 describes a series of transverse centre point loading tests on a range of reinforced plastics plates which were carried out to demonstrate the validity of the simplified design methods. The tests show that in the small deflection region the measured maximum deflections are in good agreement with those calculated from the design methods. It is also seen that for many materials the design methods may be applied with reasonable accuracy without a complete set of anisotropic elastic constants.

2. ANALYSIS OF RECTANGULAR ORTHOTROPIC PLATES

The paper is concerned with the design analysis of orthotropic plates in order to predict maximum deflections and maximum bending moments in the plate. It is thus assumed that the flexural rigidities are known, either from measurement of material properties or for laminated composites by calculation in terms of ply lay-up and thicknesses. The designer may then use the calculated maximum deflections and bending moments for comparison with imposed deflection limits and given ultimate bending moments for the plate, or to predict maximum stresses for comparison with material failure stresses. Alternatively[5] deflection and stress limited plate thicknesses may be calculated and hence the plate design thickness specified.

2.1. Definition of Plate Properties

Consider a rectangular plate with sides a, b and uniform thickness h composed of orthotropic elastic material with material symmetry axes parallel to the plate edges, as shown in Fig. 1. The plate is supported at the edges and loaded by a transverse load P applied over all or part of the plate. Attention is restricted to the small deflection flexural behaviour of the plate, with maximum transverse deflections assumed to be less than the plate thickness. The main stiffness parameters needed to characterise the flexural properties of orthotropic plates are the flexural rigidities D_1, D_2, D_3. Let x, y be rectangular co-ordinates as indicated in Fig. 1 with the x-axis lying along the side of length a. D_1 and D_2 are the plate flexural rigidities in

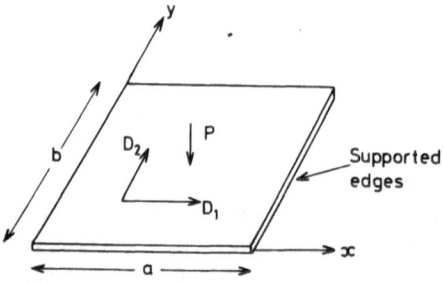

FIG. 1.   Schematic diagram of rectangular orthotropic plate under flexural load.

the x and y directions respectively, defined per unit plate width, and D_3 is defined in terms of the plate shear properties, see below.

In the case of a plate composed of a homogeneous orthotropic material, the flexural rigidities may be related to the material elastic constants and the plate thickness h. An orthotropic plate has two orthogonal symmetry directions in the plane of the plate and its elastic response is characterised by four independent elastic constants. When the two symmetry directions are parallel to the axes of a rectangular plate, it is termed specially orthotropic and this is the situation considered here. Let E_1, E_2 be the plate Young's moduli in the x and y directions, respectively, G the in-plane shear modulus and v_{12}, v_{21}, the plate Poisson's ratios. We adopt the standard convention that v_{12} is the strain in the y direction per unit strain in the x direction, when the plate is loaded in the x direction. Of these five elastic constants only four are independent since the Poisson's ratios are connected by the equation

$$v_{12}E_2 = v_{21}E_1 \tag{1}$$

In terms of these elastic constants and the plate thickness the flexural rigidities are defined by

$$D_1 = \frac{E_1 h^3}{12(1 - v_{12}v_{21})} \qquad D_2 = \frac{E_2 h^3}{12(1 - v_{12}v_{21})}$$

$$D_{12} = v_{12}D_2 = v_{21}D_1$$

$$D_3 = D_{12} + Gh^3/6 \tag{2}$$

An important class of specially orthotropic plates are glass or carbon fibre reinforced plastics with undirectional fibres aligned in the x or y directions. Short fibre reinforced thermoplastics plates may also contain aligned fibres as a result of flow during moulding. We may refer to E_1/E_2 or D_1/D_2 as the plate anisotropy ratio for these materials. A second class of specially orthotropic materials consists of composites with balanced woven reinforcement and warp or weft directions parallel to the x axis of the plate. For such materials the anisotropy ratio $D_1/D_2 = 1$ but they are not isotropic since the shear modulus G is independent of the principal Young's modulus E_1 and does not satisfy the isotropic relation $G = E/2(1 + v)$. In this case three elastic constants characterise the material, $E_1 (= E_2)$, G, v_{12} ($= v_{21}$) and $D_1 = D_2 \neq D_3$. For laminates consisting of different reinforcement types or for unidirectional cross-ply materials the plate flexural rigidities depend on the laminate construction so that the relations (1) and

(2) no longer hold. Such plates may still be characterised in terms of flexural rigidities D_1, D_2, D_3 but further formulae are required to define the flexural rigidities in terms of individual ply moduli and thicknesses. These additional relations may be obtained from laminated plate theory[6] or to a good approximation from simplified analyses.[5]

A detailed analysis of orthotropic plates[1,5] shows that the key material parameters for design calculations are not the flexural rigidities D_1, D_2, D_3 but the rigidity ratios D_1/D_2 and D_3/D_2 along with a Poisson's ratio v_{12}. In order to find appropriate values for the rigidity ratios of reinforced plastics,

TABLE 1

Typical values of moduli and rigidity ratios for fibre reinforced plastics

Material	E_1 (GPa)	E_2 (GPa)	G (GPa)	v_{12}	D_1/D_2	D_3/D_2
CSM/polyester	8	8	3	0·32	1	1
WR/polyester	15	15	4	0·15	1	0·67
Glass fabric/polyester	25	25	4	0·17	1	0·49
UD glass/polyester	40	10	4	0·3	4	1·04
UD Kevlar/epoxide	76	8	3	0·34	9·5	1·08
UD carbon/epoxide	148	10	4	0·31	14·8	1·09

we list in Table 1 some typical modulus data for a range of materials, along with values of D_1/D_2 and D_3/D_2 calculated from eqns (2). The materials data[7,8] refer to GRP materials with chopped strand mat (CSM), woven rovings (WR) and glass fabric reinforcement and to unidirectionally (UD) reinforced glass/polyester, Kevlar/epoxide and Type II carbon/epoxide composites. It is assumed that the UD fibres are aligned in the x direction in the plate so that $D_1/D_2 > 1$ for these materials. For the UD composites the flexural rigidity ratio D_1/D_2 is in the range 4–15, yet the ratio D_3/D_2 lies between 1·04–1·09, with v_{12} between 0·3–0·34. For such materials, we may simplify the presentation of design data by taking $v_{12} = 0·3$, setting $D_3/D_2 = 1$ which is the isotropic limit, and considering the single material parameter D_1/D_2. From the definitions (2), $D_1/D_2 = E_1/E_2$ is the longitudinal to transverse modulus ratio, or anisotropy ratio of the composite. Use of this single materials parameter for the presentation of design data on orthotropic plates means that there is no need to specify the anisotropic elastic constants in detail. This is an important consideration since properties such as G and v_{12} are not well documented for most reinforced plastics.

For the woven fibre reinforcement $D_1/D_2 = 1$, assuming equal fibres in the warp and weft direction, whilst D_3/D_2 varies from 0·49 to 0·67 and v_{12} from 0·15 to 0·17. Thus for woven reinforcement we may take $D_1/D_2 = 1$, $v_{12} = 0·15$ and consider the single variable parameter D_3/D_2. In this case values $D_3/D_2 < 1$ signify the departure from isotropy of reinforced plastics with woven or cross-ply reinforcement. The data in Table 1 show that $D_3/D_2 \simeq 0·5$ is relevant to woven GRP materials, and it is probable that lower values may be needed for woven carbon fibre laminates and cross-ply materials. For CSM/polyester materials, we see that $D_1/D_2 = D_3/D_2 = 1$, hence they are isotropic and since $v_{12} \simeq 0·3$, standard design tables[4] may be used for the design of CSM/polyester plates.

2.2. Maximum Deflections and Bending Moments

The flexural behaviour of orthotropic elastic plates may be determined from the transverse deflection function w over the plate. Load distribution in the plate is defined by the bending moments M_x, M_y, in the x and y directions respectively, and the plate twisting moment M_{xy}, all measured per unit width of plate. Flexure and shear stresses in the plate are derived from the bending and twisting moments. In order to obtain the detailed deflection and moment distribution it is necessary to solve the governing orthotropic plate equations. These are fourth order partial differential equations for w whose complete solution usually requires numerical computation. The full equations, boundary conditions and a numerical solution method are described elsewhere[1,5] and some particular solutions are presented in detail. To facilitate such analysis a BASIC computer program has been developed[2] suitable for use on a minicomputer and can be made available to interested users.

There will be many situations, particularly for high technology applications, where a detailed stress and deflection analysis is required and it will be necessary for the designer to have access to finite element analysis or specialised programs.[2] However, where GRP or Kevlar/epoxy materials are used in medium technology applications such as storage tanks, boat hulls and vehicle body panels the designer may not require detailed stress analysis. In these situations the need is for general design rules, backed up by design charts and design formulae. Such an approach is necessary in this case because there may be incomplete information on material properties so that the elastic constants E_1, E_2, G, v_{12} required by a full analysis may not be known. Knowledge of loading and boundary conditions may also be imprecise so that exact computed solutions are not possible.

Even for high technology applications the simplified design approach has its uses for the initial selection of materials and panel thickness based on idealised geometry. A full finite element analysis may then be used at a later stage to analyse design details such as ribs or fixing points. In the remainder of this paper we use the detailed analysis[1] to generate simplified design information valid for a wide range of orthotropic plates.

The main requirements for the design analysis of plates is to determine the maximum plate deflection and the maximum bending and twisting moments. Consider an orthotropic rectangular plate subjected to a transverse load P, as depicted in Fig. 1. A detailed analysis shows that the maximum deflection w may be expressed in the general form

$$w = \alpha Pa^2/D_2 \tag{3}$$

where α is referred to as the plate *stiffness parameter*. For orthotropic plates α is a dimensionless parameter which depends on plate geometry ratio a/b, the flexural rigidity ratios D_1/D_2, D_3/D_2, the plate edge conditions and the loading type and position. If the plate contains a free edge, α depends on Poisson's ratio v_{12}, but is independent of v_{12} for combinations of clamped and simply supported edges. It may also be shown that the maximum bending and twisting moments in the plate are expressible in the form

$$M_x = \beta_1 P \qquad M_y = \beta_2 P \qquad M_{xy} = \beta_{12} P \tag{4}$$

where $\beta_1, \beta_2, \beta_{12}$ are *stress concentration parameters* which again depend on a/b, D_1/D_2, D_3/D_2, v_{12} and the plate edge and loading conditions. Study of the full solution for orthotropic plates shows that the twisting moment M_{xy} is usually significantly lower than the bending moments M_x, M_y. Thus β_1, β_2 are expected to be the critical stress parameters for orthotropic plate design and we concentrate here on design information for α, β_1, β_2.

For plates subjected to a uniform pressure p, the design formulae corresponding to (3) and (4) are more conveniently expressed as

$$w = \alpha pa^4/D_2 \tag{5}$$

$$M_x = \beta_1 pa^2 \qquad M_y = \beta_2 pa^2 \qquad M_{xy} = \beta_{12} pa^2 \tag{6}$$

The design parameters α, β_1, β_2 again depend on the variables a/b, D_1/D_2, D_3/D_2, v_{12}.

For isotropic plates $D_1 = D_2 = D_3 = D$ and it is commonly assumed that $v = 0.3$, so that the design parameters depend only on plate geometry a/b

and loading conditions. Roark and Young[4] (Table 26) give an extensive table of values for α and β (the maximum of β_1, β_2) for isotropic elastic plates, showing the effect of varying a/b, plate edge and loading conditions. This table is widely used for design calculations on isotropic plates. The main objective of the present paper is to supplement these standard results with design information on the influence of material anisotropy on the plate stiffness and stress parameters. In order to provide information valid for a wide range of orthotropic materials, we must select values of the material parameters D_1/D_2, D_3/D_2 and v_{12} which are representative of the main reinforced plastics materials. We note that the α and β parameters are independent of the absolute values of flexural rigidities D_1 or D_2, which are only required when calculating plate deflections from eqn. (5). This simplifies considerably the presentation of design data for orthotropic plates since a wide range of materials can be covered by a few values of the ratios D_1/D_2, D_3/D_2. Following the discussion in Section 2.1 we present design data for UD reinforced plates by assuming $D_3/D_2 = 1$ and taking the anisotropy ratio D_1/D_2 as the main design parameter. Design data for woven reinforcement and cross-ply plates are presented by assuming $D_1/D_2 = 1$ and taking D_3/D_2 as the design parameter.

2.3. Design Data for UD Reinforcement

The dependence of the stiffness and stress parameters α, β_1, β_2 on D_1/D_2 will now be considered, with $D_3/D_2 = 1$ and $v_{12} = 0.3$ assumed to be constant. We consider orthotropic plates under both uniform pressure loading and centre point loads. The plates are assumed to be simply supported on all four edges with aspect ratios a/b in the range $0.25 \leq a/b \leq 4$. Corresponding design data on square plates with fully clamped edges have been presented in a previous paper.[5] Figures 2 and 3

TABLE 2

Dependence of stiffness parameter α on D_1/D_2 for rectangular plates under centre point load with simply supported edges ($D_3/D_2 = 1$, $v_{12} = 0.3$)

D_1/D_2 \ a/b	4	2	1	0.5	0.25
1	0.00107	0.00417	0.0116	0.0167	0.0173
2	0.000975	0.00379	0.00933	0.01111	0.0113
4	0.000875	0.00338	0.00676	0.00704	0.00715
7	0.000796	0.00298	0.00481	0.00475	0.0049
10	0.000753	0.0027	0.00385	0.0038	0.0039
15	0.0007	0.00237	0.00292	0.00285	0.00298

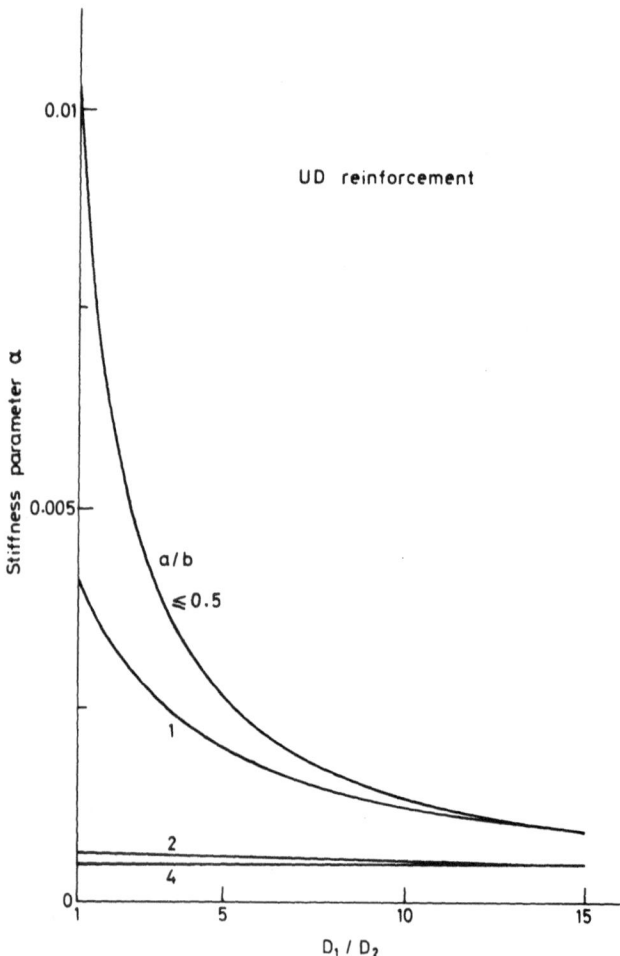

FIG. 2.   Dependence of stiffness parameter α on D_1/D_2 for simply supported rectangular plates under uniform pressure ($D_3/D_2 = 1$, $\nu_{12} = 0.3$).

give design charts for α, β_1, β_2 for plates under uniform pressure loading with anisotropy ratios D_1/D_2 in the range 1–15. Table 2 lists values of α for several values of a/b and D_1/D_2 in the same range for centre point loaded plates. We have not plotted the stiffness parameter α in the point loaded case because the design curves are very similar in appearance to Fig. 2. We do not present here values of the stress parameters β_1, β_2 for the point loaded case. The bending moments M_x, M_y have a log singularity at a concentrated load[3,4] and hence the computed values of β_1, β_2 are dependent on the assumed size of the region under concentrated load. Thus

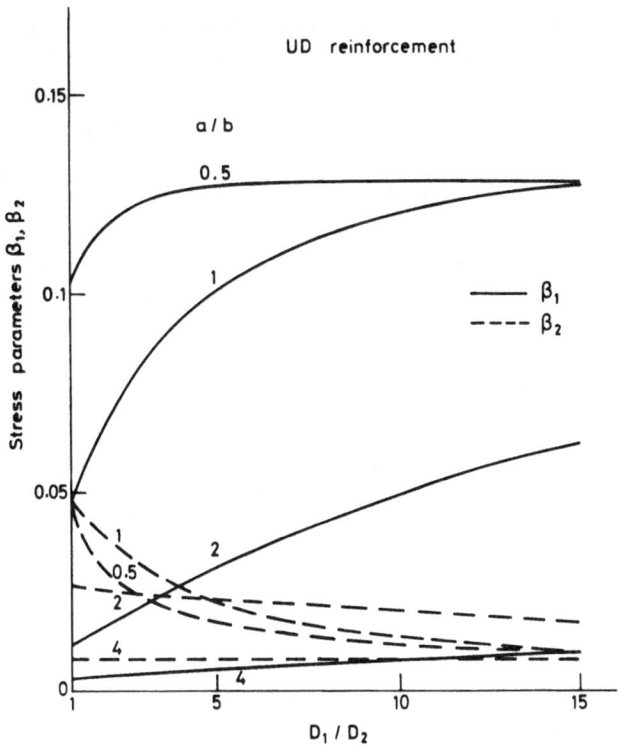

FIG. 3. Dependence of stress parameters β_1, β_2 on D_1/D_2 for simply supported rectangular plates under uniform pressure ($D_3/D_2 = 1$, $\nu_{12} = 0.3$).

to a certain extent the actual numerical values are arbitrary. Some typical computed values of β_1, β_2 are given elsewhere[5] based on a plate span/load diameter ratio of 30 and these show similar features to Fig. 3.

For plates in which $a/b \leq 1$ Fig. 2 shows significant reduction in the stiffness parameter α from the isotropic value at $D_1/D_2 = 1$. This indicates a smaller maximum deflection w and hence a stiffer plate as D_1/D_2 increases. Plates of higher aspect ratio with $a/b = 2$ or 4 are considerably stiffer. However this is brought about by the geometry change and we see that increasing the anisotropy ratio of the plate material has little influence on plate stiffness. It is interesting to note that an increase of 15 times in the plate material stiffness ratio D_1/D_2 leads to an increase in effective plate stiffness of about 5 times for square plates, of about 11 times for $a/b = 0.5$ and a negligible increase for $a/b = 4$. This demonstrates the effectiveness of placing fibre reinforcement across the shortest direction to enhance the stiffness of rectangular plates.

In Fig. 3, β_1 refers to the maximum bending moments in the x or reinforcement direction, whilst β_2 is the stress parameter in the y direction transverse to the fibres. For simply supported plates these maximum moments are at the plate centre which is thus in a biaxial stress state. Figure 3 shows that for orthotropic plates β_1 increases and β_2 falls as D_1/D_2 is increased, indicating higher bending moments and hence stresses in the fibre direction and lower bending moments transverse to the fibres. The influence of geometry is again seen to be more critical for $a/b \le 1$.

It is apparent from Figs 2 and 3 that the change in plate properties with anisotropy ratio is most significant for lower values of D_1/D_2 and falls off for $D_1/D_2 > 10$. It follows that significant errors may arise in plate stiffness and strength calculations if isotropic values for α, β_1 are used, even for moderately anisotropic materials such as UD GRP in which D_1/D_2 is in the range 2–5. We see that use of isotropic values for α, β_1, with flexural rigidity D assumed to be equal to the transverse rigidity D_2, is very conservative for calculating maximum plate deflections and transverse bending moment. However, the longitudinal bending moment is significantly underestimated since the isotropic analysis takes no account of load redistribution along the reinforcement direction.

2.4. Design Data for Woven Reinforcement

The dependence of the stiffness and stress parameters α, β_1, β_2 on D_3/D_2 is considered here, with $D_1/D_2 = 1$ and $v_{12} = 0 \cdot 15$ assumed to be constant. We consider the same geometry and loading cases as in Section 2.3. Figures 4 and 5 give the computed design charts for α, β_1, β_2 for simply supported plates with a/b ratios of 1, 2 and 4 for D_3/D_2 in the range $0 \cdot 4$–1. Note that Fig. 4 is on a log-linear plot and because of the material symmetry we no longer need to consider the cases $a/b < 1$ separately.

TABLE 3

Dependence of stiffness parameter α on D_3/D_2 for rectangular plates under centre point load with simply supported edges $(D_1/D_2 = 1, v_{12} = 0 \cdot 15)$

D_3/D_2 \ a/b	1	2	3
1	0·0118	0·002 12	0·000 271
0·8	0·013 1	0·002 26	0·000 286
0·6	0·0146	0·002 43	0·000 304
0·4	0·0165	0·002 63	0·000 325

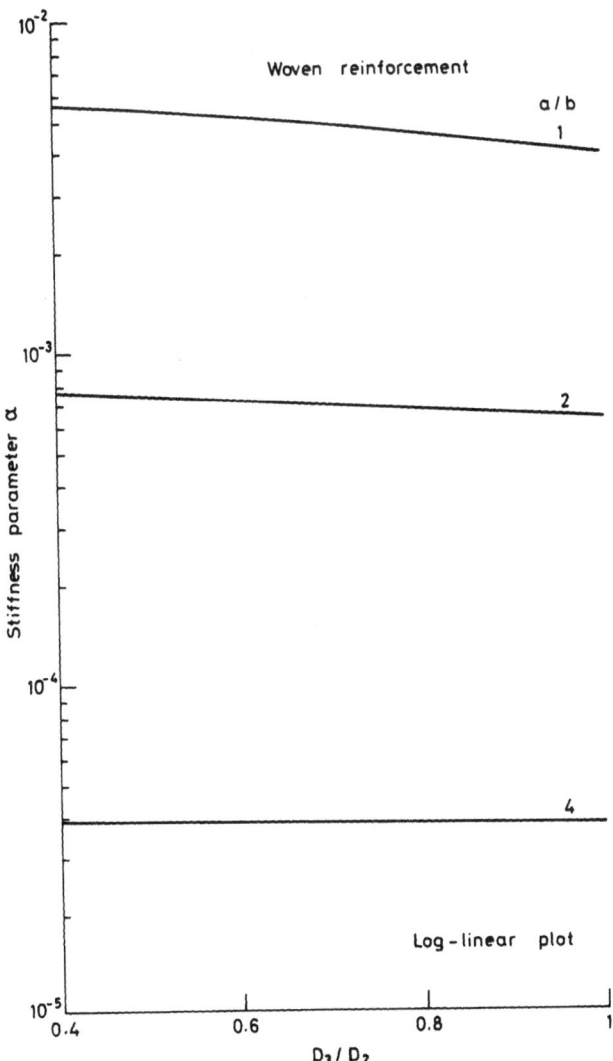

FIG. 4. Dependence of stiffness parameter α on D_3/D_2 for simply supported rectangular plates under uniform pressure ($D_1/D_2 = 1$, $\nu_{12} = 0.15$).

Table 3 gives corresponding data for the stiffness parameter α for centre point loaded plates. We see that α takes its lowest value at $D_3/D_2 = 1$ and increases as D_3/D_2 is reduced. This indicates higher deflections and hence lower stiffness for this class of orthotropic materials, compared with isotropic materials. This reduction in stiffness is accompanied by an

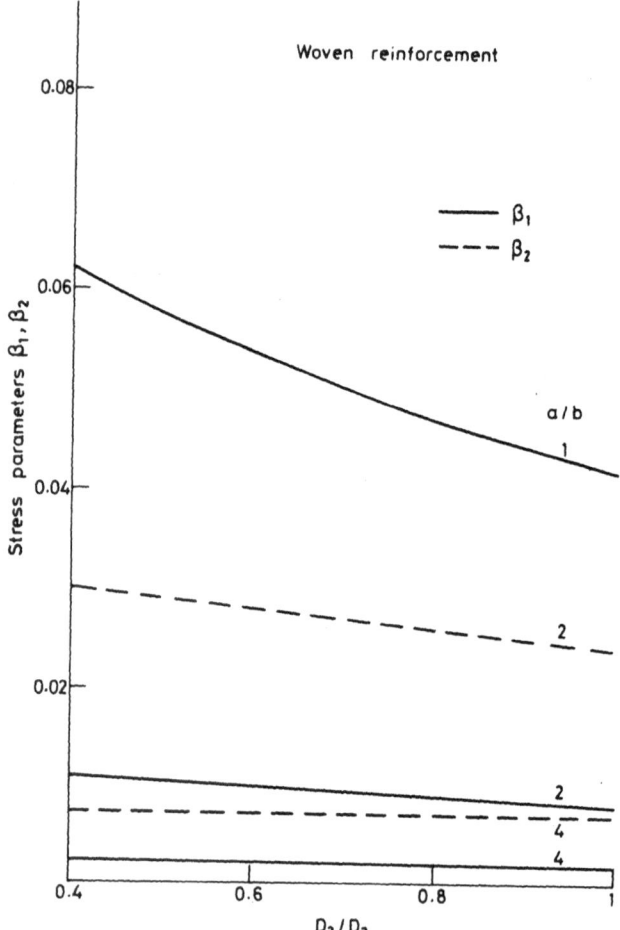

FIG. 5. Dependence of stress parameters β_1, β_2 on D_3/D_2 for simply supported rectangular plates under uniform pressure ($D_1/D_2 = 1$, $v_{12} = 0.15$).

increase in the bending stress parameters β_1 and β_2. Plates with higher aspect ratios $a/b = 2$, 4 are considerably stiffer than the square plate. This is due to the change in geometry and it is apparent that the influence of material anisotropy becomes less significant as the aspect ratio increases. Thus for $a/b = 4$ the isotropic value for α may be used for this type of orthotropic plate. In a similar manner the increases in maximum bending moment β_1 for $D_3/D_2 < 1$ are also small. It is apparent from the design curves that for orthotropic plates with woven types of reinforcement the changes in stiffness resulting from anisotropic effects are much lower than for UD reinforcement. For square plates plate stiffness falls by about 30 % as D_3/D_2 is reduced from 1 to 0·4. This is accompanied by an increase in maximum bending moment of about 50 %. The effects are even lower as a/b is increased. It is apparent that, particularly for $a/b > 1$, isotropic values of α, β_1, β_2 may be adequate for design calculations, especially for GRP materials such as woven roving/polyester where a typical value for the rigidity ratio is $D_3/D_2 = 0·7$.

3. APPROXIMATE METHODS FOR PLATE ANALYSIS

3.1. Derivation of Simplified Design Formulae

As a supplement to the design data given in Section 2 based on computed solutions to the orthotropic plate equations, we present here some approximate design formulae for the stiffness parameter α and stress parameters β_1, β_2 based on energy methods. The basis of the method[9] is to assume a form for the plate deflection function which satisfies the edge conditions and which contains one or more variable parameters. These parameters are then chosen so as to minimise the potential energy in the plate. As the number of parameters increases the resulting approximate solution tends to the exact solution of the plate equations. Instead of seeking such an exact solution, we assume a form for the plate deflection containing only a single free parameter. The method then gives a simple but approximate formula for the maximum deflection in the plate and hence for the plate stiffness parameter α. The validity of the resulting design formulae is assessed by comparison with the computed solutions presented above.

Uniform pressure loading

We adopt the notation for plate geometry and plate properties used in Fig. 1 and consider the case of a simply supported rectangular plate subjected to a uniform pressure p. We assume that the deflected shape of the plate $w_1(x, y)$ may be approximated by the function

$$w_1(x, y) = 256wxy(a - x)(b - y)(a^2 + ax - x^2)(b^2 + by - y^2)/25a^4b^4 \quad (7)$$

which has been chosen to satisfy the simply supported edge conditions:

$$w_1 = 0 \qquad \frac{\partial^2 w_1}{\partial x^2} = 0 \qquad x = 0, a$$

$$w_1 = 0 \qquad \frac{\partial^2 w_1}{\partial y^2} = 0 \qquad y = 0, b$$

The numerical factor is included so that the unknown parameter

$$w = w_1(a/2, b/2) \tag{8}$$

is equal to the maximum centre point deflection in the plate. For a specially orthotropic plate with simply supported edges, the potential energy V in the plate takes the form[9,10]

$$V = \frac{1}{2} \int_0^a \int_0^b \left[D_1 \left(\frac{\partial^2 w_1}{\partial x^2} \right)^2 + 2D_3 \left(\frac{\partial^2 w_1}{\partial x \partial y} \right)^2 + D_2 \left(\frac{\partial^2 w_1}{\partial y^2} \right)^2 \right] dx\, dy$$

$$- \int_0^a \int_0^b p w_1 \, dx\, dy \tag{9}$$

On substituting the assumed deflection (7) into (9) and integrating we find that

$$V = 12 \cdot 4[D_1 + 2D_3(a/b)^2 + D_2(a/b)^4] b w^2/a^3 - 0 \cdot 41 pabw \tag{10}$$

The condition for the minimum potential energy $dV/dw = 0$ gives an equation for the unknown centre point deflection w, from which we obtain

$$w = 0 \cdot 0165 p a^4 [D_1 + 2D_3(a/b)^2 + D_2(a/b)^4]^{-1} \tag{11}$$

Comparison with eqn. (5) shows this to be in the form of the design formula for the maximum deflection in a pressure loaded plate, with the stiffness parameter α given by

$$\alpha = n_1 [D_1/D_2 + 2(D_3/D_2)(a/b)^2 + (a/b)^4]^{-1} \tag{12}$$

where

$$n_1 = 0 \cdot 0165$$

The bending moments m_x, m_y in the plate are defined in terms of the deflection function w_1 by the relations[10]

$$m_x = -\left(D_1 \frac{\partial^2 w_1}{\partial x^2} + D_{12} \frac{\partial^2 w_1}{\partial y^2} \right) \qquad m_y = -\left(D_{12} \frac{\partial^2 w_1}{\partial x^2} + D_2 \frac{\partial^2 w_1}{\partial y^2} \right) \tag{13}$$

Approximate formulae for the maximum bending moments M_x, M_y, which for simply supported plates are at the plate centre, follow on substituting (7) into (13) and setting $x = a/2$, $y = b/2$ and take the form

$$M_x = 9 \cdot 6w(D_1/a^2 + D_{12}/b^2) \qquad M_y = 9 \cdot 6w(D_{12}/a^2 + D_2/b^2) \tag{14}$$

Reference to the definitions (5) and (6) and using the relation $D_{12} = v_{12}D_2$ gives the following formulae for the stress parameters β_1, β_2,

$$\beta_1 = n_2\alpha[D_1/D_2 + v_{12}(a/b)^2] \qquad \beta_2 = n_2\alpha[v_{12} + (a/b)^2] \qquad (15)$$

where $n_2 = 9\cdot6$.

Centre point loading

We consider a simply supported rectangular orthotropic plate subjected to a centre point load P. Again we assume the deflection function $w_1(x, y)$ defined in (7) which satisfies the plate edge conditions.

The second term in the potential energy in eqn. (9) representing the work done by the point load now becomes equal to Pw and the condition $dV/dw = 0$ gives for the maximum plate deflection

$$w = 0\cdot0403Pa^3b^{-1}[D_1 + 2D_3(a/b)^2 + D_2(a/b)^4]^{-1} \qquad (16)$$

Comparison with the plate design formula leads to the following formula for the stiffness parameter α for a point loaded plate

$$\alpha = m_1(a/b)[D_1/D_2 + 2(D_3/D_2)(a/b)^2 + (a/b)^4]^{-1} \qquad (17)$$

where $m_1 = 0\cdot0403$.

We do not give corresponding formulae to (15) for the stress parameters in the point loaded case. The stresses have a log singularity at a concentrated load as the loading radius reduces to zero, which is not obtainable from an expression for w_1 in the form (7). A reasonable estimate for β_1, β_2 may be obtained by replacing the numerical factor n_2 in (15) by an expression containing $\log(r/a)$ where r is the loading radius, but this is not considered further here.

3.2. Comparison with Computed Design Charts

These design formulae for α, β_1, β_2 are of simple form and if accurate their use is preferable to the design charts Figs 2–5. In this section we assess the validity of the formulae and show that their accuracy can be improved by adjusting the numerical factors.

Uniform pressure loading

The simplest check on the formulae is to compare them with exact solutions for isotropic plates[4] which correspond to the $D_1/D_2 = 1$ values in Figs 2–5. On doing this it is found that for pressure loading α is within 2% of the isotropic value for square plates, in error by 5% for $a/b = 2$ and by 14% for $a/b = 4$. It follows that the formulae may need improving and this

TABLE 4

Adjusted numerical coefficients for use with the formulae (12), (15), (17)

a/b	Pressure load		Point load
	n_1	n_2	m_1
1	0·0162	9·1	0·0464
0·5, 2	0·0157	8·9	0·0522
0·25, 4	0·0145	9·4	0·0773

is achieved by modifying the numerical factor n_1 from its value 0·0165 so that the formula is exact in the isotropic limit. It is apparent that such modifications will depend on the plate geometry a/b and the appropriate values of n_1 are listed in Table 4.

Using these corrected α values, the errors in the stress parameters β_1, β_2 were assessed in the isotropic limit and found to be in the range 4–13% depending on a/b ratio. Again these were corrected by changing n_2 from its value of 9·6 and the new values are also listed in Table 4.

Equations (12) and (15) are the proposed design formulae for the stiffness and stress parameters for a pressure loaded orthotropic plate, with the n_1, n_2 values listed in Table 4. For a/b ratios other than those tabulated, estimates for n_1, n_2 must be made by interpolation or extrapolation.

A detailed comparison of (12) and (15) with the design charts shows that the formulae are good approximations over a wide range of D_1/D_2, D_3/D_2 and a/b ratios. The errors increase with D_1/D_2 and as the plate aspect ratio increases. Table 5 compares computed values of α, β_1, β_2 with those obtained from the design formulae for a rigidity ratio $D_1/D_2 = 10$. Typical errors in α range from 2% for $a/b \geq 1$ to about 10% for $a/b = 0.25$. The

TABLE 5

Comparison of exact computed and approximate design formula values for pressure loaded plates, with $D_3/D_2 = 1$, $D_1/D_2 = 10$

a/b	α		β_1		β_2	
	Exact	Approx.	Exact	Approx.	Exact	Approx.
4	0·00005	0·000049	0·0079	0·0068	0·008	0·0075
2	0·00048	0·00046	0·049	0·046	0·019	0·018
1	0·00123	0·00125	0·12	0·12	0·014	0·015
0·5	0·00133	0·00149	0·13	0·13	0·012	0·007
0·25	0·0013	0·00143	0·13	0·13	0·012	0·005

largest stress parameter, usually β_1, was accurate to within 6%, but greater errors are seen in the numerically much smaller stress parameter, usually β_2. We conclude that the simplified formulae for the design parameters of pressure loaded plates provide a means of determining these parameters to a reasonable degree of accuracy, certainly for moderate anisotropy ratios $D_1/D_2 < 10$ and for plates which are approximately square.

Point loaded plates

The same procedure is adopted here to adjust the numerical factor m_1 in (17) and appropriate values for m_1 are given in Table 4. In the case of square plates this formula for α is compared directly with computed α values from Table 2 in Fig. 6. The agreement is seen to be good over the whole range $1 \leq D_1/D_2 \leq 15$ and particularly for moderate anisotropy ratios < 10. A comparison with other a/b ratios shows good agreement for moderate anisotropy except in the case $a/b \leq 0.5$. Reference to Table 2 in this case shows that α is almost independent of a/b for $a/b \leq 1$ and $D_1/D_2 > 5$, although this is not predicted by (17). In these cases it is suggested that for a better approximation the formula (17) is used with $a/b = 1$ when $D_1/D_2 > 1$ and $a/b \leq 1$.

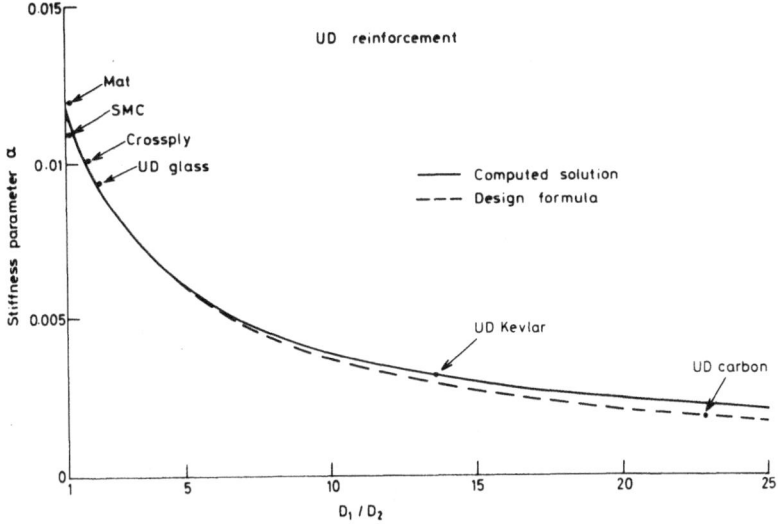

FIG. 6. Stiffness parameter α for simply supported square plates under centre point load—comparison of computed design curve with approximate design formula (17) and experimental data.

4.   EXPERIMENTAL TESTS ON COMPOSITE PLATES

4.1. Materials

A range of materials have been tested to explore the effect of variations in the D_1/D_2 and D_3/D_2 ratios on the value of the plate stiffness parameter α for a centre point loaded square plate. Press moulded samples were selected in all cases so that the sheet has two smooth surfaces and a minimum variation in sheet thickness. The materials used are listed in Table 6 along with the percentage fibre content by volume.

TABLE 6

	Material description (volume fraction of reinforcement)	Manufacturers coding
(a)	Sheet moulding compound (SMC) ($\sim 30\%$)	Freemans Chemicals
(b)	Glass fibre chopped strand mat/epoxy ($\sim 42\%$)	Permali Gloucester (33 ME)
(c)	Glass fibre fabric/epoxy ($\sim 38\%$)	Permali Gloucester (22 FE)
(d)	Crossplied aligned glass fibre/epoxy ($\sim 65\%$)	Permali Gloucester (XEB/6)
(e)	Unidirectional glass fibre/epoxy ($\sim 65\%$)	Permali Gloucester (XEB/5)
(f)	Unidirectional Kevlar 49 aramid fibre/epoxy ($\sim 60\%$)	Fothergill-Rotoway
(g)	Unidirectional high modulus carbon fibre/epoxy ($\sim 60\%$)	Fothergill-Rotoway

4.2. Test Methods

A square plate 215 mm × 215 mm was cut for each material except for the SMC, where it was necessary to cut a 145 mm × 145 mm sample so that a thin central area associated with the sprue could be positioned in a less critical area. Each plate was then tested in three modes:

(a)   Line loaded along the plate midline on a 200 mm span three point-flexure jig to obtain the flexural rigidities D_1 and D_2, using two simply supported and two free edges.

(b)   Loaded at opposite corners in a plate twist test, with free edges, to obtain in-plane shear modulus G.

(c)   Centre point loaded by a 25 mm or 12·5 mm diameter spherical indentor while simply supported on all four sides on a 200 mm square test frame, in order to obtain plate stiffness data for comparison with predictions of the design method for centre point loaded plates.

In each case a crosshead speed of 1 mm/min was used and the central deflection of the plate was continuously measured by a clip gauge extensometer attached to a spring loaded plunger resting against the lower face of the plate. Duplicate tests were undertaken in each case by turning over the plate, reversing the upper and lower faces of the plate in the test jig. In point load tests it was necessary to apply a small deadweight load to the perimeter of the plate to ensure good contact with the loading jig. The deflection per unit load was obtained from the initial portion of the load–displacement curve, so that maximum deflection/thickness ratio $w/h < 0.5$ in both the point loaded and plate twist tests.

4.3. Experimental Results

The data from line loaded plates were analysed to obtain values for the flexural rigidities D_1 and D_2 in the principal directions in the plate using the equation

$$D_1 = \alpha_1 a^3 P/12bw \tag{18}$$

with corresponding equation for D_2. Here w is the loading line deflection and P the total line load. The design constant α_1 takes the value $\alpha_1 = 0.259$ and has been computed using the general plate analysis program.[2]

The plate twist tests were analysed to obtain the in-plane shear modulus G, using the equation[11]

$$G = 3Pa^2/8wh^3 \tag{19}$$

where w is the plate central deflection and P the corner loads. The measured values of G with literature values of v_{12} from Table 1 were used to calculate D_{12} and D_3 using eqns (2).

Rather than compare measured and calculated deflections, the centre point load flexure tests were used to calculate the plate stiffness parameter α using eqn. (3) expressed in the alternative form

$$\alpha = D_2 w/a^2 P \tag{20}$$

where w is the loading point deflection and P the centre point load. The experimental values of D_1, D_2, D_3 for these materials are given in Table 7 along with the ratios D_1/D_2 and D_3/D_2 needed for the design charts and design formulae. The table also lists experimental values of the stiffness parameter α obtained from the centre point load tests and for comparison, calculated values of α obtained from the approximate design formula (17).

TABLE 7
Measured plate properties and comparison of predicted and experimental values of α for point loaded plates

Material	Thickness (mm)	v_{12} assumed	Expt G MN/m²	Expt D_1 N.m	Expt D_2 N.m	Calc. D_3 N.m	D_1/D_2	D_3/D_2	α (Point test) Expt	α (Point test) Calc. eqn. (17)
Sheet moulding compound	2·44	0·3	4·03	13·2	13·0	13·7	1·01	1·05	0·0109	0·0113
Glass fibre mat/epoxy	7·67	0·32	5·83	717·4	667·3	652·5	1·08	0·98	0·0119	0·0115
Glass fibre fabric/epoxy	3·75	0·17	3·94	102·4	87·9	49·7	1·16	0·56	0·0141	0·0141
Cross-plied aligned glass fibre/epoxy	1·74	0·15	13·73	21·6	13·0	14·0	1·7	1·08	0·0101	0·009 55
Unidirectional glass fibre/epoxy	3·52	0·3	9·27	190·4	88·6	93·4	2·15	1·06	0·0094	0·008 8
Unidirectional HMS carbon fibre/epoxy	3·11	0·31	4·00	373·7	16·4	22·1	22·8	1·35	0·0017	0·001 75
Unidirectional Kevlar 49/epoxy	3·02	0·34	1·80	155·3	11·4	10·0	13·6	0·88	0·0031	0·002 84

4.4. Comparison of Design Parameters with Measured Values

On Fig. 6 measured values of the stiffness parameter α for centre point loaded, simply supported square plates of SMC, mat and cross-ply GRP, and UD glass, Kevlar and carbon reinforced epoxide plates are compared with the theoretically determined design values. In plotting the data on Fig. 6 it is assumed that the ratio $D_3/D_2 = 1$ so that the only materials data needed is the principal rigidity ratio D_1/D_2 for the composite plates. The figure shows close agreement between the measured and predicted α values, indicating the validity of the design method for predicting plate stiffness behaviour and hence from (3) the plate maximum deflections. Calculated values of α for these orthotropic plates obtained from use of the design formula (17) are also shown in Table 7 for comparison with the measured values listed there. When using the design formula it is not necessary to assume $D_3/D_2 = 1$ and the measured value of D_3/D_2 may be used since this ratio occurs explicitly in the formula.

Also included in the table, but not on Fig. 6 are the centre point load stiffness data for the glass fabric reinforced plate. Here $D_3/D_2 = 0.56$ and it is not appropriate to use the design chart for UD reinforcement shown in Fig. 6. We see from Table 7 that the design formula also gives good agreement with data on this type of woven reinforcement. The measured and predicted stiffness parameters for the composite plates tested are seen to be within 5 % for the balanced GRP plates and within 10 % for the UD reinforced plates. These are acceptable levels of agreement for design calculations on composite plates. We note that the cross-ply GRP plate consists of a 0/90/90/0 construction which is not symmetric to flexural loads in the x and y directions. Thus it is found that $D_1/D_2 = 1.7$ and hence it is possible to include the centre point load stiffness data for the cross-ply plate on the UD design curve (Fig. 6).

5. CONCLUDING REMARKS

(1) The paper gives a simplified design method for determining stiffness and strength properties of rectangular composite plates under flexural loading. The method is based on formulae for maximum deflections and bending moments combined with design charts or approximate design formulae for a plate stiffness parameter α and flexural stress parameters β_1, β_2. These design parameters depend on plate material properties through the flexural rigidity ratios D_1/D_2 and D_3/D_2.

(2) The design method is illustrated by providing design charts and

formulae for simply supported, rectangular orthotropic plates under pressure or centre point loads. The approximate design formulae are derived from variational methods. They are found to give good agreement with the design charts, are simpler to use and more general. Other edge and load conditions would require the generation of further design charts from computed solutions to the orthotropic plate equations[1,2] or the development of further approximate formulae.

(3) Although the design data presented are specific to pressure and point loaded plates with simply supported edges, some general observations on the significance of material anisotropy on plate properties can be made. Thus for plates with UD reinforcement the change in plate stiffness with anisotropy ratio is most significant for D_1/D_2 in the range 1–5 and becomes less as D_1/D_2 increases. The extent of the stress redistribution along the fibre direction is also significant at low anisotropy ratios. Errors would be involved in using conventional isotropic design formulae even for mildly anisotropic materials with $D_1/D_2 = 2$. However for woven reinforcement with $D_1/D_2 = 1$ the isotropic design formulae can be used to a fair approximation.

(4) Experimental tests were carried out on a range of GRP, Kevlar and CFRP plates. First the basic plate flexural rigidity data D_1, D_2, D_3 were measured. Then to test the design methods, centre point load tests on square simply supported plates were carried out. Experimentally measured and design values of the stiffness parameter[1] α were found to be in good agreement.

(5) An important limitation of the design method is that it is restricted to small deflections, in which plate deflections are less than the plate thickness. However it can be shown[3,9] that design procedures based on small deflection theory are conservative at larger deflections, when deflections and stresses are lower than those predicted by the small deflection theory.

ACKNOWLEDGEMENTS

The authors wish to thank A. Woolf for the computations used to generate the design charts and D. G. Gladman for assistance with the experiments.

REFERENCES

1. JOHNSON, A. F. and WOOLF, A., Deflection and stress analysis of anisotropic plates in flexure, NPL Report DMA(A)41, 1982.

2. WOOLF, A., An interactive minicomputer program for plate design analysis, NPL Report DMA(A)51, 1982.
3. TIMOSHENKO, S. and WOINOWSKY-KRIEGER, S., *Theory of Plates and Shells*, New York, McGraw-Hill, 1959.
4. ROARK, R. J. and YOUNG, W. C., *Formulas for Stress and Strain*, 5th edition, New York, McGraw-Hill, 1975.
5. JOHNSON, A. F., Design analysis of GRP plates in flexure. *BPF Reinforced Plastics Congress* (1982), 155–162, British Plastics Federation, London.
6. TSAI, S. W. and HAHN, H. T., *Introduction to Composite Materials*, Westport, Technomic, 1980.
7. JOHNSON, A. F., *Engineering Design Properties of GRP*, London, British Plastics Federation, 1978.
8. *A Guide to High Performance Plastics Composites*, London, British Plastics Federation, 1980.
9. MANSFIELD, E. H., *The Bending and Stretching of Plates*, London, Pergamon, 1964.
10. HEARMON, R. F. S., *Applied Anisotropic Elasticity*, Oxford University Press, 1961.
11. PURSLOW, D., The shear properties of unidirectional carbon fibre reinforced plastics and their experimental determination, RAE Technical Report 76093, 1976.

23

Design Principles for Plastic Structures

LAJOS GARAY

Epitestudomanyi Intezet Budapest XI,
David Ferenc utca 6, 1502 Budapest, Hungary

ABSTRACT

The COMECON standard, Design Principles for Plastic Building Structures, puts special emphasis on the differences characterizing the design of plastic structures, compared to the design of traditional ones. For the analysis of the ultimate limit state, the 'stress based' or the 'strain based' method can be used. The first is characterized by the design stress, dependent on specified life and on specified load combination. The second is characterized by the design strain and by the time- and temperature-dependent compliance. For proper analysis, there are given principles for estimation of the action process of loads and effects.

1. INTRODUCTION

The basic principles of estimation of structural safety, i.e. the way or methodology of structural analysis, were standardized by the COMECON in 1976. Since that time similar principles have been elaborated for concrete, reinforced concrete, steel and aluminium structures. For plastic structures the acceptance of basic principles is planned at the end of this year. The main features of these principles are summarized in the following.

Plastics—including composites, using plastic as matrix or as reinforcement—are the youngest structural materials. They are new structural materials, and are not substitutes for any other traditional ones. In consequence of their particular (special) properties, in some cases they are indispensible, in other cases they are potential rivals of traditional materials. So, they appear in increasing volume in the building industry too.

The designer of a building construction has to verify that the load-carrying structure satisfies the requirements prescribed by the authorities. The first criterion of the verification is the proper material model and quantitative data, that will simulate the real behaviour of the material. The second criterion is the properly prescribed design actions (loads and effects), that enable analysis with the proper material model.

2. THE MATERIAL MODEL

Plastics are polymers, consequently they differ fundamentally from silicates and metals. The strength of silicates and metals—in usual environmental circumstances—can be assumed as constant values. It is not so in the case of structural plastics: 'the change in time of strength and of deformability, caused by mechanical stresses from loads and effects and from environmental circumstances, must be taken into account' (quoted from the draft of the standard). Consequently, the material data have to be completed by time and environment dependent values, which characterize the viscoelastic properties of plastics.

The change in time of material properties, caused by the stressing process, can be divided into two parts by the proportional limit, below which the material properties change in linear proportion with the isochronous intensity of the stress, i.e. the rules of the theory of linear viscoelasticity are valid. Exceeding this limit, the load-carrying capacity of the material is not exhausted, but the characteristic data of material properties will change irreversibly. The proportional limit signifies the starting point of the accumulation of microstructural changes and/or damages, consequently that of the ceasing of the rheological time invariancy. The strain value at the proportional limit is practically constant, that means independent of the action process (of time, of stress and of temperature).

According to the Principles, the strength of the plastics must be characterized:

—by the action process (time, stress, temperature) dependent strength value,

—or by the constant strain value of the proportional limit and by the time and temperature dependent compliance.

These two modes of characterization allow the use:

—of the stress-based and
—of the strain-based methods of the structural analysis.

The 'stress-based' method:

—is built on the failure stress, and uses a high safety factor for the material;
—keeps the analysis of the action process-dependent stresses within the standardization by prescribing specified design stresses;
—prescribes moduli of deformation independently of failure stress for controlling deformation, i.e. the serviceability limit states.

All these result in a usual, traditional method of structural analysis. The designer controls directly the limit states, and takes into account the so-called 'static fatigue process' through the use of prescribed design stresses, leaving 'the load carrying process' out of consideration.

The 'strain-based' method:

—is based on the proportional limit, and uses a low safety factor for the material;
—trusts the designer with the analysis of the loading process;
—applies consistent data for the control of the ultimate and the serviceability limit states.

All these follow the principles of the theory of viscoelasticity. The designer controls the entire viscoelastic process as a consequence of the action process (of loads, of temperature, of humidity, etc.). He finds by calculation the moment of occurrence of the limit states, and also their character.

The use of both methods is allowed. National standards prescribe—according to traditions, or to characteristics of materials and of construction—which method should be used.

The use of the 'stress-based' method is simple, namely the designer works with design stresses, given according to the specified life of the structure and to the specified action combinations. He does not analyse any processes. So this method is advantageous in cases when the material is used for a defined purpose and for a standard specified life; further, when the deformations have no definitive role in the load-carrying capacity of the structure. Yet the control of stability raises difficulties in the use of this method, namely the loss of stability is a process, controlled by the co-action of the stress and the strain processes.

The use of the 'strain-based' method is laborious, but it can be applied in any case. The safety of the structure can be controlled at each moment of the specified life. So, the entire process of the loss of stability can be followed, till the proportional limit is reached. The instantaneous safety of the structure can be controlled through comparing the measured strain data to the design values. All these lead to a more developed, complex way of thinking, which differs from traditional practice, and through which the particular properties of plastics can be better exploited, and it helps in plastic-like design.

3. ACTIONS

The Principles interpret the design values of the standard load types.

'—the dead and the permanent loads act during the entire specified life with their design values;
—the short time loads act during the entire specified life with varying intensity. The maximum limit of the intensity is the design value of the load.
—the accidental actions occur once and shortly during the entire specified life. The maximum limit of its intensity is the design value of the load.'

To determine or to estimate 'the intensity changes of the load or effect, it is advisable to state them in the form of a stochastic or deterministic, continuous or discrete process, based on appropriate statistical data'. So far these are the Principles. The appropriate statistical data in most cases are not available, consequently the estimation, based on technical considerations, is unavoidable; but such estimations lead to better and more economical design; although they usually err considerably for the sake of safety, namely, for calculation of the viscoelastic response of the structure, the knowledge of the action process is unavoidable (e.g. recovery analysis in case of the quasi-static periodic actions).

4. DISCUSSION AND CONCLUSIONS

In practice, the 'stress-based' and the 'strain-based' design methods, when taking into account the action process of loads and effects, lead to conformable results. In general, the exploitation of the fracture strength of plastics is impossible: the requirements of stability, of rigidity, of corrosion resistance, decide the structural dimensions. In such cases, exceeding the proportional limit is not advisable. The knowledge of the fracture strength gives an idea about the limit of the load-carrying capacity, and this, combined with the knowledge of the 'load-carrying process' till the proportional limit is reached, leads to a better and confident estimation of the structural safety. And this results in economical use of material.

These two different design methods do not influence the principles of the control of limit states. The 'stress-based' method keeps the traditional form. It does not require analysis of the process, that leads to limit state. It requires no more than to take into account the specified life. The 'strain-based' method clears all features of the load-carrying process through the analysis of the deformation process. The designer will know that the limit state is not the consequence of the occurrence of a particular stress state, but of the action process of stresses and environmental circumstances, and that the limit state belongs to a distinguished moment of the action process.

Considering the solution of the design task, the 'strain-based' method contains the 'stress-based' method, as a special case.

ACKNOWLEDGEMENTS

The author expresses his thanks to Mrs H. Doubravszky (Hungary), to Mr L. Skupin (Czechoslovakia), to Mr G. Ackermann (German Democratic Republic) and to Mr S. B. Yermolov (Soviet Union) for their creative cooperation.

BIBLIOGRAPHY

'*Stress-based*' *design method:*
DOMKE, H. and RÜBBEN, A., Einfluss des tatsächlichen Lastablaufes auf Kriechverhalten und Festigkeit von Kunststoff-Konstruktionen. *Plasti-construction*, **6** (1976), 5, 173–180.
RÜBBEN, A., Grundlagen zur Berechnung tragender Kunststoff-konstruktionen. *Verbindungstechnik*, **10** (1978), 4, 21–25.
'*Strain-based*' *design method:*
MENGES, G. and ROSKOTHEN, H. J., Neue einfache Dimensionierungsmöglichkeiten bei glasfaserverstärkten Kunststoffen. *Kunststoff-Rundschau*, **9** (1972), 472–487.
OWEN, M. J., Progress toward a safe-life design method for glass reinforced plastics under fatigue loading. Paper for *Reinforced Plastics Congress*, 1976, Brighton.

MENGES, G., Dimensionierung von Kunststoffteilen auf Basis von kritischen Deformationen. *Kunststoffe-Plastics*, **8** (1977), 15–25.

ROBERTS, R. C., Design strain and failure mechanism of GRP in a chemical environment. *Reinforced Plastics Congress* (British Plastics Federation 1978), pp. 145–151.

NORWOOD, L. S. and MILLMAN, A. F., Strain limited design criteria for reinforced plastic process equipment. *Composites*, January (1980), 39–45.

GARAY, L., Design requirements for load-carrying plastic structures, Paper for *ICP/RILEM/IBK International Symposium Plastics in Material and Structural Engineering*, Prague (1981), pp. 175–181.

24

An Evaluation of the Impact Properties of Carbon Fibre Reinforced Composites with Various Matrix Materials

Aa. Stori and E. Magnus

Central Institute for Industrial Research,
Forskningsv. 1, P.B. 350, Blindern, Oslo 3, Norway

ABSTRACT

Instrumented impact testing of laminates of continuous-carbon-fibre-reinforced thermosetting-materials (epoxy and PI) and thermoplastics (PSO, PES, PEI, PEEK and PC) with 50% fibre has shown that the fracture energies of the thermoplastic laminates are essentially higher than the fracture energies of laminates prepared from commercial epoxy/carbon-fibre.

Laminates made from solution-impregnated prepregs gave better fibre impregnation than those prepared from melt-impregnated prepregs. The moulding temperatures were chosen on the basis of melt-rheology tests to yield equivalent viscosities at low shear-rates for all the thermoplastics. The double-notched specimens with fibre orientations [0], [90], [0, 90], and [±45] relative to the long dimension were milled for impact testing. 'Flexural' stresses and moduli were calculated by treating the impact test as a case of high rate three-point bending.

INTRODUCTION

Thermoplastic fibre composite materials have received increasing attention during the past 5–10 years. Almost every thermoplastic material from conventional polyolefines to the more exotic temperature-resistant thermoplastics (e.g. polysulphone) can be reinforced with glass, aramide or carbon fibres in order to produce a greatly improved property profile.

Carbon fibres show their greatest potential when used in combination with engineering and exotic thermoplastics.[1-6]

Composite materials are often supplied in sheet form. These are heated to a temperature above the softening point of the thermoplastic, and formed in a match-die press. Cycle times are reported to be around 20–30 s. Since the matrix material is a thermoplastic, rejects can be recycled, which leads to a low percentage of waste.

The composite's temperature resistance depends on the thermoplastic material used. Several of the more exotic thermoplastics form carbon fibre composites with better temperature resistance than the traditional epoxy/carbon fibre composites.

The impact properties of carbon fibre reinforced thermoplastics are reported to be better than the impact properties of carbon fibre reinforced epoxy.[6] The aim of this work is to study possible differences in impact properties of carbon fibre composites with different matrix materials (both thermosetting and thermoplastic) and to compare impact properties resulting from two prepreg manufacturing methods, viz. melt impregnation and solution impregnation.

MATERIALS

The thermosetting and thermoplastic matrix materials used are given in Table 1. The two epoxy resins were reinforced with unidirectional Grafil HM-S carbon fibres, while the remaining matrix materials were reinforced with Grafil XA-S unidirectional and bidirectional fibre fabrics.

TABLE 1
Thermosetting and thermoplastic matrix materials

Abbreviation	Material	Manufacturer
	Thermosetting	
Epoxy 914C	Temperature-resistant epoxy	Ciba Geigy
Epoxy 920C	Impact-resistant epoxy	Ciba Geigy
PI	Polyimide (Kerimide 601)	Rhône Phoulenc
	Thermoplastic	
PSO	Polysulphone (Udel 1700)	Union Carbide
PES	Polyethersulphone (Victrex 300P)	ICI
PEI	Polyetherimide (ULTEM 1 000)	General Electric
PEEK	Polyetheretherketone (PEEK)	ICI
PC	Polycarbonate (Makrolon 6555)	Bayer

EXPERIMENTS

Melt Rheology

When moulding a thermoplastic composite, correct temperature and pressure are essential for a void-free and well impregnated result. The thermoplastics were therefore characterized by melt rheology in order to obtain information about the flow properties during impregnation and pressing of the laminates. The dependence of melt viscosity on temperature and shear-rate was established using a Rheometrics rheogoniometer fitted with parallel plates.

Preparation of Laminates

Both unidirectional, [0], and bidirectional, [0, 90], laminates were produced. Their thickness was 2·0–2·2 mm.

The epoxy resins were supplied as carbon fibre prepregs which were compression-moulded into laminates according to the manufacturers' specifications. The volume fraction of fibres in the epoxy laminates was 60 % and 50 % for 914C and 920C respectively.

Carbon fibre prepregs with PI, PSO, PEI and PC were prepared by impregnating the fibres with a solution of the polymer, in N-methyl-2-pyrrolidone. After evaporating the solvent, the prepregs were stacked with additional polymer film in order to obtain the desired fibre content. Laminates were then produced by compression moulding. Total exposure time (including temperature conditioning of the laminate) was 10 min, for 5 min of which maximum pressure (6 GPa) was applied. The temperature was selected according to the melt-rheological data, so that all the thermoplastics attained the same melt viscosity at low shear-rate ($0·1 \, s^{-1}$). All the laminates were cooled while under pressure and removed at room temperature. Table 2 gives the moulding parameters together with the volume fraction of fibres (V_f).

Laminates based on PI prepregs were moulded according to the manufacturers' specifications (i.e. starting at 120 °C, increasing to 180 °C over 30 minutes, and curing for 60 min at 180 °C. The pressure was kept constant at 5 MPa through this treatment). A selection of these laminates was post-cured at 250 °C for 24 h. These are termed 'cured' below. The PI-laminates that were not post-cured are termed 'uncured'.

Carbon fibre/PEEK prepregs were prepared by melt-impregnating the fibres under pressure. Prepregs and PEEK-films were then stacked in the mould, and fused under pressure. Conditions used for melt-impregnation and moulding of laminates were the same as the conditions for the

TABLE 2
Moulding parameters

Matrix	Pressure (MPa)	Moulding time (min)	Temperature (°C)	V_f
PSO	6	10	275	~50
PES	6	10	305	~50
PEI	6	10	335	~50
PEEK	6	10	385	~50
PC	6	10	245	~50
PI	5	$\begin{cases} 30 \\ 60 \end{cases}$	$\begin{array}{c} 120 \rightarrow 180 \\ 180 \end{array}$	~50

lamination of the solution-impregnated prepregs as given in Table 2. A selection of PES and PSO laminates was also produced according to the same procedure. These samples are termed 'melt-impregnated'.

Impact Testing

Impact specimens were milled from the laminates. The specimens were of the double-notched type (*see* Fig. 1), with dimensions 65 × 13 mm. The V-notch had a root radius of 0·25 mm and a depth of 2 mm. Each double-notched specimen was placed in the impact tester as indicated in Fig. 2. In this way the impact initiates fracture on an undamaged part of the specimen. The effect of the double notch is a well defined fracture path and a reproducible fracture area. The impact specimens had fibre orientations of [0], [90], [0, 90] and [±45] relative to the long dimension of the test specimen.

The velocity of the pendulum at the point of contact with the specimen is 3·16 m/s, and the equivalent kinetic energy is in the order of 100 J. The

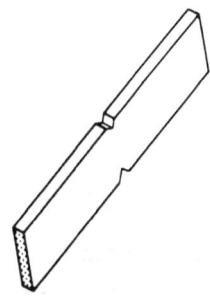

FIG. 1. The double-notched specimen used for the instrumented impact testing.

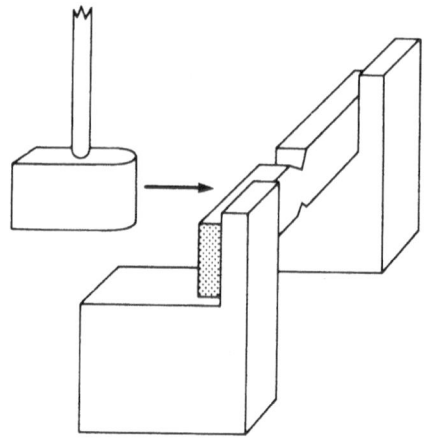

Fig. 2. The specimen is placed in the supports in such a way that the motion of the pendulum is parallel to the notches in order to initiate fracture on an undamaged part of the laminate (where no surface area is removed).

energy absorbed by the specimen is 0·1–3 J, so velocity changes of the pendulum during fracture can be neglected.

The impact pendulum was equipped with a force transducer, a charge amplifier and a transient analyser. The stress–strain relationship through the impact and the energy absorbed by the specimen were recorded (see Figs 3(a) and 3(b)). This type of impact testing can be treated as a special case of three-point bending carried out at very high strain rates. The stress–strain curve recorded by the transient analyser provides information for calculating flexural stress and modulus for the composite under investigation.

Maximum flexural stresses in three-point bending can be calculated according to ASTM D 790,

$$S = \frac{3PL}{2bd^2} \tag{1}$$

and the flexural modulus,

$$E_B = \frac{L^3 P}{4Dbd^2} \tag{2}$$

where P = load (N); L = support span (m); d = specimen thickness (m); b = specimen width (notch bottom to notch bottom) (m); D = deflection (m).

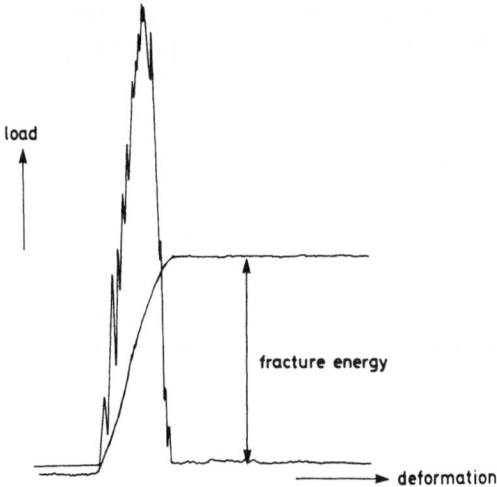

FIG. 3a. The recorded stress–strain relationship during the fracture of a solution impregnated [0] PSO laminate.

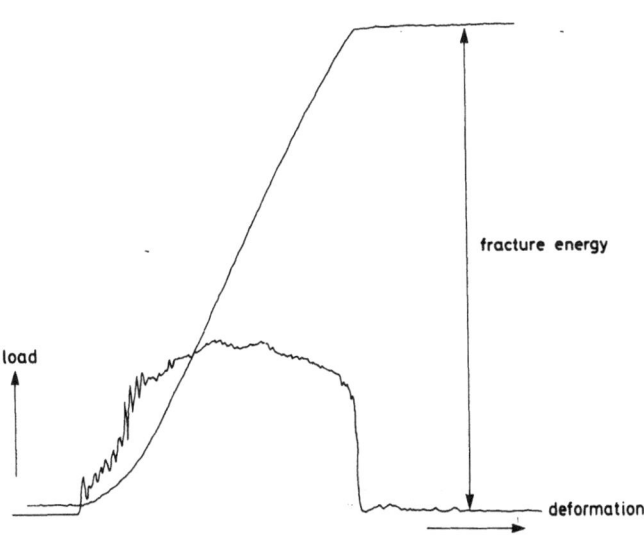

FIG. 3b. The recorded stress–strain relationship during the fracture of a solution impregnated [±45] PSO laminate.

Due to the presence of the notches, these measurements will not yield true stresses and moduli. However, since specimen dimensions were the same for all material combinations, the results provide a good basis for a comparative study of the parameters of interest.

RESULTS

In Fig. 4 the viscosity of PES is plotted as a function of shear-rate at eight temperatures from 260 to 385 °C. Corresponding measurements were also carried out for the other thermoplastics. During moulding of a laminate the

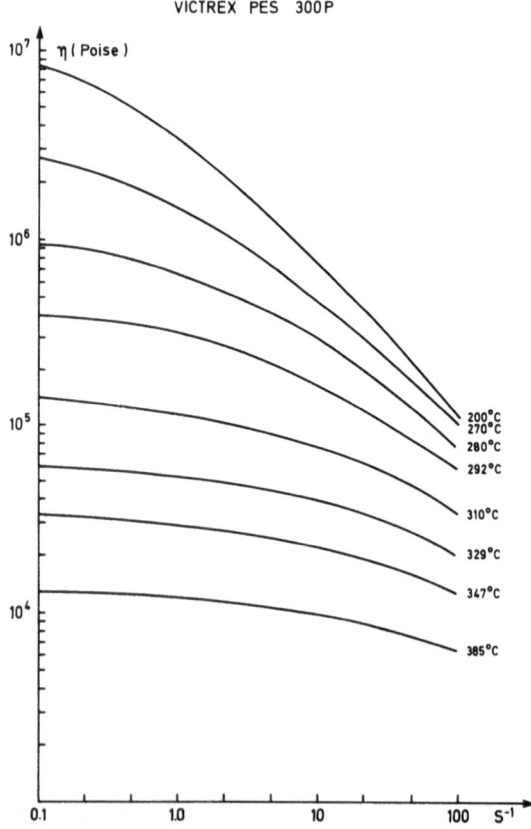

FIG. 4.   Melt viscosity for polyethersulphone (PES) measured at different temperatures and shear-rates. Note that the melt viscosity is more temperature dependent at low shear-rates than at higher ones.

material is subjected to low shear-rates. At these shear-rates the melt viscosity is considerably more temperature-dependent than at higher shear-rates, as shown in Fig. 4. Figure 5 gives viscosity as a function of temperature for PEEK, PEI, PES, PSO and PC. On the basis of these results, processing temperatures were selected to give the thermoplastics melt viscosities of 170 kilopoise at shear-rates of $0 \cdot 1 \, s^{-1}$. The moulding temperatures are listed in Table 2.

Scanning electron micrographs (SEM) of fracture surfaces in PSO show that the fibres were sufficiently impregnated when the fibres were pre-impregnated from solution (Figs 6, 7 and 8), but were insufficiently impregnated in the melt-impregnated samples (Fig. 9). The difference in pull-out of fibres during fracture for the two impregnation methods is

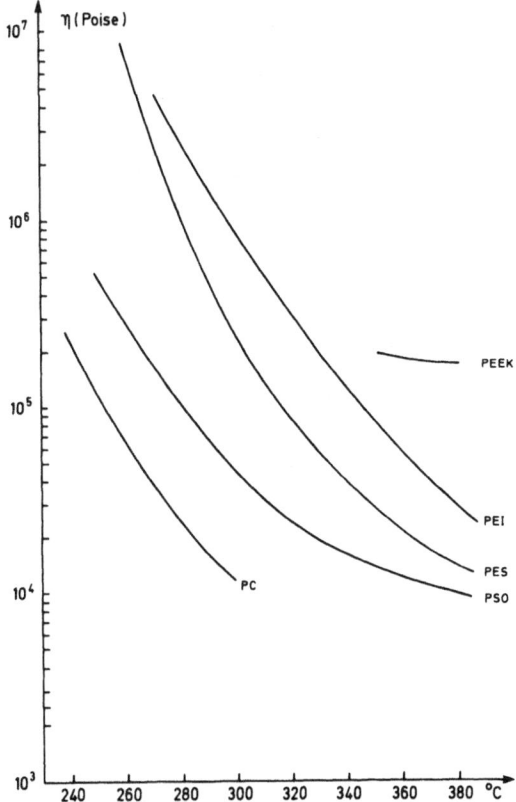

FIG. 5. Melt viscosity for thermoplastics as a function of temperature. The shear-rate is $0 \cdot 1 \, s^{-1}$.

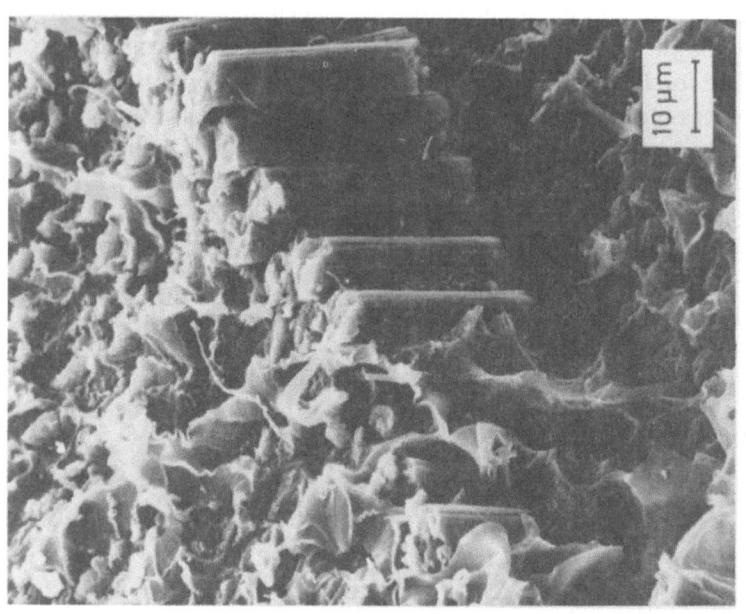

FIG. 7. Detail from Fig. 6 showing that the carbon fibres are well impregnated.

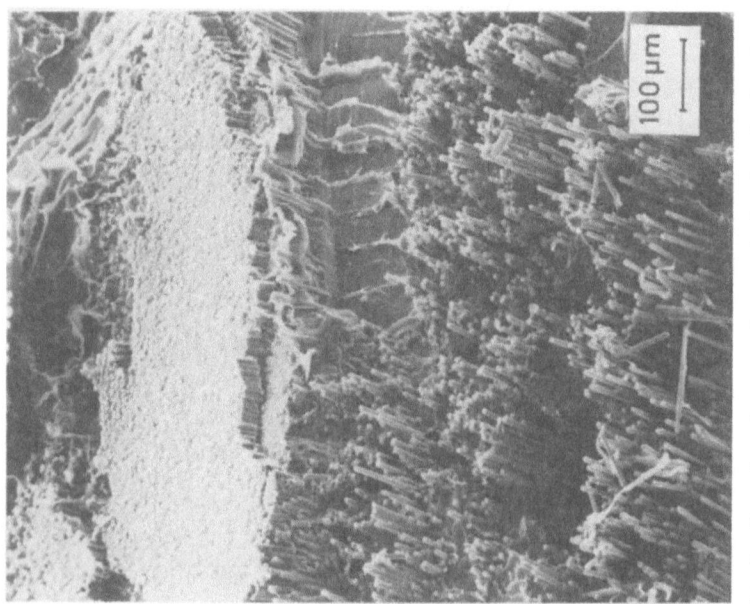

FIG. 6. Scanning electron micrograph showing details of the fracture of a solution-impregnated PSO [0] laminate. The upper part of the micrograph is nearer the compressed zone of the specimen, and the fracture here differs from the lower part where tensile stresses dominate.

FIG. 9. Scanning electron micrograph of melt-impregnated PSO [90] laminate. The fibres easily pull out during fracture.

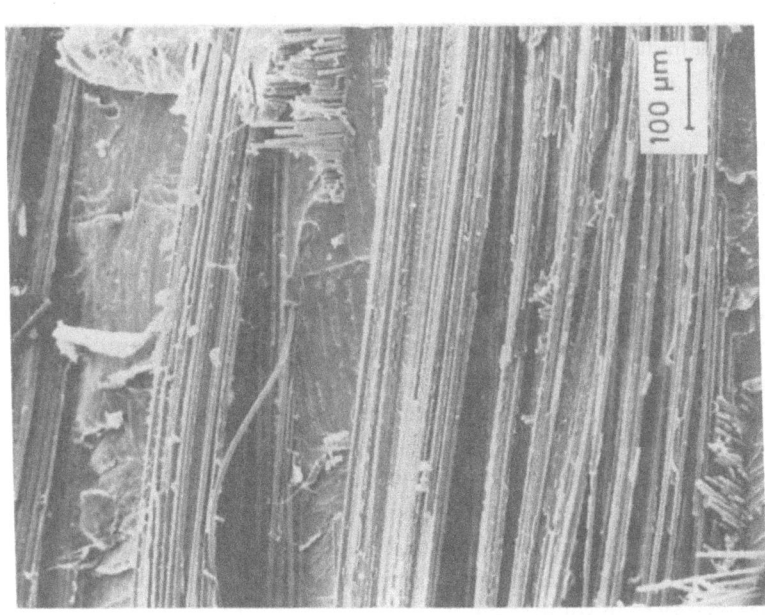

FIG. 8. Scanning electron micrograph of solution-impregnated PSO [90] laminate.

FIG. 10.   Test specimens of PSO [±45] laminates. The solution-impregnated laminate (left)
shows a brittle fracture while the melt-impregnated laminate (right) shows extensive fibre pull-
out.

compared in a larger scale in Fig. 10. Further indication of solution-
impregnated prepregs being the best prepreg manufacturing method is
found in the impact results (Figs 11–16).

Figures 11 and 12 are histograms of the measured fracture energy for the
various material combinations, impregnation processes and fibre orientations. They show that there is considerable difference between
conventional epoxy/carbon composites and the non-epoxy composites.

Maximum fracture energy is found for the [±45] specimens, followed by
[0] and [0, 90] specimens. As expected [90] specimens exhibited the lowest
values for the fracture energy. There is some degree of correlation between
the fracture energy for the pure matrix materials and the fracture energy
found for the corresponding fibre composite. An exception is solution-
impregnated PC. Pure polycarbonate did not break under impact testing,
while the carbon fibre reinforced laminate required significantly lower
fracture energy than solution-impregnated PES and PSO and slightly less
than PEI.

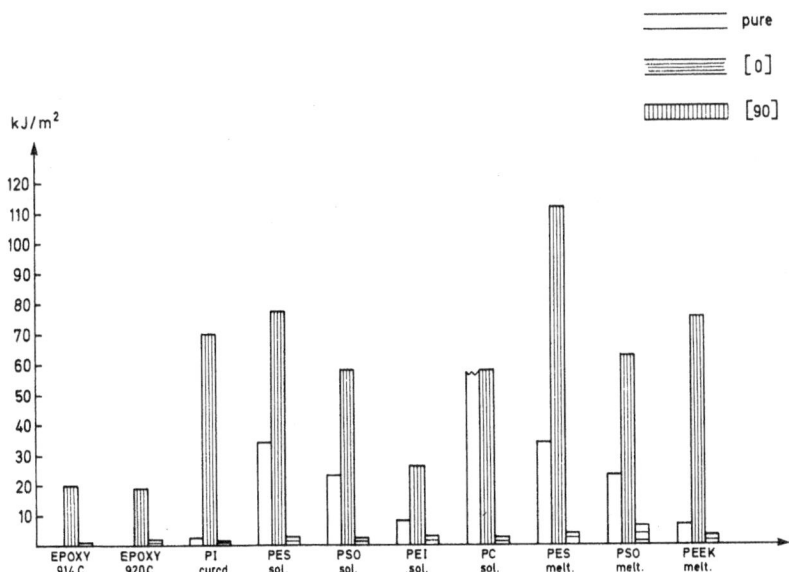

FIG. 11.   Measured fracture energies from instrumented impact testing. 'Sol.' refers to solution-impregnated laminates and 'melt' to melt-impregnated laminates. Pure PC did not break completely under testing.

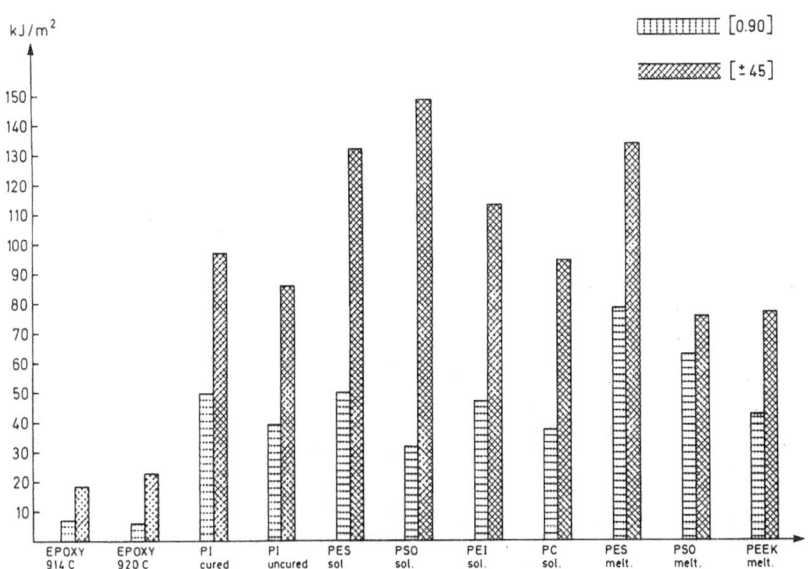

FIG. 12.   Measured fracture energies from instrumented impact testing for [0,90] and [±45] laminates.

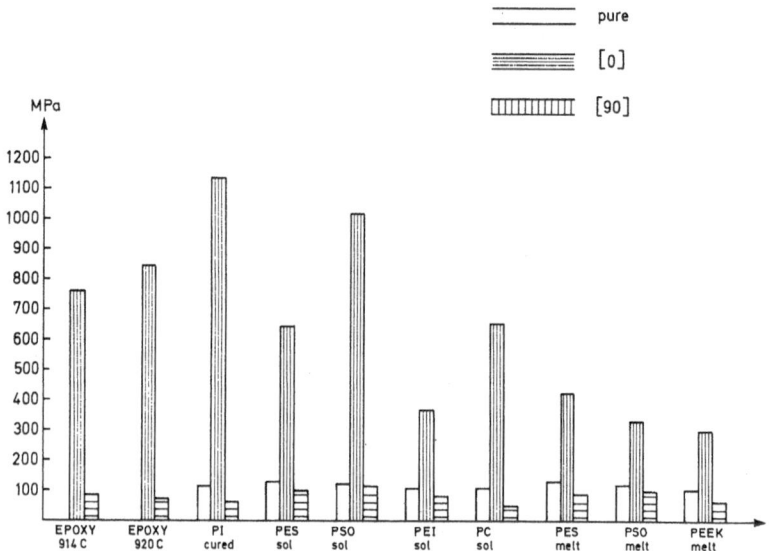

FIG. 13. Maximum 'flexural' stresses calculated using eqn. (1) and the stress–strain relationships recorded during the impact.

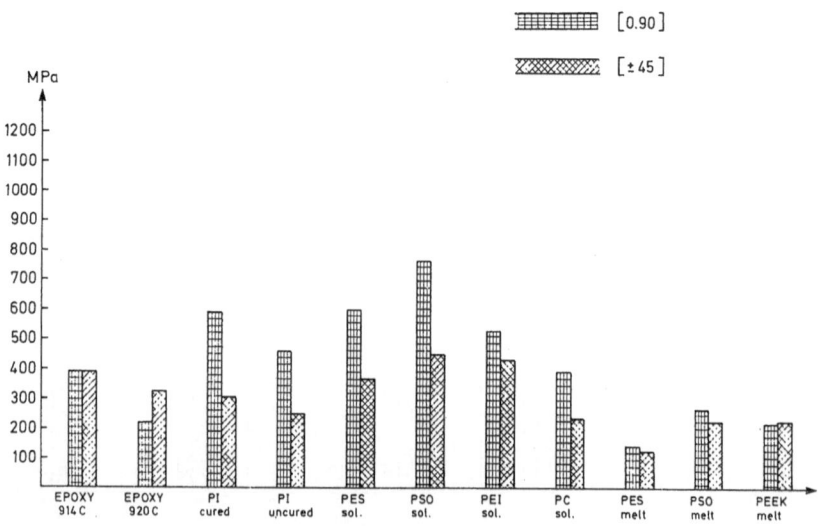

FIG. 14. Maximum 'flexural' stresses calculated using eqn. (1) and the stress–strain relationships recorded during the impact.

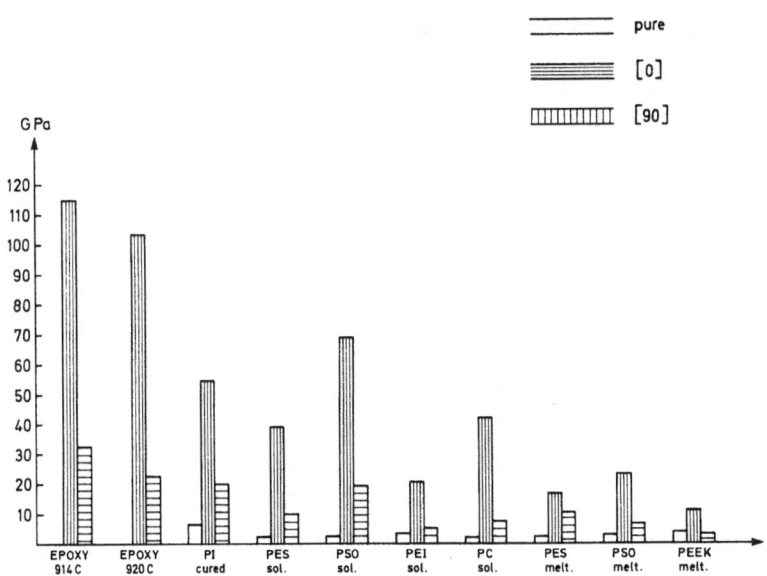

FIG. 15.   'Flexural' moduli calculated using eqn. (2) and the stress–strain relationships recorded during the impact.

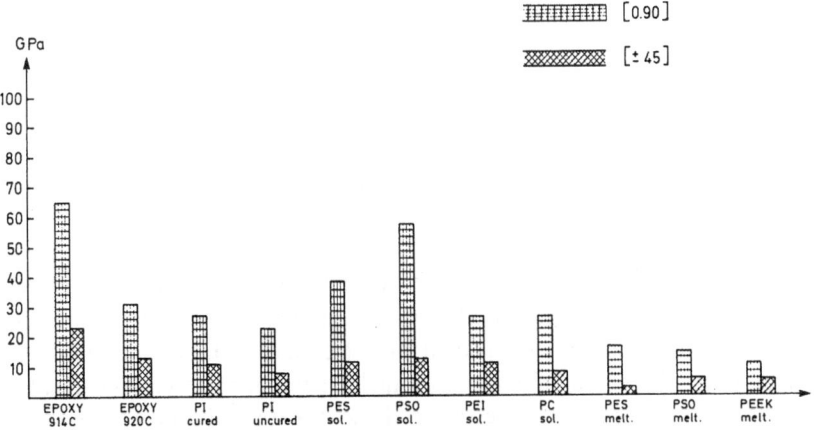

FIG. 16.   'Flexural' moduli calculated using eqn. (2) and the stress–strain relationships recorded during the impact.

Figures 13 and 14 show maximum 'flexural' stresses calculated according to eqn. (1). For [0] and [90] laminates the PI and the solution-impregnated PSO yielded a higher maximum stress than the two epoxy-based laminates investigated. For [0, 90] fibre direction, all the solution-impregnated specimens yielded a higher maximum stress than the two epoxy-based laminates. This behaviour is a result of the tendency of epoxy-composites to delaminate during deformation. This tendency was not observed for the solution-impregnated specimens, possibly due to the ability of thermo-plastics to bear higher strains before fracture. Also, in the solution-impregnated specimens the carbon fibres are concentrated in bundles to a greater extent than in epoxy laminates.

'Flexural' stresses for the melt-impregnated specimens are far lower than for their solution-impregnated equivalents. An explanation for this is found in the poorer impregnating characteristics of the melt-impregnation methods, so that fewer fibres act as stress bearers under load.

The 'flexural' modulus can be calculated according to eqn. (2). The results are given in Figs 15 and 16. Epoxy laminates tested at [0] and [90] directions have the highest 'flexural' moduli. Of the remaining laminates, solution-impregnated PSO yields the highest 'flexural' modulus. At [0, 90] and [±45] directions the highest values are found for epoxy 914C, but also the solution-impregnated PSO have a high 'flexural' modulus in these directions. The melt-impregnated composites yielded comparatively low values for all fibre directions. As stated above this method has poorer impregnating characteristics than the solution method, and therefore a lower effective volume fraction of fibres.

DISCUSSION AND CONCLUSION

Melt Rheology

The laminates were prepared with the intention of studying how their impact properties varied with the matrix materials. In order to keep the moulding parameters constant, i.e. to subject all the thermoplastic laminates to reasonably equivalent melt treatments, moulding temperatures were chosen on the basis of melt rheological tests. The results here are therefore not necessarily those for the optimum moulding conditions, but for a combination of temperature and low shear-rate that yields equivalent viscosities for all the thermoplastics.

Commercial Epoxy Laminates versus Thermoplastic Laminates

The experimental results clearly show that the fracture energies of the laminates are essentially higher than the fracture energies of commercial epoxy/carbon-fibre laminates. The solution-impregnated thermoplastic laminates were, however, not observed to have more fibre pull-out than the epoxy laminates. Increased fracture energies of the thermoplastic laminates appear, therefore, to be due to their greater ability to bear higher strains before fracture, rather than to a changed fracture mechanism. Another reason for the high fracture energy of the thermoplastic laminates is that the fibres are more uniformly distributed in the epoxy laminates than in the thermoplastic laminates.

The calculated 'flexural' stresses are higher for the thermoplastic laminates than for the epoxy laminates. The main reason for this is the tendency of carbon-fibre-reinforced epoxies to delaminate when deformed at high strain rates. The values obtained in the present study for epoxy materials are lower than would be measured at conventional 'flexural' tests with slower deflection rates.

Among all the laminates tested, the calculated 'flexural' moduli for [0] and [90] laminates are highest for the epoxy laminates. This is probably due to the more uniform fibre distribution in these laminates, which causes the stress to be transferred to a greater fraction of fibres at an early stage. Among the [0, 90] laminates only epoxy 914C, with 60% fibres, had a higher 'flexural' modulus than solution-impregnated PSO laminates with 50% fibres.

Melt-impregnated versus Solution-impregnated Laminates

The results clearly indicate that melt-impregnation yields poorer fibre impregnation than impregnation by a polymer solution. A lower degree of impregnation favours fibre pull-out during fracture, whereas in contrast a high degree of impregnation yields fracture by fibre breakage. For this reason laminates prepared by the two methods exhibited roughly equivalent fracture energies. The poorer impregnation of the melt-impregnating method results in the relatively low 'flexural' stresses and moduli found for specimens prepared by this method.

PSO/Carbon-fibre Laminates

On the basis of fracture energies, 'flexural' stresses and moduli, PSO is the most favourable matrix material for the laminates manufactured in this work. The fracture energies for the PSO laminates are from 3 to 7 times higher than the epoxy laminates, and the 'flexural' stresses are 1–2 times higher. 'Flexural' moduli are higher than the epoxy 920C laminates and comparable to epoxy 914C laminates at [0, 90] fibre directions.

ACKNOWLEDGEMENT

This work was supported by the Royal Norwegian Council for Scientific and Industrial Research (NTNF).

REFERENCES

1. LIND, D. J. and COFFEY, V. J., A method of manufacturing composite material, *Brit. Pat.* 1 485 586 (1977).
2. MURPHY, D. J. and PHILLIPS, L. N., Thermoplastics materials, *Brit. Pat.* 1 570 000 (1980).
3. HOGGATT, J. T. and LAAKSO, J. H., Fabrication of low-cost primary aircraft structures with thermoplastic composite, *Eighth SAMPE National Symposium*, **8** (1976), 459–478.
4. MAXIMOVICH, M. G., Evaluation of selected high-temperature thermoplastic polymers for advanced composite and adhesive applications, *ASTM STP 617* (1976).
5. SHOROKHOV, V. M., NOVIKOVA, O. A., LIPATOV, YU. S. and BEZRUK, L. I., Thermoplastics reinforced with continuous glass-fibre, *Int. Polym. Sci. Technol.*, **8**, No. 10 (1981), 56–58.
6. HOGGATT, J. T., Aerospace application of graphite reinforced thermoplastic composites, *Conference on Advanced Composite Technology*, El Segundo, Calif., March (1978).

25

Thermal Cycling and Vacuum Baking Effects on the Thermal Expansion of Graphite Fiber Reinforced Composite Laminates

MELVIN S. HENRIKSEN

*Beckman Instruments, 200 South Kraemer Blvd,
Brea, California 92624, USA*

PAUL D. ARTHUR and GERARD C. PARDOEN

*School of Engineering, University of California,
Irvine, California 92717, USA*

and

HARI DHARIN

*Mechanical Engineering Department, University of California,
Berkeley, California 94720, USA*

ABSTRACT

The need for dimensionally stable materials in aerospace applications has prompted interest in fiber reinforced composites. Unfortunately thermal cycling and moisture absorption can cause significant dimensional changes in these materials. Some graphite fibers have a negative coefficient of thermal expansion (CTE), so it is possible to construct structures with very low or even zero overall CTE. This paper reports experimental and numerical studies on the effect of thermal cycling, matrix microcracking and moisture desorption on the longitudinal CTE of cross-ply graphite fiber-reinforced composite cylinders. A tentative identification of the responsible mechanisms is presented.

BACKGROUND

The increased importance of graphite/epoxy as an engineering material can be attributed to its orthotropy, low thermal expansion, and favorable stiffness and strength-to-weight ratios. The material orthotropy allows laminae of differing orientations to be combined to achieve a completely anisotropic structure. The negative thermal expansion coefficient of some graphite fibers permits a zero coefficient of thermal expansion, useful where dimensional stability over long time and changing environments is required.

Problems with composites result from exposure to environments after cure and viscoelastic stress relaxation. Moisture absorption by the epoxy matrix can cause dimensional changes and alter the elastic modulus or CTE.

As cured laminated composites cool to room temperature, stresses develop at the laminae and fiber–matrix interfaces due to the different constituent expansion characteristics. The lower temperatures in space can further increase these stresses. A space vehicle moving between sun and shadow will produce stress cycling which may cause the material to fatigue. Previous investigators have shown that cool down and thermal cycling can induce microcracking in the epoxy matrix and fiber–matrix disbonding.[1-3] Disbonding at the fiber–matrix interface tends to cause an increase in longitudinal (fiber direction) CTE.[4] Microcracking has also been found to alter laminate tensile strength, interlaminar shear strength, flexure strength and stiffness and to alter the overall static dimensions.[1] Hence, initially stable composites could become unstable in an actual space application. Residual thermal stresses can cause structure warpage or induce microcracking in the matrix which, in turn, may degrade some material properties. Thus investigations of moisture and thermal effects on graphite/epoxy laminates are of great practical importance.

SCOPE OF PRESENT WORK

This paper presents the effect of thermal cycling and moisture desorption on the longitudinal CTE of cross-ply composite cylinders and an identification of the responsible mechanisms. A numerical study examined the effects of matrix microcracking on longitudinal CTE and on the stress distribution throughout the laminate.

EXPERIMENTAL STUDIES

Experimental work was done to establish the change in microcrack density and longitudinal CTE with increasing numbers of thermal cycles for two different types of graphite fibers using the same matrix resin. Vacuum baking provides the effect of moisture loss on the longitudinal CTE of individual specimens. Vacuum baked graphite/epoxy tubes were thermally cycled to measure the overall longitudinal CTE. The tubes were subsequently examined to identify mechanisms responsible.

The two particular graphite fiber reinforced composites chosen for testing were (1) a medium modulus fiber, Hercules HMS, in a Fiberite 934 epoxy resin matrix, and (2) a high modulus fiber, Celanese GY-70, in the same resin. These fibers bracket a significant range of moduli. Both fiber–resin composites were cured for 2 h at 350 °F. Nominal material properties for HMS and GY-70 tapes are given in Table 1.

The cylindrical test samples are tubes with a nominal diameter of 2·5 in and an 8 ply $(0/90/0/90)_s$ lay-up. Each ply thickness is 0·005 in for a total tube wall thickness of 0·040 in. The samples used for vacuum baking were 5 in. in length and the thermal cycling samples were 4·5 in. in length.

One each of the HMS and GY-70 tubes were baked in vacuum for four days at 250 °F. The remaining two tubes were thermal cycled a total of 30 times from -120 °C to 100 °C. Initial CTE measurements were taken of the uncycled and unbaked tubes. CTE measurements were then taken after baking and after the first, fifth, tenth and thirtieth cycle. Trimmings from each of the two cycled tubes were also cycled for optical and scanning electron microscope observation. Thermal cycling of the test samples was done in an automatic temperature controlled chamber using electrical resistance heating. Cooling was provided by liquid nitrogen. Longitudinal CTE's of the |tubular samples were obtained with a Hewlett-Packard helium–neon laser dilatometer using an interferometer for differential length measurements. The test sample was heated with electrical resistance heating tape wound outside in a helix. Tape voltage was controlled with a

TABLE 1
Nominal material properties

	$E_1 \times 10^6$	$E_2 \times 10^6$	v_{12}	$G_{12} \times 10^6$	$\alpha_1 \times 10^{-6}$	$\alpha_2 \times 10^{-6}$
HMS tape	25	0·85	0·3	0·65	$-0·3$	14
GY-70 tape	38	0·9	0·25	0·7	$-0·6$	17·5

TABLE 2
CTE vs number of thermal cycles (in/in °F)

Number of thermal cycles	CTE HMS/934	CTE GY-70/934
0	0.18 ± 0.04	-0.004 ± 0.044
1	0.34 ± 0.07	-0.012 ± 0.058
5	0.47 ± 0.09	-0.076 ± 0.076
10	2.05 ± 0.17	0.356 ± 0.071
30	1.21 ± 0.53	-0.509 ± 0.049

Variac voltage regulator. The sample temperature was monitored by four thermocouples on the outside of the tube. Tube temperature was increased from ambient in three increments to 30°F above ambient. The tube was allowed to come to temperature for 30 min prior to taking measurements.

The CTE values were calculated from plots of the measured linear displacements of the samples versus the change in temperature as shown in Table 2.

Examination of cycled samples of the HMS and GY-70 material using optical and scanning electron microscopy revealed that matrix micro-cracking was present but that these defects were produced primarily during the first thermal cycle. Subsequent cycling did not noticeably alter the crack density. These observed cracks are on the interior ply of the $(0/90/0/90)_s$ material and it is assumed that this crack density is typical of the crack densities in the other plies (Table 3).

TABLE 3
Measured crack densities in thermal cycled material

Material	Number of thermal cycles	Microcrack density (per inch)
HMS	0	3
HMS	1	27
HMS	10	26
HMS	30	26
GY-70	0	4
GY-70	1	23
GY-70	17	21

Significant matrix microcracking in the 90° plies tends to decrease the effects of those plies on the overall laminate CTE in the 0° direction such that the laminate CTE approaches the 0° laminate CTE value.[5] Examination of the crack density data indicates that microcracking occurred only after the first thermal cycle. The CTE variation with thermal cycling for HMS indicates a gradual increase in CTE with increasing cycles, a marked increase near 10 cycles and then a subsequent decrease. This behavior cannot be explained by microcracking but may be attributed to fiber–matrix disbonding, followed by subsequent ply delamination. As will be indicated it is possible that ply delamination may be initiated by transverse microcracking due to the high stress levels present at the laminae interfaces surrounding the crack tips. Fiber–matrix disbonding induced by methods other than thermal cycling has previously been shown to cause an increase in CTE,[4] resulting from a decrease in the fiber effect on the composite expansion. High shear stresses at the fiber–matrix boundary induced by temperature changes have been described by Adams.[6] Delamination (especially if adjacent to transverse microcracks) tends to decrease the longitudinal CTE as will be shown numerically.

The CTE of the GY-70/934 thermal cycled sample shows much the same pattern of variation as the HMS though shifted downward and of lesser magnitude. No significant shift in CTE after the five thermal cycles can be attributed to the observed matrix microcracking. The subsequent behavior again may be explained by fiber–matrix separation and ply delamination. Values of the CTE's for the vacuum-baked samples are as presented in Table 4.

TABLE 4
CTE's of vacuum-baked samples (in/in °F)

Vacuum baked	HMS/934	GY-70/934
Before	0.62 ± 0.12	0.03 ± 0.05
After	0.40 ± 0.09	-0.09 ± 0.06

These variations of CTE are attributed to measurement uncertainty.

The suggested mechanisms responsible for CTE variation with time make it difficult to isolate any one failure mechanism (such as matrix microcracking) and observe its effects experimentally. We attempt, however, to display the isolated effect on CTE of matrix microcracking or ply delamination by numerical modeling.

NUMERICAL STUDIES

Using the NASTRAN finite element program, several numerical models were developed and used to study matrix microcrack and ply delamination effects on overall laminate CTE and to determine the stress distribution around these microcracks and delaminations. A rectangular, four node, plane stress element with orthotropic properties was used.

The laminate modeled was a 4-ply symmetric$|(0/90)_s|$flat plate as shown in Fig. 1. Due to symmetry about the x and y axes, only a quarter plate is used. Each ply was assumed to be homogeneous, and the apparent material properties are those listed previously for HMS tape. Stress relief cracks in the 90° ply were postulated and inserted in the model. An exaggerated diagram of the crack model is shown in Fig. 1 in which models of 1, 3, 5 and 7 cracks were examined.

Previous numerical modellers have suggested that there are high interlaminar stresses at the plate edges under uniaxial loading.[7,8] Assuming a similar stress state at the crack edges, delamination between the 0° and 90° plies was postulated at the crack tips. Models of cracks with delaminations at the tips were examined for the same series of cracks as for those without delaminations. A finer mesh was used near the cracks/delaminations. The initial crack and delamination widths were set at zero. The models were initially in equilibrium, and a static stress case was developed by applying a temperature decrement of 1 °F. The output for

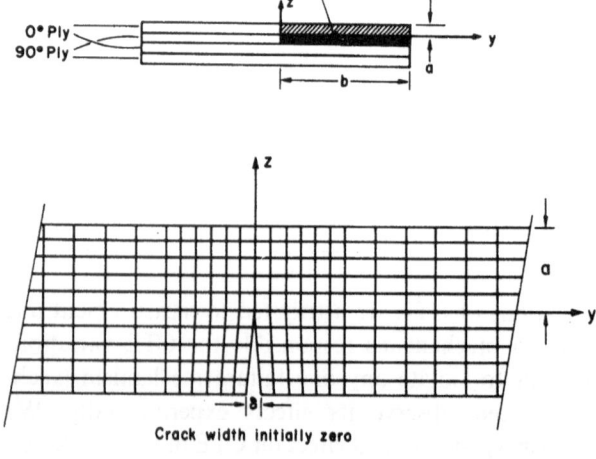

Fig. 1.   Finite element model of four-ply laminate.

each of the models described consists of nodal displacements in both the y and z directions, the average normal and average shear stresses (per degree Fahrenheit) in each of the elements, and the total strain energy of the laminate.

CTE Variation along Plate, No Cracks

The finite element laminated plate model allows one to examine the variation of CTE along the plate length in the y direction. The displacement output for the uncracked finite element model shows that the longitudinal displacement vs height is linear and perpendicular to the centreline until approximately $y/b = 0.90$, whereas closer to the plate end, the displacement is non-linear. Thus Kirchhoff's hypothesis does not hold true at the plate end, and we would expect a deviation in CTE from that given by laminated plate theory. Using the rule of mixtures equation and material properties defined earlier, the value for the longitudinal CTE is

$$\text{CTE} = \frac{\alpha_{0^\circ} E_{0^\circ} + \alpha_{90^\circ} E_{90^\circ}}{E_{0^\circ} + E_{90^\circ}} = 0.17 \, \text{m in/in} \, {}^\circ\text{F} \tag{1}$$

We use the simplified equation ($v_{12} = 0$) since the plane stress finite element model does not take v_{12} into consideration. This mixture value is exactly what the finite element model reports in the interior of the plate (*see* Fig. 2).

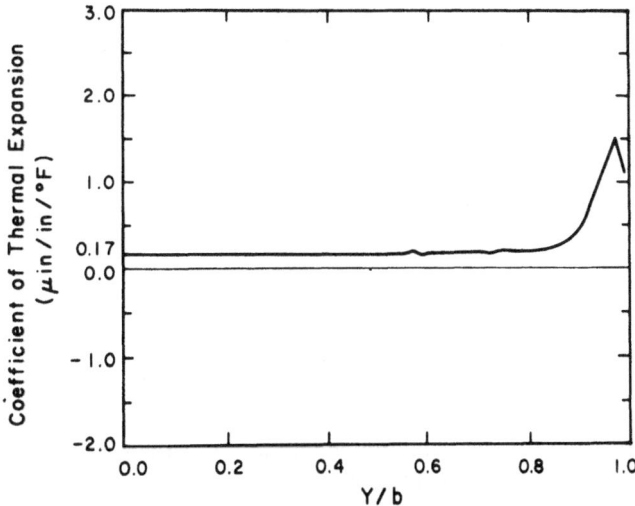

FIG. 2. Finite element CTE versus plate length for no cracks or delaminations.

For another averaged value of CTE we compute the average of each element over the total plate length:

$$CTE_{avg} = \sum_{i}^{n} \frac{CTE_i L_i}{L} L \qquad (2)$$

where n = number of elements; L_i = length of ith element; CTE_i = CTE of ith element = $\Delta L_i / L_i \Delta T$; ΔL_i = change in length of the ith element; ΔT = change in temperature; L = total length of plate.

Using the definition of CTE_i, we have:

$$CTE_{avg} = \sum_{i}^{n} \frac{\Delta L_i L_i}{L_i \Delta T L} = \frac{1}{L} \frac{\Delta L}{\Delta T} \quad \text{as expected} \qquad (3)$$

This numerical model yields a CTE value of $0.26\ \mu in/in\ °F$ at the $0°/90°$ interface. This value is larger than the 'interior' value of 0.17 and is recognized as an average of the finite element values which increase toward the laminate edge.

CTE Variation with Cracks

CTE values versus number of transverse microcracks and/or delaminations are presented in Fig. 3. The plate and displacements are shown in Fig. 4 for microcracks and for microcracks with delaminations. The plate end configuration remains unchanged, and the cracks and/or delaminations cause a translation of the plate end to the right as the failures increase. From Fig. 3 we see that delaminations alone have no effect on the laminate CTE, but the CTE decreases linearly with increasing numbers of cracks.

An elementary approach for CTE variation with ply delamination is to consider two plates with different material properties laminated together with a gap in the lower plate. The CTE along lengths L_1 and L_2 is described by

$$\alpha' = \frac{\alpha_1 E_1 + \alpha_2 E_2}{E_1 + E_2} \cdot \qquad (4)$$

The CTE along the delamination length d is just the CTE of material 1, hence the overall thermal strain is:

$$\Delta^1 = \alpha' \Delta T (L_1 T L_2) + \alpha_1 \Delta T(d) \qquad (5)$$

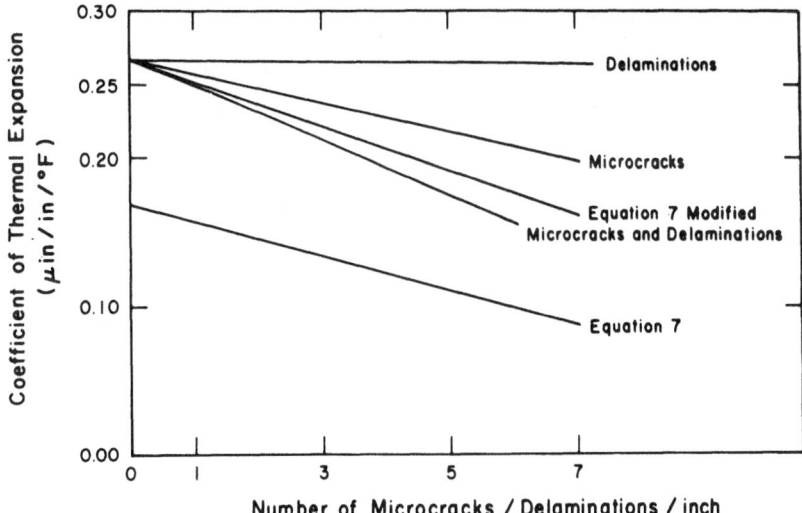

FIG. 3.   Centerline CTE versus number of microcracks per inch.

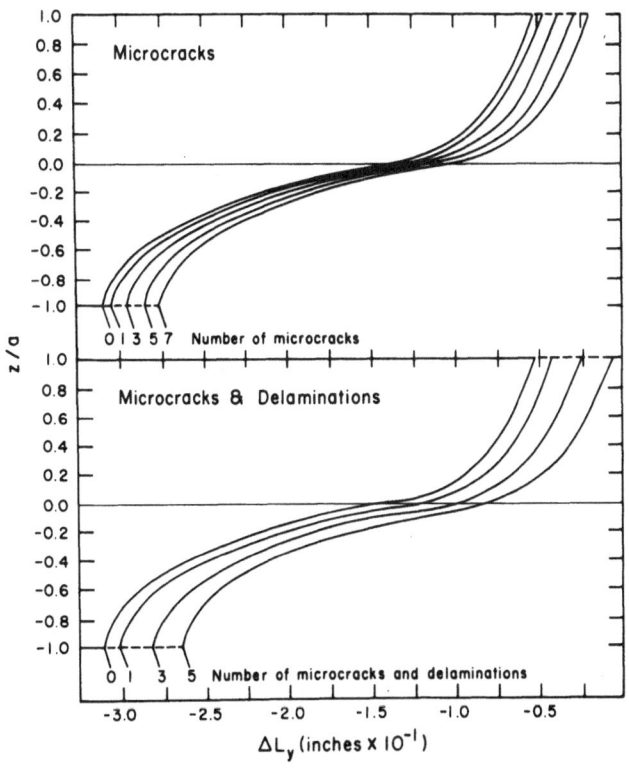

FIG. 4.   Deflection of plate end in y direction versus plate height.

Noting that $L = L_1 + L_2 + d$ and that $\alpha = \Delta L/\Delta TL$, we have:

$$\Delta L = \alpha \, \Delta TL = \alpha' \, \Delta T(L - d) + \alpha_1 \, \Delta Td$$

$$\alpha = \alpha' \, \frac{L - d}{L} + \alpha_1 \, \frac{d}{L}$$

$$\alpha = \alpha' + d/L(\alpha_1 - \alpha') \tag{6}$$

If d is the sum of n separate delaminations such that $d = \sum_{i=1}^{n} d_i = nd_i$, eqn. (6) becomes:

$$\alpha = \alpha' + \frac{nd_i}{L} (\alpha_1 - \alpha') \tag{7}$$

This model is plotted in Fig. 5 (dotted) for comparison with the numerically generated graphs. Here, we have used $\alpha' = 0.17 \, \mu\text{in/in} \, °F$. $\alpha_1 = -0.3 \, \mu\text{in/in} \, °F$, $d_i/L = 0.005$. These values of α_1 and d_i/L are consistent with the numerical model and α' is that derived from eqn. (4).

We have already seen that due to edge effects, the initial CTE as predicted by eqn. (4) is less than the numerically predicted CTE. The numerical value for CTE in eqn. (7) is designated 'Equation 7 modified' in Fig. 3, and lies between the numerically generated curves for microcracks alone and that for microcracks with delaminations.

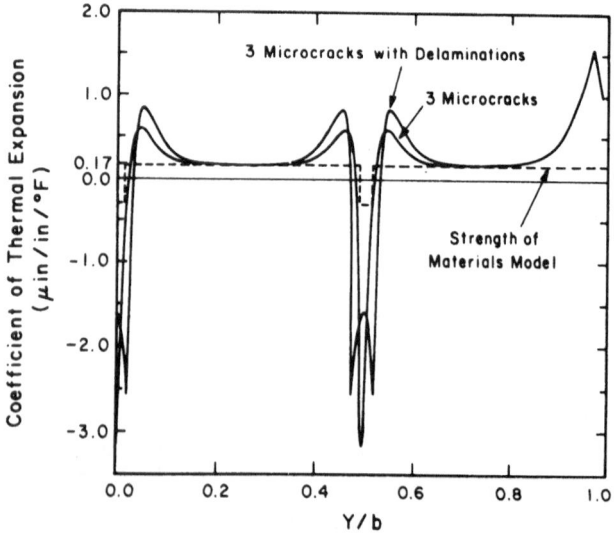

FIG. 5.   CTE versus plate length.

A typical variation of CTE along the plate length for cracks and delaminations is presented in Fig. 5. At the crack or delamination edges the CTE becomes negative, rises sharply to high positive values and returns to the laminated plate theory prediction of $0 \cdot 17 \, \mu$in/in °F well away from the crack edge. Comparing the CTE around microcracks with and without delaminations, it is seen that the delaminated CTE attains a higher positive value, and the rise is shifted away from the crack. The overall resultant decrease in CTE, however, is brought about by the greater length of negative localized CTE near the crack tip.

Stress Distribution

The numerical model of the uncracked laminate near the $0°/90°$ laminae interface is presented in Fig. 6. The figure gives the distribution just above the interface in the $0°$ ply as well as describing the stresses just below the interface in the $90°$ ply.

The normal stress σ_z and the shear stress τ_{yz} increase from zero as the

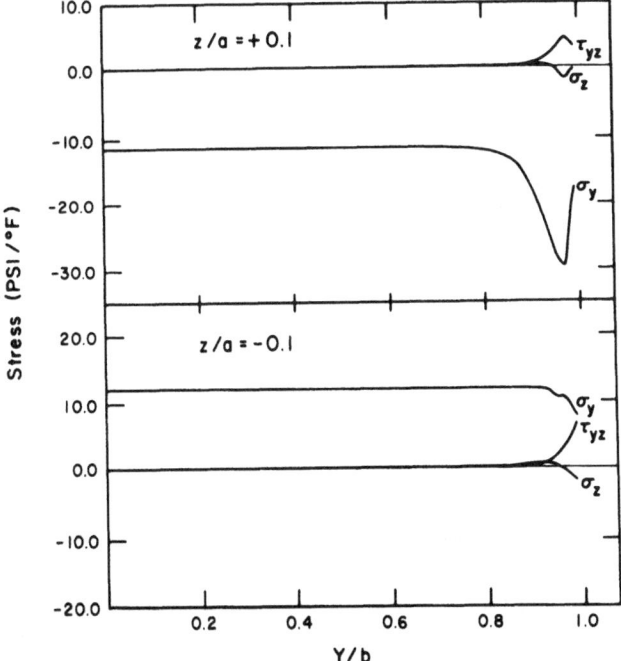

Fig. 6. Stress distribution (no micro failures).

edge is approached. At the interior of the plate σ_y agrees well with the value
of

$$\alpha_y = \frac{E_2 E_1}{E_2 + E_1} \Delta T(\alpha_2 - \alpha_1) \tag{8}$$

Note that with cooling the $90°$ ply stress is tension and in the $0°$ ply
compression. At the plate edges σ_y deviates from this value. Although not
displayed here the calculated stresses near the $90°/90°$ interface at
$z/a = -0.9$ compare favorably with the results of Wang and Grossman.[9]

When a transverse microcrack is introduced into the $90°$ ply at $y/b = 0$,
the internal stress distribution about the crack area is radically altered.
Figure 7 shows σ_y, σ_z, and τ_{yz} at $z/a = -0.1$. As expected, σ_y switches signs
to become positive (tension) in the $0°$ ply near the interface at the crack tip.
The introduction of the crack in the $90°$ ply has created a free edge which
cannot support a normal stress. Thus the drop in σ_y in the $90°$ ply near the
crack tip. This change in the distribution of σ_y with the crack is seen in the
thickness plot of normal stress through the plate (Fig. 8). Returning to
Fig. 7, we see that where σ_z and τ_{yz} were previously zero near $y/b = 0$

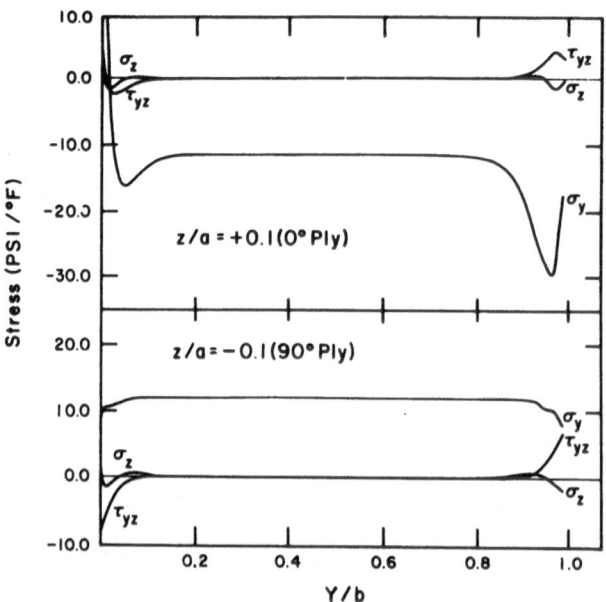

FIG. 7.   Stress distribution (microcrack at $y/b = 0$).

(Fig. 6), they have now been raised to significant levels. Note that very near the crack tip σ_z is a high tensile stress. The distribution of these stresses through the laminate thickness is shown in Fig. 8. Given a high enough temperature, these interlaminar shear and tensile stresses will rise to levels capable of causing delaminations between 0° and 90° plies.

Introducing such a delamination at the crack tip yields a stress distribution near the lamina interface as shown in Fig. 9. Stress distributions through the plate near the delamination tip shows that this redistribution

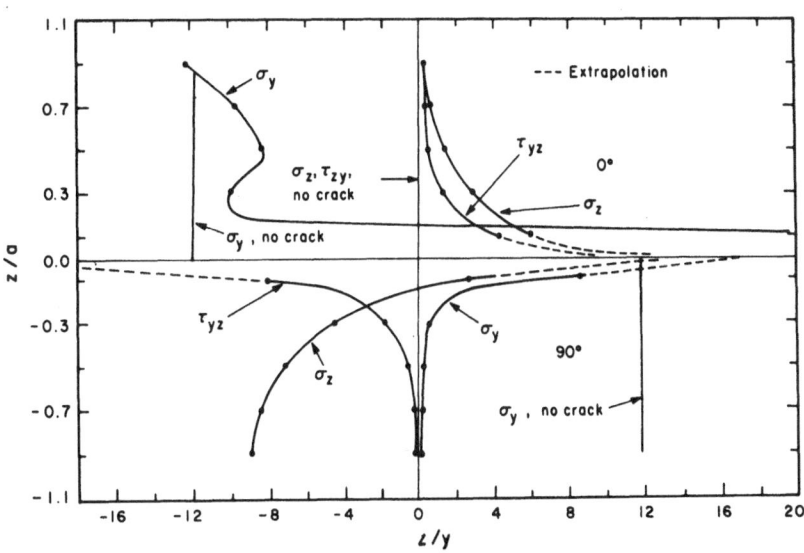

FIG. 8.   Stress distribution at $y/b = 0.0025$ (with and without microcrack at $y/b = 0$).

consists of a shifting of the normal stress curve σ_y laterally along the y axis away from the crack tip to a position corresponding to the delamination tip with a slight increase in the already higher than normal compressive stress near the failure tip. The shear stress τ_{yz} has increased slightly at the delamination tip from its previous value at the crack tip, but the normal stress σ_z has decreased by almost tenfold, no longer exhibiting the sharp tensile spike at the failure tip. Despite the higher shear stress, the lowering of σ_z may indicate a more stable failure than the undelaminated crack tip since graphite epoxies tend to be stronger in shear than in tension.[10] This is difficult to analyze quantitatively because of the need to interpolate stress values between opposite sides of the lamina interface.

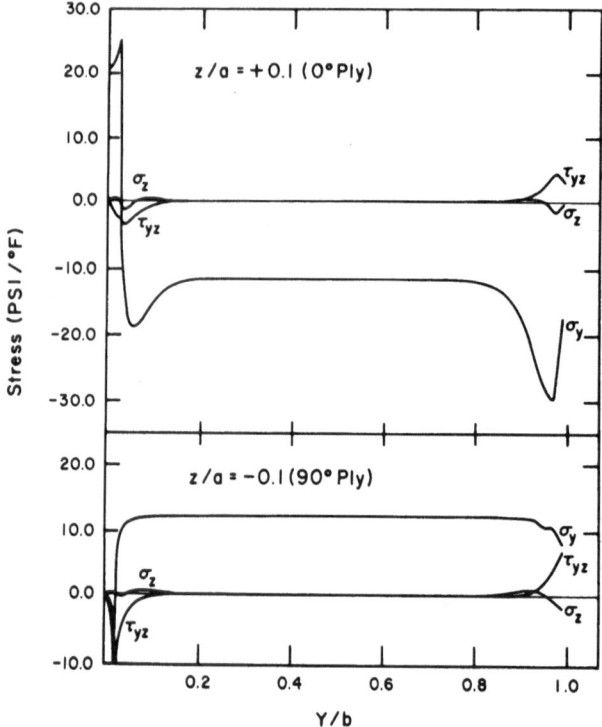

FIG. 9.   Stress distribution (microcrack at $y/b = 0$; delamination tip at $y/b = 0$).

CONCLUSIONS

The variation of CTE has been measured with cross-ply composite cylinders constructed of HMS/934 and GY-70/934 graphite epoxies. The CTE increased up to 10 thermal cycles with a decrease at 30 cycles. It is proposed here that the initial increase is due to fiber–matrix debonding as previously described.[3,4] This variation on CTE cannot be directly attributed entirely to transverse matrix microcracking. Matrix microcrack density is seen to change only after the first thermal cycle, with no change in subsequent cycles. The observed subsequent decrease in CTE may be attributed to ply delamination between the 0° and 90° plies induced by high interlaminar stresses present at the tips of matrix microcracks.

High interlaminar stresses at the microcrack tips were confirmed by a numerical finite element analysis. Delamination at the crack tip was found to decrease the normal inter-laminar stress τ_{yz}. The numerical model confirmed previous investigations into thermally induced edge stresses.

Kirchhoff's hypothesis was seen to hold only away from failure and edge effects, hence rendering the laminated plate theory derivation of laminate CTE inaccurate for other than very long uncracked plates. For shorter plates edge effects become important in altering the CTE from the laminated plate theory equation. Failures such as matrix microcracks and delaminations also detract from the accuracy of this relation.

REFERENCES

1. KIRLIN, R. L. and PNYCHON, G. E., Dimensional stability investigation—graphite/epoxy trust structure, *Natl SAMPE Symp. Exib. Proc. 24th, Enigma of the Eighties: Environ., Econ., Energy*, San Francisco, CA, May 8–10 (1979).
2. EXELUM, S. A., NEUBERT, H. D. and WOLF, E. G., Microcracking effects on dimensional stability, *Natl SAMPE Symp. Exhib. Proc. 24th, Enigma of the Eighties: Environ., Econ., Energy*, San Francisco, CA, May 8–10 (1979).
3. MAZZIO, V. F. and MEHAN, R. L., Effects of thermal cycling on the properties of graphite–epoxy composites, *Composite Materials: Testing and Design (Fourth Conference)*, ASTM STP617, American Society for Testing and Materials, pp. 466–480 (1977).
4. MAROM, G. and GERSHON, B., Interfacial bonding and thermal expansion of fibre-reinforced composites, *J. Adhesion*, **7**, 195–201 (1975).
5. CAMAHORT, J. L., RENNHACK, E. H. and COONS, W. C., Effects of thermal cycling environment on graphite/epoxy composites, *Envjronmental Effects on Advanced Composite Materials*, ASTM STP 602, American Society for Testing and Materials, pp. 37–49 (1976).
6. ADAMS, D. F., Temperature- and moisture-induced stresses at the fiber/matrix interface in various composite materials, *Natl SAMPE Symp. Exhib. Proc. 24th, Enigma of the Eighties: Environ., Econ., Energy*, San Francisco, CA, May 8–10 (1979).
7. PIPES, R. B. and PAGANO, N. J., Interlaminar stress in composite laminates under uniform axial extension, *J. Composite Materials*, **4**, 538–548 (1970).
8. HERAKOVICH, C. T., NAGARKAR, A. and O'BRIEN, D. A., Failure analysis of composite laminates with free edges, *Modern Developments in Composite Materials and Structures*, pp. 53–66 (1979).
9. WANG, A. S. D. and GROSSMAN, F. W., Edge effects on thermally induced stresses in composite laminates, *J. Composite Materials*, **11** (July) 300 (1977).
10. JONES, R. M., *Mechanics of Composite Materials*, McGraw-Hill (1975).

26

Testing Organic Composite Insulators for Fusion Magnets

H. Becker, A. M. Dawson, P. G. Marston and D. B. Montgomery

Plasma Fusion Center, Massachusetts Institute of Technology, Cambridge, Massachusetts 02139, USA

ABSTRACT

Descriptions are presented of problems associated with acquiring mechanical property data on organic composite electrical insulators for fusion magnets. Magnet environments are identified and are shown to differ from test conditions because of simulation difficulties, foremost among which is the reactor radiation field. Other problems involve the development of reliable test procedures for such properties as interlaminar shear. Examples are presented. Recent advances in test procedures are summarized and the survivabilities of newer composites are described.

A description is presented of efforts to develop test methods that approach the magnet environment more closely. An outline is presented of the current US six-year program in this area.

INTRODUCTION

Organic insulator survivability under radiation plus mechanical load can be reduced to a small fraction of the unirradiated value, leading to structural failure and loss of insulation capability. Insulators also must satisfy electrical and chemical requirements. Furthermore, the structural loads always are applied at full design value as for pressure vessels.

Magnet design details vary widely (Fig. 1). The conductors for a superconducting fusion magnet may be monolithic in character with

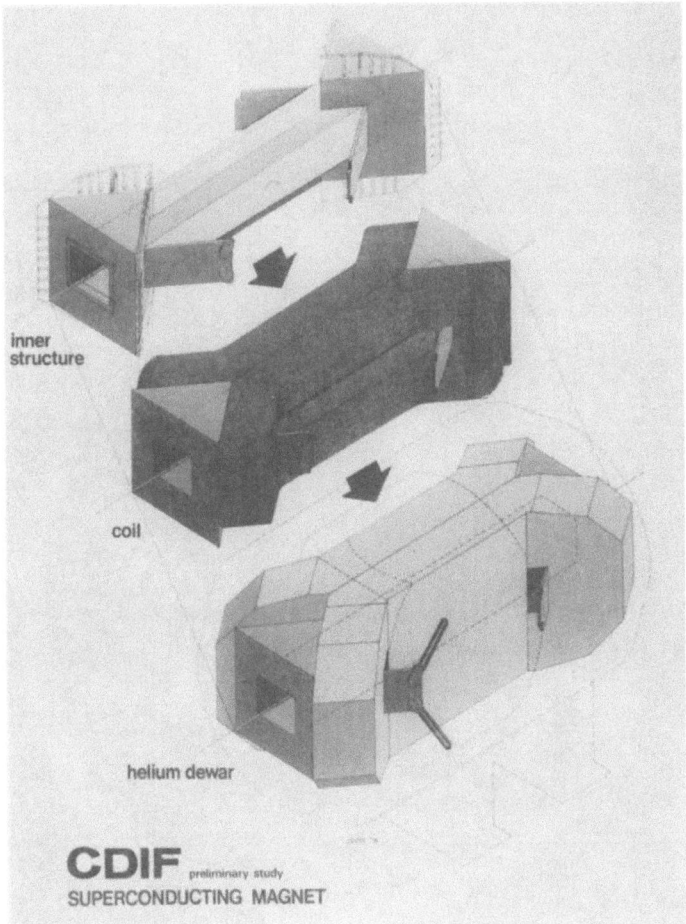

FIG. 1. Magnet configurations. (A) Artist's conception of a rectangular saddle coil superconducting MHD magnet. In this design the individual conductors are placed in grooves in machined G-10 plates, corresponding to the grooved plate end shown in Fig. 2(A).

various surface characteristics to maximize the heat transfer to the bath of liquid helium that surrounds them. They also may be fine-wire cables.

Insulation may be by simple interturn spacers leaving conductor edges exposed to the coolant or may take the form of glass tape wrapped around the conductor and epoxy impregnated. Other conductors may lie in an insulating and load-bearing subplate that contains grooves for the conductor and channels for enhanced contact between helium and conductor. Some insulation configurations are depicted in Fig. 2.

FIG. 1(B).   The yin-yang superconducting magnet pair for the Mirror Fusion Test Facility at Lawrence Livermore National Laboratory represents another complex winding geometry. Here the coils are shown enclosed in their cryostats. Figure 2(B) shows both the layer-to-layer G-11 insulation and the turn-to-turn 'button' insulation in the magnet winding.

The variety of configurations leads to a corresponding variety of mechanical, electrical, thermal, chemical and radiative loads creating a set of problems in finding and characterizing an insulating material (or a family of materials) to meet those load requirements.

Some of the leading candidates are epoxy/glass and polyimide/glass composites. All exhibit approximately the same thermal expansion as the steels, coppers, aluminums and superconductors used in magnets.

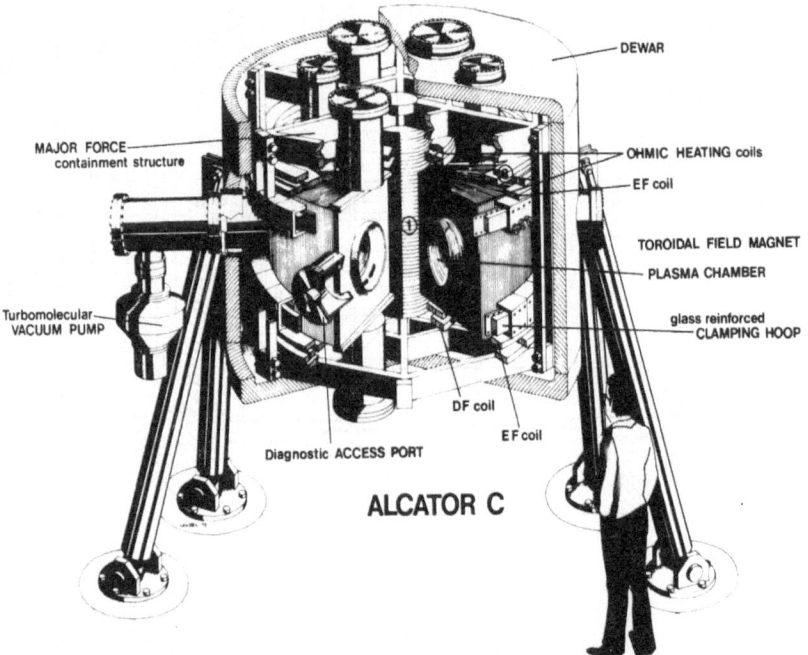

FIG. 1(C). The Alcator C Tokamak fusion experiment has toroidal field coils made of wedge-shaped plates of high purity copper interleaved with steel reinforcing plates and sheets of insulation forming triplets from which the magnet is assembled. Figure 2(C) shows the triplet detail.

The material characterization process involves conducting tests that simulate the magnet environment. Since no fusion reactor exists, such data must be considered tentative. In fact, the first self-sustaining fusion reactor will be, perforce, a materials testing reactor.

A summary is presented of fusion magnet environments in which insulators must survive. The disparities in the environments of the fusion reactor magnet and existing test arrangements are delineated. The relevant environmental parameters are identified (stress, voltage, radiant fluence, etc.) and the status of testing is summarized.

In order to portray the difficulty of acquiring property data of relevant materials, examples are presented to show how the test details affect the numerical values. A brief discussion of testing standards including a critique of one ASTM procedure is included.

A description is presented of steps that are being taken, by agencies in the USA, to define a testing program that will yield pertinent design data within reasonable cost and time for composite insulators.

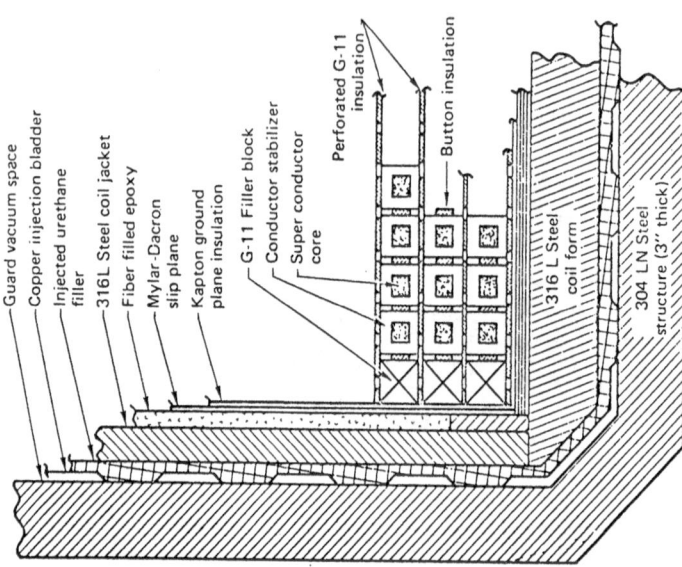

Guard vacuum space
Copper injection bladder
Injected urethane filler
316L Steel coil jacket
Fiber filled epoxy
Mylar-Dacron slip plane
Kapton ground plane insulation
G-11 Filler block
Conductor stabilizer
Super conductor core
Perforated G-11 insulation
Button insulation
316 L Steel coil form
304 LN Steel structure (3" thick)

FIG. 2(B). In this complex magnet insulation occurs as perforated sheets between coil layers, as 'buttons' between magnet coil turns and as filler blocks and spacers.

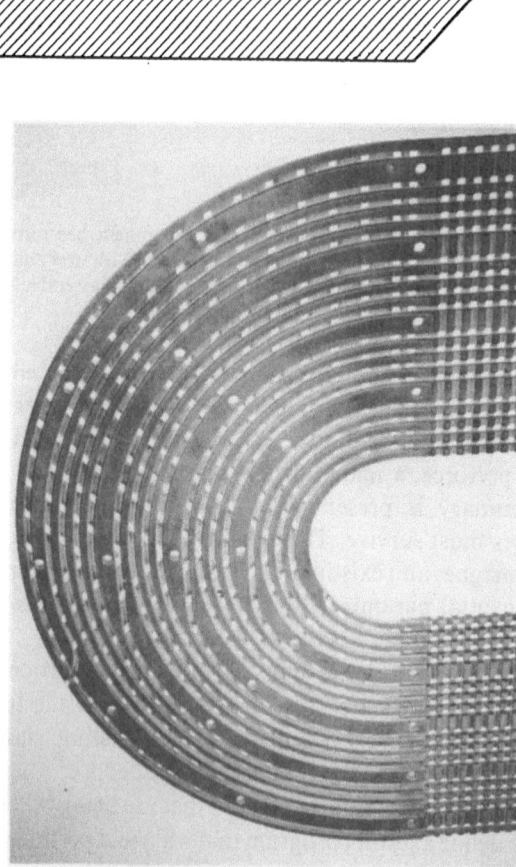

FIG. 2. Insulator configurations. (A) These machined G-10 plates have grooves on the upper surface into which superconductor is wound. The grooves on the lower surface provide abundant space for helium circulation.

FIG. 2(C). Sheets of copper, steel and composite insulator are used to form this triplet. Many of these are interleaved to form the toroidal field coil of the tokamak shown schematically in Fig. 1(C).

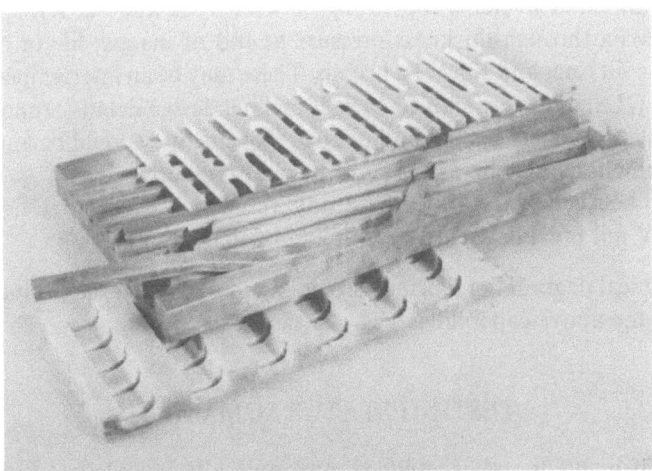

FIG. 2(D). The insulators shown here were developed for a commercial-scale MHD magnet system. The 50 kA conductor shown in the center layer is a composite of a copper stabilizing support plate with transposed superconductor. The insulation is stamped or machined to allow adequate helium circulation for cooling.

In the past, all insulation materials testing for fusion magnets was conducted after the samples were removed from the radiation source. Data provided in Ref. 1, for example, were obtained using that approach. (In-pile experiments had been performed decades ago in connection with the US nuclear-powered airplane program but the work was confined to structural metals.) A program is now underway to load insulation specimens in-pile. That effort will be described.

Three recent advances to be discussed are: (1) the development of standardized forms of epoxy/fiberglass that are designed to furnish consistent properties, batch to batch; (2) the orientation of cloth layers perpendicular to the direction of peak compressive stress thereby attaining maximum strength for a laminated sheet insulator of any organic composite system; (3) the use of polyimide and S2 glass instead of epoxy and E/glass to achieve a composite system that displays an important gain in intrinsic strength.

MAGNET ENVIRONMENT

Fusion magnet environments require insulation to withstand the following:

10^8 Gy,* or equivalent, of gammas and neutrons for superconducting magnets (at 1·8–5 K) and as much as 10^9 Gy for conventional magnets with minimal shielding (operating between 77 K and 350 K).

270 MPa through-thickness pressure at end of magnet life in bucking post split rings and 135 MPa in coils. There may be an interlaminar shear load (ILS) of the order of 30 MPa if magnet design details cannot avoid problems such as friction and relative movement. It could be applied in conjunction with the 135 MPa of normal pressure.

15 V between coil turns for steady resistance of current flow and 500 V/mil breakdown strength at end of magnet life.

Mechanical degradation during assembly and operation should not cause loss of the above capabilities.

TESTING STATUS SUMMARY

It is reasonable to ask whether an unequivocally reliable test procedure exists for determining any mechanical property of an organic laminate.

* A gray (Gy) is a measure of absorbed radiation equal to one joule per kilogram of mass.

TABLE 1

Comparison of reactor and testing environments for insulators

(a) FUSION REACTOR (all conditions act together)

---

1. Nuclear radiation flux at low gamma/neutron ratios
2. Lorentz (and preload) normal pressures, interlaminar shears and in-plane bending plus membrane loads
3. 10^3 to 10^6 cycles. Static load duration—months
4. Liquid helium temperature (constant) for some. Pulse starting at RT or LN_2 temperature for others
5. 14 MeV spike at moderate fluence

(b) TEST CONDITIONS (all testing done after application of total fluence)

---

1. Mostly RT irradiation, some cryogenic
2. Mostly static, some cyclic, little creep testing, no combined load tests
3. Most testing at RT, some at 77 K, some at 4 K
4. High gamma/neutron ratios
5. 14 MeV spike at low fluence applied alone and not added to other irradiation either before or after 14 MeV dose

---

Data are available to demonstrate the dependence of strength values on specimen types and proportions, types of end supports, the effect of contact friction at the loading head and the influence of load rate, to name some of the more obvious factors.[2-4]

Tables 1–4 constitute a review of the parameters relevant to testing organic insulators for fusion reactors. Examples of the impact of various

TABLE 2

Experimental parameters involved in insulator evaluation

Radiation dose

Spectrum, energy, dose rate, time

Current materials

Epoxy or polyimide matrices;
E, S or S2 glass reinforcements

Specimen

Composition (layered cloth, fiber mat),
fabrication parameters (pressure, temperature, time),
size, shape, orientation

Test

Temperature, thermal gradients, current,
voltage, stress type, magnitude, frequency, fluid

TABLE 3

Irradiated insulator electrical, physical and chemical testing status (some on composites, some on unfilled polymers, none under stress)
× Some testing conducted

Material parameter measured	Irradiation temp., K						Test temp., K		
	Gammas			Neutrons					
	300	77	4	300	77	4	300	77	4
Superconductivity		×		×					×
Contraction	←	→		←	→				
Surf. resist.	×						×		
Dielectric	×	×		×			×		
Breakdown		×		×			×		
Thermal K	×						×		
Gas									
Corrosion									

test parameters on measurement of mechanical properties are presented in Tables 5 and 6 and in subsequent figures.

It is clear from Table 1 that there is disparity between the insulator environments in fusion reactors and the environments in which tests have been conducted. Parameters of interest appear in Table 2. The outstanding problems can be seen in the broad view of the testing status for irradiated

TABLE 4

Irradiated insulator mechanical testing status (some on composites, some on unfilled polymers, all after irradiation)
× Some testing conducted

Type	Irradiation temp., K						Test temp., K		
	Gammas			Neutrons					
	300	77	4	300	77	4	300	77	4
‖ Tension	×						×	×	
⊥ Tension									
‖ Compression	×	×	×	×			×	×	×
⊥ Compression	×	×	×	×			×	×	×
‖ Shear	×			×			×		
Interlaminar shear (ILS)									
Combinations									
Creep									
Relaxation									
Erosion									

organics (the ×'s in Tables 3 and 4) which indicates that only a relatively small portion of the total field has been covered.

Most irradiation doses have involved predominantly gammas.[3] The absorbed neutron fluence energy was less than 1/10 of the gamma energy. Tests are now in progress to bring the relative magnitudes into closer agreement with the reactor ratios, which are of the order of 1 to 3, gammas to neutrons. It also is not clear that mechanical damage can be evaluated on absorbed energy alone for gammas and neutrons. In addition, it is conceivable that the application of load during irradiation may reveal behavior different from that observed during loading after irradiation. This last situation also is being examined as described below.

One of the biggest holes in the data bank is the absence of information on irradiated interlaminar shear (ILS) specimens. A potential problem may be debonding between laminae after irradiation at 4 K and subsequent warmup to room temperature.[5] Electrical properties, including surface breakdown ('tracking') at 4 K *in situ*, also need attention. Furthermore, the effect of mechanical damage on electrical integrity is not known. The largest unexplored territory is the effect of in-pile loading on strength.

IMPACT OF TEST CONDITIONS ON DATA

Introduction.

Examples have been chosen of structural tests conducted on composites. They illustrate the effect of some of the test parameters on numerical values of the measured 'mechanical properties'. The quotes emphasize the fact that the term actually arises from assigning to a material the numerical results of measurements made on a structure. The examples emphasize the need for conducting a 'materials properties' test in a manner that truly duplicates the conditions in the device being designed or improper conclusions may be drawn about the usefulness of the material in that application.

Compression Strength

The point to be made in Table 5, and in Figs 3 and 4, is that compressive strength perpendicular ($\perp$) to the layers of a cloth composite is greater than parallel ($\parallel$) strength.[6-8] Ordinary G-10 disks under $\perp$ loading withstood approximately the same stress level as Spaulrad® and Norplex® rods, both $\parallel$, although the G-10 fluence was $1\frac{1}{2}$ orders of magnitude greater. Norplex rods $\perp$ also have exhibited excellent performance as have Spaulrad disks.

TABLE 5
Comparison of compression test data (References 6–8)

Rods: $D = 6 \cdot 4$ mm, $L = 12 \cdot 7$ mm; 10^8 Gy* $\gamma + 8 \cdot 7 \times 10^{16}$ nvt**

Static loads at 77 K

G-10CR	‖	55 MPa (avg. of 3)
G-11CR	‖	70
Spaulrad	‖	360
Norplex	‖	360
Norplex	⊥	890

Disks: $D = 11$ mm, $t = 0 \cdot 5$ mm: Cyclic loads ⊥

G-10	RT	$3 \cdot 8 \times 10^9$ Gy $+ 10^{20}$ nvt	460 cycles (F) at 345 MPa
G-10	77 K	$3 \cdot 8 \times 10^9$ Gy $+ 10^{20}$ nvt	30 000 cycles (S) at 345 MPa
DGEBA + OCA + S glass	RT	$2 \cdot 3 \times 10^9$ Gy $+ 2 \times 10^{19}$ nvt Survived 30 000 cycles at 310 MPa	
KERIMID 601 + S glass	RT	$2 \cdot 3 \times 10^9$ Gy $+ 2 \times 10^{19}$ nvt Survived 250 000 cycles at 640 MPa	

* Gy (gray): 1 joule of absorbed energy per kilogram.
** nvt: neutrons per square centimeter.

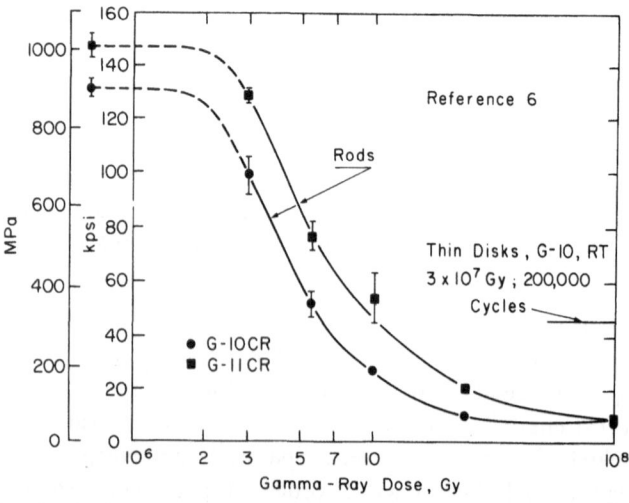

FIG. 3. Static compression strength of epoxy/E-glass laminates at 77 K.

In the most recent MIT tests, Spaulrad disks resisted 640 MPa for 350 000 cycles at room temperature after irradiation to approximately 3×10^9 Gy.

Large advances can be made by using S2 glass in a polyimide matrix and by designing magnet details to take advantage of the great strength available in thin sheet laminates when they are loaded in $\perp$ compression. Furthermore, fiber mat composites appear to offer a useful alternative since they may not be subject to ILS problems observed in layered composites.

Figure 3 displays data for G-10CR and G-11CR.[6] These are two products developed under a standardization program for insulation materials.[9] They have essentially the G-10 composition of epoxy and E-glass. However, they are produced under fabrication control designed to yield good G-10 properties with high uniformity. The materials standardization program is an advance derived from the US insulation development program since it fosters a greater degree of reliability in composite properties than has been available in the past.

The use of polyimides appears to offer another gain in insulation development for fusion reactors.[8] As may be seen in Fig. 4, they provide superior strength compared with epoxy, data for which are shown in Fig. 3.

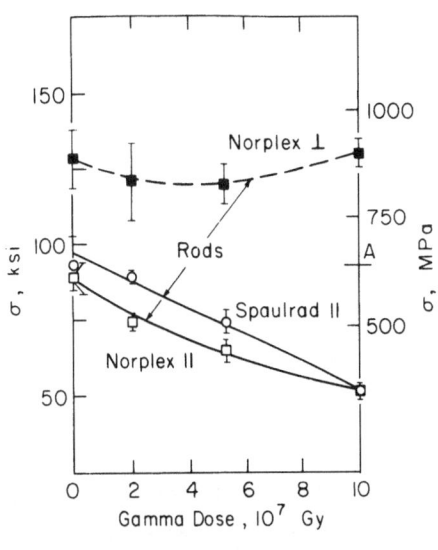

A: Disks, $\perp$, 10^9 Gy at RT
300,000 Cycles at RT,
No Failure

FIG. 4. Polyimide/S-glass at 77 K.

TABLE 6
Effect of thickness on compressive strength
(Courtesy of J. Benzinger, Spaulding Fibre Company, Inc.)

Compressive strength* 295 K	G-10CR	G-11CR	Spaulrad™ E	Spaulrad™ S
Flatwise, MPa (ksi)				
3 mm thickness	420 (60·9)	461 (66·8)	572 (82·9)	893 (129·5)
13 mm thickness	483 (70·0)	552 (80·0)	638 (92·5)	862 (125·4)
Edgewise, warp, MPa (ksi)				
3 mm thickness	375 (54·4)	396 (57·4)	333 (48·3)	471 (68·3)
13 mm thickness	414 (60·0)	469 (68·0)	446 (64·7)	539 (78·1)
Edgewise fill, MPa (ksi)				
3 mm thickness	283 (41·0)	315 (45·7)	317 (45·9)	430 (62·4)
13 mm thickness	331 (48·0)	359 (52·0)	447 (64·8)	500 (72·5)

* Average of three test specimens or more.

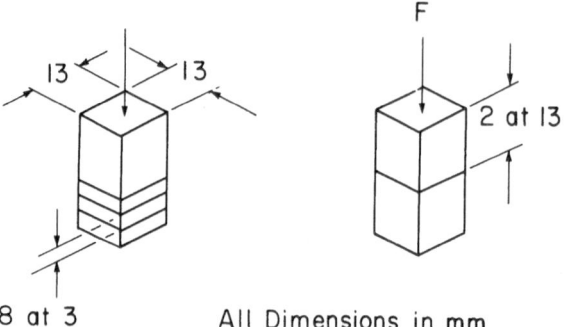

8 at 3   All Dimensions in mm

Stacking Effect on Compressive Stiffness

During the disk tests it was observed that a compressed pair failed at a lower load than single disks, which may indicate a problem arising from flatness tolerances on the disk faces. The same sort of behavior can be seen from data in Table 6. Spaulding S, flatwise ($\perp$) defied the trend.

The previous two graphs relate to strength. A similar stacking problem also can exist with compression stiffness measurements (Fig. 5). The magnet coil is made from layers of extruded copper bars 0·5 in high. In the test arrangement they were stacked to a total height of 5 in. The nominal Young's modulus at RT is 110 GPa. Since the magnet coil design stress level is 14 MPa, it is obvious that the coil stiffness during operation is only one-seventh of the material stiffness.

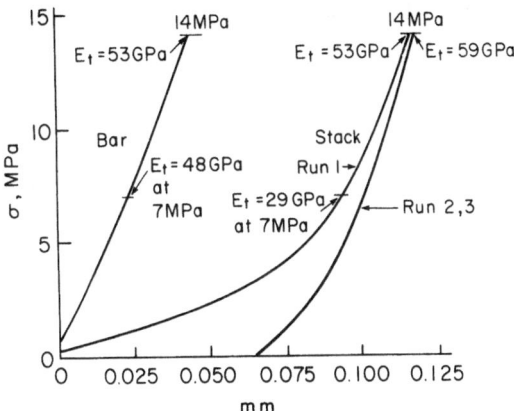

FIG. 5. Compression of stack versus bar; $H = 13$ cm. Quarter hard copper.

$\dfrac{\text{Wound tube}}{\text{Short beam shear}}$		0·8	RT
$\dfrac{\text{Wound tube}}{\text{Guillotine}}$	$\dfrac{L}{t} = 2.7$	2·1	RT
	$\dfrac{L}{t} = 7.8$	2·9	RT
Guillotine plotting extrapolation		1·3	RT
$\dfrac{\text{Guillotine theoretical extrapolation}}{\text{Guillotine plotting extrapolation}}$		0·75	RT
$\dfrac{\perp \text{ Rod in torsion}}{\text{Guillotine plotting extrapolation}}$		1	RT
		1	77 K

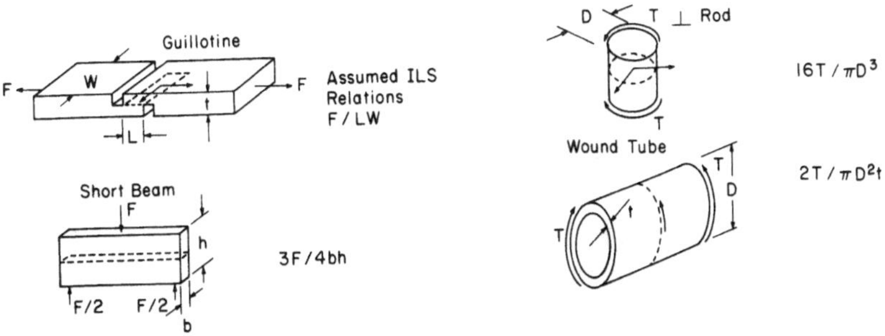

FIG. 6. ILS comparisons and tests.

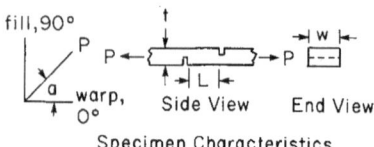

Specimen Characteristics

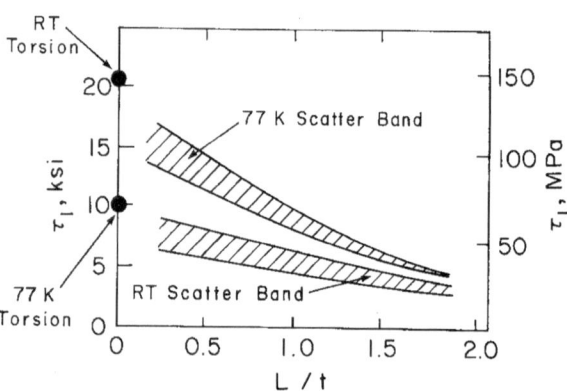

FIG. 7.　Variation of ILS stress with distance between grooves; G-10 and G-10CR. $t = 12\cdot7\,\text{mm}$.

TABLE 7

MIT program on radiation survivability of organic insulators
(organic composites—testing thin disks under $\perp$ compression)

FY'82

Design of in-pile loader for RT and 77 K testing
RT and cryogenic irradiation of polyimide/S-glass
disks. RT and cryogenic testing under static and
dynamic loads
Failure mode analysis

FY'83

Fabrication of in-pile loader
RT and 77 K in-pile testing
Design of in-pile loader for 4 K testing
Failure mode analysis

FY'84

Fabrication of 4 K in-pile loader
RT and cryogenic in-pile testing
Failure mode analysis

Interlaminar Shear Strength

Interlaminar shear strength determination is a complex problem.[10-13] Figure 6 indicates some of the test methods and specimen shapes used to attempt ILS determination for a number of layered composites. Comparisons of the results are presented in ratio form in the figure. The guillotine plotting extrapolation (Fig. 7) involved plotting measured strength as a function of L/t (see sketch on figure) and extending the plot to

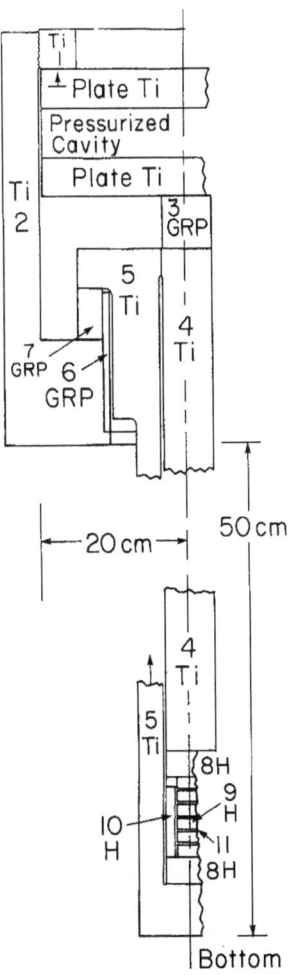

FIG. 8. Schematic of load generator for in-pile mechanical testing. (1, reaction loading ring; 2, reaction sleeve; 3, heat trap; 4, loading rod; 5, reaction sleeve; 6, radiation shield; 7, heat trap; 8, load spreader; 9, pad; 10, guide sleeve; 11, test specimen. Materials: Ti, titanium alloy; GRP, G-10CR; H, hardened steel or WC.)

$L/t = 0$. The potential for this procedure is indicated in the agreement (shown on Fig. 6 to one significant figure) of the extrapolation and torsion test strengths.

In-pile Loading

An attempt is being made to evaluate the influence of in-pile loading on insulator strength. It is important to make that assessment since it approaches the magnet environment more closely than post-irradiation testing, which has been employed on all insulator radiation survivability testing conducted to date.

The MIT program for in-pile insulator testing is outlined in Table 7. Figure 8 depicts a concept for an in-pile loader currently under construction. It would be capable of producing 700 MPa static or cyclic face pressure on disks 11 mm in diameter. It also avoids the use of complex mechanical load trains.

The loader will be made from titanium alloy since the strength is great enough to keep the structure relatively small thereby minimizing heat generation when it is in the radiation source. Otherwise, cooling problems could defeat the attempt to achieve in-pile loading.

OUTLINE OF US SIX-YEAR PROGRAM PLAN

The US Department of Energy has a program plan for developing materials and test methods to produce insulators for fusion reactors by FY88 at which time a design will be selected for the US magnetic confinement fusion Engineering Test Reactor.

The goal of the planned program is to develop organic insulation systems engineering capability by the start of the design of the selected fusion system. The program would provide data for designs of insulation systems to be used in cryogenic and conventional magnets for tokamak and mirror fusion devices.

The program now underway will first resolve the outstanding questions concerning insulator behavior. There then would be a transition period during which 'best' materials would be identified and standardized. Test methods also would be standardized. The last several years would be focused on verification testing of insulation under simulated fusion reactor conditions and on the development of fusion magnet insulation systems design procedures which will dovetail with the final ETR design selection. A broad view of the schedule is shown in Table 8.

TABLE 8
US insulation development schedule

Item	Category	Fiscal year					
		83	84	85	86	87	88
1	Resolve current outstanding problems		————————				
2	Standardize testing		————————				
3	Standardize material composites and fabrication procedures		————————				
4	Develop design data base			————————————			
5	Conduct verification testing of candidate insulation systems				————————		
6	Develop insulation system design procedures for fusion magnets			————————————————————			

REFERENCES

1. VON DE VOORDE, M. M., *Selection Guide to Organic Materials for Nuclear Engineering*, CERN 72-7, May (1972).
2. EREZ, E. A., MARSTON, P. G. and THOMPSON, J. B., Some mechanical properties of G-10, GRP at low temperature, paper presented at *NBS–DOE Workshop, Materials at Low Temperatures*, October 24–26, Vail, Colorado (1978).
3. COLTMAN, JR, R. R., KLABUNDE, C. A., KERNOHAN, R. M. and LONG, C. J., *Radiation Effects on Organic Insulators for Superconducting Magnets*, ORNL/TM-7077, November (1979).
4. BRECHNA, H., *Effect of Nuclear Radiation on Organic Materials; Specifically Magnet Insulations in High-energy Accelerators*, SLAC Report No. 40 (1965).
5. KLABUNDE, C. E. and COLTMAN, JR, R. R., Debonding of epoxy from glass in irradiated laminates, paper presented at the *Symposium on Radiation Damage Analysis for Fusion Reactors*, TMS-AIME, 82-246, October 24–28, St. Louis, Missouri (1982).
6. COLTMAN, JR, R. R. and KLABUNDE, C. E., The strength of G-10CR and G-11CR epoxies after irradiation at 5 K by gamma rays, *J. Nucl. Mater.*, **113** (1983, in press).
7. COLTMAN, JR, R. R., Organic insulators and the copper stabilizer for fusion reactor magnets, *J. Nucl. Mater.*, **108, 109**, 559–571 (1982).
8. EREZ, E. A. and BECKER, H., Radiation damage in thin sheet fiberglass insulators, *Nonmetallic Materials and Composites at Cryogenic Temperatures* (Hartwig and Evans, eds), Plenum Press (1982).
9. KASEN, M. B., SCHRAMM, R. E. and KRIZ, R. D., Effect of cryogenic temperatures on the mechanical performance of glass-fabric-reinforced epoxy and polyimide matrix laminates, *ICEC 9-ICMC Conference*, May 11–14, Kobe, Japan (1982).

10. BECKER, H. D. and EREZ, E. A., A study of interlaminar shear strength at cryogenic temperatures, *Adv. Cryogen. Eng. Mater.*, **26**, 259 (1980).
11. MARKHAM, M. F. and DAWSON, D., *Composites*, **6**(4), 73 (1975).
12. MCKENNA, G. B., MANDELL, J. F. and MCGARRY, F. J., *Interlaminar Strength and Toughness of Fiberglass Laminates*, Department of Civil Engineering Report R72–76, Massachusetts Institute of Technology, Cambridge, Massachusetts (1972).
13. CHAIO, C. C. and MOORE, R. I., *Evaluation of Interlaminar Shear Test for Fiber Composites*, Report UCRL-51766, Lawrence Livermore Laboratory, Livermore, California (1975).

27

A Comparison of Plain and Double Waisted Coupons for Static and Fatigue Tensile Testing of Unidirectional GRP and CFRP

P. T. CURTIS and B. B. MOORE

Materials and Structures Department, Royal Aircraft Establishment, Farnborough, Hants GU14 6TD, England

ABSTRACT

The static and fatigue tensile performance of unidirectional GRP and CFRP plain coupons and coupons waisted in width and thickness was studied. The static strengths obtained were good in all cases except for the double waisted GRP coupons. The longest fatigue lives for the CFRP were obtained with plain coupons since the double waisted coupons all failed prematurely in the machine grips. For the GRP the double waisted coupons gave the longest fatigue lives, but the values for the plain coupons were as good as previous values for double waisted coupons. Quantities of unreacted solid hardener were found to be present in the GRP coupons. Explanations for these observations are discussed.

1. INTRODUCTION

The mechanical testing of fibre composite materials has proved to be significantly different from the testing of conventional structural materials. This is principally because of the anisotropy of composites, the materials having good strength and stiffness in the fibre direction but poor strength and stiffness perpendicular to the fibres and low shear strength along the fibres. In structures, this problem is usually overcome by using laminates with fibres in several directions, but there are applications where it is beneficial to position most of the fibres in one direction. In addition, a knowledge of the unidirectional properties is essential for design purposes and for a full understanding of the material behaviour. Because unidirectional fibre composites are weak in shear, it is often difficult to introduce load into them and problems are frequently encountered with longitudinal shear failure at waists or changes of section. This is particularly so in the fatigue testing of waisted fibre composite coupons, where resin or fibre/resin interface degradation, due to the cyclic loading, leads to an inability of the coupon to sustain the shear loads at waists.[1]

Different establishments have used various types of coupon for the fatigue testing of unidirectional composites, but two basic types predominate: firstly a plain coupon, of 1–2 mm thickness and 10–25 mm width, with end plates of either aluminium alloy or GRP bonded to the coupon; secondly a coupon with some form of a waist within the gauge length to limit the maximum stress to a gauge length in the middle of the coupon, if possible keeping the shear stress below the critical value. Neither of these designs is ideal, the plain coupon being susceptible to failure at the end plates due to the associated stress concentration and the waisted coupon being susceptible to shear failure from the waist.

Many variations of the waisted coupon have been proposed to try and reduce the susceptibility to shear failure. Currently favoured for static testing is a coupon with a waisted region having a 1-m radius in the thickness, the large radius being chosen to minimize the shear stress and any stress concentration, but machining of this coupon has proved to be a difficult task.[2] The more commonly used coupon with a 125-mm radius,[3] although suitable for static testing of most fibre composites (excluding those with very low shear strengths), is not suitable for fatigue testing since longitudinal shear cracks develop more readily under cyclic loading from the shoulders of the waist causing the coupon to fail in shear, without fibre failure. In an attempt to reduce the susceptibility to shear failure, coupons waisted in both the thickness and the width have been used,[4] so that the minimum cross-section is substantially less than the cross-sectional area of the unwaisted part.

In this work the effect of coupon geometry on the static and fatigue tensile properties of fibre composites was studied. The work involved tensile static testing and fatigue testing, at long life-times and low stresses, of GRP and CFRP coupons of both plain parallel-sided design and double waisted design. Tensile strengths and fatigue lives were compared and failure mechanisms in the GRP were studied using optical microscopy. There was insufficient material for a similar microscopic examination to be made of the CFRP.

2. EXPERIMENTAL

The GRP coupons consisted of Equerove E-glass fibres in Ciba-Geigy
epoxy resin type BSL913 and the CFRP coupons consisted of Courtaulds
XA-S fibres in the same resin. Laminates 1 mm, 2 mm or 7 mm thick were
made from unidirectional pre-impregnated warp sheet and these were
cured in a press-clave for 2 h at 90 °C followed by 1 h at 120 °C. The volume
fraction of fibres in the materials was approximately 60 %.

Two types of coupon were studied, a plain coupon shown in Fig. 1 and a
coupon waisted in both the thickness and the width shown in Fig. 2. The
plain coupons were cut from laminates 1 and 2 mm thick and the double
waisted coupons from laminates 7 mm thick. The 1-m radii profiles of the
double waisted coupons were ground using a fine grinding wheel, but the
reject rate was high. Aluminium alloy end plates, 40 mm long, were bonded
to the ends of the parallel-sided coupons using an epoxy resin based
adhesive, type Hysol EA-9309/2, cured at room temperature. Because

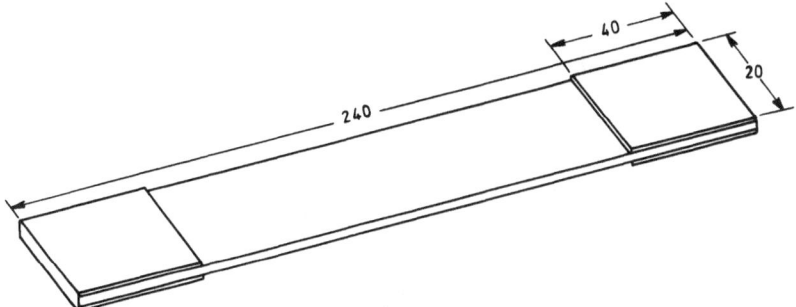

FIG. 1.   Plain test coupon (not to scale) (dimensions in mm).

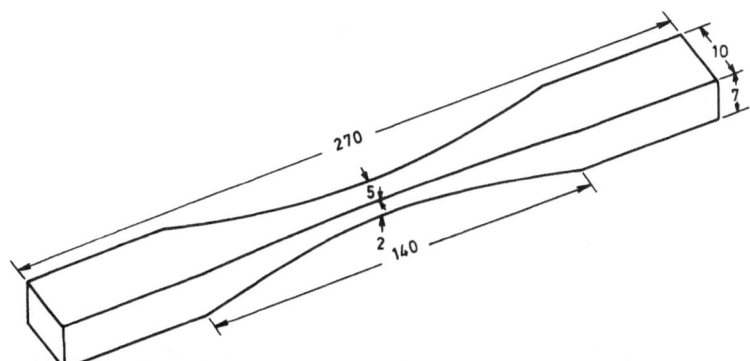

FIG. 2.   Double waisted test coupon (not to scale) (dimensions in mm, radius of
waists = 1 m).

problems were encountered during fatigue with debonding of the end plates during the tests, the adhesive was post-cured for 18 h at 65 °C. In addition the adhesive BSL403, found to be better in fatigue, was used to bond the end plates to some of the coupons.

Static tests were performed in position control at a ram movement rate of 2·5 mm/min. Fatigue tests were performed in load control at a frequency of 25 Hz, although a few tests were carried out at 5 Hz and 10 Hz to investigate the effect of frequency. Fatigue tests were performed by sinusoidal loading between zero load and a maximum tensile load ($P \pm P$). The load P was selected to induce failure at long lives (10^4–10^7 cycles). A fan was used to blow cool air over the GRP coupons during the fatigue tests. This helped to reduce the heating that occurs during the fatigue of GRP due to the relatively large dynamic strain range and the poor thermal conductivity of the material.

A few of the GRP coupons were unloaded after approximately 10^6 cycles but before final fatigue failure, and examined macroscopically for evidence of damage. In addition, sections were cut from the coupons, mounted and polished, and examined using an optical microscope. Areas of typical damage were recorded photographically.

3.   RESULTS

3.1. Static Tests

Average values of the static strengths (from five tests) are given in Table 1 for both the GRP and CFRP coupons. Photographs of typical failures are

TABLE 1

Static strengths of the unidirectional GRP and CFRP coupons

Coupon	Static strength (MPa) GRP	Static strength (MPa) CFRP
Plain 1 mm thick	$1\,283 \pm 45$	$1\,719 \pm 81$
Plain 2 mm thick bonded with Hysol EA-9309/2	$1\,173^{*}$	$1\,628 \pm 67$
Plain 2 mm thick bonded with BSL403	—	$1\,877 \pm 61$
Double waisted	$1\,029 \pm 71$	$1\,697 \pm 87$

Note: Mean values from five tests are quoted; $\pm$ are standard deviation. The volume fraction of fibres was nominally 60 %.

* The ends debonded from three coupons before failure. These values were rejected.

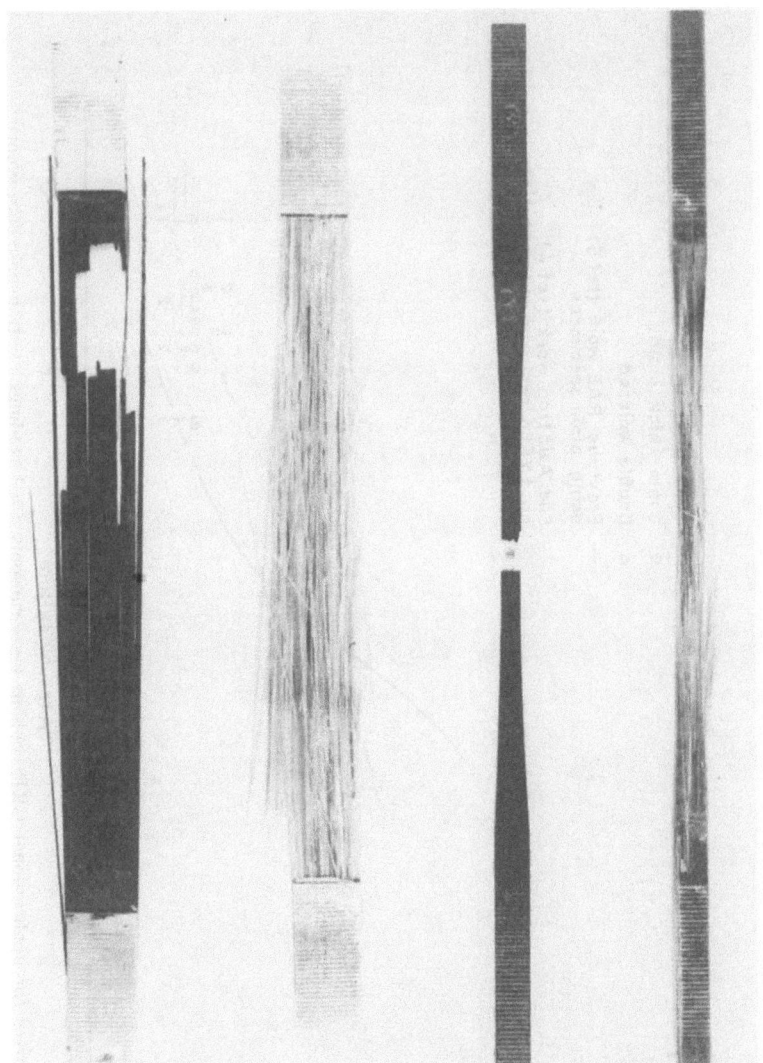

FIG. 3.   Static failures in unidirectional GRP and CFRP composites.

given in Fig. 3. The plain GRP coupons split the entire length between the end plates and failed mainly at the end tags in the 'brush-like' mode typical of GRP. The plain CFRP coupons failed by a combination of fibre fracture across the coupons and longitudinal splitting parallel to the fibres. A few of the CFRP coupons showed evidence of failure at the end tags, but this was not always the case, as in the GRP. The double waisted GRP and CFRP

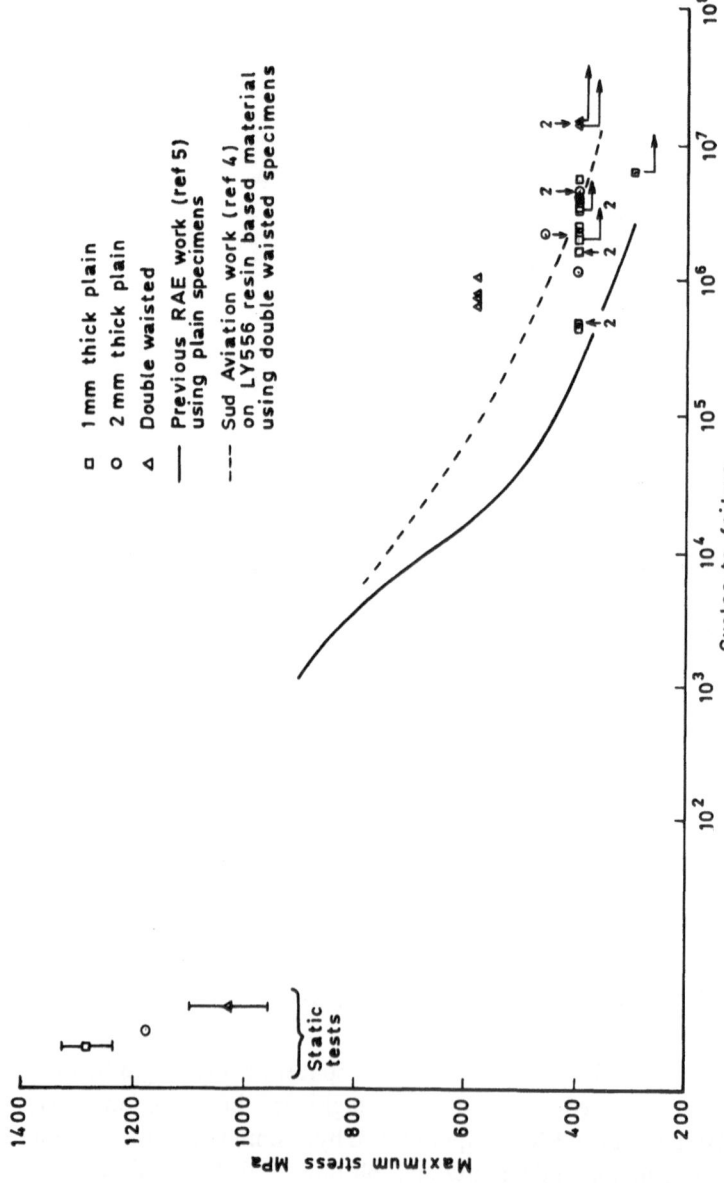

FIG. 4. Fatigue curves for unidirectional GRP coupons ($\rightarrow$ = run-outs, $\overset{2}{\downarrow}$ = 2 readings, $\overset{\circ}{\downarrow}$ = transpose data point).

coupons failed within the waisted regions by a combination of fibre fracture and longitudinal splitting, the splitting being more apparent in the GRP coupons. This splitting extended just to the gripped portions of the coupons. In the CFRP, splitting occurred before final failure at approximately 90 % of the UTS.

3.2. Fatigue Tests

Fatigue data are presented as diagrams of peak stress versus log number of cycles in Figs 4 and 5. Some problems were experienced with debonding of the end plates during the fatigue tests, especially on the 2-mm thick plain CFRP coupons. This was lessened to some extent on all coupons (both GRP and CFRP) by post-curing the end plate adhesive overnight at 65 °C, although premature failure still occurred in some of the 2-mm thick coupons by debonding of the end plates. Those coupons in which the adhesive BSL403 was used to bond the end plates suffered no problems with end plate debonding. Varying the test frequency between 5 Hz and 25 Hz had no significant effect on the fatigue lives of the plain coupons.

Typical fatigue failures are shown in Fig. 6. The plain CFRP coupons 1 mm thick exhibited only limited longitudinal splitting along the coupon lengths prior to failure, but this was more extensive in the 2-mm thick coupons. The plain CFRP coupons failed in a manner similar to that of the static coupons, but the double waisted CFRP coupons developed extensive shear cracks from the waists quite early in the fatigue tests. These propagated back to the gripped portion of the coupons, resulting in premature, short life failure by pulling a narrow plug out of the gripped end. During these tests, as a result of the shear cracking, the compliance of the coupons increased by a factor of 2.

The plain GRP coupons all showed considerably more damage than the CFRP coupons before ultimate fatigue failure. Relatively early in the tests a few damaged areas were observed which appeared dark in transmitted light (light in reflected light). In previous work[5] this damage was more extensive, but was shown to be associated with defects such as kinks in the fibre tows. The number of these damaged areas increased as the tests proceeded, and typical examples are shown in Fig. 7. Further fatigue cycling resulted in the development of longitudinal splits, usually linking up the damaged regions. These splits sometimes extended into the end tags. Longitudinal splits frequently developed from the cut edges of the coupons and these also propagated back into the gripped region. Fatigue failures were similar in appearance to the static failures, but were less 'brush-like' (e.g. compare Figs 3 and 6). As in the static failures the glass fibres appeared to have failed

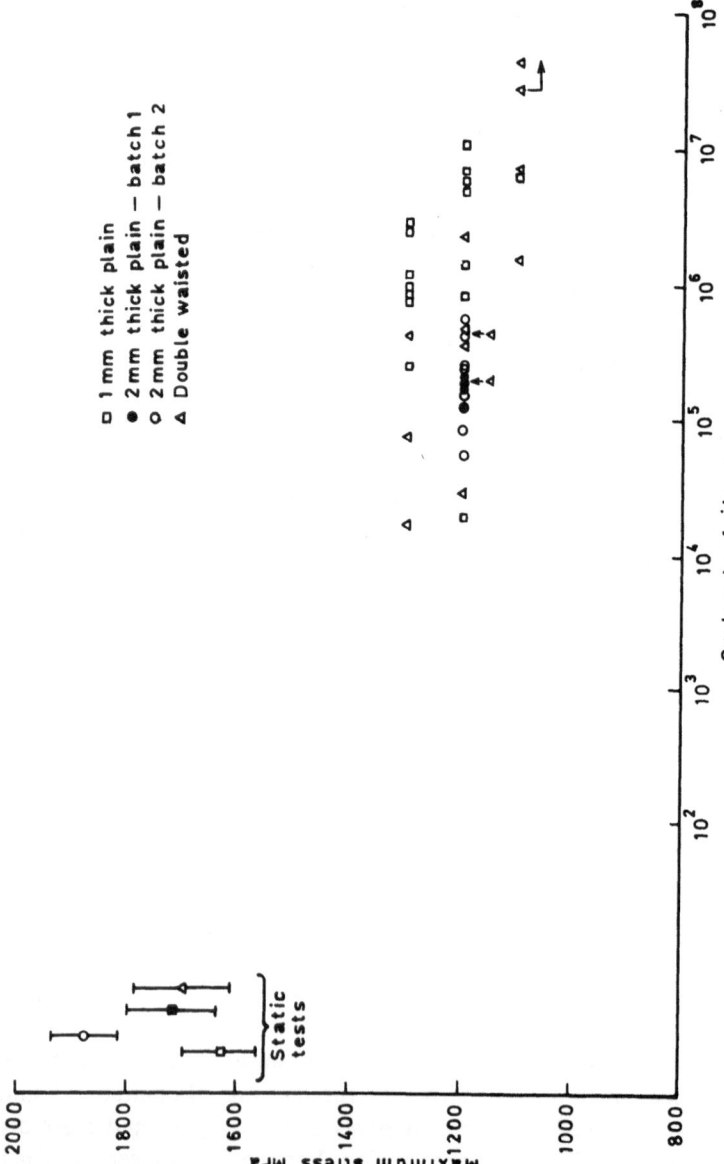

FIG. 5.   Fatigue curves for unidirectional CFRP coupons ($\rightarrow$ = run-outs, $\downarrow$ = transpose data points).

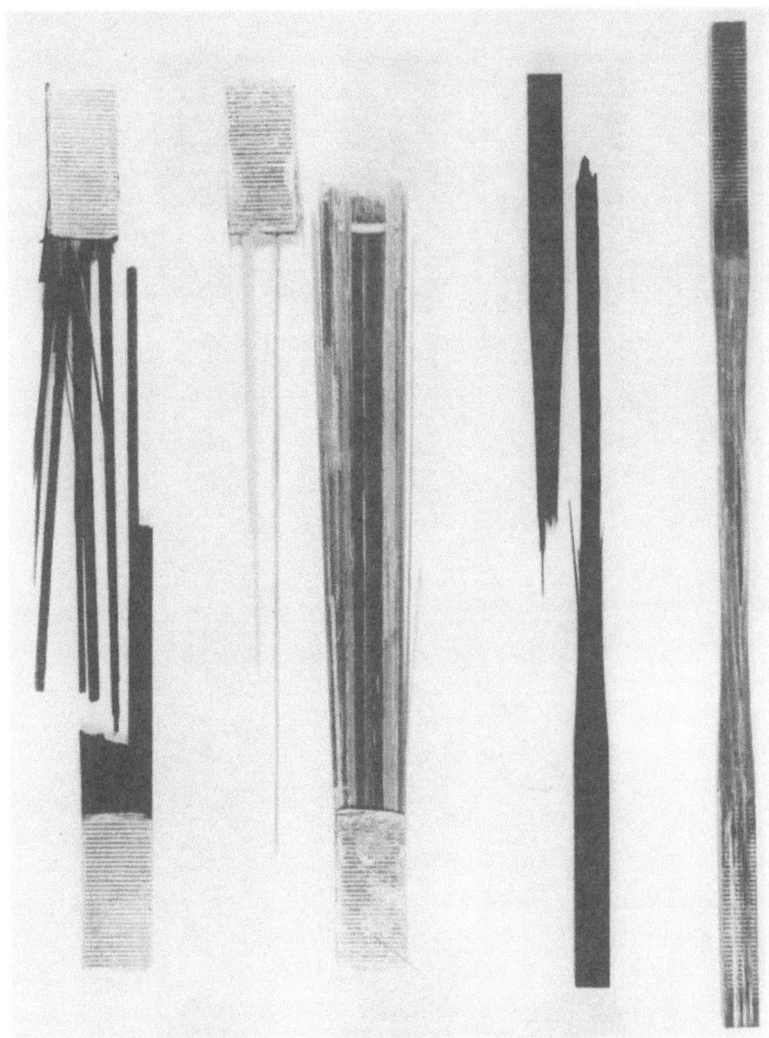

FIG. 6.   Fatigue failures of unidirectional GRP and CFRP composites.

at the end plates. The failure of the double waisted GRP coupons was different from that of the CFRP coupons, failing mainly within the gauge length although in some cases part of the section failed in the grips. Longitudinal shear cracks did form during these tests, and areas of damage were observed in the gauge length similar to those seen during the tests on the plain coupons. Final fatigue failure occurred mainly within the gauge

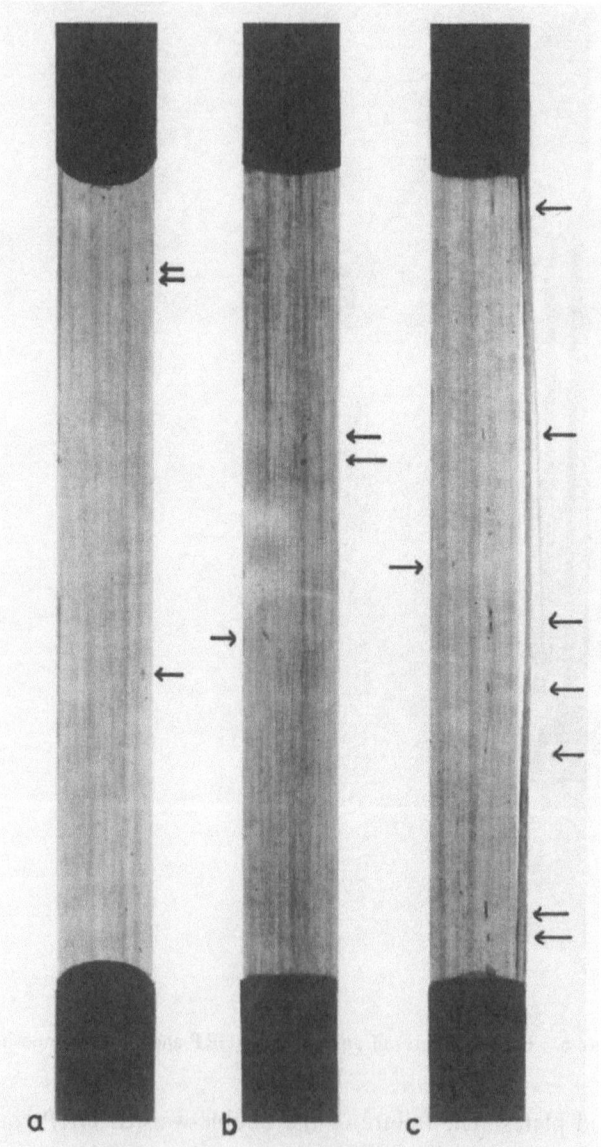

FIG. 7. GRP fatigue coupons in transmitted light showing dark damaged areas. Typical areas of damage are arrowed. (All 25 Hz, $200 +/- 200$ MPa.) (a) 2×10^6 cycles; (b) 2×10^6 cycles; (c) $3 \cdot 3 \times 10^6$ cycles.

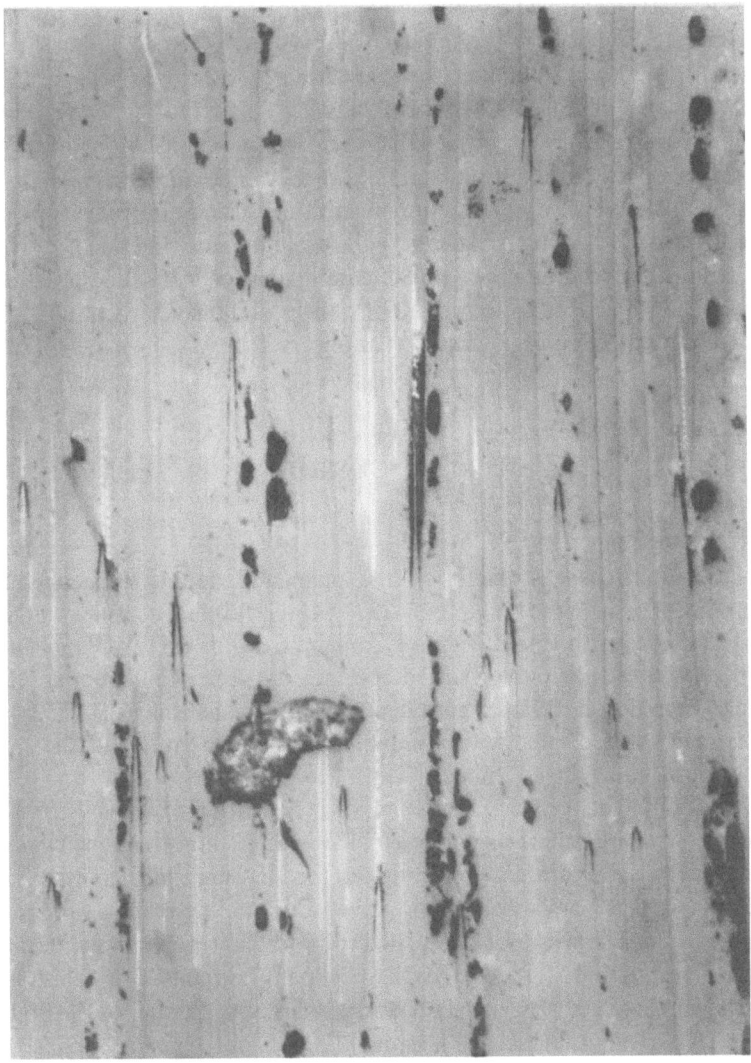

FIG. 8.   Fibre/resin debonding in fatigued GRP coupon (200 + / − 200 MPa, 2 × 10⁶ cycles).
Typical areas of debonding, which appear light, are indicated. The black areas are solid
hardener.

length by a combination of fibre fracture and longitudinal splitting, and there was no evidence of the shear plug pull-out failure observed in the CFRP. A typical failed coupon is depicted in Fig. 6.

Optical microscopy of the damaged areas showed them to be due to resin cracking and fibre debonding as shown in Fig. 8. Microscopy of the GRP coupons also revealed many black areas, shown in Fig. 8. Repeated washing and polishing confirmed that these were not polishing artifacts but pieces of unreacted dicyanodiamide hardener. The hardener was visible both on the polished surface and below the surface by refocusing the microscope and using polarized light to eliminate the main surface reflection. The occurrences of unreacted solid hardener and fatigue damage appeared to be unconnected and no instances of damage originating from accumulations of solid hardener were observed.

4. DISCUSSION

4.1. Static Properties

The static tensile strengths of the plain and double waisted CFRP coupons were all in the range 1628–1877 MPa. The plain coupons with end plates bonded with EA-9309/2 gave similar strengths (1719 MPa) to those (1697 MPa) of the double waisted coupons, but the plain coupons 2 mm thick bonded with BSL403 gave significantly greater values (1877 MPa). This may be associated with a reduction in the magnitude of the stress concentration at the end plates.

In the GRP the plain coupons 1 mm thick were the strongest (1283 MPa) followed by the 2-mm thick coupons (1173 MPa) but the difference was not statistically significant. The lower values for the 2 mm thick coupons may well have been associated with the end plate debonding problems encountered. The GRP double waisted coupons, however, had significantly lower static strengths (1029 MPa) than the plain coupons. From this result it could be inferred that the static stress concentration due to the double waist was greater than that at the end of the tags of the plain coupons. The low strengths of the double waisted GRP coupons could also be associated with misaligned fibres terminating at the surface of the waisted region.

It can be concluded from the data on static strengths that both the plain and the double waisted coupons are suitable for the static tensile testing of CFRP and GRP although the plain coupon may give slightly greater results.

4.2. Fatigue Properties

The fatigue tests showed that for GRP the reduction in fatigue strength, after cycling to large numbers of cycles, compared with the static strength, was considerably greater than for CFRP. The fatigue strength was in the range 400–600 MPa after approximately 10^6+ |cycles, compared with 1100–1300 MPa for CFRP after 10^4–10^7 cycles. This is primarily because of the greater tensile strains in the GRP than in the CFRP for similar stresses, which lead to more rapid resin and fibre/resin interface degradation, although the glass fibres themselves do also degrade in fatigue. At long life-times, the maximum strains corresponding to the maximum applied stresses were similar for both CFRP and GRP, being in the range 0·8–1 %, implying that failure was strain limited. Since both materials had the same resin matrix, failure was probably determined by the ability of the resin to withstand the fatigue strains. This agrees with the optical microscopy observations on GRP coupons which showed significant resin damage prior to failure. Consequently the use of high strain fibres in composites might not lead to improved fatigue properties unless a different resin matrix was used.

The fatigue lives of the plain GRP coupons 1 mm thick were similar to those reported by Sud Aviation for GRP using a double waisted coupon with a similar fibre/resin combination[4] and also greater than fatigue lives found in previous work on nominally identical material[5] (*see* Fig. 4). This increase in fatigue life is thought to be due to an improvement in material quality due to better moulding techniques.

The fatigue lives of the double waisted GRP coupons were greater than those for the plain coupons (approximately 1 order of magnitude greater) and also greater than previous values using the double waisted specimens.[4] This was probably because the GRP plain fatigue coupons failed at the end of the tags due to the associated stress concentration.

In contrast to the GRP, the fatigue lives of the CFRP plain coupons 1 mm thick were greater than for the double waisted coupons which were similar to the values for the plain coupons 2 mm thick. The CFRP double waisted coupons probably behaved differently from the GRP coupons because of the greater fibre modulus in the CFRP, which leads to greater shear stresses at the fibre ends in the waisted region. This causes shear cracks to initiate from the waists of the CFRP coupons at lower fatigue lives than in the GRP. The shear cracks then propagate into the gripped portions of the coupons and cause premature failure with reduced apparent fatigue lives. In the CFRP coupons 2 mm thick the shorter lives were associated with premature end plate debonding and were not a result of material or coupon geometry effects. The fatigue lives for both the plain GRP and CFRP coupons 2 mm thick were slightly lower than those for the 1-mm thick coupons, due to the problem of end plate debonding encountered when testing the thicker plain coupons. This appears to have been due to the inadequacy of the original adhesive (EA-9309/2) chosen to bond the end plates to the coupons. A considerable hysteresis heating effect in the end plate adhesive was noticed which caused the end plates to debond and premature failures to occur within the grips. Post-curing the adhesive substantially reduced the problem and in subsequent tests, in which the adhesive BSL403 was used, no problems were encountered.

4.3. Failure Processes in GRP

The mode of fatigue failure in the GRP was shown to be a result of an accumulation of areas of fibre/resin debonding and resin degradation and probably also fibre failure, although the latter was not observed since it is very difficult to detect. This caused longitudinal splits to form along the length of the coupons which propagated into the gripped ends. This may have limited the ability of the material to redistribute the load from areas with broken fibres to other areas of the coupons. Eventually the damaged material was unable to support the applied load and the coupons failed. In the plain coupons, longitudinal splits were sometimes formed at the coupon edges which may have further reduced the load bearing capacity of the material and could serve to partly explain the lower fatigue lives in these coupons.

4.4. The Effect of the Unreacted Solid Hardener

Optical microscopy revealed considerable quantities of unreacted dicyanodiamide hardener in the GRP coupons. This could have been a result of the cure employed which involved an initial period at 90 °C at which temperature the hardener does not fully dissolve in the resin and would thus be unable to react completely with it. After 2 h at this temperature the resin is at an advanced stage of cure, so that even after raising the temperature to 120 °C, it was probably too viscous and the diffusion processes too slow, to permit the hardener to dissolve further and become involved in additional curing of the resin. Thus on cooling, some dicyanodiamide hardener would remain as discrete particles within the resin. However, this material would have to be compared with GRP material fully cured at 120 °C to confirm that this was the case.

Whether the unreacted hardener has an effect on the mechanical properties of the composite is of considerable importance. The evidence in this work suggests that damage occurring during the tests is unlikely to be associated with the accumulations of dicyanodiamide but this does not mean that they were not involved in the failure process of the material. Indeed other work has suggested that accumulations of solid hardener between the prepreg layers do influence the behaviour of angle-ply laminates, acting as stress concentrators and reducing the laminate strength.[5] To confirm whether the solid hardener affects the strength of unidirectional composites it would be necessary to examine the failure processes in different resin systems.

5. CONCLUSIONS

For the measurement of static tensile strength, both plain and double waisted coupons gave satisfactory results, although slightly greater values were obtained for the plain coupons, particularly those in which the end plates were bonded with BSL403.

In terms of achieving the greatest fatigue lives, the double waisted coupon proved to be slightly superior for the GRP and the plain coupon slightly superior for the CFRP. However, the fatigue lives for GRP recorded with plain coupons were as long as previous tests with double waisted coupons and better than for past work on plain coupons, implying that acceptable design allowables can be obtained from fatigue tests on plain coupons.

The CFRP plain coupons gave greater fatigue lives because the double waisted coupons failed by shear in the grips. In the GRP the double waisted coupons gave longer fatigue lives because the stress concentration at the ends of the tags reduced the lives of the plain coupons. Longitudinal edge splits also made parts of the plain coupons ineffective and increased the local stress, leading to reduced fatigue lives.

The adhesive used to bond the end plates to the coupons was unsuitable for this particular fatigue application and for future use the adhesive BSL403 is recommended, since this has proved successful.

The initial failure process in the GRP was identified as local fibre/resin debonding and resin cracking, these areas increasing in number and eventually leading to longitudinal shear cracking. Eventually the accumulation of interfacial and resin damage, together with longitudinal splitting and probably also fibre fracture, lead to fatigue failure of the material.

Fatigue failure at long lives in both materials may have been strain limited and dependent on the fatigue resistance of the resin. This implies that the use of high strain fibres in composites might not lead to improved fatigue properties unless a different matrix resin was used.

Optical microscopy revealed considerable quantities of unreacted solid hardener present in the GRP. This may be a consequence of the cure schedule used, but further work is needed to confirm this. The evidence in this work suggests that damage occurring in unidirectional GRP is not associated with the accumulations of unreacted hardener.

ACKNOWLEDGEMENT

The authors wish to thank Mr C. Evans and his colleagues at Westland Helicopters Ltd for supplying the fibre composite laminates used in this work.

REFERENCES

1. STURGEON, J. B., Fatigue testing of carbon fibre reinforced plastics. RAE Technical Report 75135 (1975).
2. DOOTSON, M., Standard specimens and test methods for unidirectional carbon fibre reinforced plastics. BAe Report TN4440 (1976).
3. STURGEON, J. B., Specimens and test methods for carbon fibre reinforced plastics. RAE Technical Report 71026 (1971).
4. Sud Aviation Report DTH/ER 122-77 (1977).
5. STURGEON, J. B., The tensile fatigue of 0, ± 30 and ± 60 angle-plied glass fibre reinforced epoxy resin. RAE Technical Report 80151 (1980).

28

Sandwich Structures for Light Turrets

R. W. McLay, D. P. Tassie and W. W. Thompson

*General Electric Company, Armament & Electrical Systems Department,
Lakeside Avenue, Burlington, Vermont 05402, USA*

ABSTRACT

The development of sandwich turret structures for patrol boats is presented. Problems involving design compromise are discussed in the areas of high stiffness-to-weight ratio, corrosion control, damage tolerance, manufacturing methods, and cost control. The preliminary design of the turret is shown to involve predictions of performance through both analysis and testing. Design principles are seen to be different with material changes, which require a careful program for both analysis and test of a shipboard sandwich structure to guarantee the performance of a given design. However, it is seen that boat builder's methods and materials can be used to construct a composite turret for a Gatling gun.

INTRODUCTION

Structural foam parts have replaced wood and metal parts on small boats over the last decade. For example, PVC foam core glass laminated sandwiches are now used on boats as large as 70 foot patrol boats in hull and cabin structures.[1] Experiments have also shown polyimide foams, vinyl ester resins, epoxy resins, Kevlar, and carbon fiber cloths to be useful in these constructions. The combination of high rigidity, good strength and toughness, lightweight, ease of design, and favorable economics of manufacture, shipping, handling, and installation are important factors in this trend.

In using these sandwiches for the primary structure of a naval turret, however, certain design problems immediately become apparent. A primary performance requirement for a naval turret is high rigidity. This is necessary to guarantee accuracy in addressing threats from aircraft, missiles, and surface weapons. Sandwich structures can achieve adequate rigidity for most applications through inherently stiff structural shapes while using the basic materials of PVC foam and E-glass/polyester or vinylester resin laminates. However, for more demanding applications, higher stiffness materials can be employed, e.g. Kevlar or carbon fiber cloths. Unfortunately, these higher performance materials are also higher in cost and may introduce problems such as poor environmental resistance and the inability to fabricate certain complex shapes without special tools, vacuum bag procedures, or prepregs and special materials preparation.

PVC core E-glass laminates are known to be attractive, highly cost effective materials when used with a standard layup mold. When compared with the high cost materials, the PVC sandwiches often possess the best material property profiles; for example: high apparent flexural modulus, high flexural strength, high apparent toughness due to the memory of the core material, high impact strength, and the lowest volume cost. Furthermore, these sandwiches have good environmental resistance for ocean-going vessels and they can be fashioned into a variety of structural shapes with only minor changes in the basic mold.

In this paper we illustrate the advantages of using PVC core E-glass laminates in turret primary structural parts. For demonstration purposes, we will consider the design and fabrication of a 25-mm Gatling gun turret. We will see that the concept is practical from both a design and production standpoint. In addition, there are clear advantages in corrosion resistance for these designs.

DIMENSIONING A TURRET

To effectively dimension a sandwich part, one must know: the material properties, the imposed design loading, and the performance or failure criterion. In the case of the turret, the imposed loading is very complex and is governed by extreme cases, such as the recoil forces, the green water loading of waves breaking over the deck of the boat, the fatigue loading of the gun on the laminate, and the heat and explosive gases generated within the dome by the gun as it fires a long burst against an incoming target. Emerging specifications[2] are setting performance criteria both in the

United States Navy and in other navies worldwide. For example, the function of the turret should not be impaired with green water loadings in excess of $500\,\mathrm{lb/ft^2}$ ($7300\,\mathrm{N/in^2}$). In addition, the life of the gun system requires a comparable fatigue life in the turret, about 100 000 cycles of gun operation. Finally, the PVC foam is formed at temperatures above 160 °F (71 °C), which implies that the temperature of the mount be such that neither convection nor radiation of heat to the sandwich will cause the core to lose its stiffness during or after a long burst.

The design of a lightweight turret for a small boat also implies a constraint on the turret size. In effect, a minimum envelope is desirable for space requirements and eliminating visual obstructions which, in general, makes meeting the green water requirement easier for a given sandwich. However, the load paths in the fatigue problem are shortened with smaller effective areas and the stress levels are accordingly slightly higher. The heat transfer problem is also accentuated by the minimum envelope.

The envelope for a prototype turret test fixture is shown in Fig. 1. Implied in the preliminary design are the requirements that the system both feed and eject from below, that the gun be balanced about the center of rotation, and that a minimum radius dome be fashioned about the mount to facilitate the training and elevation of the gun. In addition, the dome and

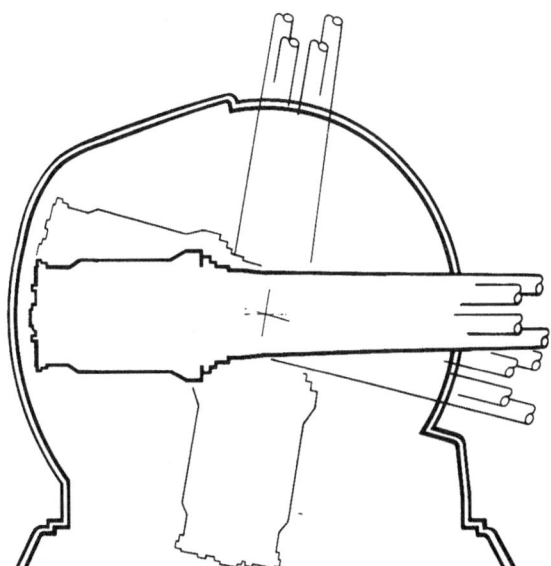

FIG. 1.   Envelope for 25-mm Gatling gun.

TABLE 1
Tension/compression tests

Material combination	Nominal thickness (in)	Tension/compression	Ultimate load (lb)
PVC/1 E-glass/Polyester	1 1/4	T	4 400
PVC/2 E-glass/Polyester	1 1/4	T	(grip pulled out)
PVC/3 E-glass/Polyester	1 1/4	T	17 000
PVC/1 E-glass/Polyester	1 1/4	C	2 410 (ultimate)
PVC/2 E-glass/Polyester	1 1/4	C	3 800 (buckled)
PVC/3 E-glass/Polyester	1 1/4	C	4 800 (buckled)
PVC/1 E-glass/Polyester	3/4	T	5 210
PVC/2 E-glass/Polyester	3/4	T	6 010
PVC/3 E-glass/Polyester	3/4	T	10 060
PVC/1 E-glass/Polyester	3/4	C	4 330 (buckled)
PVC/2 E-glass/Polyester	3/4	C	6 010 (buckled)
PVC/3 E-glass/Polyester	3/4	C	6 320 (buckled)
PVC/1 E-glass/Polyester	1/2	T	6 680
PVC/2 E-glass/Polyester	1/2	T	13 400
PVC/3 E-glass/Polyester	1/2	T	19 500

Material		Type	Load
PVC/1 E-glass/Polyester		C	1 370 (ultimate)
PVC/2 E-glass/Polyester		C	2 280 (buckled)
PVC/3 E-glass/Polyester		C	3 300 (buckled)
Polyimide/1 E-glass/Polyester		T	6 000 (debonded—core fractured)
Polyimide/2 E-glass/Polyester		T	8 300 (debonded—core fractured)
Polyimide/3 E-glass/Polyester		T	9 100 (debonded—core fractured)
Polyimide/1 E-glass/Polyester		C	4 140 (laminate buckled—debonded)
Polyimide/2 E-glass/Polyester		C	5 030 (laminate buckled—debonded)
Polyimide/3 E-glass/Polyester		C	7 640 (laminate buckled—debonded)
PVC/1 E-glass/Vinylester		T	4 930
PVC/2 E-glass/Vinylester		T	5 180
PVC/3 E-glass/Vinylester		T	6 880
PVC/1 E-glass/Vinylester		C	2 930 (buckled)
PVC/2 E-glass/Vinylester		C	4 430 (buckled)
PVC/3 E-glass/Vinylester		C	4 790 (buckled)
Polyimide/1 E-glass/Vinylester		T	3 960
Polyimide/2 E-glass/Vinylester		T	6 195
Polyimide/3 E-glass/Vinylester		T	7 150
Polyimide/1 E-glass/Vinylester		C	5 130 (laminate buckled—debonded)
Polyimide/2 E-glass/Vinylester		C	5 930 (laminate buckled—debonded)
Polyimide/3 E-glass/Vinylester		C	5 930 (laminate buckled—debonded)

associated fairing must be tight to the environment, yet have adequate ventilation to eliminate heat buildup and explosive gun gases during firing.

MATERIALS

A number of sandwich materials have been used in boat construction. Experiments continue in the use of PVC, polyimide, polycarbonate, and other foams for cores. Also, various glass and carbon fabrics have been tried in conjunction with polyester, vinyl ester, and epoxy resins. For the first prototype, it was our intention to demonstrate the function of the turret sandwich structure while at the same time reducing the production problems.

A PVC foam material was chosen for the sandwich core. It is serviceable in a range of temperature between $-40\,°C$ and $+75\,°C$. At approximately $75\,°C$, the foam loses its structural strength. Thus, it can be molded and have its properties return with a fall in temperature. It is also compatible with the polyester resin used with the majority of glass fabrics. However, in our view, its main advantage is in its toughness at failure. In effect, the laminate can be buckled into the core at failure and have 80–90% of the sandwich strength remain as the sandwich regains its original shape through material memory in the core. Thus, its damage tolerance is high, making the PVC sandwich ideal for small boat construction.

In addition to these material advantages, the PVC/polyester and E-glass sandwich has been used for over a decade in the boat industry. The materials, tools, and processes are all well known to boat builders.[3] Thus, a turret for a small boat built of this sandwich is well within the perception and experience of the industry. The materials are state of the art.

PREDICTIONS OF STRUCTURAL PERFORMANCE THROUGH TESTING

It is normal practice to institute ancillary test, full-scale firing test, and shipboard test programs for any new turret system. For the subject turret, the ancillary test program was instituted early in the design process in order to gain insight not available from material specifications. This test program consisted of tensile and compression tests on coupons and static tests of two flat panels to simulate the combined stress condition of green water loadings.

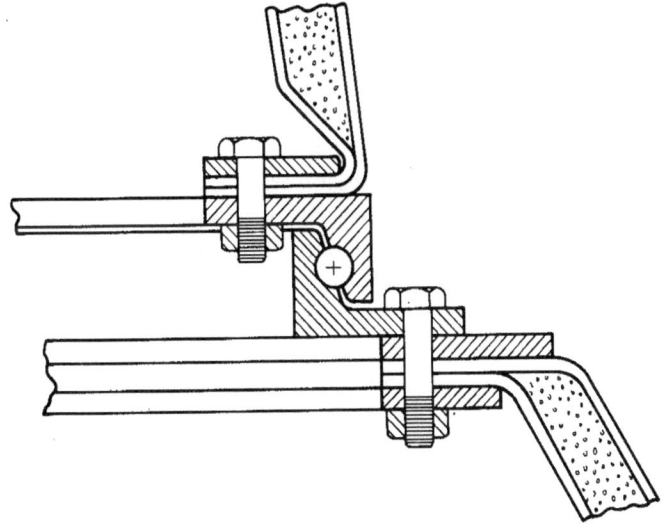

FIG. 2. Laminate/mechanical connection.

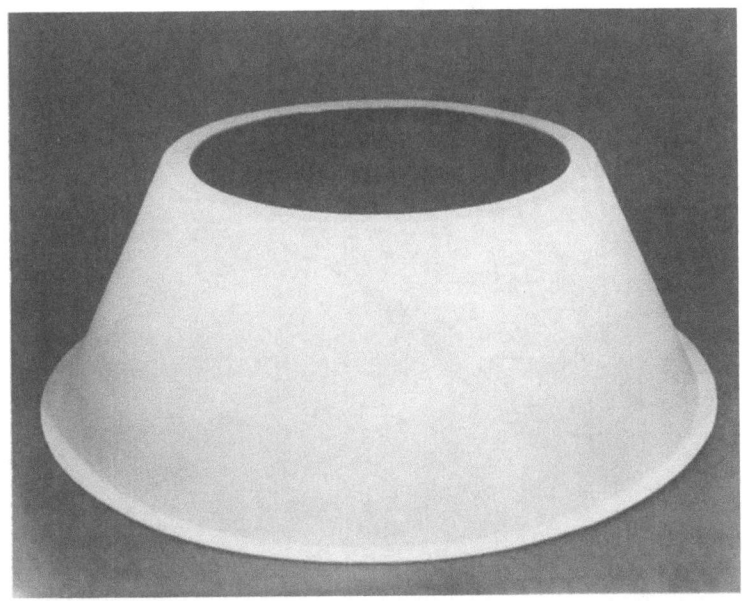

FIG. 3. Sandwich base.

In the coupon tests, panels were fabricated from various materials with nominal $\frac{1}{2}$ inch, $\frac{3}{4}$ inch and $1\frac{1}{2}$ inch (double sandwich) core thicknesses. The panels were then cut into two inch strips 18 in long and prepared for test in a Tinius Olsen testing machine. Tension tests were run using plywood inserts replacing the core foams at the specimen ends, so as to provide an adequate grip for the machine jaws. Compression tests were similarly run using the crosshead of the machine acting on vice jaws stabilizing the ends of the compression specimens. The 18-in length of these specimens was chosen to simulate the appropriate dimension of the turret base.

Results of these tests are shown in|Table 1. It is seen, for example, that all of the PVC sandwiches provide adequate strength for supporting the nominal 42-in wide turret bearing. Since space was available and since the

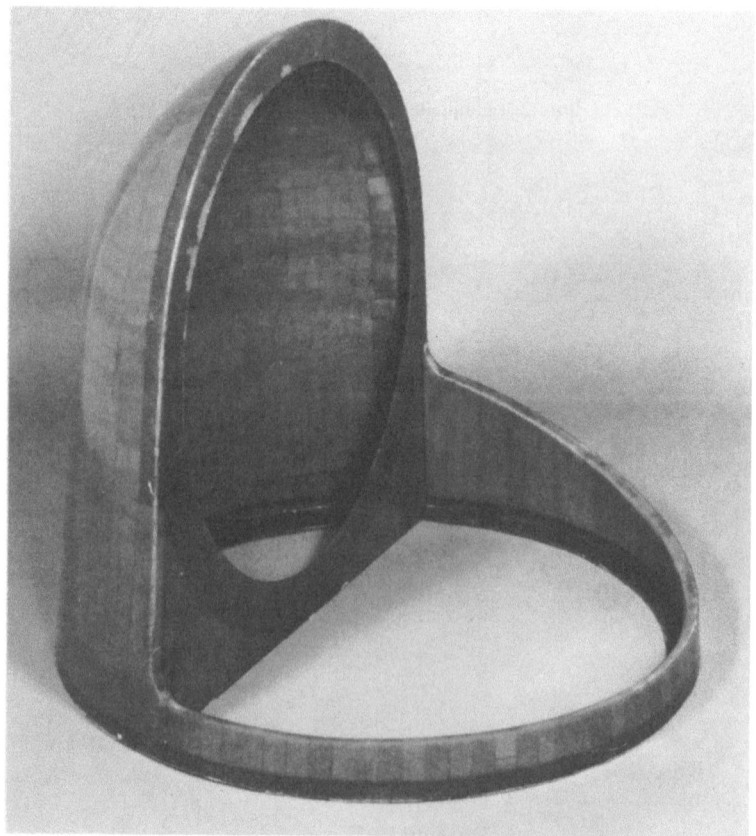

FIG. 4. Sandwich Gimbal.

nominal $\frac{3}{4}$-in PVC core showed virtually no weight penalty over the $\frac{1}{2}$-in core, it was decided to build the prototype from the thicker sandwich. The $1\frac{1}{2}$-in double sandwich, on the other hand, showed relatively low performance due to local buckling while at the same time being more difficult to fabricate in both the base and dome molds. Note that 1 E-glass, etc., refers to the laminate schedule.

The first structure then consisted of $\frac{3}{4}$-in PVC foam with polyester resin/ E-glass laminates. This construction was used for the base, half-dome, and enclosure of the turret.

DESIGN PRINCIPLES

With the basic sandwich configuration settled by testing coupons, we turned to the details of design for the turret operation.

For experimentation in the prototype test fixture, a set of off-the-shelf bearings was chosen. From the basic bearing dimensions, the sandwich structure envelope was determined and the laminate/mechanical connections[4] formed as shown in Fig. 2. In a second phase of design, a green

FIG. 5.   25-mm cannon installed in composite mount.

water skirt was employed to provide a mechanical seal between the dome and base cone of the structure. When combined with an elastomeric seal on the bearing, the skirt provided a positive seal for the metal parts of the turret.

At the time of writing this paper, the shade and fairing are being fabricated. Photographs of the primary sandwich structure are shown in Figs 3, 4 and 5. Figure 5 illustrates the compatibility of the 25-mm cannon in the mount and the system size relative to a man.

STRUCTURAL TESTING

Progressive testing took place leading up to the first firing tests on the primary structure. Tests included:

(1) Deflection and proof tests on the base with a steel 'wishbone' supporting the vertical ring and gun cradle.
(2) Deflection and proof tests on the dome and base with the wishbone tying the system together.
(3) Strain gauge instrumented coupon tests for the laminate schedule used.
(4) Strain gauge instrumented static tests on the dome and base with the wishbone tying the system together.
(5) Strain gauge instrumented static tests on the dome and base with the wishbone removed.

The setup for test Series 1 is shown in Fig. 6. These tests showed that the base structure was adequate to support a 5200 lb firing load plus a dead weight of 500 lb. Deflection measurements indicated that the base bearing rotated 0.04×10^{-3} radians, negligible relative to the accuracy requirements of 1×10^{-3} radians.

In the second set of tests, a 6000 lb firing load was applied to the wishbone, dome, and base to illustrate the structural integrity of the system. In the process, the mount stiffness was determined to be approximately 60 000 lb/in, well within the requirements dictated by experience.

In the third series of tests, the coupons were tested in tension in an Instron machine. The results showed that failure of a uniaxially loaded laminate occurred at approximately 5000 μin/in.

The strain gauge instrumentation on the dome for the fourth and fifth series of tests is shown in Fig. 7. From experience, the areas at the front,

FIG. 6. Static test setup for base.

FIG. 7. Strain gauge rosettes mounted on Gimbal sandwich.

top, and back of the dome, where the laminate leaves the PVC foam and enters the bearing clamps, were instrumented with strain gauge rosettes. A fourth area, in the middle of the sandwich, was also rosette instrumented. The results in both the fourth and fifth series of tests showed that the strains in the laminate were approximately $\frac{1}{4}$th that required for failure in the tensile test specimens of series 3 at a simulated static firing load of 5000 lb. Since the actual peak firing load, from experience, did not exceed 3000 lb, it was felt that the structure was adequate for both strength and deflection for the 25-mm cannon. In the process, the stiffness of the all-sandwich mount was found to be 42 000 lb/in, still within the experience criterion for mount stiffness.

FIRING TESTS

Various schedules were established for firing tests. Initially, the wishbone was left in the structure. After 1 shot, 5 shots, 10 shots, and several 20 shot bursts at various rates with inspections, it was felt that the system was ready for the all-laminate test. Accordingly, the wishbone was removed and a

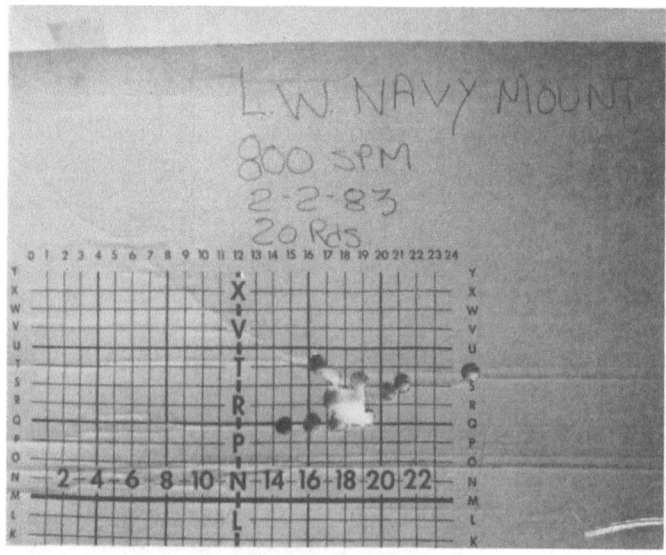

FIG. 8. Twenty round burst target at 800 shots/min.

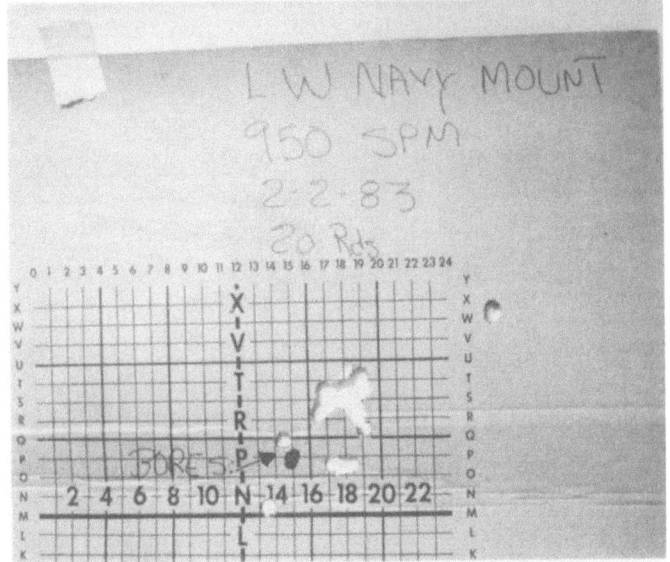

FIG. 9. Twenty round burst target at 950 shots/min.

single shot fired from the fixture. The laminate was then thoroughly inspected using a strong light. Finding no damage, we loaded the magazine with the first five-shot burst and tested the sandwich structure with its true loads. To date we have fired 134 rounds from the all-laminate structure and noted no structural changes in either the base or dome.

To illustrate the effectiveness of the mount, we have included Figs 8 and 9 showing the hits on targets placed at 1000 in from the muzzle. We hasten to note that accuracy will be improved when a better class of bearings is inserted in the mount.

MANUFACTURING ANALYSIS

Manufacturing at General Electric—Burlington completed a functionally similar naval gun mount system (Fig. 10) in 1980. The system is presently undergoing evaluation by the US Navy mounted on the aft deck of a 65 foot patrol craft.

The prototype development costs of this system are known and a production cost analysis has been completed. This will provide a convenient basis on which a weighted comparison of both prototype development costs and projected production costs of the new system can be

FIG. 10.   Metal turret on patrol boat.

made. At the present time, our experience during the initial concept phase gives encouragement that production costs will be competitive between the proposed turret system and the metal turret.

The load-carrying structure of the system presently under evaluation is aluminium, heavy gauge sheet, extruded structural shapes and machined sand castings. The structure assembly employs riveting, welding and AN type hardware.

The new composite structure fabrication parallels the technique presently used in boat hull manufacture—a oneside, open mold, made from a wood pattern. The mold construction is a fiberglass hand layup against the pattern. The molds made would have to be considered a temporary type with a limited useful life.

A typical schedule of composite layup would be as follows:

Gel coat
$\frac{3}{4}$ oz mat
38 oz cloth
$\frac{3}{4}$ oz mat
38 oz cloth
$\frac{3}{4}$ oz mat

$\frac{3}{4}$ in thick closed cell PVC foam

$\frac{3}{4}$ oz mat

38 oz cloth

$\frac{3}{4}$ oz mat

38 oz cloth

$\frac{3}{4}$ oz mat

Both the mat and cloth are 'E' glass. Room temperature cure resins, polyesters and vinyl esters have been used. Preference at this point would be the vinyl esters which appear to wet-out and work-up somewhat easier than the polyesters. The vinyl esters also exhibit improved physical properties over the polyesters.

Costs incurred to date and projected costs indicate that the composite structure program is viable and should be pursued to completion.

CONCLUSIONS

PVC core E-glass laminate turret structures appear to be feasible for naval turrets. They are adequate structurally for 25-mm Gatling cannon at various rates of fire. Structures of this type are compatible with fabrication techniques used in the boat industry for over a decade. They are competitive in price with metal turrets and offer advantages in corrosion resistance as well as ease and flexibility of manufacturing techniques. There are pronounced time and cost savings in prototype development. Finally, through an adequate testing program it is possible to design the laminate to yield a given stiffness for the mount.

REFERENCES

1. JOHANNSEN, T. J., Airex cored fiberglass for construction of larger recreational power boats, *Symposium on Design and Construction of Recreational Power Boats*, Department of Naval Architecture and Marine Engineering, Ann Arbor, Michigan (1979).
2. MUSSELMAN, K. A., United States Naval Surface Weapons Center, Dahlgren, VA, Private communication.
3. BIRD, J. and ALLAN, R. C., The development of improved FRP laminates for ship hull construction, *Composite Structures* (edited by I. H. Marshall), Applied Science Publishers, London, pp. 202–223 (1981).
4. *National Fisherman*, October, p. 58 (1982).

29

A Study of the Buckling of Some Rectangular CFRP Plates

T. J. Craig and D. J. Dawe

*Department of Civil Engineering, University of Birmingham,
PO Box 363, Birmingham B15 2TT, England*

ABSTRACT

An investigation of the buckling of moderately thick, rectangular CFRP plates is described. The plates are subjected to uniform uniaxial compression, with their loaded edges being clamped and with simple supports provided just inboard of the unloaded edges. The investigation comprises both experimental and theoretical work. Details of the test procedure are given first and these relate to the buckling of four laminated plates, two of the cross-ply and two of the angle-ply type. Brief description is then given of a finite strip method of calculating the buckling loads which is based on the use of Mindlin plate theory and hence takes account of transverse shear effects. Finally, the measured and calculated buckling loads are recorded and compared.

INTRODUCTION

There is increasing interest nowadays in the performance of carbon fibre reinforced plastic (CFRP) plates as structural components. These laminated plates exhibit a high strength-to-weight ratio which makes their use attractive in a range of circumstances, particularly in the aircraft industry. Since such plates are frequently called upon to operate in circumstances where they are subjected to an in-plane loading there naturally arises a need to predict their buckling loads. This paper is concerned with studying the buckling behaviour of rectangular CFRP plates, both experimentally and analytically. In particular, the laminated plates under consideration here are relatively thick and there appears to have been little work reported in the literature on the buckling of such plates. The present work represents a preliminary study, undertaken in order to compare the results of tests on a limited number of specimens with the predictions of a recently-developed numerical approach.

The experimental investigation concerns CFRP laminated plates, of aspect ratio approximately three, which were subjected to a uniform uniaxial compressive load acting in the direction of the longer sides. The laminates were clamped along their loaded edges and simply-supported just inboard of their unloaded edges. Two plates of nominal thickness 4 mm and two of 5 mm were tested, one of each type being a cross-ply and one an angle-ply plate. Buckling loads for each plate were determined by a procedure described later.

For moderately thick CFRP plates it is likely that the effects of transverse shear deformation will be of considerable significance, to an extent which depends upon the relative thickness and upon the ratio of extensional stiffness to transverse shear stiffness of a particular plate. Classical plate theory does not include transverse shear effects, of course, and so the analysis used in the present investigation employs Mindlin plate theory which does include these effects. The particular numerical approach adopted is the finite strip method which has been developed recently for the analysis of isotropic Mindlin plates and is here extended to include specially orthotropic laminates. The nature of Mindlin plate theory is such that the form of assumed displacement fields for use in the finite strip approach is fundamentally different when using this plate theory rather than the classical one.

Comparison of the experimental and theoretical buckling loads is made at the end of the paper.

TEST SPECIMENS

The specimens tested in this investigation were fabricated from unidirectional prepreg tape. This was supplied by CIBA-GEIGY and consisted of high tensile 'Grafil' carbon fibre embedded in the 'Fibredux 914' epoxy resin. The tape was laid up to form 32- and 40-layer specially orthotropic quasi-isotropic plates. The 32-layer laminates had a thickness of 4 mm and the following stacking sequences $[(90/0)_8]_s$ and $[(\pm 45/0/-45/90/+45/0/90)_s]_2$; stacking sequences for the two 40-layer

TABLE 1
Experimental measurements and results

Specimen	Thickness		Young's modulus E_L	Experimental buckling load (kN)
	Nominal (mm)	Actual (mm)		
32-layer, cross-ply	4	3·78	128·7	80
32-layer, angle-ply	4	4·04	124·7	130
40-layer, cross-ply	5	5·21	129·1	190
40-layer, angle-ply	5	5·21	119·4	240

laminates were $[(90/0)_{10}]_s$ and $[(\pm 45/0/90/0/-45/90/0/+45/90)_s]_2$, each of these laminates being 5 mm thick. The laminates were cured in an autoclave using standard techniques.

The overall length of all the laminates was 300 mm and the width was 110 mm. When mounted in the test rig the distance between the simple-supports was 100 mm, giving a 5 mm overhang at each support.

Due to the curing techniques used no control could be exerted over the final thickness of each laminate. There was a variation in thickness over the area of each laminate of around 5%: the average actual thicknesses are given with the results in Table 1.

TEST APPARATUS AND PROCEDURE

The test specimens were clamped along the loaded edges and simply-supported just inboard of the other two edges. These boundary conditions were provided by the test rig shown in Fig. 1. The test rig consisted of four main parts, two end blocks and two side assemblies. The end blocks were machined from solid steel and slotted. The laminate was bonded into the slots in the end blocks using epoxy resin, and so the edges were clamped firmly in place. The epoxy resin was allowed to dry with the application of a small axial load to ensure that the laminate was properly bedded into the end block. Each side assembly consisted of three pieces, comprising a main support constructed from aluminium channel section, and two knife edges. The use of channel section for the side assemblies gave the test rig stability and a high degree of rigidity. The knife edges were bolted to the channel section, and slots allowed the knife edges to move, and so accommodate a variety of laminate thicknesses. The knife edges only provided restraint out

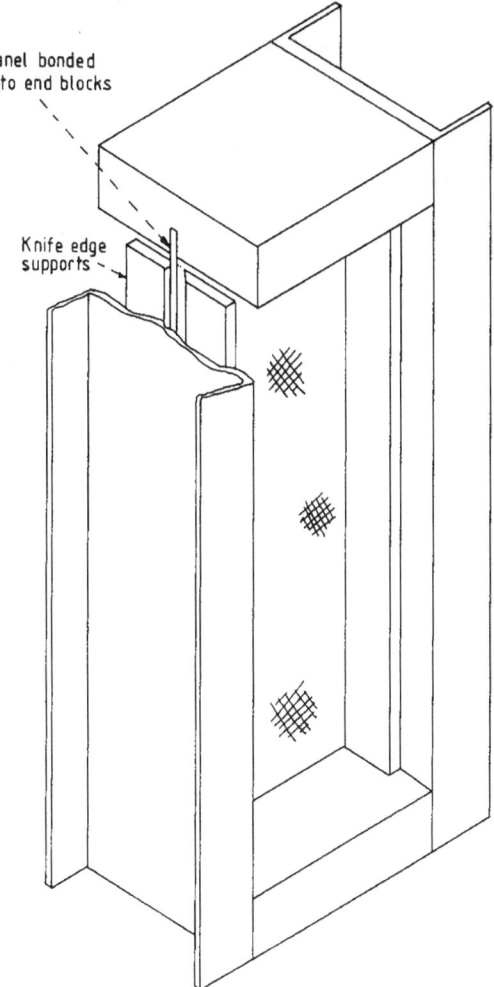

Panel bonded
into end blocks

Knife edge
supports

FIG. 1.   Test rig.

of plane, so the laminate was free to 'wave' in plane. Further consideration is given to the boundary conditions later on.

Electrical resistance strain gauges were used to measure the surface strain of the laminates during the tests. The strain gauges were mounted on the laminates according to the scheme in Fig. 2, in a 'back to back' configuration, and close to a point where it was anticipated a buckle peak (maximum out-of-plane deflection) would occur.

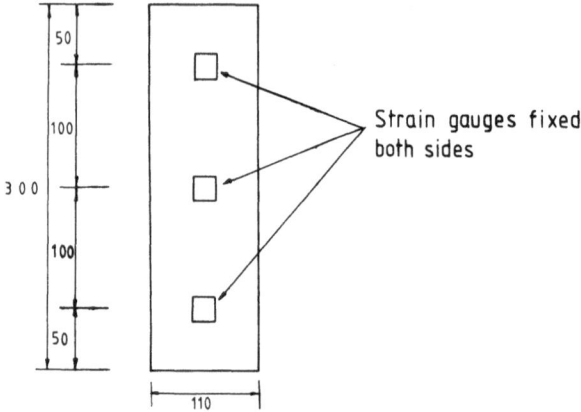

Fig. 2.   Location of strain gauges.

Dial gauges were located at the same points as the strain gauges after the rig had been positioned in the testing machine.

For the buckling tests a Schenk hydraulic testing machine with computer control was used. The strain gauges were scanned by a data logger at pre-determined load levels, and the dial gauge readings were taken manually at the same load levels. The specimens were loaded with a constant rate of straining until failure occurred.

REMARKS ON BUCKLING TESTS

Before looking in detail at the experimental results it is pertinent to discuss briefly here some of the problems encountered in buckling tests.

One such problem is that it can often be difficult to ascertain the precise boundary conditions at the edges of a plate in an experimental study, and this makes comparison with a theoretical analysis difficult. In the present work clamped boundary conditions were achieved at the loaded edges of the specimens by bonding the laminates into slots in the end blocks of the test rig. No movement of the laminates in the end blocks was detected during tests and hence the boundary conditions at the loaded edges were as follows:

$$u = v = w = 0$$

The simple-supports local to the unloaded edges were provided by means of knife edge supports which restricted movement in the out-of-plane (z, w)

direction. The laminate was free to move in plane (hence $u \neq v \neq 0$) and so the boundary condition at a simple support was simply $w = 0$. The friction between the knife edges and the specimens was assumed to be very small compared to the total applied load and so could be ignored.

Another well-known major difficulty associated with buckling tests is that associated with the presence of initial imperfection, and the effect of this can only be reduced by careful experimental technique. Initial imperfection can cause a considerable reduction in the initial buckling load, which is itself often difficult to locate. In practice, out-of-plane deflections occur as soon as the plate is loaded, due to imperfections, and often nothing apparently drastic happens when the initial buckling load is reached and passed. Detection of the point of initial buckling from test results can be tackled in a number of ways. One of these which has been used extensively is the Southwell plot. In this procedure the out-of-plane deflection w is plotted against w/P (where P is the applied in-plane load) and a straight line should be found in the region where w is less than the thickness of the plate, but is not too small. The inverse of this slope gives the buckling load. However, it is often difficult to obtain a straight line, and small changes in slope give rise to large changes in the estimated buckling load. In alternative approaches load-end displacement or load-surface strain plots could be used to determine buckling loads from tests since both show a reversal at the buckling load. However both methods are highly sensitive to imperfections, and the reversal often occurs over a wide range of loads. A better procedure is based on examining the reversal of compressive membrane strain which occurs at initial buckling, at a point on the plate local to a point of maximum out-of-plane deflection (i.e. on a buckle peak). The reversal occurs rapidly and over a small range of load, giving a well defined buckling load. The compressive membrane strain, ε_m, can be calculated from the readings ε_1 and ε_2 taken from two back-to-back strain gauges simply as follows

$$\varepsilon_m = \frac{\varepsilon_1 + \varepsilon_2}{2}$$

EXPERIMENTAL RESULTS

A full summary of experimental results is given in Table 1. A stress/load versus compressive membrane strain plot is given for the 32-layer, cross-ply laminate in Fig. 3.

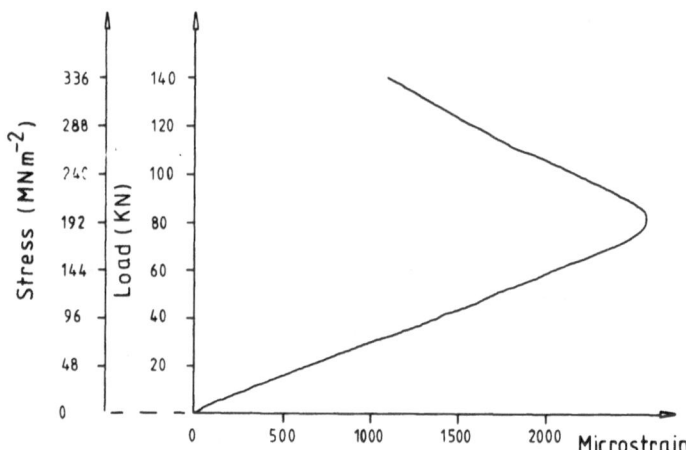

FIG. 3.   Stress load–membrane strain plot for the 32-layer cross-ply plate.

The values of the Young's modulus, E_L, given in Table 1 were calculated from the values of compressive membrane strain at the centre of each laminate. Using the method of least squares a straight line was fitted to the linear part of the stress–strain graphs. The value of E_L for the whole laminate can be determined from the slope of this line.

It can be seen clearly from the load–membrane strain graph for the 32-layer, cross-ply laminate that membrane strain reversal occurs very rapidly over a small range of load. The relationship between load and membrane strain appears to be linear both before and after buckling. It is apparent from Fig. 3 that seeking the point of reversal of membrane strain provides a relatively simple method of estimating accurately the buckling load of a plate: this was found to be so for all the plates tested.

NUMERICAL ANALYSIS

The laminates tested in the experimental study described above have been analysed by the finite strip technique. This technique is a specialization of the finite element and Rayleigh–Ritz methods wherein the structure under study, in this case a plate, is divided up into strips which span the plate in one direction and are connected to each other at their longitudinal edges. The ends of a strip form part of the boundary of the plate. This displacement field of a strip is represented in the form of a series comprising products of special functions in the longitudinal direction, that satisfy the geometric boundary conditions at the ends of the strip, and simple polynomial functions across the strip. Both the number of series terms and the order of the crosswise polynomials can be varied without undue difficulty.

The laminates to be analysed are moderately thick, and so transverse shear effects may well be significant and are included in the calculations. This is achieved by abandoning the classical plate theory and instead adopting the plate theory of Mindlin[1] in which the u, v and w displacement components at a general point are expressed as

$$u(x, y, z) = z\psi_x(x, y)$$

$$v(x, y, z) = z\psi_y(x, y) \tag{1}$$

$$w(x, y, z) = w(x, y)$$

Here ψ_x and ψ_y are the rotations of cross sections in the x and y directions respectively (*see* Fig. 4). The displacement assumptions of eqn. (1) imply that lines initially straight and normal to the middle surface remain straight and unchanged in length during the deformation process but are not constrained to remain normal to the deformed middle surface. Thus the rotations are not simply equal to the slopes of the middle surface as in classical plate theory but depend also upon the transverse shear strains: hence ψ_x and ψ_y are independent reference quantities within Mindlin plate theory. The displacement assumptions give rise to uniform transverse shear strain distributions which are corrected for by the introduction of a shear connection factor.

The application of the finite strip method, in conjunction with Mindlin

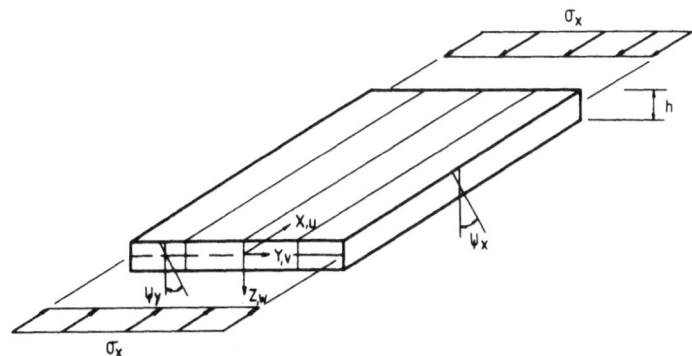

FIG. 4. The thick finite strip.

plate theory, to the vibration and buckling of thick, isotropic plates has been discussed in detail elsewhere.[2,3] It is now only necessary to describe briefly the extension of this work to cover the buckling of specially orthotropic, laminated plates.

The analysis of laminated plates can take two forms. Firstly, each layer can be analysed separately and whole-plate equations can be generated by applying inter-layer equilibrium and continuity conditions. Alternatively, and much more economically, the plate can be analysed as a whole, assuming there to be a common angle of shear through the thickness of the laminate as is assumed in Mindlin plate theory. Although this produces discontinuous transverse shear stresses within the plate, accurate results can be obtained for problems in which it is the overall plate behaviour that is important, such as occurs in the buckling problem. This has been demonstrated by Sun[4] who has shown that excellent predictions can be obtained using this approach, provided that the transverse shear rigidities of the different layers are not too dissimilar. Hence the assumptions of Mindlin plate theory are regarded as a suitable basis for the numerical study of the laminated plates considered here.

The displacement field for a finite strip (shown in Fig. 4) in the context of Mindlin plate theory is expressed as follows, in terms of the three independent reference quantities w, ψ_y and ψ_x (Refs 2 and 3):

$$
\begin{Bmatrix} w \\ \psi_y \\ \psi_x \end{Bmatrix} = \sum_{i=1}^{r} \begin{bmatrix} W_i(x) & 0 & 0 \\ 0 & W_i(x) & 0 \\ 0 & 0 & \Psi_i(x) \end{bmatrix} \begin{bmatrix} \phi_n(y) & 0 & 0 \\ 0 & \phi_n(y) & 0 \\ 0 & 0 & \phi_n(y) \end{bmatrix} \begin{Bmatrix} A_1 \\ A_2 \\ \vdots \\ A_{3n+3} \end{Bmatrix}_i
$$

(2)

Here $\phi_n(y) = [1, y, \ldots, y^n]$ represents a polynomial interpolation of order n across the strip and the A_i are generalized coefficients which can be related to the strip degrees of freedom $\mathbf{d}_{ni}$ for the ith series term. These freedoms are located at $n + 1$ reference lines across the strip and comprise values of w, ψ_x and ψ_y at each line: Fig. 4 shows a strip with five reference lines, corresponding to $n = 4$. $W_i(x)$ and $\Psi_i(x)$ are the mode shapes of vibration of appropriate Timoshenko beams, that is are unidirectional functions describing the lateral deflection and cross-sectional rotation, respectively, in such modes. These functions, which are recorded in Ref. 2 vary with the relative thickness of the plate (beam) as well as with the boundary conditions, and satisfy the appropriate Mindlin plate boundary conditions for simply-supported and clamped edges.

During buckling the bending strain energy set up per unit middle-surface area of a specially orthotropic plate is

$$\Delta U = \frac{1}{2}\left[A_{44}\left(\psi_y^2 + 2\psi_y\left(\frac{\partial w}{\partial y}\right) + \left(\frac{\partial w}{\partial y}\right)^2\right) + A_{55}\left(\psi_x^2 + 2\psi_x\left(\frac{\partial w}{\partial x}\right) + \left(\frac{\partial w}{\partial x}\right)^2\right)\right.$$

$$\left. + D_{11}\left(\frac{\partial \psi_x}{\partial x}\right)^2 + D_{22}\left(\frac{\partial \psi_y}{\partial y}\right)^2 + 2D_{12}\left(\frac{\partial \psi_x}{\partial x}\right)\left(\frac{\partial \psi_y}{\partial y}\right) + D_{66}\left(\frac{\partial \psi_x}{\partial y} + \frac{\partial \psi_y}{\partial x}\right)^2\right] \tag{3}$$

where

$$D_{ij} = \int_{-h/2}^{h/2} Q_{ij}^m z^2 \, dz \qquad i,j = 1, 2, 6 \tag{4}$$

and

$$A_{ij} = \int_{-h/2}^{h/2} Q_{ij}^m k_{ij}^2 \, dz \qquad i,j = 4, 5 \tag{5}$$

in which Q_{ij}^m are the reduced elastic constants of the mth layer and k_{ij}^2 are shear correction factors.

The loss of potential energy of the in-plane membrane stress σ_x^0 during the buckling process is the integrated sum of the stress and its corresponding second-order strain. Assuming that buckling deformations are sufficiently small that the membrane stress is unchanged during buckling leads to the following expression for the loss of potential energy per unit area of the middle surface,

$$\Delta V_g = \frac{h}{2}\,\sigma_x^0\left(\frac{\partial w}{\partial x}\right)^2 + \frac{h^3}{24}\,\sigma_x^0\left\{\left(\frac{\partial \psi_x}{\partial x}\right)^2 + \left(\frac{\partial \psi_y}{\partial x}\right)^2\right\} \tag{6}$$

To solve the plate buckling problem the elastic stiffness matrix $\mathbf{K}_n$ and the geometric (or initial-stress) stiffness matrix $\mathbf{L}_n$ need first to be established for the individual finite strip (corresponding to crosswise interpolation order n). The formation of these matrices is described at length for plates of isotropic material in Refs 2 and 3 and need not be considered in any detail here when extension is made to anisotropic material. The derivation of $\mathbf{K}_n$ and $\mathbf{L}_n$ is based, of course, on the substitution of the expression for the displacement field into expressions for the strain energy per unit area (eqn. (5)) and the loss of potential energy per unit area (eqn. (6)), respectively, with subsequent integration over the area of the strip middle surface. The whole-plate elastic stiffness matrix $\bar{\mathbf{K}}$ and the whole-plate geometric stiffness matrix $\bar{\mathbf{L}}$ can be obtained from the $\mathbf{K}_n$ and $\mathbf{L}_n$ of individual strips by the usual direct stiffness method. The matrix equation governing the plate buckling problem, after applying any necessary displacement boundary conditions at reference lines parallel to the x axis, becomes the familiar eigenvalue statement

$$(\bar{\mathbf{K}} - f\bar{\mathbf{L}})\bar{\mathbf{d}} = 0 \tag{7}$$

where $\bar{\mathbf{d}}$ is the complete column matrix of plate degrees of freedom and f is a load factor. The solution of this equation gives the critical values of the load factor and the associated mode shapes of buckling. The critical values are upper bounds to the exact Mindlin-theory values.

For the analysis of the particular plates used in the experimental study a finite strip with three internal nodal lines (5 in total) is used, giving a quartic interpolation of each of w, ψ_x and ψ_y across the strip. The loaded ends of the plate are clamped, and so the longitudinal displacement series is composed of the modes of a clamped–clamped vibrating Timoshenko beam with three terms in the series being used. The laminate is divided up into three strips as shown in Fig. 5. The full set of boundary conditions applied in the finite strip model are that $w = \psi_x = \psi_y = 0$ at the clamped plate edges, and that $w = 0$ at the line supports inboard from the longitudinal edges, located to correspond to the position of the knife edges in the test rig. The shear correction factors used in the analysis are calculated by a method proposed by Whitney.[6]

The material properties of the laminates used in the calculation are derived from those obtained during the tests (including E_T which due to the quasi-isotropic nature of the laminates can be considered to be equal to E_L

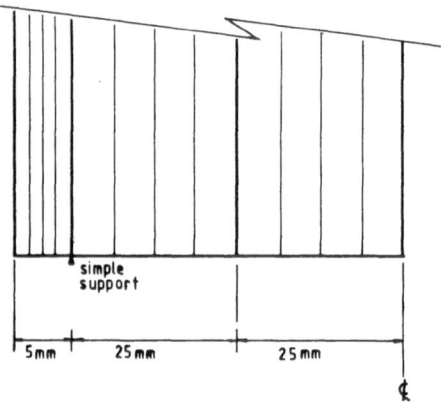

FIG. 5.   Finite strip layout.

TABLE 2
Numerical results using the finite strip method

| Specimen | Calculated buckling loads (kN) | |
	Based on nominal thickness	Based on actual thickness
32-layer, cross-ply	111	94
32-layer, angle-ply	139	145
40-layer, cross-ply	214	241
40-layer, angle-ply	246	277

which was measured in the tests), and those obtained from manufacturers' literature. The transverse shear rigidities were unavailable, but values typical of the type of fibre composite used are,

$$G_{yz} = 5\,\mathrm{GNm}^{-2} \qquad G_{xy} = G_{xz} = 6\,\mathrm{GNm}^{-2}$$

The major Poisson's ratio is 0·27.

The buckling loads calculated using the finite strip method are given in Table 2. Two sets of results are given, one set for the nominal thickness of the laminates (*see* Table 1) and one set for the actual thickness of the laminates (*see* Table 1). The 'actual' thicknesses have been calculated from an average of thickness measurements taken over the whole area of a laminate.

DISCUSSION OF RESULTS AND CONCLUSIONS

On the whole, reasonable agreement is evidenced between experimental (Table 1) and numerical (Table 2) buckling loads. The results for the angle-ply plates compare rather closer than do those obtained for the cross-ply plates. The worst comparison is between the 40-layer, cross-ply experimental and 'actual' thickness numerical results.

There are clearly a number of sources of possible differences between the experimental and numerical results, and in the main these relate to deficiencies in the test results or lack of precise knowledge of the physical characteristics of the test specimens. A common problem in fibre-reinforced plastic plates is that it is not easy to manufacture plates of uniform thickness to within a very close tolerance and this was the case in the present work. This causes difficulty in deciding what average value of thickness to use in the numerical calculations, which are, of course, restricted to perfectly flat, uniform thickness plates. Again, the non-uniformity of the test plates is very likely to increase the possibility of significant error associated with equating the knife-edge supports of the tests with the simple supports of the theory. Clearly, if a closer match between experiment and theory is to be hoped for then much will depend on the quality control on thickness and flatness which is achieved in manufacturing the test plates. Another area to which more attention needs to be paid to help improve the correlation of experimental and theoretical results is the measurement of material properties which are subsequently used as data for the calculations. The values of E_L obtained during the tests show a quite wide range. These values relied on readings taken on the surface of the laminate and on the assumption of a linear distribution of strain through the laminate. Consequently the stiffness of the first few plies of each laminate may well affect the calculations significantly.

The preceding paragraph has been concerned with sources of experimental error but it is appreciated that the theoretical model is, of course, not an exact one. However, on the basis of a range of results which have not been presented here it is believed that the errors in the results obtained using the numerical method are small.

All things considered, the correlation of experimental and theoretical results achieved in this preliminary study of a small number of plates is encouraging.

ACKNOWLEDGEMENTS

The buckling tests described in this paper were conducted at the Royal Aircraft Establishment, Farnborough, and the authors are very grateful for the materials and facilities provided there, and for the support of various personnel in Structures Department. One of the authors (T.J.C.) is grateful for the financial support of the SERC during the period of the investigation reported in this paper.

REFERENCES

1. MINDLIN, R. D., Influence of rotary inertia and shear on flexural motions of isotropic, elastic plates, *J. Appl. Mech.*, **18** (1951), 31–38.
2. ROUFAEIL, O. L. and DAWE, D. J., Vibration analysis of rectangular Mindlin plates by the finite strip method, *Comp. and Struct.*, **12** (1980), 833–841.

3. DAWE, D. J. and ROUFAEIL, O. L., Buckling of rectangular Mindlin plates, *Comp. and Struct.*, **15** (1981), 461–471.
4. SUN, C. T., Incremental deformations in orthotropic laminated plates under initial stress, *J. Appl. Mech.*, **40** (1973), 193–200.
5. TIMOSHENKO, S. P., On the correction for shear of the differential equation for transverse vibrations of prismatic bars, *Phil. Mag.*, **41** (1921), 744–746.
6. WHITNEY, J. M., Shear correction factors for orthotropic laminates under static load, *J. Appl. Mech.*, **40** (1973), 302–304.

30

Postbuckling Behaviour of Cylindrically Curved Panels of Generally Layered Composite Materials with Small Initial Imperfections of Geometry

Y. ZHANG

Shenyang Aircraft Company, China

and

F. L. MATTHEWS

Department of Aeronautics, Imperial College,
Prince Consort Road, London SW7 2BY, England

ABSTRACT

An analysis is presented in the current work to investigate the effect of small geometrical imperfections on the postbuckling behaviour of flat and cylindrically curved composite panels under compression. A general lay-up situation of the panels is considered in the analysis. The geometrical imperfection is assumed to have a general form of a double sine series. A pair of governing equations in the Von Karman sense is solved in conjunction with simply supported edges. Numerical results are presented for panels having different curvatures, different lay-ups and different materials. The results show that the initial geometrical imperfections in composite panels have an important influence on the postbuckling behaviour of the panels.

INTRODUCTION

The effect of initial geometrical imperfection on the buckling behaviour of thin plates and shells has been the subject of many investigations in the literature. Research on isotropic and orthotropic plates and shells[1-6] indicates that small initial deviation from perfect geometry has significant influence on the behaviour of such structural members. Small imperfections cause flat members to bend when axial compression is applied and curved members to collapse under a level of compressive loading much lower than the critical value predicted for perfect members.

Naturally, it may also be expected that geometrical imperfections have some influence on the buckling and postbuckling behaviour of structural members of generally laminated composite materials. It seems that there is no analysis available on this subject.

In the authors' previous investigations[7,8] an analysis for the postbuckling behaviour of cylindrically curved panels of composite materials is presented. A completely general lay-up situation is considered in the analysis. The theory is based on the virtual displacement and virtual force principles and both bifurcational and non-bifurcational types of behaviour are studied.

In the current work, however, the effect of geometrical imperfection is included. The initial imperfection in the panels is assumed to have the form of a double sine series. Computation is carried out for panels with different curvatures, different lay-ups and different materials. The effect of imperfection shapes is also discussed in the examples.

GOVERNING SYSTEM

A cylindrically curved panel on a rectangular base is assumed to consist of composite layers with arbitrary orientations of orthotropic axes with respect to the generators of the panel. The mid-surface radius of the panel is R, the thickness h, the length a and the arc b (*see* Fig. 1). The panel thickness is assumed small compared with the other dimensions of the panel so that the Kirchhoff hypothesis is valid. The constitutive relations of the panel may be written in partial inverted form, i.e.

$$\begin{Bmatrix} \boldsymbol{\varepsilon}^0 \\ \mathbf{M} \end{Bmatrix} = \begin{pmatrix} \mathbf{A}^* & \mathbf{B}^* \\ -(\mathbf{B}^*)^t & \mathbf{D}^* \end{pmatrix} \begin{Bmatrix} \mathbf{N} \\ \mathbf{k} \end{Bmatrix} \tag{1}$$

where

$$\boldsymbol{\varepsilon}^0 = [\varepsilon^0 \varepsilon_y^0 \gamma_{xy}^0]^t$$

$$\mathbf{N} = [N_x N_y N_{xy}]^t$$

$$\mathbf{M} = [M_x M_y M_{xy}]^t$$

$$\mathbf{k} = [k_x k_y k_{xy}]^t = -\left[\frac{\partial^2 w}{\partial x^2} \ \frac{\partial^2 w}{\partial y^2} \ 2\frac{\partial^2 w}{\partial x \, \partial y}\right]^t \tag{2}$$

and

$$\varepsilon_x^0 = \frac{\partial u_0}{\partial x} + \frac{1}{2}\left(\frac{\partial w_0}{\partial x}\right)^2 + \frac{\partial w_0}{\partial x}\frac{\partial \bar{w}_0}{\partial x}$$

$$\varepsilon_y^0 = \frac{\partial v_0}{\partial y} + \frac{1}{2}\left(\frac{\partial w_0}{\partial y}\right)^2 - \frac{w_0}{R} + \frac{\partial w_0}{\partial y}\frac{\partial \bar{w}_0}{\partial y}$$

$$\gamma_{xy}^0 = \frac{\partial u_0}{\partial y} + \frac{\partial v_0}{\partial x} + \frac{\partial w_0}{\partial x}\frac{\partial w_0}{\partial y} + \frac{\partial w_0}{\partial x}\frac{\partial w_0}{\partial y} + \frac{\partial w_0}{\partial y}\frac{\partial \bar{w}_0}{\partial x} \tag{3}$$

in which N_x, N_y, N_{xy} are membrane forces; M_x, M_y, M_{xy} bending moments and torque; ε_x^0, ε_y^0, γ_{xy}^0 midsurface strains; k_x, k_y, k_{xy} flexural curvatures; u_0, v_0, w_0 midsurface displacements in x, y, z directions; and $\bar{w}_0$ initial transverse deflection of the panel. Superscript t denotes the transpose of a matrix.

A^*, B^*, D^* are all 3×3 square matrices. The elements of these matrices are the reduced stiffness coefficients.

The governing equations including the effect of boundary moments are obtained from the virtual displacement and virtual force principles in the previous work,[9] i.e.

$$\int_0^a \int_0^b \left\{ \frac{\partial^2 M_x}{\partial x^2} + \frac{\partial^2 M_y}{\partial y^2} + 2\frac{\partial^2 M_{xy}}{\partial x\,\partial y} + \frac{N_y}{R} + N_x\frac{\partial^2 w_0}{\partial x^2} \right.$$

$$\left. + N_y\frac{\partial^2 w_0}{\partial y^2} + 2N_{xy}\frac{\partial^2 w_0}{\partial x\,\partial y} \right\} \delta w_0\,\mathrm{d}x\,\mathrm{d}y$$

$$+ \int_0^b \left[M_x\delta\left(\frac{\partial w_0}{\partial x}\right) \right]\Big|_{x=0}^{x=a} \mathrm{d}y + \int_0^a \left[M_y\delta\left(\frac{\partial w_0}{\partial y}\right) \right]\Big|_{y=0}^{y=b} \mathrm{d}x = 0 \tag{4}$$

$$\int_0^a \int_0^b \left\{ \frac{\partial^2 \varepsilon_x^0}{\partial y^2} + \frac{\partial^2 \varepsilon_y^0}{\partial x^2} - \frac{\partial^2 \gamma_{xy}^0}{\partial x\,\partial y} - \left(\frac{\partial^2 w_0}{\partial x\,\partial y}\right)^2 \right.$$

$$+ \frac{\partial^2 w_0}{\partial x^2}\frac{\partial^2 w_0}{\partial y^2} + \frac{1}{R}\frac{\partial^2 w_0}{\partial x^2} + \frac{\partial^2 w_0}{\partial x^2}\frac{\partial^2 \bar{w}_0}{\partial y^2}$$

$$\left. + \frac{\partial^2 w_0}{\partial y^2}\frac{\partial^2 \bar{w}_0}{\partial x^2} - 2\frac{\partial^2 w_0}{\partial x\,\partial y}\frac{\partial^2 \bar{w}_0}{\partial x\,\partial y} \right\} \delta\phi\,\mathrm{d}x\,\mathrm{d}y = 0 \tag{5}$$

where ϕ is the stress function defined by

$$N_x = \frac{\partial^2 \phi}{\partial y^2} \qquad N_y = \frac{\partial^2 \phi}{\partial x^2} \qquad N_{xy} = -\frac{\partial^2 \phi}{\partial x\,\partial y} \tag{6}$$

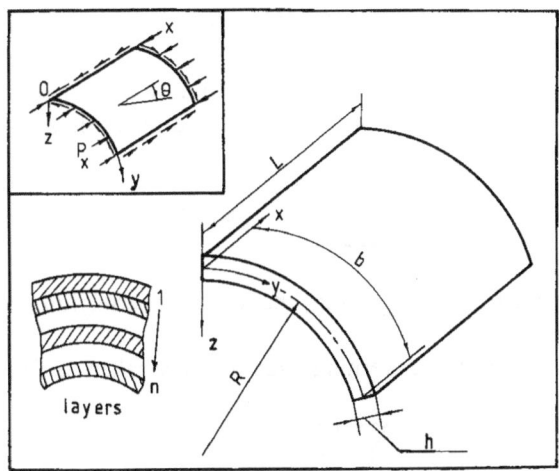

Setting (1), (2), (3) into the eqns (4) and (5) and using the following, and other, parameters:

$$\xi = \frac{x}{a} \qquad \eta = \frac{y}{b} \qquad \beta = \frac{a}{b}$$

$$w = \frac{w_0}{h} \qquad F = \frac{\phi}{A_{22}h^2} \qquad K_R = \frac{b^2}{Rh} \qquad \bar{w} = \frac{\bar{w}_0}{h}$$

the governing system in non-dimensional form is obtained.

The membrane forces and moments are also non-dimensionalized as

$$\begin{Bmatrix} N_\xi \\ N_\eta \\ N_{\xi\eta} \end{Bmatrix} = \frac{b^2}{A_{22}h^2} \begin{Bmatrix} N_x \\ N_y \\ N_{xy} \end{Bmatrix} \quad \text{and} \quad \begin{Bmatrix} M_\xi \\ M_\eta \\ M_{\xi\eta} \end{Bmatrix} = \frac{b^2}{A_{22}h^3} \begin{Bmatrix} M_x \\ M_y \\ M_{xy} \end{Bmatrix} \tag{7}$$

The panel under consideration is subjected to uniform edge compression P_x per unit width. The simply supported boundary condition is taken in non-dimensional form as follows:

$$w = 0 \qquad M_\xi = 0 \qquad \frac{\partial^2 F}{\partial \eta^2} = -\bar{N}_x \qquad \frac{\partial^2 F}{\partial \xi \partial \eta} = 0 \qquad \text{at } \xi = 0, 1$$

$$w = 0 \qquad M_\eta = 0 \qquad \frac{\partial^2 F}{\partial \xi^2} = 0 \qquad \frac{\partial^2 F}{\partial \xi \partial \eta} = 0 \qquad \text{at } \eta = 0, 1 \tag{8}$$

where $\bar{N}_x = P_x b^2 / A_{22} h^2$ is the non-dimensional edge compression.

SOLUTION

The geometrical imperfection is assumed to have a form described by the following double sine series in terms of non-dimensional coordinates

$$\bar{w} = \sum_{m=1}^{\infty} \sum_{n=1}^{\infty} \bar{w}_{mn} \sin m\pi\xi \sin n\pi\eta \qquad (9)$$

With the coefficients $\bar{w}_{mn}$ given, the initial imperfection in the panel is prescribed.

The solutions to the governing equations are taken as follows:

$$F = -\bar{N}_x \eta^2/2 + \sum_{m=1}^{\infty} \sum_{n=1}^{\infty} F_{mn} X_m(\xi) Y_n(\eta)$$

$$w = \sum_{p=1}^{\infty} \sum_{q=1}^{\infty} [w_{pq}^{(1)} \sin p\pi\xi \sin q\pi\eta + w_{pq}^{(2)} \sin^2 p\pi\xi \sin^2 q\pi\eta] \qquad (10)$$

in which the characteristic beam functions $X_i(\xi)$ and $Y_i(\eta)$ are defined by

$$X_i(\xi) = \cosh \alpha_i\xi - \cos \alpha_i\xi - \gamma_i(\sinh \alpha_i\xi - \sin \alpha_i\xi)$$
$$Y_i(\eta) = \cosh \alpha_i\eta - \cos \alpha_i\eta - \gamma_i(\sinh \alpha_i\eta - \sin \alpha_i\eta) \qquad (11)$$

where α_i and γ_i are constants which with 15 significant digits may be found in the previous work.[9] The functions $X_i(\xi)$ and $Y_i(\eta)$ are orthogonal functions. Their properties are also discussed in Ref. 9.

It is found that the assumed function F satisfies the boundary conditions (8) but the assumed function w satisfies only the zero deflection conditions. The effect of unbalanced moments on the boundaries has been, however, considered in eqn. (5). So we simply set the solution (10) into the non-dimensional forms of (4) and (5).

After much tedious algebraic manipulation, details of which are given in Ref. 9, a system containing an infinite number of equations is obtained. In each equation there are an infinite number of terms. To obtain numerical results it is necessary to truncate the series solutions (10). In the current work the first M^2 terms in the series for F and the first $2M^2$ terms of the series for w are taken (i.e. m, n, p, q = 1, 2, ..., M). In that case the resultant algebraic system contains $3M^2$ equations, which may be written in matrix form. These equations contain the column matrices $\{F\}$, $\{w^{(1)}\}$

and $\{w^{(2)}\}$, which are the matrices to be determined, the elements of which are the coefficients F_{mn}, $w_{pq}^{(1)}$, $w_{pq}^{(2)}$ in expressions (10) respectively. Also included is the column $w^{(0)}$ which contains the components $\bar{w}_{mn}$ of initial imperfections (9). This system may be simplified by solving for $\{F\}$ from one equation and substituting the result into the other two. The resultant equations are finally written as

$$[(D^{(1)}) + K_x(D^{(2)})]\{w^{(1)}\} + [(D^{(3)}) + K_x(D^{(4)})]\{w^{(2)}\}$$

$$= K_x(D^{(5)})\{w^{(0)}\} + K_x\{D^{(6)}\} \qquad (12)$$

$$[(E^{(1)}) + K_x(E^{(2)})]\{w^{(1)}\} + [(E^{(3)}) + K_x(E^{(4)})]\{w^{(2)}\} = K_x(E^{(5)})\{w^{(0)}\}$$

$$(13)$$

where

$$K_x = \frac{P_x b^2}{E_t h^3}$$

In (12) and (13) the coefficient matrices $[D^{(i)}]$ and $[E^{(i)}]$ are related in various ways to the panel dimensions, the layer properties and the columns $\{w^{(0)}\}$, $\{w^{(1)}\}$ and $\{w^{(2)}\}$. Since the matrices $[D^{(1)}]$, $[D^{(3)}]$, $[E^{(1)}]$ and $[E^{(3)}]$ contain quadratic terms of $w_{pq}^{(1)}$, $w_{pq}^{(2)}$, and $w_{pq}^{(0)}$ eqns (12) and (13) represent a set of non-linear (cubic) algebraic equations.

With initial imperfections given, together with the properties of composite materials and geometry of the panel, the columns $\{w^{(1)}\}$ and $\{w^{(2)}\}$ may be solved from eqns (12) and (13).

COMPUTATION AND RESULTS

An iterative scheme based on the modified Newton method is used to solve the system (12) and (13). The starting point of the iterative procedure is the undeformed state of the panel (zero load and zero deflection). In the examples below the first component $w_{11}^{(1)}$ of the deflection function in (10) is prescribed to increase step-by-step. For the first few steps, the step length of the prescribed component of the deflection is kept relatively small because the load–deflection curves are usually very steep at the beginning of the loading process. With every increase of one step in the prescribed variable, the iterative procedure produces the solution for a new point on the behaviour path. After three points are obtained an extrapolation scheme is introduced to create a better estimate for succeeding points. For every

TABLE 1
Elastic constants for calculations

	Boron/epoxy	Carbon/epoxy	Glass/epoxy
E_l (GN/m^2)	206·9	206·9	53·8
E_t (GN/m^2)	20·7	5·2	17·9
v_{lt}	0·3	0·25	0·25
G_{lt} (GN/m^2)	5·2	2·6	8·9

point on the path, $\{F\}$, the stress function, the deflection function and the membrane forces and moments are obtained from the relevant equations.

The computing results are summarized in Figs 2 to 10. The properties of the composite materials used in the calculations, appropriate to boron/ epoxy, carbon/epoxy and glass/epoxy are given in Table 1. E_l and E_t are the tensile modulus in or perpendicular to the filament direction, v_{lt} and G_{lt} are the Poisson's ratio and shear modulus respectively.

In the examples a four-term approximation for the stress function F and an eight-term approximation for the deflection w are used. The length a and the midsurface arc b of the panel are taken both equal to 250 mm and the thickness h equal to 2·5 mm.

Figure 2 is obtained for four-layer panels of $(\pm 45°)_s$ boron/epoxy composite materials with different curvatures. The dotted lines indicated the behaviour of perfect panels and solid lines show the results of imperfect panels in which the first component w_{11} of the imperfection expression (9) is equal to 0·5, and the other components vanish. It can be seen that, because of the initial imperfection, the behaviour of the panels under compression is not bifurcational. Bending occurs as soon as the load is applied. This is true both for flat plates and curved panels. For flat plates with geometrical imperfection the load–central deflection relation is still monotonic. This means that the equilibrium state is always unique, corresponding to a certain level of load. For curved panels, however, there may be more than one equilibrium state which corresponds to the same load level. This means that snap-through may occur in such panels. It is seen from the figure that the net deflection of these imperfect panels may be smaller than that of the perfect ones, but the total deflection (net deflection plus initial imperfection) is always larger than that of the perfect panels (for perfect panels net deflection is equal to the total deflection). This phenomenon was pointed out by Yamaki[3,4] for isotropic plates. Now it is found also to be true for laminated curved panels. It is seen in all the following figures of the current work.

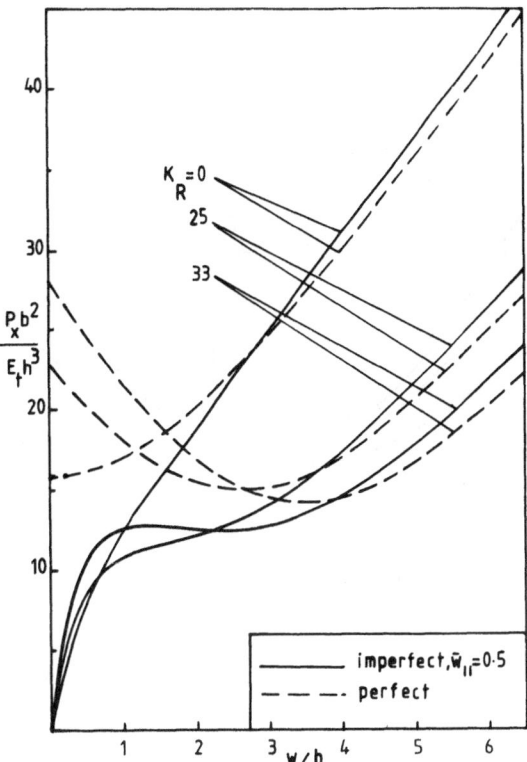

FIG. 2.   Variation of central deflection with axial compression of imperfect boron/epoxy panels; ($\pm45°$)$_s$ lay-up, different curvatures.

Figures 3, 4 and 5 illustrate the effect of the magnitude of the imperfection on the postbuckling behaviour of the panels. These panels are assumed to have only an imperfection of the form $\bar{w}_{11} \sin \pi\xi \sin\pi\eta$ (non-dimensional form). Figure 3 is obtained for four-layer ($\pm45°$)$_s$ boron/epoxy panels having the curvature parameter $K_R = 50$. Figure 4 is for four layer ($\pm45°$)$_s$ glass/epoxy flat plates ($K_R = 0$), and Fig. 5 is for four layer ($\pm45°$)$_s$ carbon/epoxy panels with $K_R = 25$. All these results obtained for panels of different materials and different curvatures indicate that bending occurs in imperfect panels immediately the load is applied, and the bending effect increases with increasing magnitude of the imperfection. The snap-through type of collapse may occur for imperfect panels at a much lower level of load than that of the bifurcation point for perfect panels. That is, the initial imperfections cause early collapse of the curved panels. On the other hand, when the initial imperfections in the curved panels are large, the

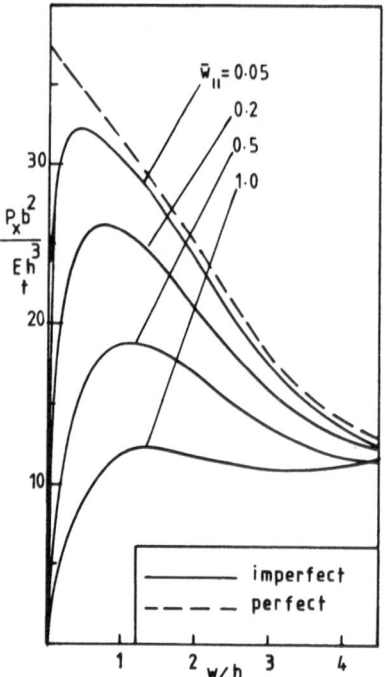

FIG. 3. Variation of central deflection with axial compression of imperfect boron/epoxy panels; ($\pm 45°)_s$ lay-up and $K_R = 50$.

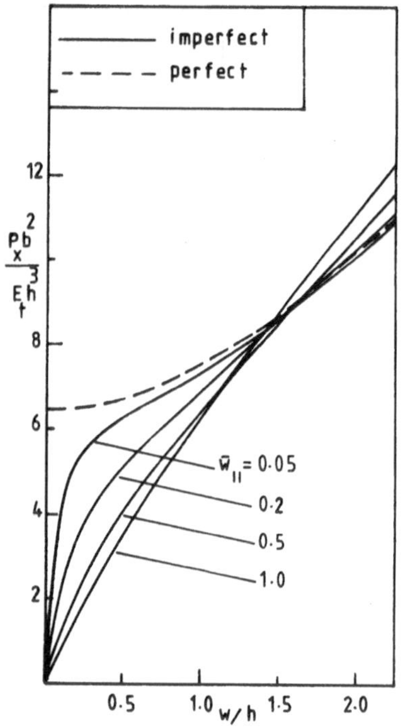

FIG. 4. Variation of central deflection with axial compression of imperfect glass/epoxy flat plates; ($\pm 45°)_s$ lay-up.

load–deflection curves become very flat and snap-through may no longer happen in the loading process.

Figure 6 demonstrates the effect of the coupling terms D_{16} and D_{26} on the postbuckling behaviour of imperfect panels. The load–central deflection curves in the figure are calculated for panels with symmetric boron/epoxy layers with $K_R = 25$. It is obvious that the largest D_{16} and D_{26} terms appear in the panel having unidirectional layers at 45° and they decrease with increasing number of alternate layers at $\pm 45°$ in the panel. It is found that the coupling terms D_{16} and D_{26} degrade the panel performance, both for perfect panels and imperfect ones under compression, unless the deflection is very large.

Figure 7 is obtained for boron/epoxy panels with antisymmetric angle layers. The effect of the coupling terms B_{16} and B_{26} can be seen clearly in this figure. Panels having two-layer antisymmetric arrangements have the lowest performance, no matter whether the panels are perfect or imperfect.

Figure 8 shows the effect of different shapes of imperfection. It can be

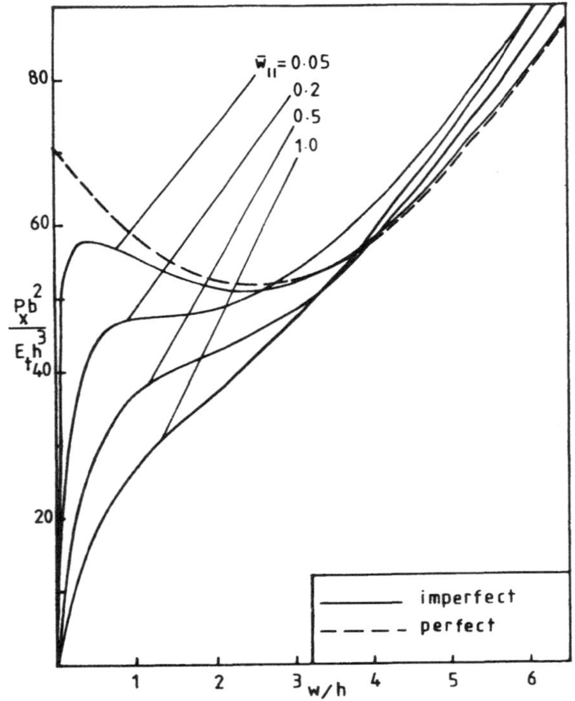

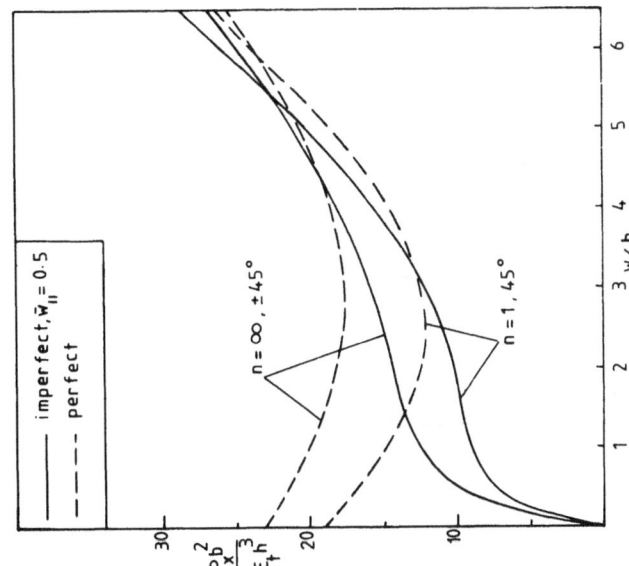

FIG. 6. Variation of central deflection with axial compression of imperfect boron/epoxy panels; different lay-ups and $K_R = 25$.

FIG. 5. Variation of central deflection with axial compression of imperfect carbon/epoxy panels; $(\pm 45°)_s$ lay-up and $K_R = 25$.

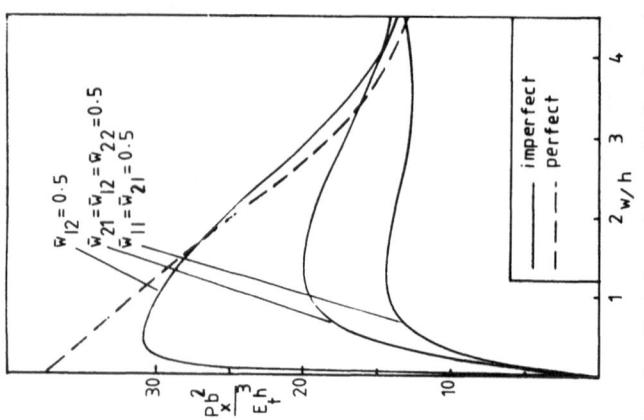

FIG. 8.   Variation of central deflection with axial compression of imperfect boron/epoxy panels; $(\pm 45°)_s$ lay-up and $K_R = 50$.

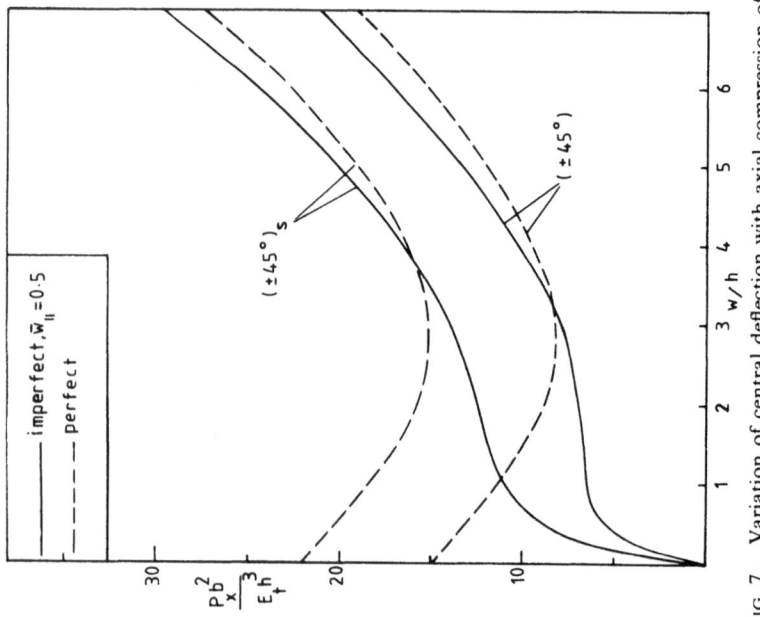

FIG. 7.   Variation of central deflection with axial compression of imperfect boron/epoxy panels; antisymmetric lay-up and $K_R = 25$.

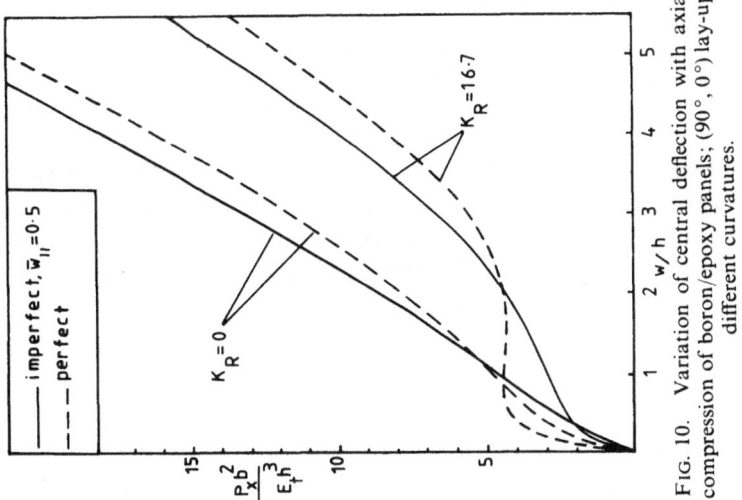

FIG. 10. Variation of central deflection with axial compression of boron/epoxy panels; (90°, 0°) lay-up, different curvatures.

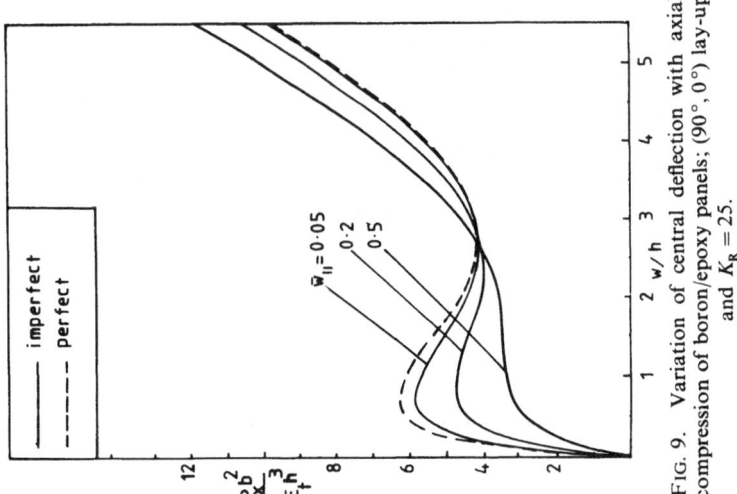

FIG. 9. Variation of central deflection with axial compression of boron/epoxy panels; (90°, 0°) lay-up and $K_R = 25$.

seen that the sensitivity of the panel to the different imperfection shapes is quite different. For this four-layer $(\pm 45°)_s$ boron/epoxy panel with the curvature parameter $K_R = 50$, the components $\bar{w}_{11}$ and $\bar{w}_{21}$ of imperfection have more influence than $\bar{w}_{12}$ and $\bar{w}_{22}$.

Figures 9 and 10 are obtained for boron/epoxy panels with unsymmetric crossply layers at $(90°, 0°)$. As discussed in the previous work,[8] even for panels with perfect geometry under compression, bending of the panel is induced by the unsymmetric crossply layers. The dotted lines show the behaviour of such perfect panels. For the imperfect geometry the bending effect of the panel is even more severe. It follows that the unsymmetric crossply arrangement and geometrical imperfection both reduce the resistance of the curved panel against collapse.

CONCLUSIONS

From the above examples, the following conclusions may be summarized:

(1) Initial geometrical imperfection in a composite panel has an important influence on the behaviour of the panel under in-plane loading. Bending of the panel is induced by such imperfections immediately the load is applied. A panel with geometrical curvature is even more sensitive to initial imperfections. It may greatly reduce the resistance of such panels against collapse.

(2) The coupling terms D_{16}, D_{26} or B_{16}, B_{26} degrade the panel performance both for perfect panels and imperfect panels.

(3) Laminated composite panels have different sensitivity to different types of imperfection shape.

(4) Panels having certain unsymmetric lay-ups which, even when perfect, bend under in-plane loading, will experience increased bending when imperfect.

REFERENCES

1. DONNELL, L. H. and WAN, C. C., Effect of imperfection on buckling of thin cylinders and columns under axial compression, *J. appl. Mech., Trans. ASME*, **17**, 1950, 73–83.
2. COAN, J. M., Large deflection theory for plates with small initial curvature loaded in edge compression, *J. appl. Mech., Trans. ASME*, **18**, 1951, 143–151.
3. YAMAKI, N., Postbuckling behaviour of rectangular plates with small initial curvature loaded in edge compression, *J. appl. Mech., Trans. ASME*, **26**, 1959, 407–414.

4. YAMAKI, N., Postbuckling behaviour of rectangular plates with small initial curvature loaded in edge compression (continued), *J. appl. Mech., Trans. ASME*, **27**, 1960, 335–342.

5. KHOT, N. S., Postbuckling behaviour of geometrically imperfect composite cylindrical shells under axial compression, *AIAA J.*, **8**, 1970, 579–581.

6. BOOTON, M., Buckling of imperfect anisotropic cylinders under combined loading, UTIAS Report No. 203, CNISSN 0082-5255, Institute for Aerospace Studies, University of Toronto, Aug. 1976.

7. ZHANG, Y. and MATTHEWS, F. L., Postbuckling behaviour of curved panels of generally layered composite materials, *J. Comp. Struct.*, **1**, August 1983, also presented at Ninth U.S. Congress of Applied Mechanics, Ithaca, New York, U.S.A., June 1982.

8. ZHANG, Y. and MATTHEWS, F. L., Bending behaviour of laminated panels under in-plane loading, to be published.

9. ZHANG, Y., Buckling and postbuckling behaviour of generally laminated composite panels, Ph.D. Thesis, Imperial College, University of London, Oct. 1982.

31

The Instability of Composite Channel Sections

W. M. BANKS and J. RHODES

*Department of Mechanics of Materials, University of Strathclyde,
Glasgow G1 1XJ, Scotland*

ABSTRACT

At previous conferences the authors have examined the buckling and postbuckling behaviour of reinforced plastic plates subject to in-plane loading, both theoretically[1] and experimentally.[2] That work has been extended to cover the buckling and postbuckling behaviour of orthotropic box sections.[3] The present contribution extends this work further to examine the instability of orthotropic channel sections. The sections are considered as a series of linked plates with rotationally restrained unloaded edges or rotationally restrained and free unloaded edges. These conditions have been considered for the plates alone. The linking procedure enables the instability of the section to be evaluated.

After buckling the linked plates are given a common end displacement. The moments and slopes at each edge are related to this and combined in such a way as to ensure that equilibrium and compatibility are satisfied at the plate edges, using an iterative procedure. Thereafter the relevant postbuckling stiffnesses and coefficients are obtained.

NOTATION

a, b Plate dimensions in x and y directions respectively

D_{11}, D_{22} Flexural rigidity of plate per unit width for bending about the y and x axes respectively, given by

$$D_{11} = E_{11}t^3/12(1 - v_{12}v_{21}) \qquad D_{22} = D_{11}E_{22}/E_{11}$$

E_{11}, E_{22} Modulus of elasticity in the x and y directions respectively
e Ratio of buckle half wavelength to plate width
G_{12} Elastic shear modulus in x-y plane
K Elastic buckling coefficient for orthotropic plates
t Plate thickness
$Y(y)$ Deflections across buckled plate
v_{12}, v_{21} Poisson's ratio in the x and y directions respectively
σ_{cr} Critical buckling stress
w Out-of-plane deflections of the plate

INTRODUCTION

The increasing use of reinforced plastics in structural applications has led to the need to examine their behaviour when subjected to compressive loading. Their low elastic modulus coupled with a high strength makes instability a major problem area. To increase and exploit applications to composite structures, this problem will need to be understood and appropriate steps taken at the design stage.

The purpose of the present paper is to make a contribution in this direction and to permit an understanding of the instability of composite channel sections. The sections considered are fabricated from glass reinforced plastic (GRP) with unidirectional orientation. However, provided the properties of the composite are known, the method could be applied to other materials.

METHOD OF ANALYSIS

A typical channel section is shown in Fig. 1. In Fig. 1(a) the overall problem is presented. The channel is subjected to a unidirectional load which will cause buckling. The method of analysis is indicated in Fig. 1(b). The section is effectively considered as a series of linked plates. The unloaded edges on the flanges are rotationally restrained and free while on the web the unloaded edges are both rotationally restrained. This latter case has been studied in detail in Ref. 3, where a box section was considered, and all the appropriate equations given. In this contribution the governing equations for rotationally restrained/free plates will first be presented and then the linking procedure invoked to obtain the analysis of the column.

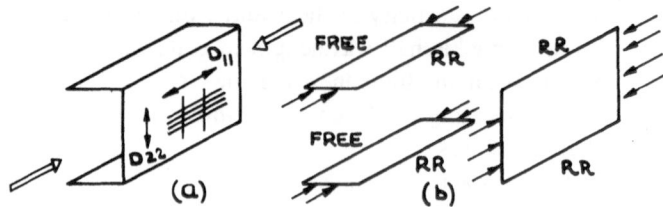

FIG. 1.   Channel analysis.

PLATE EQUATION

The behaviour of a single composite plate is first considered, the fundamental problem being shown in Fig. 2. The plate is uniformly compressed on the loaded ends which are considered to be simply supported. The unloaded edges are elastically restrained against rotation on one edge and free on the other. Also the unloaded edges are stress free while on the loaded ends the shear stress is considered to be zero.

The Buckling of Rotationally Restrained/Free Plates

The detailed analysis of this plate using a semi-energy approach is given in Ref. 4. The results for the buckling of plates with various rotational restraints on one edge varying from the simply supported case to the fixed case were obtained and comparison with existing solutions for particular values showed excellent agreement and gave confidence in the results.

At buckling the deflection for the plate shown in Fig. 2 can be taken in the form

$$w = Y(y)\cos\frac{\pi x}{eb} \tag{1}$$

where e is a measure of the buckle half wavelength in the x direction and is introduced to enable the effect of changing the buckle wavelength to be

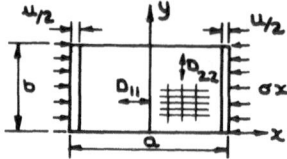

FIG. 2.   Plate analysis.

studied. The function $Y(y)$ are polynomials satisfying the boundary conditions on the unloaded edges.

The critical buckling stress for the plate can be written in the form

$$\sigma_{cr} = \frac{K\pi^2 \sqrt{(D_{11}D_{22})}}{b^2 t} \qquad (2)$$

where K is defined as the elastic buckling coefficient and is a function of the rotational restraint on the plate unloaded edge and the buckle half wavelength.

It has been shown[1] that the coefficient of restraint at the plate edge can be written as

$$R = \frac{\alpha b}{D_{22}} \qquad (3)$$

where α is an elastic constant.

Using the above approach, it is possible to obtain the buckling coefficients for a large number of different plates, i.e. with different aspect ratios and different restraints on the unloaded edge. Relatively simple expressions governing the plate buckling problem can then be obtained. The variation of K with e for a range of different R values is shown in Fig. 3 for a typical unidirectional GRP plate. When $R = 0$ the plate is simply supported, while for high positive values of R the plate can be considered as fully fixed. Note that for negative values of R the plate buckling is being assisted and hence the buckling coefficient is lower than that for a simply supported plate.

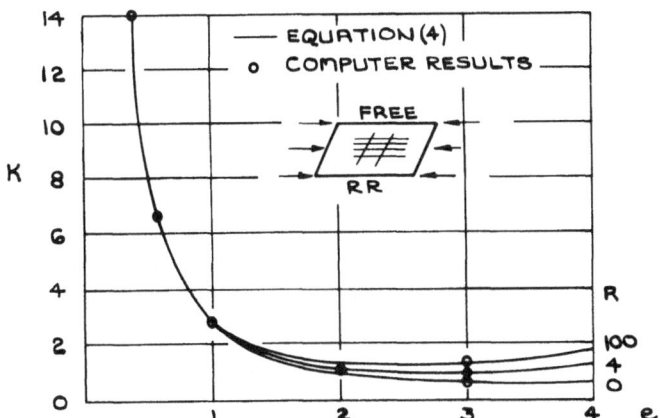

FIG. 3. Comparison of computer and derived results for K–e curves.

The variation of K with e was obtained, as indicated in Ref. 5 in the form

$$K = \frac{K_0 + RQ_1 K_\infty}{1 + RQ_1} \tag{4}$$

where K_0 is the coefficient for a simply supported/free orthotropic plate and is given by

$$K_0 = 0 \cdot 4398 + \frac{2 \cdot 235}{e^2} \tag{5}$$

and K_∞ is the coefficient for a clamped/free plate and is given by

$$K_\infty = 0 \cdot 5439 + 0 \cdot 064\,83 e^2 + \frac{2 \cdot 2313}{e^2} \tag{6}$$

and Q_1 is a function of e given by

$$Q_1 = \frac{1 - 1 \cdot 018\,25 e}{4 \cdot 2048 - 4 \cdot 350\,27 e} \tag{7}$$

Thus, knowing the value of e, the value of K could be obtained from eqn. (4) for any particular value of R. Figure 3 also gives a comparison of results obtained using eqn. (4) with those obtained directly from the computer. The agreement is seen to be excellent and eqn. (4) can therefore be claimed to describe accurately the buckling coefficient for this type of plate.

The Postbuckling of Rotationally Restrained/Free Plates

It is necessary in the postbuckling range to obtain expressions for both relative stiffness at buckling and edge slope coefficient.

Relative stiffness at buckling

Expressions for the ratio of postbuckling to prebuckling stiffness E^*/E were obtained in a similar fashion to that described above for the buckling coefficients. The variation of E^*/E with e for a range of R is given in Fig. 4. The full lines given on this graph are those obtained from the derived equations. Values from the computer output are given at selected points. The form of the expression for the simply supported free plate, i.e. with $R = 0$ is taken as

$$\frac{E_0^*}{E} = 1 - \frac{2}{3 + Ae/(1 + Be)} \tag{8}$$

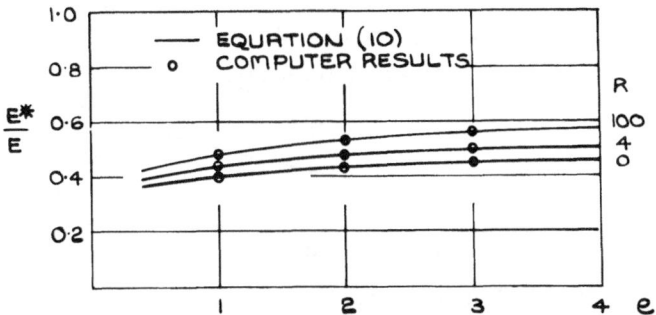

FIG. 4. Variation of relative stiffness with R and e.

By a curve fitting process, A was found to be 0·4526 and B to be 0·4567. For the fixed plate the form of the expression was altered slightly and was found to be of the form

$$\frac{E^*_\infty}{E} = 1 - \frac{2}{3 + 1·22e(1 + 0·4618e)} \tag{9}$$

To take account of the variation with R the general form

$$\frac{E^*}{E} = \frac{\dfrac{E^*_0}{E} + Q_2 R \dfrac{E^*_\infty}{E}}{1 + Q_2 R} \tag{10}$$

was used. Q_2 had to be taken in the form

$$Q_2 = \frac{0·4142e}{e + 1·6} \tag{11}$$

Edge slope coefficient

To satisfy compatibility conditions between linked plates it is necessary to know how the edge slope varies after buckling. It was shown in (3) that the edge slope could be written in the general form

$$\theta \frac{b}{t} = \bar{\theta}(\bar{u} - k)^{1/2} \tag{12}$$

where $\bar{\theta}$ is the edge slope coefficient and depends on e and R. As before the values of $\bar{\theta}$ were plotted for various values of e and R. The results are shown in Fig. 5.

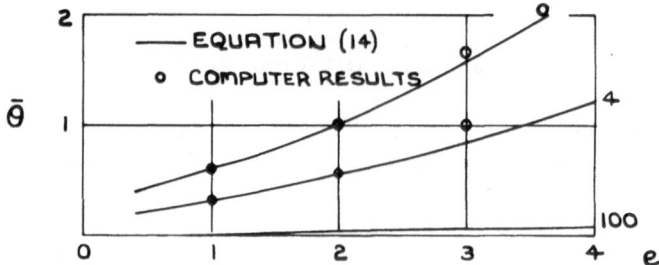

FIG. 5. Variation of edge slope coefficient with R and e.

For the simply supported free case curve fitting produced an equation for $\bar{\theta}_0$ of the form

$$\bar{\theta}_0 = 0 \cdot 1 + 0 \cdot 5e - \frac{0 \cdot 013\,25}{e - 0 \cdot 3896} \qquad (13)$$

As the restraint increases on one edge the value of $\bar{\theta}$ reduces there. This led to an expression of the form

$$\bar{\theta} = \frac{\bar{\theta}_0}{1 + Q_3 R} \qquad (14)$$

with Q_3 given as

$$Q_3 = \frac{0 \cdot 2605e}{0 \cdot 4312 + e} \qquad (15)$$

A comparison of the computer and derived results are also given in Fig. 5.

CHANNEL SECTION ANALYSIS

As indicated earlier the above plate analysis together with that presented in Ref. 3 can be used as a basis for evaluating the behaviour of a channel section. The plates are linked together in such a way that the boundary conditions between adjacent plates are satisfied. The concepts involved in the instability of a section are indicated on Fig. 6. The original angle between the plates is maintained during buckling. In addition the wavelengths of the buckles which occur in all plates simultaneously are the same.

The method of analysis can be indicated by considering two adjacent

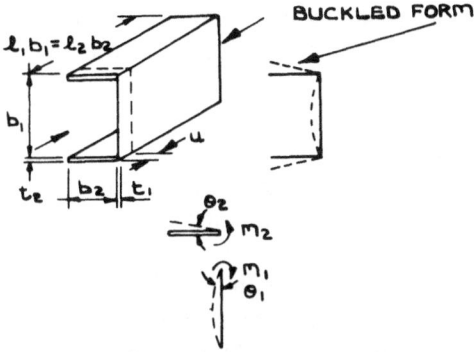

FIG. 6.   Concepts involved in buckling of sections.

plate elements as shown in Fig. 6. The compatibility and equilibrium conditions for the plate edges require that

$$\theta_1 = \theta_2 \quad \text{and} \quad M_1 = M_2 \tag{16}$$

In addition the buckle wavelength for each plate is the same, i.e.

$$e_1 b_1 = e_2 b_2 \tag{17}$$

Introducing the coefficient of restraint defined earlier in eqn. (3) and remembering that α is given by

$$\alpha = \frac{M}{\theta} \tag{18}$$

gives the ratio restraint on plates 1 and 2 using eqn. (16) as

$$R_2 = -R_1 \frac{b_2}{b_1} \left(\frac{t_1}{t_2}\right)^3 \tag{19}$$

The difference in signs of R_1 and R_2 arise because moment M_2 tends to reduce the rotation θ_2 while moment M_1 tends to increase the rotation θ_1.

The subsequent buckling and postbuckling analysis is identical to that given in (3), except that eqn. (30) of that paper has to be replaced by

$$\bar{P} = \bar{P}_1 + 2\bar{P}_2 \tag{20}$$

APPLICATION TO A SPECIFIC CASE

The above equations were derived for a glass reinforced plastic channel with the following general properties.

$$E_{11} = 30\,\text{GN/m}^2 \qquad E_{22} = 6\,\text{GN/m}^2 \qquad G_{12} = 5\,\text{GN/m}^2 \qquad \nu_{12} = 0.33$$

The results of course could be derived for any reinforced plastic composite provided the fundamental mechanical properties of the material were known.

The Buckling of a GRP Channel

The application of the buckling analysis permitted Fig. 7 to be drawn. This shows the variation in the minimum buckling coefficient with b_2/b_1 for various thickness ratios.

It is seen that as b_2 increases, with the other dimensions held constant, the value of the buckling stress rises and then falls away. Initially the flange is stiffening the web but then contributes adversely to the buckling of the section. The stiffening effect, as expected, is more marked with the thicker flange. Also as t_2 reduces in thickness the buckling stress for the section reduces as anticipated for the same b_2/b_1 ratio.

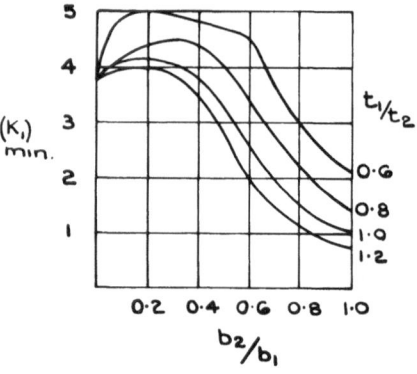

FIG. 7.   Minimum buckling coefficient for GRP channels.

The Postbuckling Behaviour of a GRP Channel

The postbuckling behaviour for three representative GRP channels is given in Figs 8–10. The channel details are given at the top of the figures. The value of e was chosen to give the minimum buckling load. The load–end displacement curves are given in Fig. 8. Figure 9 shows the reduction in panel loads after buckling for channel 1. Δk is the reduction in load of the plate compared with that for an unbuckled plate with the same deflection. It can be seen that for channel 1, panel 1 buckles first with a corresponding reduction in load. As b_2 is increased it is found as expected that panel 2 buckles first followed by panel 1.

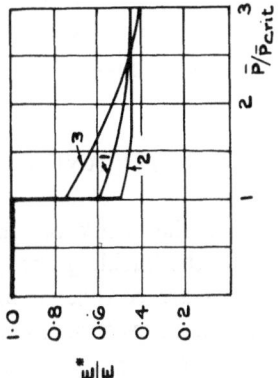

Fig. 10. Variation of postbuckling stiffness.

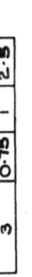

CHANNEL No	b_2/b_1	t_2/t_1	e
1	0·25	1	1·5
2	0·50	1	2·0
3	0·75	1	2·5

Fig. 9. Reduction in plate loads after buckling for channel 1.

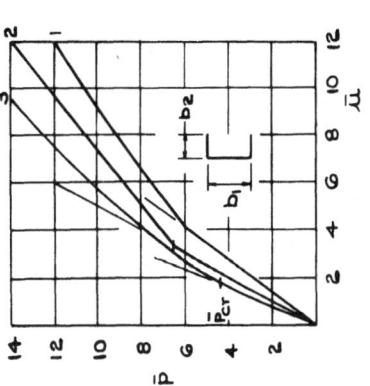

Fig. 8. Load–end displacement curves.

The variation of postbuckling stiffness for each section was obtained from Fig. 8 and is plotted in Fig. 10. It can be seen from this figure that after a sudden reduction in stiffness there is a more gradual fall off. For channel 2 the reduction is almost complete at buckling.

SUMMARY AND CONCLUSIONS

This paper extends earlier work on glass reinforced plastic plates to examine theoretically the buckling and postbuckling behaviour of GRP sections. The sections are considered as a series of linked orthotropic plates. Application is made to channel sections in particular and the critical loads and postcritical behaviour predicted for particular geometries.

The market for reinforced plastic products is continually expanding. This is leading to the structural application of composites in, for example, the aircraft industry. This in turn means that problems solved earlier for isotropic systems need to be re-examined and analysed for the new materials. The work presented in this paper is a contribution in that direction.

REFERENCES

1. BANKS, WILLIAM M., HARVEY, JAMES M. and RHODES, JAMES, The nonlinear behaviour of composite panels with alternative membrane boundary conditions on the unloaded edges, *Proc. 2nd Int. Conf. Composite Materials*, Toronto, April 1978.
2. BANKS, WILLIAM M., Experimental study of the nonlinear behaviour of composite panels, *Proc. 3rd Int. Conf. Composite Materials*, Paris, 1980.
3. BANKS, WILLIAM M. and RHODES, JAMES, The postbuckling behaviour of composite box sections, in: *Composite Structures*, Proc. 1st Int. Conf. Composite Structures (ed. I. Marshall), Applied Science Publishers, 1981, pp. 402–414.
4. BANKS, WILLIAM M., A contribution to the geometric nonlinear behaviour of orthotropic plates, Ph.D. Thesis, University of Strathclyde, 1977.
5. BANKS, WILLIAM M. and RHODES, JAMES, The buckling behaviour of reinforced plastic box sections, *Proc. R.P. Congress*, 1980, Brighton, November.

32

Analysis of a Hybrid, Unidirectional Buffer Strip Laminate*

LOKESWARAPPA R. DHARANI

*Department of Engineering Mechanics, University of Missouri—Rolla,
Rolla, Missouri 65401, USA*

and

JAMES G. GOREE

*Department of Mechanics and Mechanical Engineering,
Clemson University, Clemson, South Carolina 29631, USA*

ABSTRACT

*A method of analysis capable of predicting accurately the fracture behavior
of a unidirectional composite laminate containing symmetrically placed
buffer strips is presented. As an example, for a damaged graphite/epoxy
laminate, the results demonstrate the manner in which to select the most
efficient combination of buffer strip properties necessary to inhibit crack
growth. Ultimate failure of the laminate after crack arrest can occur under
increasing load either by continued crack extension through the buffer strips
or the crack can jump the buffer strips. For some typical hybrid materials it
is found that a buffer strip spacing-to-width ratio of about four to one is the
most efficient.*

INTRODUCTION

One of the major difficulties in designing an advanced composite structure
such as an aircraft to comply with current safety regulations, is meeting the
damage-tolerant (fail-safe) requirements. One very promising method of

* This work was supported by the Fatigue and Fracture Branch, Materials
Division, NASA-Langley Research Center under Grant NSG-1297.

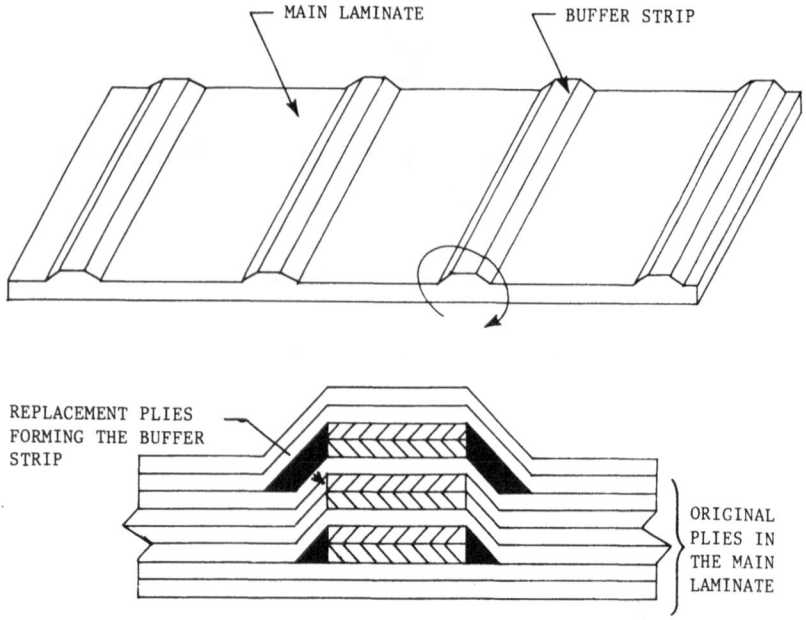

FIG. 1.   A typical buffer strip laminate.

constructing a damage-tolerant composite laminate is to use hybrid, embedded stringers (buffer strips) as a crack arrest mechanism. A typical laminate is shown in Fig. 1 and the geometry assumed for the present study is given in Fig. 2. Two fundamental differences are seen between the real construction and the model; first the model is assumed to consist of only zero degree (parallel to the load) fibers and second, it contains an initial central crack between two buffer strips and two half-planes rather than a

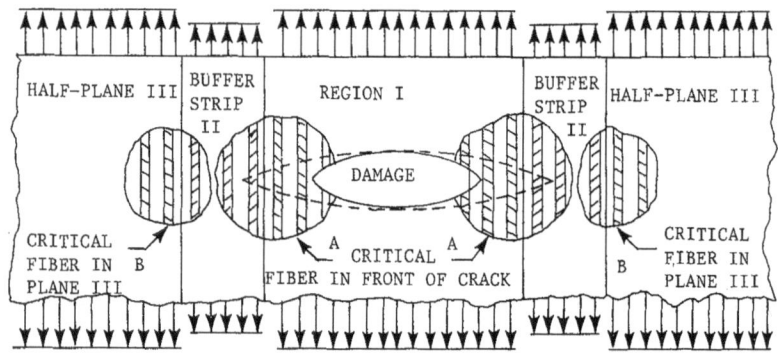

FIG. 2.   Geometry of a symmetric buffer strip.

periodic array of buffer strips. It is felt that much of the characteristic behavior can be represented by the unidirectional laminate, as a dominant portion of the load is carried by these fibers. A primary function of the angle plies in Fig. 1 is to prevent longitudinal matrix splitting in a brittle matrix such as epoxy. This is accounted for to some degree in the present solution by allowing the matrix to support large strains without splitting.

This work is an extension of the studies presented by the authors in Refs 1, 2 and 3 and is the latest solution developed in an attempt to understand the damage tolerant behavior of a buffer strip laminate. The intent is to be able to estimate the remote stress required to fail the hybrid unidirectional laminate of Fig. 2. The fibers and matrix are assumed to be linearly elastic and the failure criterion is simple tension failure of the fibers. The classical shear-lag model is used to represent the shear stress distribution between adjacent fibers. From the previous work[1] it is known that for a single material laminate, without matrix yielding and splitting, the most highly stressed fiber is the first unbroken one directly in front of the notch. The significant question in this study is, if the first fiber in front of the notch breaks at a given applied stress will the next fiber require a higher or lower applied stress to also fail or will a fiber break at some other location at an even lower stress? That is, is the crack growth stable or unstable, and how does this behavior depend on materials and geometry? The shear-lag model and the assumptions and simplifications made in the analysis are somewhat restrictive but the ability of these simple models to represent actual laminate response has been found[1] to be very good. It is then felt that the results given in this paper are a good indication of the behavior of a buffer strip laminate.

The initial studies using the shear-lag model to analyze notched unidirectional laminates were given by Hedgepeth[4] and Hedgepeth and Van Dyke.[5,6] The work of Refs 1, 2 and 3 extends these methods up to the present treatment. Experimental investigations concerning buffer strip laminates are discussed by Eisenmann and Kaminski;[7] Hess, Huang and Rubin;[8] Avery and Porter;[9] Verette and Labor;[10] and Poe and Kennedy.[11] Because of the limited space allowed for this presentation much of the background details and development must be referred to these papers.

FORMULATION

The fundamental solution needed in the analysis of this problem is the case of a unidirectional half-plane with broken fibers and matrix splitting as

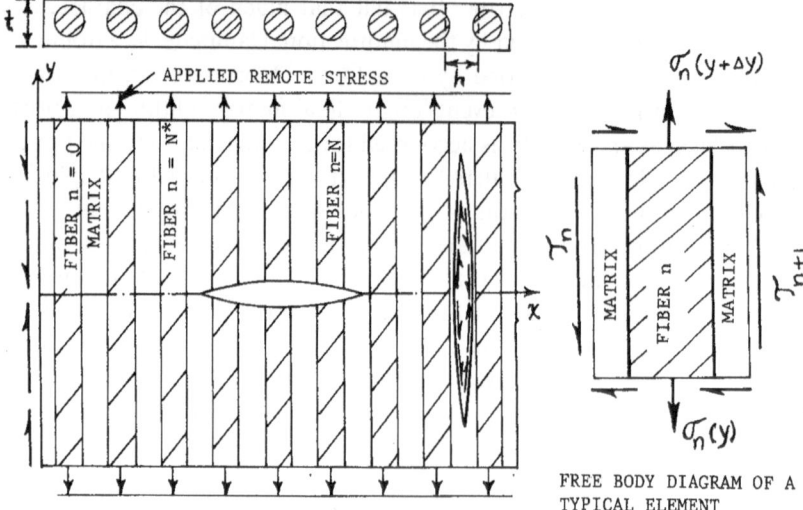

FIG. 3.   Unidirectional half-plane with broken fibers.

shown in Fig. 3. This basic solution will be developed first and then by taking appropriate combinations of particular forms of this result, the complete solution will be presented.

A unidirectional array of parallel fibers with an arbitrary number of broken fibers in the form of a notch and a longitudinal split in the matrix is shown in Fig. 3. The laminate is subjected to prescribed shear stresses $\tau_a(y)$ along the free edge and $\tau_b(y)$ along the split, and a remote uniform tensile strain in the axial direction. Fiber breaks occur along the x-axis (axis of symmetry) and, since the loading is symmetric, only the upper half of the laminate is considered in the analysis.

The fibers are taken to be of much higher strength and extensional stiffness than the matrix and all of the axial load is assumed to be carried by the fibers with the matrix transferring load by shear stresses as given by the classical shear-lag assumption.[4] The axial fiber stress, $\sigma_n(y)$, and matrix shear stress, $\tau_n(y)$, are then given by the simple relations

$$\sigma_n(y) = E_F \frac{dv_n(y)}{dy} \quad \text{and} \quad \tau_n(y) = \frac{G_M}{h} [v_n(y) - v_{n-1}(y)] \quad (1)$$

where $v_n(y)$ is the axial displacement of the fiber n at the location y, E_F is the Young's modulus of the fiber, G_M is the equivalent matrix shear modulus and h is a shear transfer distance. Because of the interference between fibers it is unlikely that G_M will be the homogeneous matrix shear modulus or h the actual fiber spacing. It is pointed out by Dharani and Goree[3] that these values can be determined experimentally for a given laminate. Batdorf[12] also discusses this question in considerable detail.

By virtue of the shear-lag assumption the longitudinal and transverse equilibrium equations become decoupled and the fiber axial displacements and stresses can be obtained without solving the transverse equilibrium equation. Therefore, only the equilibrium equation in the longitudinal (axial) direction will be considered. With reference to the free-body diagram of a typical fiber–matrix region shown in Fig. 3, the equilibrium equations in the longitudinal direction are given by

$$\frac{A_F}{t}\frac{d\sigma_0(y)}{dy} + \tau_1(y) - \tau_a(y) = 0 \qquad \text{for fiber } 0$$

$$\frac{A_F}{t}\frac{d\sigma_n(y)}{dy} + \tau_{n+1}(y) - \tau_n(y) = 0 \qquad \text{for fiber } n$$

$$\frac{A_F}{t}\frac{d\sigma_{NW}(y)}{dy} + \tau_b(y) - \tau_{NW}(y) = 0 \qquad \text{for fiber } NW \text{ when } y \leq l$$

and

$$\frac{A_F}{t}\frac{d\sigma_{NW+1}(y)}{dy} + \tau_{NW+1}(y) - \tau_b(y) = 0 \qquad \text{for fiber } NW+1 \text{ when } y \leq l$$

$$(2)$$

Using the stress–displacement relations, eqn. (1), in the above equilibrium equations, the following set of differential-difference equations is obtained:

$$\frac{A_F E_F h}{G_M t}\frac{d^2 v_0}{dy^2} + v_1 - v_0 = \tau_a(y)$$

$$\frac{A_F E_F h}{G_M t}\frac{d^2 v_n}{dy^2} + v_{n+1} - 2v_n + v_{n-1} = 0$$

$$\frac{A_F E_F h}{G_M t}\frac{d^2 v_{NW+1}}{dy^2} - v_{NW} + v_{NW-1} = -\tau_b(y)$$

and

$$\frac{A_F E_F h}{G_M t}\frac{d^2 v_{NW+1}}{dy^2} + v_{NW+2} - v_{NW+1} = \tau_b(y) \qquad (3)$$

Noting the coefficient of the second derivative term in the above equations, the following changes in the variables are suggested: let

$$y = \sqrt{\frac{A_F E_F h}{G_M t}}\, \eta \qquad \sigma_n = \sigma_\infty \bar{\sigma}_n = E_F \frac{dv_n}{dy} \qquad \text{and} \qquad v_n = \sigma_\infty \sqrt{\frac{A_F h}{E_F G_M t}}\, V_n$$

$$(4)$$

Algebraic manipulation then gives

$$\sigma_n = \sigma_\infty \frac{dV_n}{d\eta} \qquad \tau_n = \sigma_\infty \sqrt{\frac{G_M A_F}{E_F h t}}\,(V_n - V_{n-1}) \qquad \text{and} \qquad l = \sqrt{\frac{A_F E_F h}{G_M t}}\, \beta$$

$$(5)$$

where η, β, $\bar{\sigma}_n$ and $V_n(\eta)$ are non-dimensional.

By making use of Fourier transform techniques[3] the resulting differential-difference equilibrium equations can be written in the form of a single differential equation given by

$$\frac{d^2 \bar{V}(\eta, \theta)}{d\eta^2} - \delta^2 \bar{V}(\eta, \theta) = \bar{\tau}_a(\eta) \cos\left(\frac{\theta}{2}\right) + \langle \eta - \beta \rangle [g(\eta) - \bar{\tau}_b(\eta)] F^2 \quad (6)$$

where

$$V_n(\eta) = \frac{2}{\pi} \int_0^\pi \bar{V}(\eta, \theta) \cos[(n + \tfrac{1}{2})\theta]\, d\theta \qquad \delta^2 = 2[1 - \cos(\theta)] = 4 \sin^2\left(\frac{\theta}{2}\right)$$

$$\bar{\tau}_a(\eta) = \sqrt{\frac{E_F t h}{A_F G_M}}\, \frac{\tau_a(y)}{\sigma_\infty} \qquad \bar{\tau}_b(\eta) = \sqrt{\frac{E_F t h}{A_F G_M}}\, \frac{\tau_b(y)}{\sigma_\infty}$$

$$F^2 = \cos[(NW + \tfrac{1}{2})\theta] - \cos[(NW + \tfrac{3}{2})\theta]$$

$$\langle \eta - \beta \rangle \begin{cases} = 1 & \text{for } \eta \le \beta \\ = 0 & \text{for } \eta > \beta \end{cases}$$

and

$$g(\eta) = V_{NW+1} - V_{NW}$$

The solution to the problem of vanishing stresses and displacements at infinity and uniform compression on the ends of the broken fibers will now be sought. The complete solution is obtained by adding the results corresponding to uniform axial strain and no broken fibers to this solution. As before[3] the solution to eqn. (6) satisfying vanishing stresses and displacements reduces to solving a set of linear algebraic equations, in terms of the unknown Fourier constants B_m, given by

$$\sum_{m=1}^{M} B_m \frac{2}{\pi} \int_0^\pi \cos\left[(N^* + m + \tfrac{1}{2})\theta\right] \cos\left[(n + \tfrac{1}{2})\theta\right] d\theta$$

$$+ \frac{2}{\pi} \int_0^\pi \cos\left(\frac{\theta}{2}\right) \cos\left[(n + \tfrac{1}{2})\theta\right] \int_0^x e^{-\delta t}\bar{\tau}_a(t)\, dt\, d\theta$$

$$+ \frac{2}{\pi} \int_0^\pi F^2 \cos\left[(n + \tfrac{1}{2})\theta\right] \int_0^x e^{-\delta s}\{g(s) - \bar{\tau}_b(s)\}\, ds\, d\theta = 1 \quad (7)$$

for all broken fibers, i.e. $n = 0, \ldots, M$.

The displacement of any fiber n at η is then given by

$$V_n(\eta) = \frac{2}{\pi} \int_0^\pi e^{-\delta\eta} \sum_{m=1}^{M} B_m \cos\left[(N^* + m + \tfrac{1}{2})\theta\right] \cos\left[(n + \tfrac{1}{2})\theta\right] d\theta$$

$$- \frac{1}{\pi} \int_0^\pi \frac{\cos(\theta/2)}{\delta} \int_0^x D(\delta, \eta, t)\bar{\tau}_a(t)\, dt \cos\left[(n + \tfrac{1}{2})\theta\right] d\theta$$

$$- \frac{1}{\pi} \int_0^\pi \frac{F^2}{\delta} \int_0^x D(\delta, \eta, t)\{g(t) - \bar{\tau}_b(t)\}\, dt \cos\left[(n + \tfrac{1}{2})\theta\right] d\theta \quad (8)$$

where $\qquad\qquad D(\delta, \eta, t) = e^{-\delta|\eta - t|} - e^{-\delta(\eta + t)}$

SYMMETRIC BUFFER STRIP LAMINATE

Since the laminate shown in Fig. 2 is symmetric about the x and y axes, only the upper right quadrant will be considered. Figure 4 shows the three distinct regions of the laminate. Regions I and II are finite width unidirectional strips with broken fibers subjected to remote tensile stresses σ_∞^I and σ_∞^{II} and varying shear stresses along the free edges. The solution of these two regions can be obtained by setting the split length equal to infinity in the basic solution obtained in the previous section. The region III is a unidirectional half-plane subjected to uniform remote tensile stress σ_∞^{III} and varying shear stress $\tau_b^{III}(y)$ along the free edge, the solution of which is obtained by setting the split length equal to zero in the basic solution of the previous section. Thus the solutions for all the three regions are known for given applied shear and axial stresses.

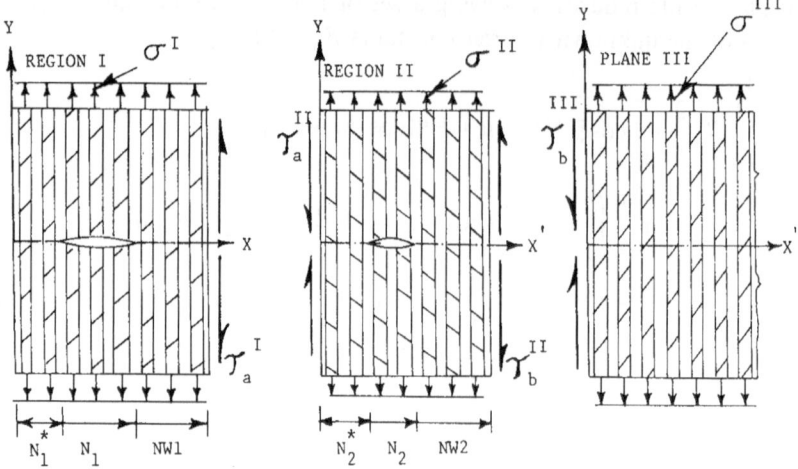

FIG. 4.   Three regions of the buffer strip laminate.

Where these regions are joined together the shear stress is unknown. But from equilibrium, the shear stresses on each of the adjacent regions must be equal at their respective interfaces. Further, as the shear stress is directly related to the distortion of the matrix from the shear-lag assumption, it follows that these stresses must be proportional to the difference in the displacement of the adjoining fibers of the adjacent regions.[3] These conditions, along with the stress boundary conditions on the broken fibers in regions I and II, are used in obtaining the solution for the entire buffer strip laminate. The superscripts I and II indicate the variables in region I and II. Further, $(G_M/h)^{i1}$ and $(G_M/h)^{i2}$ are the ratios of G_M and h for interfaces I and II, respectively. Denoting

$$f^{I}(\eta) = \bar{\tau}_a^{I}(\eta) - g^{I}(\eta) \qquad f^{II}(\xi) = \bar{\tau}_b^{II}(\xi) - g^{II}(\xi)$$

$$F^{I} = \cos[(NW1 + \tfrac{1}{2})\theta] - \cos[(NW1 + \tfrac{3}{2})\theta]$$

$$F^{II} = \cos[(NW2 + \tfrac{1}{2})\theta] - \cos[(NW2 + \tfrac{3}{2})\theta]$$

and

$$C(k) = \cos[(k + \tfrac{1}{2})\theta]$$

the governing equations for the buffer strip laminate can be given as follows:

$$\frac{2}{\pi}\int_0^\pi \left\{ \sum_{m=0}^{M_1} B_m^{I} C(m)\delta - F^{I}\int_0^x e^{-\delta t}f^{I}(t)\,dt \right\} C(n)\,d\theta = 1 \qquad (9)$$

$$\frac{2}{\pi} \int_0^\pi \left\{ \sum_{m=0}^{M_2} B_m^{\text{II}} C(N_2^* + m) \delta + G_{12} \frac{C(0)}{R_1^2} \int_0^x e^{-\delta_1 t} \bar{\tau}_a^{\text{I}}(t) \, dt \right.$$

$$\left. - F^{\text{II}} \int_0^x e^{-\delta s} f^{\text{II}}(s) \, ds \right\} C(j) \, d\theta = 1 \qquad (10)$$

for $n = 0, \ldots, M_1$ and $j = N_2^* + 1, \ldots, M_2$,

$$\bar{\tau}_a^{\text{I}}(\eta) = \frac{1}{\pi} G_{i1} \int_0^\pi \left\{ - \sum_{m=0}^{M_1} 2 B_m^{\text{I}} C(m) e^{-\delta \eta} C(NW1) \right.$$

$$- \frac{1}{\delta} \int_0^x \left\{ F^{\text{I}} D(\delta, t, \eta) f^{\text{I}}(t) C(NW1) + \frac{G_{12}}{R_1} C^2(0) D(\delta_1, t, \eta) \bar{\tau}_a^{\text{I}}(t) \right\} dt$$

$$+ R_1 \sum_{m=0}^{M_2} 2 B_m^{\text{II}} C(N_2^* + m) e^{-\delta_1 \eta} C(0) - F^{\text{II}} \frac{C(0)}{\delta}$$

$$\left. \times \int_0^x D(\delta, s, \eta/R_1) f^{\text{II}}(s) \, ds \right\} d\theta \qquad (11)$$

$$g^{\text{I}}(\eta) = -\frac{1}{\pi} \int_0^\pi \left\{ \sum_{m=0}^{M_2} 2 B_m^{\text{I}} C(m) C(NW1) e^{-\delta \eta} \right.$$

$$\left. + \frac{1}{\delta} \int_0^\infty \left\{ F^{\text{I}} C(NW1) f^{\text{I}}(t) - C^2(0) \bar{\tau}_a^{\text{I}}(t) \right\} dt \right\} d\theta \qquad (12)$$

$$g^{\text{II}}(\xi) = \frac{1}{\pi} \int_0^\pi \left\{ - \sum_{m=0}^{M_2} 2 B_m^{\text{II}} C(N_2^* + m) C(NW2) e^{-\delta \xi} \right.$$

$$+ G_{12} \frac{C(0) C(NW2)}{R_1^2 \delta} \int_0^\infty D(\delta, t/R_1, \xi) \bar{\tau}_a^{\text{I}} \, dt$$

$$\left. - \frac{1}{\delta} \int_0^\infty \left\{ F^{\text{II}} C(NW2) f^{\text{II}}(s) + C^2(0) \bar{\tau}_b^{\text{II}}(s) \right\} ds \right\} d\theta \qquad (13)$$

$$\bar{\tau}_b^{II}(\xi) = \frac{G_{i2}}{\pi} \int_0^\pi \left\{ -2 \sum_{m=0}^{M_2} B_m^{II} C(N_2^* + m) C(NW2) e^{-\delta\xi} + G_{12} \frac{C(0)C(NW2)}{\delta} \right.$$

$$\times \int_0^x D(\delta, t/R_1, \xi) \bar{\tau}_a^I(t) \, dt - \frac{1}{\delta} \int_0^x \left\{ F^{II} C(NW2) D(\delta, \xi, s) f^{II}(s) \right.$$

$$+ G_{23} \frac{C^2(0)}{R_2^2} D(\delta_2, \xi, s) \bar{\tau}_b^{II}(s) \left. \right\} ds \left. \right\} d\theta \tag{14}$$

where M_1 and M_2 are the number of broken fibers in I and II,

$$R_1 = \sqrt{\left(\frac{A_F E_F h}{G_M t}\right)^{II} \left(\frac{G_M t}{A_F E_F h}\right)^I} \qquad R_2 = \sqrt{\left(\frac{A_F E_F h}{G_M t}\right)^{III} \left(\frac{G_M t}{A_F E_F h}\right)^{II}}$$

$$\delta_1 = \delta/R_1 \qquad \delta_2 = \delta/R_2 \qquad G_{i1} = (G_M/h)^{i1} (h/G_M)^I$$

$$G_{i2} = (G_M/h)^{i2} (h/G_M)^{II} \qquad \text{and} \qquad G_{23} = (G_M/h)^{II} (h/G_M)^{III}$$

The above governing equations are of the same form as those obtained in the case of a single buffer strip laminate (Dharani and Goree[3]), except that this problem has an additional integral equation due to the finiteness of the center panel.

SOLUTION

The above six equations (9–14) contain the unknown Fourier constants B_m^I and B_m^{II} and the unknown functions $g^I(\eta)$, $g^{II}(\xi)$, $\bar{\tau}_a^I(\eta)$, and $\bar{\tau}_b^{II}(\xi)$. The solution is developed by representing the integrals containing the unknown functions using a Gauss–Laguerre[3] quadrature formula and reducing the six equations to a single system of equations having as unknowns the Fourier constants and the values of the unknown functions at specific points (quadrature points).

For any continuous, integrable function the Gauss–Laguerre quadrature formula gives

$$\int_0^\infty f(x) \, dx = \sum_{i=1}^K w_i e^{-x_i} f(x_i) \tag{15}$$

where x_i is the ith zero of the Laguerre polynomial, $L_K(x_i)$, and w_i is the corresponding weight function given by

$$w_i = x_i / [(K+1) L_{K+1}(x_i)]^2 \tag{16}$$

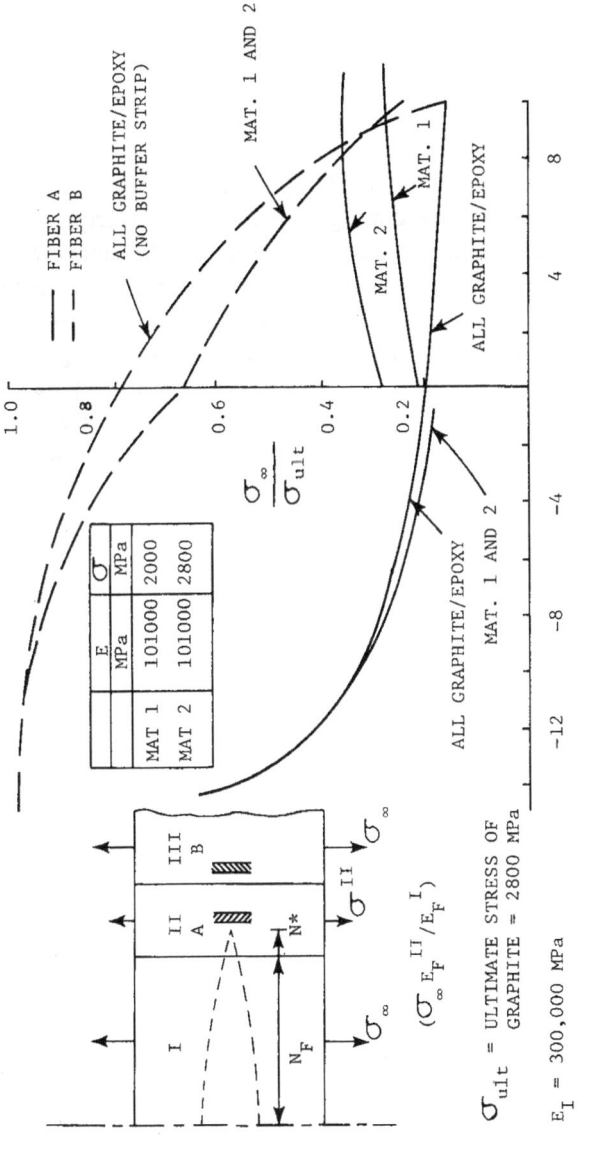

FIG. 5.   Failure stress as a function of crack length.

For the results presented in this paper, forty-five terms ($K = 45$) were taken to represent each of the four unknown functions. Computation time on the Clemson University IBM 3081-K computer was about two minutes for a typical geometry.

RESULTS

Figure 5 presents results corresponding to initial crack growth in region I, crack arrest at the interface, crack growth in the buffer strip and subsequent laminate failure. In these results all fibers are of the same cross-sectional area and in all cases the buffer strips are ten fibers wide and are thirty fibers apart. Two buffer strip materials are considered, each with the same modulus but with different ultimate stresses as shown in Fig. 5. Material 2 has properties close to that of S-glass and the parent laminate is graphite/epoxy. The solid line in Fig. 5 represents the remote stress required to initiate crack extension (failure of the first unbroken fiber in front of the notch, fiber A). The remote stress required to fail the laminate

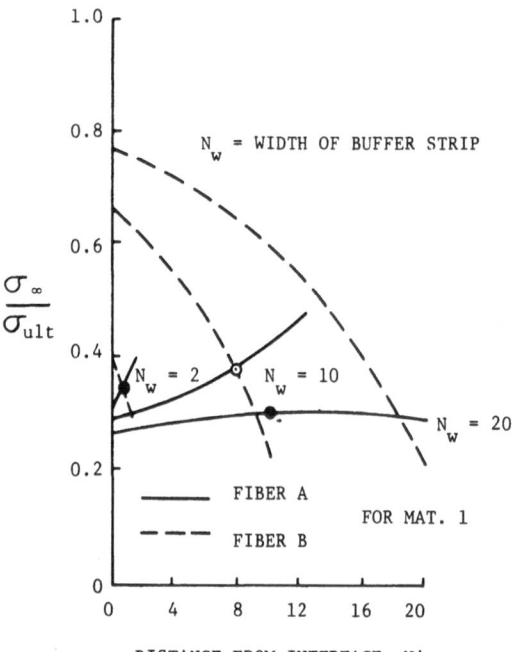

FIG. 6. Effect of buffer strip width on crack growth.

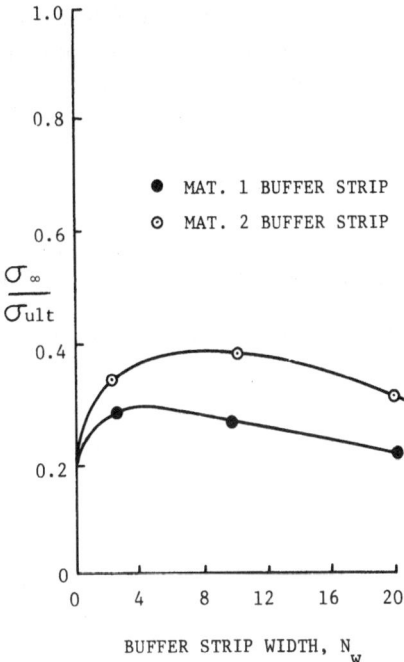

FIG. 7.   Ultimate failure stress versus buffer strip width.

catastrophically (failure of the first fiber in plane III, fiber B) is given by the broken line in Fig. 5. Both these stresses are functions of the initial crack length and decrease with increasing length. Results for an all graphite/epoxy laminate are also given. The crack growth takes place by breaking consecutive fibers from the crack tip to the interface. Then, depending on the stress level required to run the crack to the interface and depending on the buffer strip material, the crack may arrest. Both buffer strip materials require an increasing stress to continue the crack growth in the buffer strip, although material 1 will arrest a crack only if it initiates under fairly low load, i.e. initially close to the interface. For the particular lamina of Fig. 5, all fibers in the material 1 buffer strip fail before fiber B attains its failure stress, whereas for material 2 fiber B fails when there are still some fibers left unbroken, i.e. the crack jumps the buffer strip.

In Fig. 6 the effect of buffer strip width on crack growth for a fixed spacing between buffer strips of thirty fibers is given. The ultimate failure stress of the laminate as a function of buffer strip width is plotted in Fig. 7. From Fig. 7 it is seen that for material 1 the optimum buffer strip width is about 3–4 fibers and for material 2, about 8 fibers. Additional results indicate that one may think of individual fibers as groups of fibers and Fig. 7 then implies that, for a (graphite, S-glass)/epoxy hybrid laminate, the optimum aspect ratio should be about four to one.

REFERENCES

1. GOREE, J. G. and GROSS, R. S., Analysis of a uni-directional composite containing broken fibers and matrix damage, *Engng Frac. Mech.*, **13** (1979), 563–578.

2. DHARANI, L. R., JONES, W. F. and GOREE, J. G., Mathematical modeling of damage in uni-directional composites, *Engng Frac. Mech.*, **17**, No. 6 (1983), in press.

3. DHARANI, L. R. and GOREE, J. G., Analysis of a hybrid, uni-directional laminate with damage, *Proceedings of the IUTAM Symposium on Mechanics of Composite Materials*, Aug. 1982.

4. HEDGEPETH, J. M., *Stress Concentrations in Filamentary Structures*, NASA TN D-882, May 1961.

5. HEDGEPETH, J. M. and VAN DYKE, P., Local stress concentrations in imperfect filamentary composite materials, *J. Comp. Mater.*, **1** (1967), 294–309.

6. HEDGEPETH, J. M. and VAN DYKE, P., Stress concentrations from single-filament failures in composite materials, *J. Textile Res.*, **29** (1969), 618–626.

7. EISENMANN, J. R. and KAMINSKI, B. E., Fracture control for composite structures, *Engng Frac. Mech.*, **4** (1972), 907–913.

8. HESS, T. E., HUANG, S. L. and RUBIN, H., Fracture control in composite materials using integral crack arresters, *Proc. AIAA/ASME/SAE 17th Structures, Structural Dynamics, and Materials Conference*, May 1976, pp. 52–60.

9. AVERY, J. G. and PORTER, T. R., Damage tolerant structural concepts for fiber composites, *Proc. Army Symposium on Solid Mechanics, Composite Materials: The Influence of Mechanics of Fracture on Design*, AMMRC-MS-76-2, 1976.

10. VERETTE, R. M. and LABOR, J. D., *Structural Criteria for Advanced Composites*, AFFDL-TR-142, Air Force Dynamics Laboratory, 1976.

11. POE, C. C., JR and KENNEDY, J. M., An assessment of buffer strips for improving damage tolerance of composite laminates, *J. Comp. Mater. Suppl.*, **14** (1980), 57–70.

12. BATDORF, S. B., Measurement of local stress distributions in damaged composites using an electric analogue, *Advances in Aerospace Structures and Materials*, ASME, AD-03 (1982), 71–74.

33

Basic Failure Mechanisms of Laminated Composites and Related Aircraft Design Implications

R. C. SANDERS, E. C. EDGE and P. GRANT

*British Aerospace PLC, Aircraft Group, Warton Division,
Warton Aerodrome, Preston, Lancashire PR4 1AX, England*

ABSTRACT

A critical literature survey of the major failure criteria of laminated fibre reinforced composites is initially carried out. The failure modes of unidirectional material are then summarized, and related to layer failure modes in multidirectional laminates.

A critical upper bound of matrix strain is identified as being a limiting factor in layer longitudinal tensile strength, and interaction criteria are proposed for combined layer shear and longitudinal tension or compression. It is shown that reasonable knowledge of hydrothermal stresses can be important in predicting laminate failure, and that further work concerning the quantification of these is required.

A strength envelope of the cross-ply laminate is produced showing the significance of the various failure modes.

NOTATION

E Extensional modulus
G Shear modulus
Q Stiffness matrix
S Shear strength across the fibre direction
T Shear strength parallel to the fibre direction

V Volume fraction
X Strength in the fibre direction
Y Strength transverse to the fibre direction
ε Strain
γ Shear strain, when not in tensorial criterion
σ Stress
τ Shear stress, when not in tensorial criterion

Subscripts

c   compression
crit   critical
F   fibre
m   matrix

1. INTRODUCTION

Since the early 1970s the design and manufacture of large scale composite structures at British Aerospace has necessitated the development of an understanding of the basic failure mechanisms of laminated composites.

The inherent inhomogeneity and anisotropic nature of these materials represent a novel challenge to the modern aerospace designer. An appreciation of their manifold failure mechanisms is essential, and a mere extension of metals techniques in testing and design can often lead to erroneous and sometimes dangerous conclusions. Consequently it would appear that the gathering of mere numerical data will not suffice, and that it is essential to obtain at least a qualitative understanding of failure.

The ideal failure criterion is obviously one which requires a minimum of material testing but recognizes the different failure modes. The multiplicity of different possible laminate configurations make the amount of material testing potentially prohibitive, especially if extensive biaxial testing is required. Consequently it is hoped that once the catastrophic failure modes are identified, the problem of quantifying failure of any material system will become practicable.

Further advantages of understanding the basic failure mechanisms are that once the relative importance of each of the various material constituents is appreciated, the effects of modification of manufacturing techniques are better understood and any improvements necessary in the material system can be better appreciated.

2. SURVEY OF EXISTING UNNOTCHED FAILURE CRITERIA

A survey of criteria published prior to 1973 was carried out by Grant and Sanders,[1] where it was shown that the tensor polynomial criterion developed by Tsai and Wu[2] encompassed all other criteria published up to that date. Other criteria surveyed included maximum stress and strain theories, and the common quadratic interaction formulae.

In the simple maximum stress theory layer failure is assumed to occur when either of the longitudinal, transverse or shear stresses reaches its corresponding strength. Similarly the maximum strain theory assumes failure to occur when any one of the strains associated with the axes of the layer reaches its maximum.

The so-called Hill criterion has been developed from a yield criterion suggested by Hill[3] for anisotropic metals having no Bauschinger effect. Hill pointed out that this yield criterion is only applicable when the principal axes of an isotropy are the axes of reference. For a transversely isotropic material, with '1' being the fibre direction, and for the case of plane stress in the 1–2 plane, Hill's original yield criterion can be written in the form:

$$\left(\frac{\sigma_1}{X}\right)^2 + \left(\frac{\sigma_2}{Y}\right)^2 - \left(\frac{\sigma_1\sigma_2}{X^2}\right) + \left(\frac{\sigma_6}{T}\right)^2 = 1 \tag{1}$$

It is in this form that Hill's criterion is frequently used to predict failure of a fibre reinforced layer.

However, the observation has been made[1,2] that this criterion can only be applied with '1' being the fibre direction, since otherwise the material is no longer transversely isotropic. Clearly then, this criterion cannot be applied to gross strength estimation of multidirectional laminates where the out-of-plane strength is independent of in-plane strengths. Layer stresses in multidirectional laminates are obtained using classical laminated plate theory, by rotation of the laminate strain vector to the layer principal axes and multiplication of this result by the layer stiffness matrix. This transformation of strain is incompatible with the fact that the strength criterion is only valid for one set of layer axes. Additionally such a simplistic interaction between fibre and resin dependent properties seems unrealistic. In practical situations the expression, (σ_2/Y), can become unity at very low applied loads, with the layer still able to react to longitudinal load. This is often due to the fact that the transverse strength Y is commonly obtained from the simple transverse tension test where failure occurs at the first occurrence of a crack, whereas in practical laminates load will be transferred across the crack via adjacent layers, and much of the layer will still be capable of reacting to load.

An extension of the Hill criterion was developed by Hoffman,[4] to account for different properties in tension and compression, but this still suffers from the same deficiencies inherent in eqn. (1).

The coefficient $(1/X^2)$ in eqn. (1) has been replaced by $(1/XY)$, making the equation then yield identical answers when employed along or transverse to the fibre direction, but this lacks analytic foundation and many of the previous criticisms are still valid.

The first attempt to develop a failure criterion specifically for fibre reinforced composites was made by Tsai and Wu.[2] This made a minimum of assumptions, and a failure surface in the stress space of the following form was proposed:

$$f(\sigma_k) = F_i\sigma_i + F_{ij}\sigma_i\sigma_j = 1 \qquad (2)$$

It was stipulated that F_i and F_{ij} are strength tensors of the second and fourth rank respectively.

The main important feature of this criterion is its invariance, i.e. it is valid for all coordinate systems once it is valid for one coordinate system.[2] Also its interaction terms between stress components are independent material properties. There are features not possessed by other quadratic approximations such as that represented by eqn. (1). This criterion can be applied to complete laminate or individual layer failure.

However, use of this criterion to predict catastrophic laminate failure assuming the laminate to be a homogeneous whole, presents several practical problems. Categorization of the criterion for any one laminate configuration alone requires extensive testing, including biaxial loading incorporating tubular specimens. Even if the dubious assumption that the criterion is independent of laminate thickness is made, the level of expense for the designer is prohibitive. Additionally the failure surface developed will be a continuous function and as such it is doubtful that changes in failure mode and their origin will be known.

The prediction of ultimate laminate failure using this criterion to identify layer failure is potentially attractive. However, the degree of sophistication required in the laminate stress analysis techniques is at present unattainable and the possibilities of achieving this appear remote.[5] It is also worth noting that maximum stress, Hoffman and tensor polynomial criteria have been shown to yield answers for the combined loading of unidirectional material sufficiently in agreement with each other for design purposes.[6,7]

Wu and Scheublein[5] have attempted to remedy these deficiencies in the quadratic form of the tensor polynomial, in the development of a cubic polynomial criterion. Here layer failure criteria are used to predict optimal biaxial tests for evaluation of the strength tensors. However, the amount of testing required still appears to be prohibitive, and whilst the cubic form allows a more general mathematical representation of the failure surface, recognition of true failure modes is still absent.

Tennyson *et al.*[8] have analysed the failure of glass–epoxy and graphite–epoxy tubes subjected to internal pressure. The cubic tensor polynomial is shown to give a better fit to the experimental data than the quadratic form, but significantly overpredicts in the region of optimal fibre angles, i.e. 45–60°.

Nuismer, in recognizing the need for using a more realistic layer transverse stiffness in laminate stress analysis, has used an assumed model, where the stiffness eventually diminishes to zero at some ultimate strain level, considerably beyond that achieved in the simple transverse tension test.[9] Although residual stresses are neglected, and the model seems a little too artificial this approach appears to be the correct one.

Much knowledge is being gained considering the mechanisms of unidirectional failure, including the influence of residual stresses, and ultimately a comprehensive understanding of laminate failure will probably include this along with realistic laminate stress analysis, and a tensor polynomial failure criterion.

3. DEVELOPMENT OF BRITISH AEROSPACE FAILURE CRITERION

The British Aerospace understanding of failure mechanisms of laminated composites is the result of an ongoing test and analytical programme of work which started in the early 1970s. The results of this programme up to 1979 have been summarized by Sanders *et al.*[10,11]

In view of the lack of a complete model, and the present intractable nature of the problem, the present approach to quantifying and understanding laminate failure has to be essentially semi-empirical. Initially an understanding is obtained of the various modes of failure of unidirectional material subjected to uniaxial loading. It is believed that this is a necessary prerequisite to the study of layer failure under combined loading in a multidirectional laminate.

3.1. Failure Modes of Unidirectional Material

It is common knowledge that in the tensile failure of glass or carbon–epoxy, individual fibre fractures occur at levels well below the ultimate tensile strength,[12] and that this can be some 1·5 times that of the fibre bundle strength.[13] When a fibre fractures, load is transferred from the fracture site to an adjacent fibre via the matrix. Consequently the fractured fibre has a small ineffective length local to the fracture site. Ultimate failure occurs when there are sufficient fibre fractures across any one section to reduce the net section strength to that of the applied load. This section must have a critical 'band width' of length approximating to that of the fibre ineffective length such that all the fibre fractures lie within this band. Alternatively failure can occur where isolated fracture sites are linked by shear failure of the resin and/or fibre-resin bond, parallel to the fibres. It is clear that longitudinal tension strength is very much influenced by the resin and fibre–resin bond strength.

Longitudinal compression failure has received much attention in the literature since the work by Dow and Gruntfest[14] in 1960. The basic problem is that the upper bound of failure which is approximately equal numerically to the longitudinal tensile strength is very rarely achieved, the largest discrepancies occurring in adverse environmental conditions. The lower failure mode is generally considered to be one of small wavelength fibre instability, which is analogous to that of the buckling of a column on an elastic foundation. The two buckling modes illustrated in Fig. 1 are normally considered, and a simple strain energy evaluation of these was carried out by Dow and Rosen,[15] where the following two equations were derived:

for the out-of-phase mode:

$$\sigma_{1c} = 2V_F \left[\frac{V_F E_m E_F}{3(1 - V_F)} \right]^{1/2} \tag{2}$$

and for the in-phase mode:

$$\sigma_{1c} = \frac{G_m}{1 - V_F} \tag{3}$$

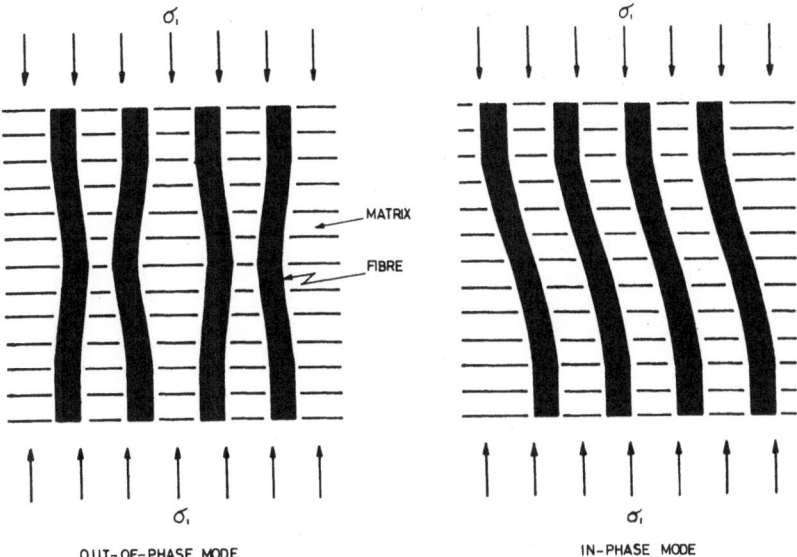

FIG. 1. Microbuckling modes.

For most composites and carbon–epoxy in particular the shear mode represented by eqn. (3) represents the lower energy condition. However, correlation between this ideal model and test is rarely achieved, and, many attempts to incorporate practical effects such as variable fibre packing density and resin modulus, fibre misalignment, fibre-resin bond defects, etc., have failed to satisfactorily explain the theory–test discrepancies, particularly when the material is subjected to adverse environmental conditions. Strictly microbuckling analyses do not predict catastrophic failure, but more probably localized yielding.

An ultimate failure mode that has been observed in failed unidirectional compression specimens is that of 'kink-band' failure, which is shown diagramatically in Fig. 2. This failure mode appears to be a form of gross, shear instability, and a simple analytical treatment of this mechanism has been carried out by Sanders and Grant.[16] Here the compressive stress at this instability level is shown to be equal to the resin shear modulus as follows:

$$\sigma_{1c} = G_m \tag{4}$$

This equation correlates much better with test results[17] than other apparently more sophisticated models. A typical resin shear modulus is $1200\,\text{N/mm}^2$ (BSL 914).

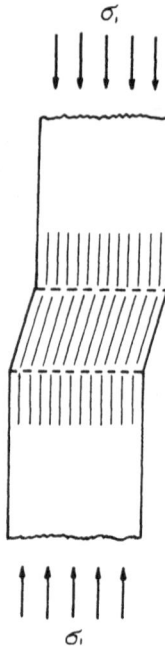

FIG. 2. Kink-band failure mechanism.

Despite the absence of a comprehensive criterion for longitudinal compression failure, it would appear certain, that in most instances, this failure mode is extremely sensitive to resin moduli, and in particular the shear modulus.

Transverse and shear failure modes are also extremely resin sensitive. In the former transverse tension strength is known to exhibit high variability,[17] to such an extent that it can be extremely difficult to obtain meaningful design values. Transverse compression failure, although again resin dominated, occurs at a much higher stress,[17] which can be three times that in tension. Clearly compression failure is not as sensitive to initial flaws and local inhomogeneities, since cracks will not propagate as readily, and in most cases load will be transferred across cracks through bearing and friction.

Shear failure of unidirectional material normally occurs parallel to the fibres through the resin and fibre–resin interface, and the integrity of the latter may be an important factor in the shear strength. An attempt to understand the nature of this was made by Sanders and Grant,[16] in experimental observations of non-linearity in shear, using $\pm 45^\circ$ coupon specimens subjected to longitudinal tension. It was concluded that breakdown of this bond was responsible for most of the non-linearity and that the nature of the bond was probably largely frictional.[16]

Failure across the fibre normally only occurs in multidirectional laminates, when the shear component parallel to the fibres is reacted by adjacent layers in alternative orientations.

3.2. Basic Failure Modes of Multidirectional Laminates

The present approach at British Aerospace to predicting failure is based upon a semi-empirical quantification of layer failure and/or delamination. This is considered to be justified by the fact that in most structural laminate configurations catastrophic layer failure is synonymous with total laminate collapse, and the fact that in structural design, laminate integrity can no longer be ensured if any layer is incapable of carrying load in the fibre direction.

One of the main effects of the laminate environment upon layer failure is to impose extraneous stresses due to both Poissson deformation and inhibition of layer free hydrothermal expansion. Additionally interlaminar stresses are created due to load transfer across cracks and other free boundaries.

The following is a summary of what are considered to be the basic layer failure modes in multidirectional laminates.

3.2.1. Layer tensile failure

The upper bound of longitudinal tensile failure obtained from the pure unidirectional test is not normally fully achieved in multiangular laminates. However, there are many cases where for normal design purposes this can be considered as a valid failure level. These normally occur in uniaxial loading of the (0, ±45) family of laminates. However, it has been observed that the presence of 90° layers in this laminate can result in somewhat lower failure strain levels, even when the applied load is pure tension on the 0° axis. In this instance the lowest failure level is obtained in the 0/90 laminate.[16]

Valid biaxial test data is difficult to achieve and is not often quoted in the literature. However, a lengthy programme of work at British Aerospace,[11] has resulted in the development of a sophisticated tubular specimen and loading rig suitable for the application of all forms of combined in-plane loading. One of the more interesting data points so far obtained from the use of this facility is the failure level of a ±45° or 0°/90° laminate subjected to in-plane hydrostatic tensile stress. Initial attempts to predict this failure level proved abortive,[18] where the laminate failed with a longitudinal fibre tensile stress some 30 % below that obtained in the pure unidirectional tension test. It is worth noting that it is this case that has resulted in the most significant test theory discrepancy, employing tensor polynomial criteria.[8]

These low failure levels observed in 0/90 laminates prompted an assessment of the role of the matrix in longitudinal tensile failure.[16] It has previously been assumed that resin cracking would never be the primary cause of failure in multiangular laminates,[10] since cracking can occur at very low load levels, and has even been observed in laminates at room temperature due to thermal stresses alone, without apparently affecting utlimate failure.

An assessment has been made of the condition of the matrix in the 0/90 laminate at ultimate laminate failure, assuming the matrix to be fully effective between cracks. If ε is the layer strain matrix, the nominal matrix stresses are given by

$$\sigma_m = Q_m \varepsilon \tag{5}$$

However, it is considered that a better assessment of matrix failure is made by an examination of strain levels. The nominal layer strains do not readily reflect the restraining influence of the fibres upon matrix Poisson deformation which can be quite high for typical epoxy resins, which have exhibited Poisson ratios of 0·4 or greater.[19]

It has been proposed that a valid assessment of gross matrix collapse can be made by examination of what has been referred to as the 'true' matrix strain, ε^T.[16] This is defined as the strain in any direction which when multiplied by the relevant extensional modulus will yield the stress, i.e.

$$\varepsilon^T = \frac{\sigma}{E} \tag{6}$$

Written in matrix notation the matrix 'true' strain can be obtained from the layer nominal strain as follows:

$$\varepsilon^T = N\sigma = M\varepsilon \tag{7}$$

where N is a diagonal matrix with

$$N_{11} = N_{22} = \frac{1}{E_m} \quad \text{and} \quad N_{33} = \frac{1}{G_m}$$

and the elements of M are thus dependent upon the matrix Poisson ratio only.

A typical uniaxial failure stress at room temperature for the 0/90 laminate manufactured from HTS/BSL 914 material, is 900 N/mm^2.[18] The maximum value of ε^T at this stress is 0·0217 and occurs along the fibre axis in the 0° layers and transverse to the fibre axis in the 90° layers. If the material is assumed to be in the dry condition, the thermal component of this strain is 0·0056. The ±45° laminate manufactured from this material has failed on test under in-plane hydrostatic tension at a level of 650 N/mm^2. Again evaluation of the 'true' matrix strain at this failure level yields a value of 0·0218.

It has been shown that the matrix plays an important role in unidirectional tensile failure, by redistributing load in areas of broken fibres. Consequently it is believed that there exists an upper bound of gross matrix strain at which the matrix ceases to become effective in this load transfer mechanism and the fibres then become effectively a 'loose bundle', and that the layer will no longer support tensile load at this strain level. It is considered that for contemporary carbon–epoxy material the 0°/90° tension test at room temperature will yield this critical strain. Clearly a proper understanding of the true nature of this failure mechanism requires detailed analytical analysis at the micromechanical level. However, in the light of present knowledge it is considered that a gross estimate of 'true' resin strain from the 0°/90° test is of great importance for design and assessment of the material system.

It should be noted that thermal strain at room temperature forms a significant portion of the total strain. At this temperature strains created by the inhibition of expansion due to absorbed moisture will relieve the thermal strains. Consequently a true assessment of strain should incorporate an allowance for net hydrothermal deformation. It has been common practice in analysis to treat moisture and thermal deformation as being analogous. However, it is extremely difficult to quantify true moisture levels, and also it is believed that there may be interaction effects between thermal and moisture created deformations. It has been considered necessary to initiate a programme of work to quantify net hydrothermal deformations, and this is at present in progress.

3.2.2. Layer compression failure

Generally it is considered that the 'kink-band' failure mechanism observed in pure unidirectional compression failure will still apply. However, in this case the failure level obtained in the pure unidirectional test will most likely constitute a lower bound since this is an instability mode, and in a multidirectional laminate this will be inhibited by support given by other layers.

Since the mode of failure is a form of shear instability one would expect that laminates exhibiting higher shear stiffness will provide higher failure levels. This has been supported by test data on multidirectional laminates, where the lower failure stresses have been observed in 0/90 laminates, and the higher failure stresses in $0/\pm45$ laminates. Indeed some of the failure stresses in the $0°$ layers in the latter have approached the longitudinal tension strength. However, the results have tended to lack consistency, and occurrence of the lower failure bound always appears to be a possibility.

At present it is extremely difficult to quantify the supporting influence of the other layers, and the failure mode is sensitive to slight eccentricity of load. Consequently the lower failure mode should always be anticipated in design.

Transverse compression failure is unlikely in structural multidirectional laminates, since the ultimate failure strain ($\simeq 3\%$) is much higher than that of other failure modes. Additionally transverse compression stress has frequently, initially, to relieve residual thermal stresses.

3.2.3. Layer shear failure

This failure mode is best observed in $\pm45°$ laminates subjected to pure longitudinal tension. The loading on each layer is almost pure shear, equal to half the applied tension stress. Transverse tension stresses exist sufficiently large to create cracks, but these have a negligible effect upon the essential aspects of shear behaviour.[16]

Local shear failure parallel to the fibres can occur at very low load levels, probably helped by the transverse tensile stresses and residual thermal stresses. However, ultimate failure normally occurs when the shear stress approximately equals the interlaminar shear strength at a longitudinal strain of approximately 3%.[16,18] Examination of failed specimens of this type has indicated both shear failure parallel to the fibres and shear failure across the fibres. It is believed that when shear parallel to the fibres occurs, this component of shear is reacted in adjacent layers across the fibres, where the fibres eventually fail as short beams in shear.[16] Ultimate failure is thus less resin dependent than initial thoughts have indicated. This failure mode is obviously dependent upon interlaminar integrity, and failed specimens have exhibited a degree of delamination.

A different level of failure some 30% higher has been observed at British Aerospace. Examination of failed specimens has suggested that failure was precipitated by matrix disintegration followed by a tearing apart of the fibres. It appears that, lacking the support of the matrix the fibres have failed in flexure and/or tension.

3.3. Delamination

For most structural laminates this is not normally a catastrophic failure mode. However, for design purposes it is advisable to have some knowledge of the loading level at which this is likely to occur.

Delamination is created by load transfer at free boundaries and the mechanisms of this have received extensive attention in the literature, the results of which are at present unclear, and much further work is still required. Away from the edges of structure, delamination is normally precipitated by tensile loading, where cracks and defects are traversed by out-of-plane load diffusion. In compression these cracks and defects become less significant, as a large proportion of load can be transferred through friction.

Catastrophic failure due to delamination has been observed in the stratified $\pm 45°$ laminate subjected to tensile loading.[16] In this laminate $(+45_4, -45_8, +45_4)$, an approximate estimate of delamination strength has been obtained, neglecting edge effects.[16]

3.4. Interaction Between Failure Modes

Test data has suggested that the influence of layer transverse loading upon longitudinal failure is negligible until the critical ultimate matrix strain is achieved in tension. Biaxial tests have shown that the major interaction effects occur between shear and longitudinal tension or compression.[16]

3.4.1. Longitudinal tension–shear interaction

Biaxial test data has suggested the possibility of significant interaction effects when the matrix strain exceeds 70 % of its critical value in tension.[16] This data which has been obtained from biaxial tests on $\pm 45°$ tubular specimens is shown in Fig. 3.

Exact analysis of this failure mode requires a detailed micromechanical analysis of resin and fibre–resin bondline stresses, and this at present appears impracticable. Hence for present purposes an interaction criterion based upon layer nominal shear strain and resin strains has been proposed[16] as follows:

$$\left(\frac{\gamma_{12}}{\gamma_{12\,\text{ult}}}\right)^2 + \left(\frac{\varepsilon_m}{\varepsilon_{m\,\text{crit}}}\right)^2 = 1 \cdot 0 \tag{7}$$

It can be seen from Fig. 3 that failures represented by eqn. (7) tend to give a conservative estimate of failure.

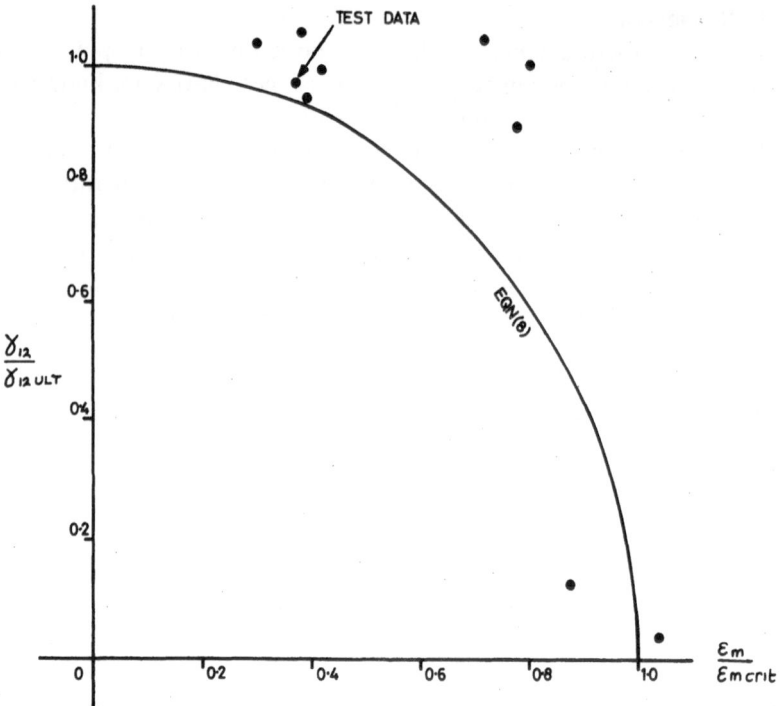

FIG. 3.   Proposed layer tension–shear interaction.

3.4.2. Longitudinal compression–shear interaction

Since in most cases layer longitudinal compression failure is a form of shear instability significant interaction between these two failure modes is not unexpected.

It is assumed that over the narrow kink-band, the fibre kink due to compression loading applies an average shear stress τ_c, which gives a local shear strain, γ_{12} ($= \tau_c/G_m$). The following simple interaction criterion has been derived:[16]

$$\frac{\tau_{12}}{T} + \frac{\sigma_1}{G_m} = 1 \cdot 0 \tag{8}$$

where absolute values of τ_{12} and σ_1 are employed.

It has been shown,[16] that in multidirectional laminates shear failure parallel to the fibres is not a valid catastrophic failure mode. Consequently use of the interlaminar shear strength, T, in this expression is not strictly correct. In multidirectional laminates the upper bound of shear strength is

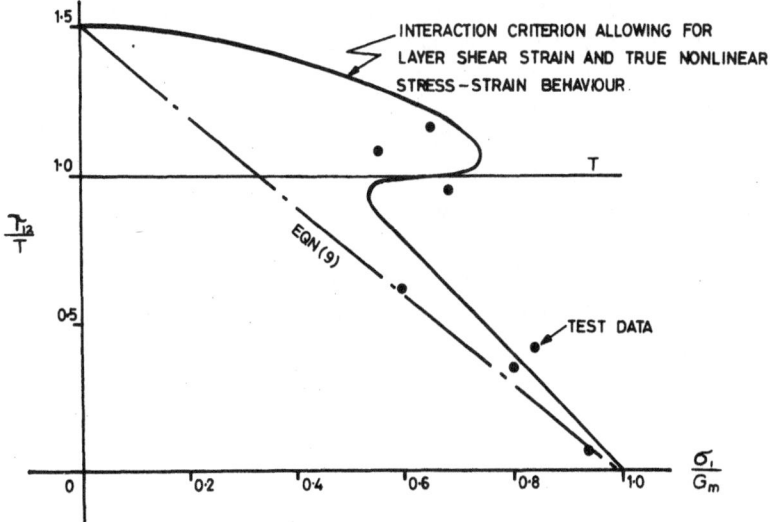

FIG. 4.   Proposed layer compression–shear interaction.

achieved by shear across the fibres, and the correct shear strength to be used here, must lie somewhere between these two limits. Consequently eqn. (8) is rewritten as follows:

$$\frac{\tau_{12}}{K \cdot T} + \frac{\sigma_1}{G_{\mathrm{m}}} = 1 \cdot 0 \qquad (9)$$

where $1 < K < S/T$.

This expression is compared with biaxial test data in Fig. 4, employing a value of K of $1 \cdot 5$. Also shown in Fig. 4 is a curve representing a more complex relationship[16] allowing for shear strain in the layer due to applied shear, and employing a true non-linear shear stress–strain relationship. It can be seen that this latter curve shows a reasonable agreement with test data, but the more simple expression gives a more conservative estimate of failure.

4.   APPLICATION OF STRENGTH CRITERIA

A technique for strength estimation of laminated composites based upon the foregoing failure criteria has been developed at British Aerospace.[16] Layer failure using these criteria is generally assumed to constitute

catastrophic laminate failure, since for most structural laminates this has
been shown to be true. Classical laminated plate theory is employed with a
limited assessment of non-linearity. This, at present, crude non-linear
capability assumes a very low transverse tensile modulus for layers loaded
in transverse tension, and use of a secant layer shear modulus equal to that
normally achieved in a tension test of $\pm 45°$ laminates at the failure strain
of unidirectional material. In addition to fibre dominated failures good
agreement has been obtained with test data, where the failures are
significantly influenced by the matrix, e.g. $90°/\pm 45°$ loaded in pure
uniaxial tension or compression.

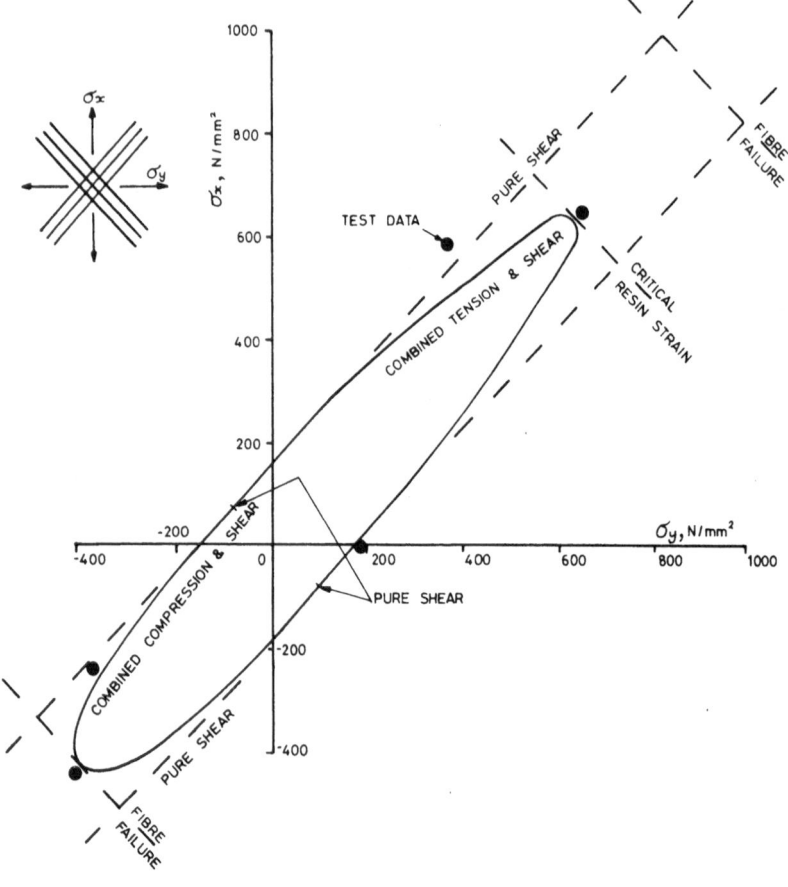

Fig. 5. Layer failure modes in the strength envelope of the $\pm 45°$ laminate.

For characterization of the various criteria it is considered that, having established the basic mechanisms, simple uniaxial testing will suffice.

In addition to basic fibre unidirectional tension and compression tests it is considered that the following key coupon tests are necessary as a basic minimum to categorize the material system:

(i)   Tension test on a $0°/90°$ laminate.
     This should yield the critical true matrix strain, $\varepsilon_{m\,crit}$.

(ii) Tension tests on both a well distributed and stratified $\pm 45°$ laminate. Here the former is intended to yield ultimate shear strength and stress–strain behaviour in shear and the latter an approximate estimate of delamination strength.[18]

In addition to these tests both the resin Poisson ratio and shear modulus, and net hydrothermal expansion characteristics of a unidirectional material are required.

An evaluation of the failure modes of the $\pm 45°$ laminate subjected to biaxial loading, using the foregoing criteria has been made. The various failure modes of this laminate are compared with test data in Fig. 5. The strength envelope for this laminate is composed of all the basic failure modes. Of particular importance is the significant strength increase from that in the uniaxial stress state to that in the combined stress state. This paradox is clearly explained by the change in failure mode from layer shear to pure longitudinal fibre, and combined longitudinal and shear failure. The ultimate longitudinal strength in plane hydrostatic tension is never fully achieved due to the high resin strains initiating total collapse of the resin-to-fibre supporting mechanism.

5.   CONCLUSIONS

A semi-empirical technique evaluating the basic failure mechanisms of laminated fibre reinforced composites has been developed. Various layer failure criteria have been proposed, and incorporated in a strength prediction technique employing classical laminated plate theory.

The more significant criteria developed include the identification of a critical matrix strain in tension and interaction between layer longitudinal and shear loading. The former indicates a crucial need for better quantification of residual hydrothermal stresses and requires further work at the micromechanical level. The layer compression mode and its interaction with shear, indicate that caution should be employed in the interpretation of laminate compression data, where high degree of shear support has been given to the compressively loaded layers.

In addition to the significantly resin influenced failure modes of compression and shear, it has been shown that tensile failure can be initiated by high resin strains, and this should be noted when new material systems or manufacturing techniques are being evaluated.

ACKNOWLEDGEMENT

The authors would like to acknowledge the fact that much of the work reported in this paper has been supported by the Ministry of Defence.

REFERENCES

1. GRANT, P. and SANDERS, R. C., *Strength Theories of Failure for Laminated Composite Materials*, British Aerospace, SON(P) 105 (Warton) Department, 1973.
2. TSAI, S. W. and WU, E. M., A general theory of strength for anisotropic materials, *J. Comp. Mater.*, **5** (1971), 58–80.
3. HILL, R., *The Mathematical Theory of Plasticity*, Oxford, 1956.
4. HOFFMAN, O., The brittle strength of orthotropic materials, *J. Comp. Mater.*, **1** (1967), 200–206.
5. WU, E. M. and SCHEUBLEIN, J. K., *Composite Materials: Testing and Design*, 3rd Conference, ASTM STP 546, 1974, pp. 188–206.
6. PIPES, R. B., Micromechanical models for the stiffness and strength of fibre composites, *Practical Considerations of Design, Fabrication and Tests for Composite Materials*, AGARD lecture series no. 124, September 1982.
7. NARAYANASWAMI, R. and ADELMAN, H. M., Evaluation of the tensor polynomial and Hoffman strength theories for composite materials, *J. Comp. Mater.*, **11** (1977), 366–377.
8. TENNYSON, R. C., MACDONALD, D. and NANYARO, A. P., Evaluation of the tensor polynomial failure criterion for composite materials, *J. Comp. Mater.*, **12** (1978), 63–75.
9. NUISMER, R. J., Predicting the performance and failure of multidirectional polymeric matrix composite laminates: a combined micro-macro approach, *3rd International Conference on Composite Materials*, Paris, 26–29 August, 1980.
10. SANDERS, R. C., RHODES, F. E. and TAIG, I. C., *Final Report on MOD Carbon Fibre Composites*, Data Compilation Contract no. K43a/63/CB 43 A2, British Aerospace, SOR(P)111 (Warton) August 1976.
11. SANDERS, R. C. and TAIG, I. C., *Final Report on MOD Carbon Fibre Composites, Basic Technology*, programme contract no. K/LR32B/2327, British Aerospace, SOR(P) 120 (Warton) September 1979.

12. ROSEN, B. W., Tensile failure of fibrous composites, *AIAA J.*, **2** (1964), 1985–1991.
13. 'T HART, W. G. J., JACOBS, F. A. and NASSETTE, J. H., *Tensile Failure of Unidirectional Composite Materials*, NLR TR 751140, NLR Amsterdam, The Netherlands, July 1975.
14. DOW, N. F. and GRUNTFEST, I. J., *Determination of Most Needed Potential Possible Improvements in Materials for Ballistic and Space Vehicles*, TIS R60SD389, General Electric Co., Space Sciences Lab., June 1960.
15. DOW, N. F. and ROSEN, B. W., *Evaluation of Filament-reinforced Composites for Aerospace Structural Applications*, annual report, NASA Contract NASW-817, October 1964.
16. SANDERS, R. C. and GRANT, P., *The Strength of Laminated Plates under In-plane Loading*, British Aerospace, SOR(P)130 (Warton), January 1982.
17. EASTHAM, J., SHAW, C. and HINNELLS, G., Final report on MOD contract no. K/LR32B/2126—*In Depth Evaluation of Unidirectional BO SC 10,000/ Fibredux 914C Carbon Fibre Composites*, British Aerospace, MDR 0182 (Warton), June 1979.
18. CURTIS, A. R. and GRANT, P., *Multi-angular Laminate Failure Prediction*, British Aerospace, 852/237 (Warton), March 1979.
19. WOOLSTENCROFT, D. H., *The Compressive Behaviour of Unidirectional Carbon Fibre Reinforced Plastic*, Ph.D. Thesis, CNAA Preston Polytechnic, England, June 1981.

34

The *J*-Integral as a Fracture Criterion for Composite Materials

B. D. Agarwal, Prashant Kumar and B. S. Patro

Department of Mechanical Engineering, Indian Institute of Technology Kanpur, Kanpur 208 016, India

ABSTRACT

Fracture behaviour of randomly oriented short glass fibre reinforced epoxy resin has been investigated. Fracture tests were conducted on single edge notched specimens and the J-integral evaluated using energy rate interpretation. Its value is found to be independent of crack length when crack length-to-specimen width ratio (a/w) is larger than 0·35. For smaller cracks, general material damage away from the crack tip also influences the energy absorbed significantly. An extrapolation method has been developed to separate the crack tip energy from the energy absorbed due to general material damage. The J-integral thus obtained is independent of crack length and specimen length and its critical value is the same as obtained for a/w > 0·35 without extrapolation. It also agrees well with the critical stress intensity factor obtained in an earlier study using the R-curve approach.

INTRODUCTION

Most of the research work on fracture of composite materials has, so far, centred around linear-elastic fracture mechanics concepts employing elastic analysis of the crack tip region.[1−12] There are practical difficulties in accurately analysing the crack tip region even for homogeneous isotropic materials and more so for heterogeneous composites. A characterization of the crack tip area by a parameter calculated without focusing attention directly at the crack tip would provide a more useful method for analysing fracture. The path independent J-integral proposed by Rice[13] is such a parameter. Its value depends on the near tip stress–strain field. However, the path independent nature of the integral allows an integration path, taken sufficiently far from the crack tip, to be substituted for a path close to the crack tip region. Therefore, the J-integral can be calculated using numerical methods more accurately compared to the stress intensity factor. Also, an experimental evaluation of the J-integral can be accomplished quite easily by considering the load deflection curves of identical specimens with varying crack lengths.

The use of the J-integral as an elastic–plastic fracture criterion has been discussed by Broberg[14] from an analytic standpoint. A justification for choosing this parameter as a fracture criterion comes from a consideration of the Hutchinson–Rice–Rosengren (HRR) crack tip model[15,16] where the product of plastic stress and strain is shown to have a $1/r$ singularity; r is a near tip crack field length parameter. For a deformation plasticity theory, McClintock[17] has demonstrated, through the crack tip plastic stress and strain equations expressed from the HRR singularity, the existence of a singularity in r whose strength is the J-integral. Thus, the J-integral may be chosen as a parameter to characterize the crack tip environment because it can be evaluated experimentally and calculated with less difficulty than the plastic stress and strain intensity factors.

Begley and Landes[18–20] discussed various aspects of using the J-integral as a failure criterion for metals. They demonstrated the applicability of the J-integral for the case of large scale plasticity at the crack tip through experimental results on an intermediate strength rotor steel for which the J-integral at failure for fully plastic behaviour was found to be equal to the linear elastic value of strain energy release rate (G) at failure for extremely large specimens. Thus, the J-integral approach eliminates the necessity of testing very large specimens.

A major limitation of the approach arises from the fact that the J-integral is path independent only when the stress–strain relation is unique. It is truly path independent for linear and non-linear elastic stress–strain laws and also for elastic–plastic behaviour under situations of monotonic loading. This rules out its application to materials which exhibit significant subcritical crack growth prior to fracture since any crack extension necessarily implies unloading near the crack tip.

It is well known that in composite materials, microcracks at the fibre matrix interface appear at very low loads due to the stress concentrations produced by the fibres lying perpendicular to the load. It is probably this unavoidability of microcracks that has deterred researchers from exploring the applicability of the J-integral as a fracture criterion for composite materials.

In the present paper, the J-integral is being developed as a fracture criterion for composite materials, based on test results. The evaluation of J is not tied down to the analytical limitations. A new set of limitations are being developed around the analytical limitations but not strictly adhering to them. The results reported in this paper on short fibre composites are very promising. Similar developments for laminates are currently underway.

EXPERIMENTAL DETAILS

The present studies were performed on randomly oriented short glass fibre reinforced epoxy resin. A chopped strand mat of glass fibres having a mass of $0.6\,kg/m^2$ and an average fibre length of 50 mm was used as the reinforcement. The matrix material was Araldite CY 230 epoxy cured with hardener HY 951. The composite plates (3 mm thick) were cast in the laboratory and cured at room temperature for at least 10 days. The cured plates exhibit a fibre volume fraction of about 36%. The single edge notched specimens were 25 mm wide and the length between grips was at least 3 times the specimen width. The initial notches were machined using a 0.2 mm thick slit cutter and their length was varied between 1.25 and 17.5 mm.

The fracture toughness tests were performed on a 10 ton MTS machine. Load and load point displacement were recorded on an X–Y recorder. All the tests were conducted in a displacement controlled mode. The data was analysed using J-integral approach.

RESULTS AND DISCUSSION

Typical load displacement (at load point) curves for specimens with different initial crack lengths are shown in Fig. 1. The tests were conducted under displacement controlled conditions so that the load displacement behaviour beyond maximum load is also clearly indicated. Specimens with small cracks fracture suddenly, causing an abrupt drop in load, whereas the specimens with larger cracks show a more gradual fracture process beyond maximum load. The behaviour is similar to that observed in metals.[18] This is because the strain energy stored during loading in a small crack length

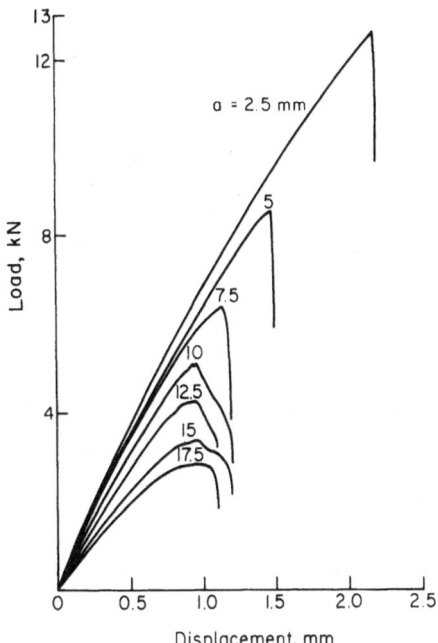

FIG. 1.   Load displacement curves for different initial crack lengths.

specimen is sufficient to cause catastrophic failure. It is not the case for longer crack length specimens.

The observed fracture load is plotted against crack length in Fig. 2 along with the fracture load that would be expected if the strength was unaffected by the crack, i.e. the fracture load obtained by multiplying the net cross sectional area and the unnotched strength. The observed fracture load is smaller than the expected indicating that the crack reduces the fracture load far greater than can be accounted for by the reduction in cross-sectional area. The extent of this influence is illustrated in Fig. 3 through the ratio of observed to expected fracture loads. The decreasing ratio indicates the increasing influence of cracks which stabilizes for cracks larger than 7·5 mm.

The fracture process becomes unstable at a displacement beyond which the load decreases monotonically. This displacement may be referred to as the critical displacement. It is plotted against the crack length in Fig. 4. Initially the critical displacement decreases with increase in crack length and remains constant for cracks larger than 7·5 mm. The initial variation in critical displacement occurs due to the significant deformations away from

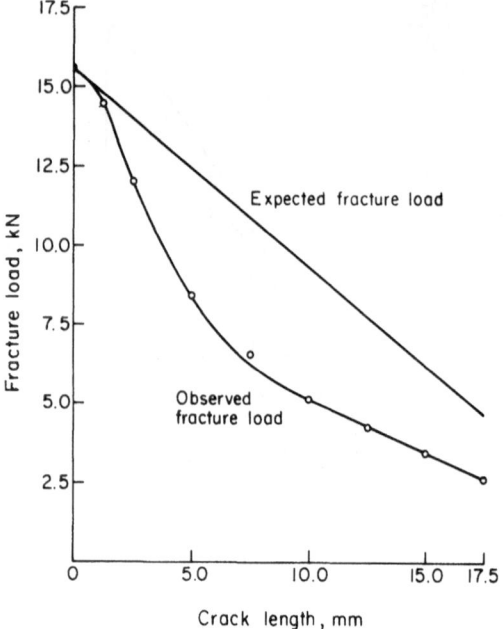

FIG. 2.   Observed and expected (based on net cross-sectional area) fracture loads for notched specimens.

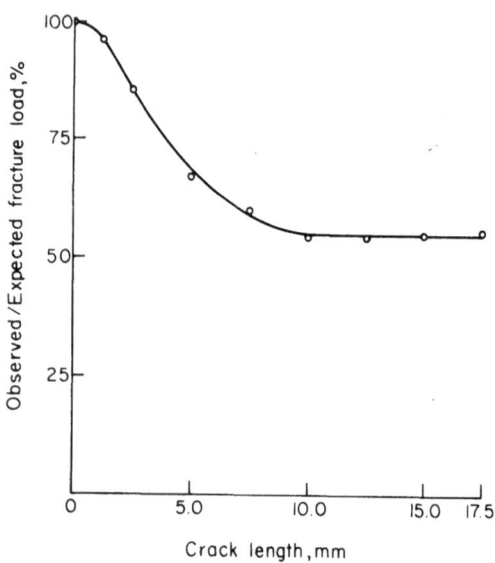

FIG. 3.   The ratio of observed to expected fracture load as a function of crack length.

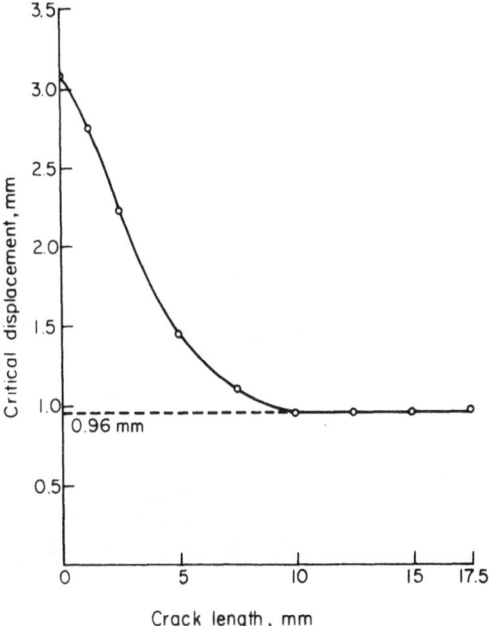

Crack length, mm

FIG. 4. Variation of critical displacements with initial crack lengths.

the crack plane because of large loads. This point will be further explained later in this section. The critical value of the J-integral is obtained corresponding to the constant critical displacement of 0·96 mm as shown in Fig. 4.

The load displacement curves can be used to obtain the value of the J-integral experimentally through its energy interpretation as follows:[19]

$$J = -\frac{\partial U}{\partial a}\bigg|_{\text{constant displacement}} \tag{1}$$

where U is the potential energy per unit thickness and a is the crack length. It may be mentioned that when displacement is kept constant for evaluating J the potential energy, U, reduces to the area under the load deflection record and is equal to the strain energy.[19] Thus, area under load displacement curves is first obtained and plotted against crack length for several displacements (Fig. 5). For a given displacement, energy absorbed by a specimen decreases as the crack length increases (Fig. 5) because smaller loads are required. The variation in energy absorbed is less for cracks shorter than 8·75 mm compared to that for longer cracks because in

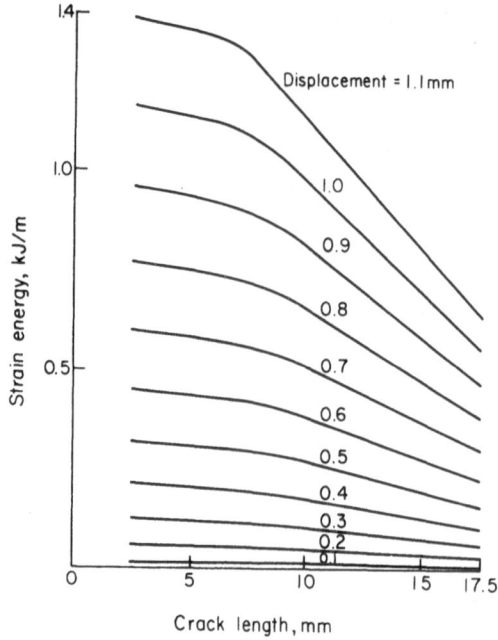

FIG. 5.   Strain energy per unit thickness of specimen for different displacements.

specimens with longer cracks the energy absorbed is essentially in the vicinity of the crack tip and is thus strongly influenced by the crack length.

The J-integral is obtained from eqn. (1) through slopes of the energy curves in Fig. 5. The J-integral is independent of crack length for cracks larger than 8·75 mm since the energy curves are straight lines in this range. The variation of J with displacement is shown in Fig. 6. The critical value of J corresponding to the critical displacement of 0·96 mm is 51·8 kJ/m². For smaller cracks, the value of the J-integral depends upon the displacement as well as the crack length because the slope of the energy curve changes with crack length (Fig. 5). The variation of J for cracks smaller than 8·75 mm has not been shown because it is not unique. However, in view of Fig. 5 it may be stated that in this region J will be smaller for a given displacement but defined for a greater range of displacement. Its apparent critical value is also expected to be larger in these cases. The applicability of the J-integral in this region is further discussed later in this section.

From the preceding discussion, it appears that when the crack is larger than 8·75 mm or when $a/w > 0·35$, the fracture behaviour is governed essentially by the crack tip environment resulting in a constant critical displacement and a unique value of the J-integral. For these crack lengths,

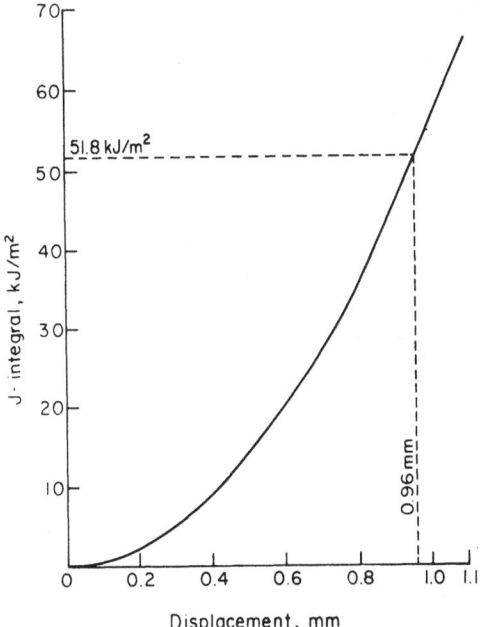

FIG. 6.   *J*-integral as function of displacement.

the fracture load is small which does not cause any general material damage away from the crack tip region. On the other hand when cracks are small ($a < 8.75$ mm or $a/w < 0.35$), the *J*-integral and critical displacement depend upon the crack length, indicating that in addition to the crack tip environment, the region away from it also influences such quantities as the energy absorbed and displacement at fracture. This may be attributed to the fact that the fracture loads are high enough to cause general material damage.

In order to study the influence of the general material damage in specimens with smaller cracks, additional specimens with varying specimen lengths were tested. The length between grips was varied from 3 to 6 times the width of the specimen. Displacement at fracture (critical displacement) is plotted against specimen length in Fig. 7. As expected the critical displacement increases with specimen length for all crack lengths. However, critical displacements for specimens with 10 and 12·5 mm long cracks are the same, which is consistent with Fig. 4. The total displacement of the specimen is the sum of the displacement in the crack tip region, which may be expected to be independent of the specimen length and displacement in the region away from the crack tip which should be a

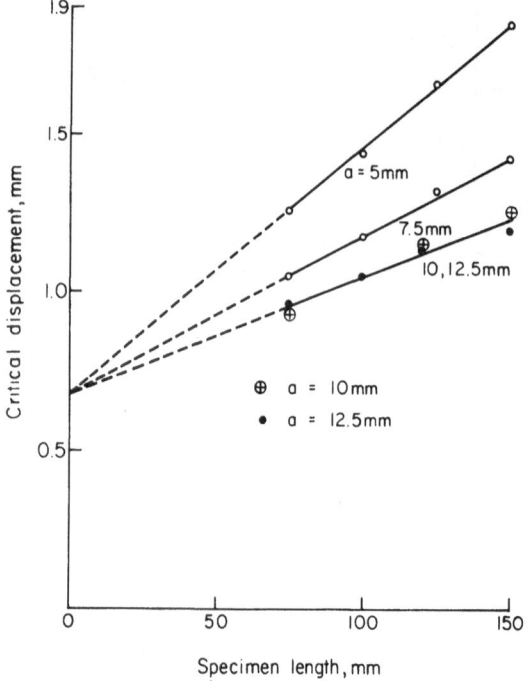

FIG. 7.   Variation of critical displacement with specimen length for different crack lengths.

function of specimen length. The intercept on the ordinate obtained through extrapolation of a straight line in Fig. 7 may be regarded as the displacement in the crack tip region alone. Interestingly, all the straight lines in Fig. 7 intercept the ordinate at the same point. This common intercept may be regarded as a critical displacement due to the presence of the crack and whose value is independent of crack length and specimen length.

Variations of energy absorbed up to fracture are shown in Fig. 8 for different crack lengths. The total energy absorbed may also be thought of as the sum of the energies absorbed in the crack tip region and the region away from it. The energy absorbed in the crack tip region should depend upon the crack length but not on specimen length whereas the energy absorbed in the region away from the crack tip does depend upon the specimen length. It is observed that when the crack length is 10 or 12·5 mm, the energy absorbed is independent of the specimen length, signifying negligible energy absorption in the region away from the crack tip. For crack lengths of 5 and 7·5 mm, the total energy absorbed increases linearly

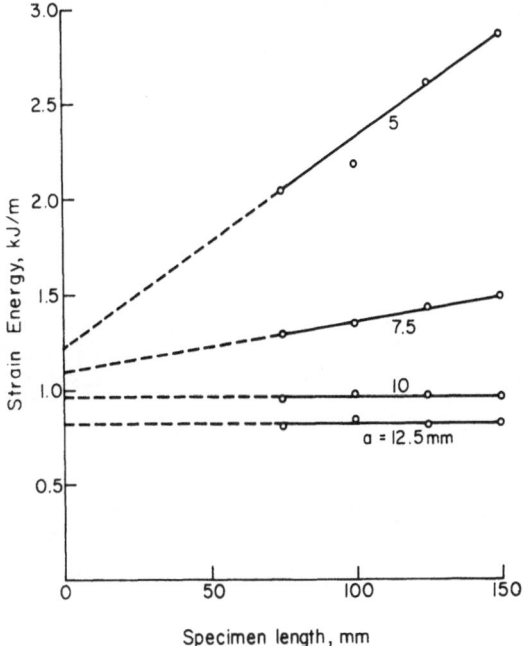

FIG. 8.   Variation of strain energy with specimen length for different crack lengths.

with the specimen length indicating a significant energy absorption in the region away from the crack tip as well. These observations are further supported by visual observation of the specimens. The damage in the specimens with 10 and 12·5 mm cracks is confined to the crack tip region whereas in specimens with smaller cracks the material damage is all over. This is illustrated in Fig. 9 through a photograph of two fractured specimens. The photograph was taken in a bright light background and therefore, the damage (opaque to light) is indicated by dark areas.

The intercept on the ordinate obtained by extrapolation of a straight line in Fig. 8, may be regarded as the energy absorbed in the crack tip region. Energy absorbed thus obtained is plotted in Fig. 10. It was explained with respect to Fig. 7 that the critical displacement due to the presence of a crack alone is independent of crack length. Thus, it may be argued that the energy absorbed for different crack lengths (Fig. 10) correspond to the same critical displacement and therefore, the slope of the straight line may be used to obtain the critical value of J, independent of crack length. The J_{critical} thus obtained is $50 \cdot 7 \, \text{kJ/m}^2$ which is close to the value $51 \cdot 8 \, \text{kJ/m}^2$ obtained earlier in Fig. 6. This is a very significant observation. This shows

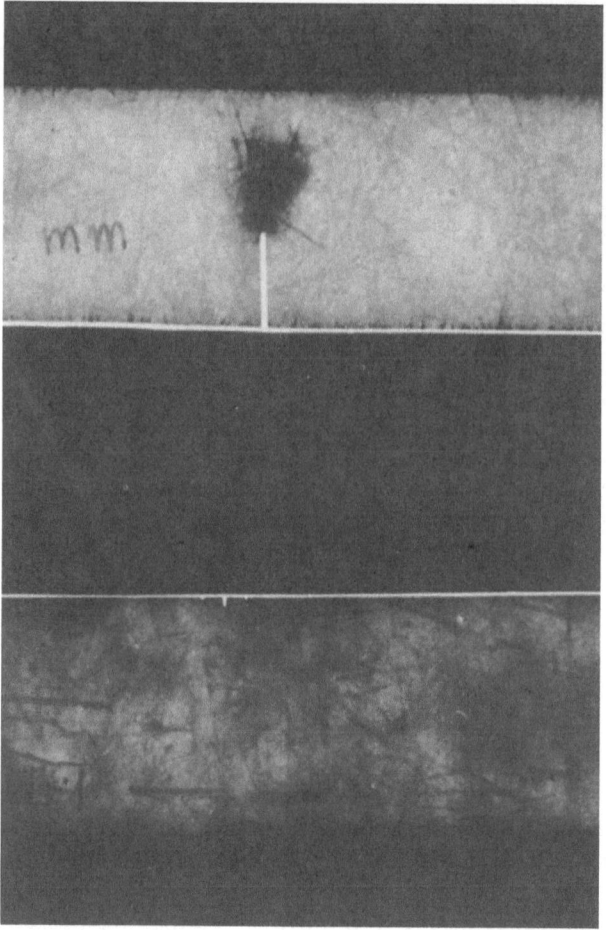

FIG. 9.   Transmitted light photograph of two specimens with different crack lengths.

that the energy absorbed at the crack tip may be isolated from that absorbed in the region away from it. Thus, a parameter independent of testing variables (i.e. crack length and specimen length) is obtained which may be used as a fracture criterion for the material.

It has been shown that the critical value of the J-integral, J_c, is related to the parameters of linear elastic fracture mechanics. For the plane stress case it is related to critical stress intensity factor in mode 1, K_c, by the following relation:[21]

$$J_c = \frac{K_c^2}{E} \tag{2}$$

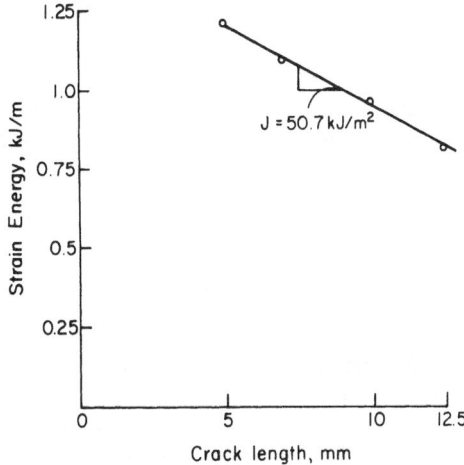

FIG. 10.   Strain energy at the crack tip for different initial crack lengths.

where E is the modulus of elasticity. The present material has an average elastic modulus equal to 11·5 GPa. Therefore, eqn. (2) gives K_c equal to 24·4 MPa$\sqrt{m}$. This value of critical stress intensity factor agrees very well with the K_c of 24·82 MPa$\sqrt{m}$ obtained by Agarwal and Giare[3] for similar glass fibre composite. This demonstrates that the present method of characterizing fracture toughness is consistent with the R-curve approach. However, the J-integral method is a lot simpler for experimental as well as analytical (computational) evaluation.

CONCLUSIONS

Fracture behaviour of a short fibre composite has been investigated. The J-integral has been evaluated using the energy rate interpretation. Its value is found to be independent of crack length when the ratio of crack length to specimen width (a/w) is larger than 0·35. For smaller crack lengths general material damage away from the crack tip also influences the energy absorbed significantly. However, an extrapolation method has been developed through which the crack tip energy may be separated from the energy absorbed due to general material damage. The J-integral thus obtained is independent of crack length and specimen length and its critical value is the same as obtained for $a/w > 0·35$ without extrapolation. Further, it also agrees well with the critical stress intensity factor obtained in an earlier study using the R-curve approach.

498   B. D. Agarwal, Prashant Kumar and B. S. Patro

ACKNOWLEDGEMENTS

This research work was sponsored by the Aeronautics Research and Development Board (Structures Panel), Government of India. The authors would also like to thank Shri B. R. Somashekar, Professor K. Rajaiah and Dr K. N. Raju for their interest in the work.

REFERENCES

1. GAGGAR, S. K. and BROUTMAN, L. J., Crack growth resistance of random fibre composites, *J. Comp. Mater.*, **9**, 1975, 216–227.
2. GAGGAR, S. K. and BROUTMAN, L. J., Strength and fracture properties of random fibre polyester composites, *Fibre Sci. Tech.*, **9**, 1976, 205–224.
3. AGARWAL, B. D. and GIARE, G. S., Crack growth resistance of short fibre composites: I—influence of fibre concentration, specimen thickness and width, *Fibre Sci. Tech.*, **15**, No. 4, 1981, 283–298.
4. AGARWAL, B. D. and GIARE, G. S., Crack growth resistance of short fibre composites: II—influence of test temperature, *Fibre Sci. Tech.*, **16**, No. 1, 1982, 19–28.
5. AGARWAL, B. D. and GIARE, G. S., Effect of matrix properties on fracture toughness of short fibre composites, *Mater. Sci. Eng.*, **52**, No. 2, 1982, 139–145.
6. AGARWAL, B. D. and GIARE, G. S., Fracture toughness of short fiber composites in modes II and III, *Engng Fract. Mech.*, **15**, No. 1–2, 1981, 219–230.
7. SIH, G. C., HILTON, P. D., BADALLIANCE, R., SHENBERGER, P. S. and VILLARREAL, G., Fracture mechanics for fibrous composites, *Analysis of the Test Method for High Modulus Fibers and Composites*, ASTM STP 521, American Society for Testing and Materials, Philadelphia, 1973, pp. 98–132.
8. MANDELL, J. F., WANG, S. S. and MCGARRY, F. J., The extension of crack tip damage zones in fiber reinforced plastic laminates, *J. Comp. Mater.*, **9**, No. 2, 1975, 266.
9. YEOW, Y. J., MORRIS, D. H. and BRINSON, H. F., The fracture behaviour of graphite/epoxy laminates, *Exp. Mech.*, **19**, No. 9, 1979, 1–8.
10. MORRIS, D. H. and HAHN, H. T., Fracture resistance characterization of graphite/epoxy composites, in: *Composite Materials Testing and Design (Fourth Conference)*, ASTM STP 617, American Society for Testing and Materials, Philadelphia, 1977, pp. 5–17.
11. AWERBUCH, J. and HAHN, H. T., K-calibration of unidirectional metal matrix composites, *J. Comp. Mater.*, **12**, 1978, 222–237.
12. BATHIAS, C., ESNAULT, R. and PELLAS, J., Application of fracture mechanics to graphite fibre-reinforced composites, *Composites*, **12**, July 1981, 195–200.
13. RICE, J. R., A path independent integral and the approximate analysis of strain concentration by notches and cracks, *J. appl. Mech.*, **35**, 1968, 379–386.

14. BROBERG, K. B., Crack growth criteria and non-linear fracture mechanics, *J. Mech. Phys. Solids*, **19**, 1971, 407–418.

15. RICE, J. R. and ROSENGREN, G. F., Plane–strain deformation near a crack tip in a power law hardening material, *J. Mech. Phys. Solids*, **16**, 1968, 1–12.

16. HUTCHINSON, J. W., Singular behaviour at the end of a tensile crack in a hardening material, *J. Mech. Phys. Solids*, **16**, 1968, 13–31.

17. McCLINTOCK, F., Plasticity aspects of fracture, *Fracture*, Chap. 2, Vol. III (ed. H. Liebowitz), Academic Press, New York, 1971, pp. 47–225.

18. BEGLEY, J. A. and LANDES, J. D., The *J* integral as a fracture criterion, in: *Fracture Toughness*, ASTM STP 514, American Society for Testing and Materials, Philadelphia, 1972, pp. 1–20.

19. LANDES, J. D. and BEGLEY, J. A., The effect of specimen geometry on J_{IC}, in: *Fracture Toughness*, ASTM STP 514, American Society for Testing and Materials, Philadelphia, 1972, pp. 24–39.

20. LANDES, J. D. and BEGLEY, J. A., Recent developments in J_{IC} testing, in: *Developments in Fracture Mechanics Test Methods Standardization*, ASTM STP 632, American Society for Testing and Materials, Philadelphia, 1977, pp. 57–81.

21. RICE, J. R., Mathematical analysis in the mechanics of fracture, in: *Fracture*, Chap. 3, Vol. II (ed. H. Liebowitz), Academic Press, New York, 1968, pp. 191–311.

35

Fracture of a Bimaterial Plate with a Crack along the Interface

E. E. GDOUTOS

School of Engineering, Democritus University of Thrace,
Xanthi, Greece

ABSTRACT

A thorough study of the fracture behaviour of a bimaterial plate with a crack along the interface was undertaken. The plate was subjected to a uniform normal stress at infinity, while its two phases were perfectly bonded along the interface. The angle of initial crack propagation into either of the two phases and the corresponding critical failure loads were determined for many material combinations of the plate by using the strain energy density criterion proposed by Sih. Finally, the fracture trajectories from the crack tip for conditions of unstable fracture were determined.

1. INTRODUCTION

In recent years composite materials have extensively been used in structural applications. The prediction of the strength of such materials within the framework of fracture mechanics is made by assuming that imperfections, in the form of holes, voids, cracks or inclusions, exist in the composite material. The existence of such imperfections near bimaterial boundaries is of particular importance due to the interaction effects which take place between the imperfections and the boundaries. Thus, the study of the strength of a composite with a crack along the interface of its constituents is of special interest. Such a problem can be used as an idealization of the eventual manufacturing flaws developed during bonding or casting processes in composites.

In the present paper the problem of fracture of a bimaterial plate resulting from a crack along the interface was studied. The stress field near to the crack tip is governed by both opening-mode and sliding-mode stress intensity factors and therefore the crack would not generally propagate along the interface. Since the crack propagation path is not known in advance fracture criteria such as the Griffith or the energy release rate criterion cannot be applied for the prediction of the critical fracture loads of the plate.

Recently Sih[1] has proposed a theory of fracture based on the field strength of the local strain energy density in an element of material ahead of the crack. According to this theory the particular stationary values of the strain energy density, as seen from the crack tip, dictate the direction and magnitude of the load required for crack extension. Strain energy density theory has been proved quite successful for the prediction of the fracture behaviour of a variety of cracked bodies.[2] This theory was used in the present work for the determination of the failure loads of a cracked bimaterial plate. Furthermore using this theory the fracture trajectories from the crack tip for unstable fracture conditions were determined.

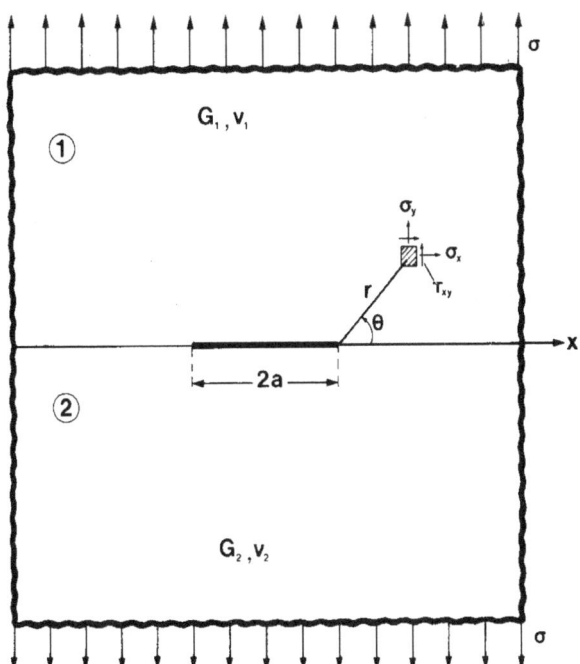

FIG. 1.   A bimaterial plate with a crack along the interface.

2. THE STRESS FIELD NEAR TO THE CRACK TIP

We consider two homogeneous, isotropic, elastic materials numbered 1 and 2 which occupy the upper $(y > 0)$ and lower $(y < 0)$ half-planes respectively (Fig. 1). The materials 1 and 2 are characterized by the elastic properties G_1, v_1 and G_2, v_2 where G_i $(i = 1, 2)$ are the shear moduli and v_i $(i = 1, 2)$ are the Poisson's ratios. The two materials are bonded to each other along the x-axis except for a finite segment of length $2a$ forming an internal crack. We suppose that the crack surfaces are free of loads and that the two media are subjected to stresses at infinity.

For this situation Erdogan[3] and Rice and Sih[4] determined the stress field in both media. The singular expression for the σ_{1x} stress component for the medium 1 $(0 < \theta < 180°)$ in the vicinity of the crack tip is given by:

$$
\sigma_{1x} = \frac{K_I}{2\sqrt{2\pi r}} \left\{ \tfrac{1}{2}\exp(-\beta(\pi - \theta)) \left[5\cos\left(\frac{\theta}{2} + \beta\log\frac{r}{a}\right) + \cos\left(\frac{5\theta}{2} + \beta\log\frac{r}{a}\right) \right. \right.
$$
$$
\left. + 4\beta\sin\theta\cos\left(\frac{3\theta}{2} + \beta\log\frac{r}{a}\right) \right] - \exp(\beta(\pi - \theta))\cos\left(\frac{\theta}{2} - \beta\log\frac{r}{a}\right) \Bigg\}
$$
$$
- \frac{K_{II}}{2\sqrt{2\pi r}} \left\{ \tfrac{1}{2}\exp(-\beta(\pi - \theta)) \left[5\sin\left(\frac{\theta}{2} + \beta\log\frac{r}{a}\right) + \sin\left(\frac{5\theta}{2} + \beta\log\frac{r}{a}\right) \right. \right.
$$
$$
\left. + 4\beta\sin\theta\sin\left(\frac{3\theta}{2} + \beta\log\frac{r}{a}\right) \right] + \exp(\beta(\pi - \theta))\sin\left(\frac{\theta}{2} - \beta\log\frac{r}{a}\right) \Bigg\}
$$

$$(1)$$

and analogous expressions valid for the other two stresses σ_{1y} and τ_{1xy}.

In the above relations r, θ are the polar coordinates of the point considered, β is a bimaterial constant given by:

$$
\beta = \frac{1}{2\pi}\log\beta_0 \tag{2}
$$

with:

$$
\beta_0 = \frac{G_1 + \kappa_1 G_2}{G_2 + \kappa_2 G_1} \tag{3}
$$

where $\kappa_i = 3 - 4v_i$ or $\kappa_i = (3 - v_i)/(1 + v_i)$ $(i = 1, 2)$ for plane strain or generalized plane stress conditions respectively, and K_I, K_{II} are the so-called stress intensity factors which are independent of the coordinates r, θ and depend on the elastic constants of the two media and the form of loading of the bimaterial plate.

For the case when both materials have the same elastic properties $\beta = 0$ and relation (1) gives the well-known expression for the stress σ_x in the vicinity of the tip of a crack in an infinite plate.

Let us now consider that the plate is subjected to a uniform uniaxial stress σ along the y-axis at large distances from the crack. For this case the K_I and K_{II} stress intensity factors for the tip $z = a$ are given by the following relations (Rice and Sih[4]):

$$K_I = \frac{\cos (\beta \log 2a) + 2\beta \sin (\beta \log 2a)}{\cosh \pi \beta} \sigma \sqrt{\pi a}$$

$$K_{II} = -\frac{\sin (\beta \log 2a) - 2\beta \cos (\beta \log 2a)}{\cosh \pi \beta} \sigma \sqrt{\pi a} \tag{4}$$

3. THE STRAIN ENERGY DENSITY CRITERION

According to the strain energy density criterion fracture of a plate originating from a pre-existing crack is governed by the minimum values of the strain energy in the vicinity of the crack tip. In plane problems the general form of the strain energy density function is given by:

$$\frac{dW}{dV} = \frac{1}{16G} [(1 + \kappa)(\sigma_x^2 + \sigma_y^2) - 2(3 - \kappa)\sigma_x\sigma_y + 8\tau_{xy}^2] \tag{5}$$

The strain energy density factor S is now defined from the relation:

$$S = r \frac{dW}{dV} \tag{6}$$

where r is the distance from the crack tip.

As it was pointed out by Sih, for a fixed distance r, S assumes some critical value, S_{cr}, which characterizes the toughness of the material. From relation (6) it is observed that specifying S_{cr} to be a constant is equivalent to requiring that $(dW/dV)_{cr}$ be a constant once the value of r has been chosen. Thus for a given value r the minimum values of S could provide the direction of crack propagation as well as the corresponding critical loads of the plate. The minimum values of S are determined from the following relations:

$$\frac{\partial S}{\partial \theta} = 0 \qquad \frac{\partial^2 S}{\partial \theta^2} > 0 \tag{7}$$

4. DETERMINATION OF FAILURE LOADS

Following the previously outlined procedure the minimum values of the strain energy density factor S defined from relation (6) were determined. For this reason the stress components σ_x, σ_y, τ_{xy} were introduced into relations (5) and (6) and further on to relations (7). Determination of S requires knowledge of the radius r of the core region surrounding the crack tip. The necessity of the introduction of the concept of the core region arises from the difficulty in the interpretation of the state of stress and displacement in the close neighbourhood of the crack tip by analytical models. Indeed, in the core region the physical behaviour is unknown and cannot be incorporated into a mathematical model. The material being highly strained may become inhomogeneous and generally not conducive to modelling.

The radius r of the core region was given in the present work the following values $r/a = 10^{-4}$, 10^{-3}, 10^{-2} and 10^{-1}. These values of r/a are generally outside the critical region surrounding the crack tip in which the stresses present the well-known oscillatory behaviour (Williams[5]). Equating the minimum values of S with its critical values S_{1cr} and S_{2cr} for materials 1 and 2 the critical values of the applied stress σ_{cr} for crack extension were determined.

Figure 2a presents the variation of the quantity $\sigma_{cr}^t(a/16G_1S_{1cr}^t)^{1/2}$ versus G_2/G_1 for $v_1 = v_2 = 0.2$ and $r/a = 10^{-4}$, 10^{-3}, 10^{-2} and 10^{-1} when the bimaterial plate is subjected to a tensile stress σ. σ_{cr}^t is the critical value of σ for crack extension and S_{1cr}^t is the critical value of S for material 1 under tensile loads. It was found that for the case under consideration the crack always extends into material 1 and therefore for the calculation of σ_{cr}^t the constants G_1 and S_{1cr}^t of material 1 should be introduced into relations (5) and (6). Similarly, Figs 2b, 3a and 3b present the same results under tensile applied stresses for the following combinations of the values v_1 and v_2 of the Poisson's ratios of materials 1 and 2: $v_1 = v_2 = 0.4$ for Fig. 2b, $v_1 = 0.2$, $v_2 = 0.5$ for Fig. 3a and $v_1 = 0.5$, $v_2 = 0.2$ for Fig. 3b. It was found that for Figs 2b and 3a the crack extends into material 1, while for Fig. 3b it extends into material 2. Thus for the calculation of σ_{cr}^t in Fig. 3b the constants G_2 and S_{2cr}^t were introduced.

In Figs 2a and 2b the values of the quantity $\sigma_{cr}(a/16G_1S_{1cr}^t)^{1/2}$ for $G_2/G_1 = 1$ are equal to 0·645 and 1·118 respectively. These values are equal to those found previously (Sih[1]) for the case of a crack in an infinite plate subjected to tensile loads. From Figs 2 and 3 is established the trend of the increased loading permitted as the radius of the core region r becomes

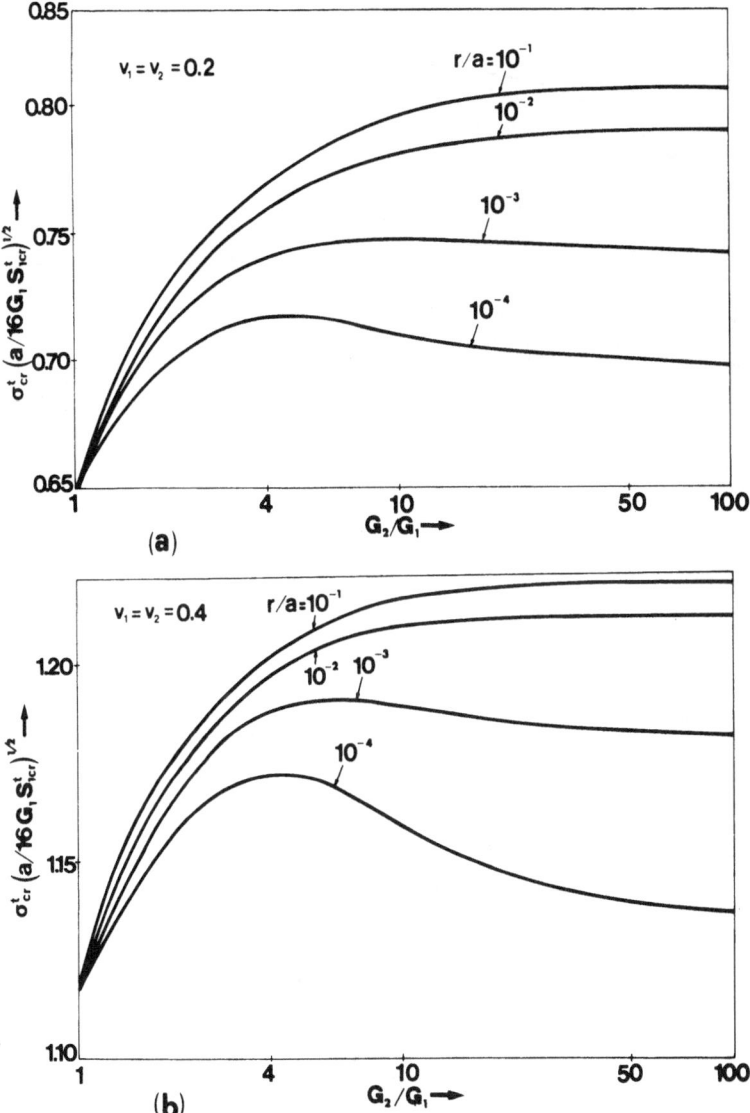

FIG. 2. Variation of the normalized critical stress of fracture $\sigma_{cr}^t (a/16 G_1 S_{1cr}^t)^{1/2}$ versus G_2/G_1 for $v_1 = v_2 = 0.2$ (a) and $v_1 = v_2 = 0.4$ (b). The plate is subjected to a tensile stress and the radius r of the core region takes the values $r/a = 10^{-4}, 10^{-3}, 10^{-2}$ and 10^{-1}. The crack always extends into material 1.

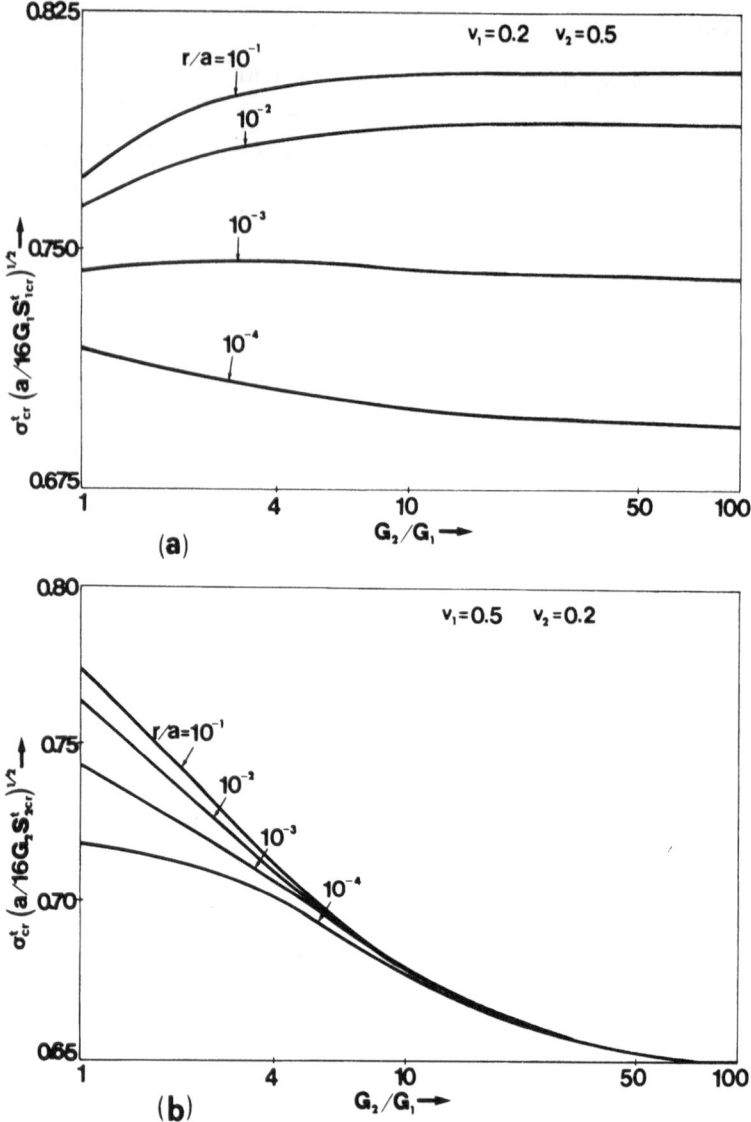

FIG. 3.   Variation of the normalized critical stress of fracture $\sigma_{cr}^t(a/16G_{1,2}S_{1,2cr}^t)^{1/2}$ versus G_2/G_1 for $v_1 = 0\cdot2$, $v_2 = 0\cdot5$ (a) and $v_1 = 0\cdot5$, $v_2 = 0\cdot2$ (b). The plate is subjected to a tensile stress and the radius r of the core region takes the values $r/a = 10^{-4}, 10^{-3}, 10^{-2}$ and 10^{-1}. The crack extends into material 1 for case (a) and into material 2 for case (b).

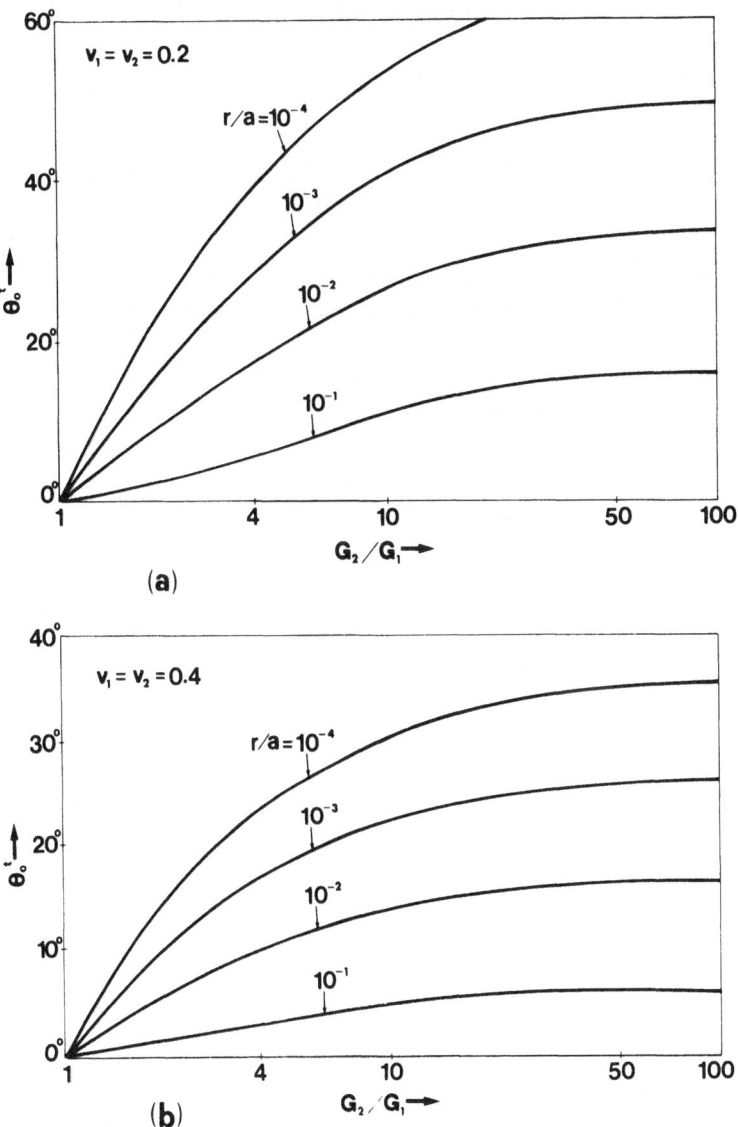

FIG. 4. Variation of the crack extension angle θ_0^t versus G_2/G_1 for $v_1 = v_2 = 0.2$ (a) and $v_1 = v_2 = 0.4$ (b). The plate is subjected to a tensile stress and the radius r of the core region takes the values $r/a = 10^{-4}$, 10^{-3}, 10^{-2} and 10^{-1}.

larger. These figures enable the determination of the critical stress for initiation of crack growth provided that the radius of the core region r is known. r represents a material constant and could be determined experimentally.

5. CRACK EXTENSION ANGLES—FRACTURE TRAJECTORIES

According to the strain energy density criterion the polar angle θ_0 at which the value of S is a minimum gives the value of the crack extension angle. Figure 4 presents the variation of the crack extension angle θ_0^t for tensile applied loads versus G_2/G_1 for $r/a = 10^{-4}$, 10^{-3}, 10^{-2}, 10^{-1} and $v_1 = v_2 = 0.2$ (Fig. 4a) and $v_1 = v_2 = 0.4$ (Fig. 4b). It is shown from Fig. 4 that for $G_2/G_1 = 1$ the crack extension angle is zero, which reflects the well-known result that a crack in a plate subjected to a uniform tensile stress perpendicular to the crack axis extends along its own plane.

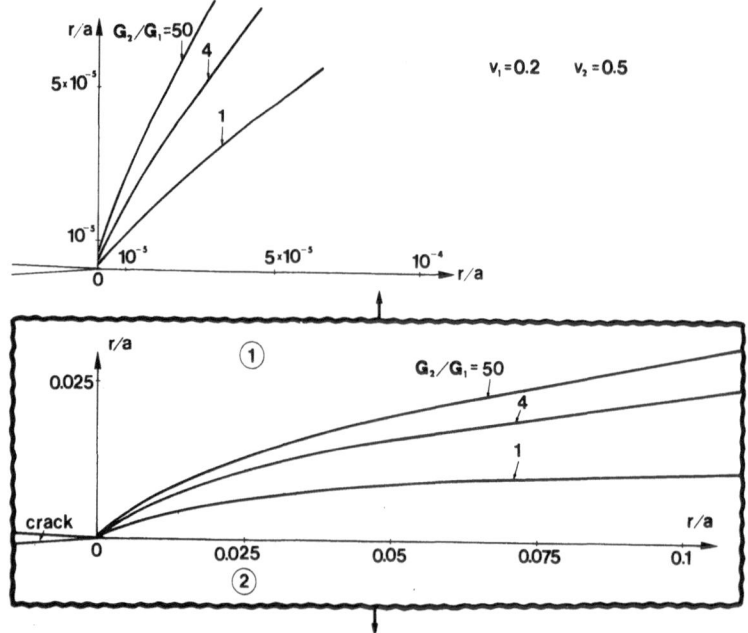

FIG. 5. Fracture trajectories for a bimaterial plate subjected to tensile loads with $v_1 = 0.2$, $v_2 = 0.5$ and $G_2/G_1 = 1$, 4 and 50. The detailed form of the trajectories in the close neighbourhood of the crack tip is shown in the upper left part of the figure.

Using Fig. 4 the fracture trajectories can be determined. Indeed, for brittle fracture behaviour with the fracture growth sudden and unstable it does not seem unreasonable to assume that the fracture path can be determined from the conditions before fracture initiation takes place. In such cases the material does not have time to redistribute the stresses during the fracture process and the fracture path can be predetermined with high accuracy. A sound justification in favour of such behaviour comes from the case of brittle fracture of blunt notches or line cracks at low angles of loading where the loads required to initiate fracture are relatively high. In such cases there exists a high density energy field near and far from the initial crack extension area which will be responsible for the rapid fracture when other physical resources exist to inhibit the fracture from propagating.

Figure 5 presents the fracture trajectories for $v_1 = 0.2$, $v_2 = 0.5$ and $G_2/G_1 = 1, 4, 50$ when the bimaterial plate is subjected to tensile loads. In the upper left side of this figure the detailed form of fracture trajectories very close to the crack tip is shown. It is observed that as the distance from the crack tip increases the crack tends to become parallel to its initial direction, which is in accordance with the results of Wang *et al.*[6]

6. CONCLUSIONS

The brittle fracture behaviour of a bimaterial plate with a crack along the interface subjected to a uniform uniaxial stress perpendicular to the crack axis was investigated. The critical loads for initiation of crack extension as well as the fracture trajectories were determined by using the strain energy density criterion. The dependence of the fracture characteristic quantities on the material properties of the two phases of the bimaterial plate was established. From the whole study the following main conclusions may be derived by taking $G_2 > G_1$:

(i)   For tensile loads and $v_1 \leq v_2$ the crack extends into material 1, while for $v_1 > v_2$ it extends into material 2. Quite the contrary happens for compressive loads.

(ii) The critical stress for crack extension increases as the radius r of the core region surrounding the crack tip also increases. Determination of r which is a material constant enables the immediate calculation of the critical stress.

(iii) The critical stress for crack extension is always larger for compressive than for tensile applied loads.

(iv) The crack extension angle θ_0^t for tensile loads decreases and tends
to zero as the radial distance r/a from the crack tip increases. For
compressive loads the angle θ_0^c tends to $\pm 180°$ as r/a increases.
(v) The fracture path for both tensile and compressive loads tends to
become parallel to the crack axis as r/a increases. However, the
crack extends in opposite directions for tensile and compressive
loads.

REFERENCES

1. SIH, G. C., Strain-energy-density factor applied to mixed mode crack problems,
    Int. J. Fract., **10** (1974), 305–321.
2. GDOUTOS, E. E., *Mixed-Mode Crack Growth* (ed. G. C. Sih), Martinus Nijhoff
    (in press).
3. ERDOGAN, F., Stress distribution in bonded dissimilar materials with cracks,
    ASME Trans., J. appl. Mech., **32** (1965), 403–410.
4. RICE, J. R. and SIH, G. C., Plane problems of cracks in dissimilar media, *ASME
    Trans., J. appl. Mech.*, **32** (1965), 418–423.
5. WILLIAMS, M. L., The stresses around a fault or crack in dissimilar media. *Bull.
    Seism. Soc. Amer.*, **49** (1959), 199–204.
6. WANG, T. T., KWEI, T. K. and ZUPKO, H. M., Tensile strength of butt-joined
    epoxy–aluminum plates. *Int. J. Fract. Mech.*, **6** (1970), 127–137.

36

The Effects of Laminate Thickness on the Fracture Behavior of Composite Laminates

C. E. Harris and D. H. Morris

Engineering Science and Mechanics Department,
Virginia Polytechnic Institute and State University,
Blacksburg, Virginia 24061, USA

ABSTRACT

The relationship between both fracture toughness and the development of crack-tip damage as a function of specimen thickness was investigated. Fracture toughness was determined experimentally using center-cracked tension specimens for $[0/\pm 45/90]_{ns}$ and $[0/90]_{ns}$ laminates, where ns means multiple layers with the same repeated sequence and symmetric about the midplane. Laminate thicknesses ranged from 8 plies to 96 plies. As with isotropic metals, fracture toughness was found to decrease with increasing specimen thickness and asymptotically approached a lower bound. The crack tip damage in the $[0/\pm 45/90]_{ns}$ laminate does not appear to be a function of thickness. However, there are differences in the damage of the $[0/90]_{2s}$ and $[0/90]_{8s}$ laminates and associated differences in fracture toughness.

INTRODUCTION

In order to design laminated composite structural components the design engineer must be able to assess size effects. He must have thick laminate data or be assured by the research engineer that thin laminate data can be extrapolated to thick laminate applications. A large body of fracture toughness data exists for thin (6–16 plies) laminated composites but relatively little data exists for thick (90–120 plies) laminates. Also, numerous studies have been conducted to determine the fracture characteristics of thin laminates. These studies have led to the development of several mathematical models to predict fracture. However, the validity of the conclusions reached from these studies has not been demonstrated nor have the predictive models been shown to be applicable to thick laminates. In order to address this issue a fracture test program is being conducted at Virginia Tech in which laminate thickness is the primary variable. This paper presents some of the early program results. Experimental results are presented herein that compare the fracture toughness and crack-tip damage development at various thicknesses for several laminate types.

The relationship between fracture toughness and specimen thickness has been established for isotropic materials. Fracture toughness is a decreasing function of increasing specimen thickness. A lower bound toughness value corresponding to a condition of plane strain at the crack-tip is asymptotically approached. Discussions of this relationship and the implications thereof can be found in references such as ASTM STP 410[1] and STP 463.[2] The crack-tip state-of-stress is directly related to the relationship between the size of the plastic zone at the crack-tip and the specimen thickness. If the specimen thickness is large relative to the size of the crack-tip plastic zone, sufficient constraint in the interior will be present to produce plane strain conditions. Plane stress conditions prevail when the specimen is too thin to provide constraint in the interior. Using a failure criterion such as the von Mises Criterion or Tresca Criterion it can be shown that the plane stress plastic zone size is much greater than the plane strain plastic zone size.[3] After much research and review the ASTM decided that the specimen thickness should be 50 times greater than the radius of the plane strain plastic zone to ensure plane strain conditions.[2] Using this requirement and an expression for the plastic zone size obtained for linear elastic fracture mechanics and the von Mises or Tresca Criterion, the following limitation on specimen thickness, B, and plane strain fracture toughness, K_{IC}, was obtained[2]

$$B \geq 2 \cdot 5 (K_{IC}/\sigma_{ys})^2$$

where σ_{ys} is the yield strength of the isotropic material.

Several investigations of the fracture toughness of laminated composites have reported specimen thickness as a test variable.[4-7] However, the range of specimen thickness values was limited and most probably did not result in a plane strain fracture condition. None of these investigations reported any significant variation in toughness as a function of thickness. Assuming that the above plane strain fracture toughness thickness requirement applies to laminated composites, the thickness would have to be 0·35–0·40 in (8·89–10·16 mm) to ensure plane strain conditions. (More will be said about this in a later section.) Using this thickness range as a guide specimens were tested from 8 plies (0·0395 in or 1·00 mm) to 96 plies (0·473 in or 12·0 mm). The test results are reported and discussed herein.

MATERIAL PROPERTIES

Graphite/epoxy (T300/5208) laminate panels were prepared by a tape layup and autoclave curing process. The laminate stacking sequences were $[0/\pm45/90]_{ns}$ and $[0/90]_{ns}$ where ns means multiple layers with the same repeated sequence and symmetric about the midplane. The stress analysis method (described in the next section) for calculating fracture toughness required laminate stiffness properties. These laminate properties were computed from basic lamina properties by standard laminate equations such as those given by Jones.[8] The following basic lamina properties were experimentally measured:

$$E_{11} = 20\cdot1 \times 10^6 \text{ psi } (138\cdot6 \text{ GPa})$$
$$E_{22} = 1\cdot56 \times 10^6 \text{ psi } (10\cdot76 \text{ GPa})$$
$$v_{12} = 0\cdot318$$
$$G_{12} = 0\cdot867 \times 10^6 \text{ psi } (5\cdot98 \text{ GPa})$$

where the subscript 1 refers to the fiber direction and subscript 2 denotes the direction perpendicular to the fibers.

EXPERIMENTAL PROCEDURES

Fracture Tests

All fracture tests were conducted using center-cracked tension panels. All specimens were 2 in (51 mm) wide and 8 in (203 mm) long. The crack length-to-width ratio, $2a/\omega$, and specimen thickness were test variables. For each laminate type the specimens were oriented such that the specimen and laminate panel thickness direction coincided and the machined slot was perpendicular to the 0° fiber direction. Specimens from the $[0/\pm45/90]_{s}$ and $[0/90]_{2s}$ laminates were tested at four crack length-to-width ratios,

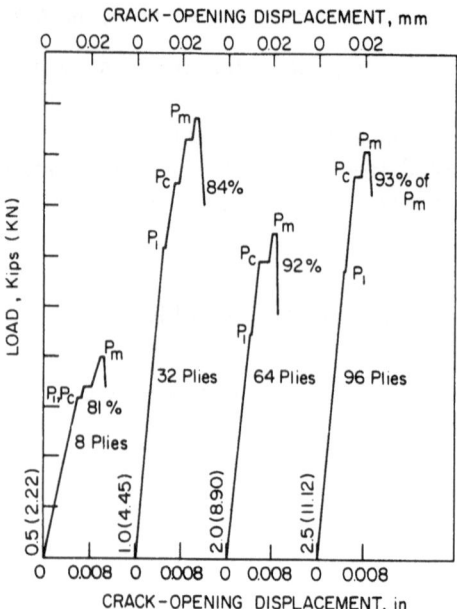

FIG. 1.   Records of typical load versus crack-opening displacement for the $[0/\pm45/90]_{ns}$ laminate at various thicknesses.

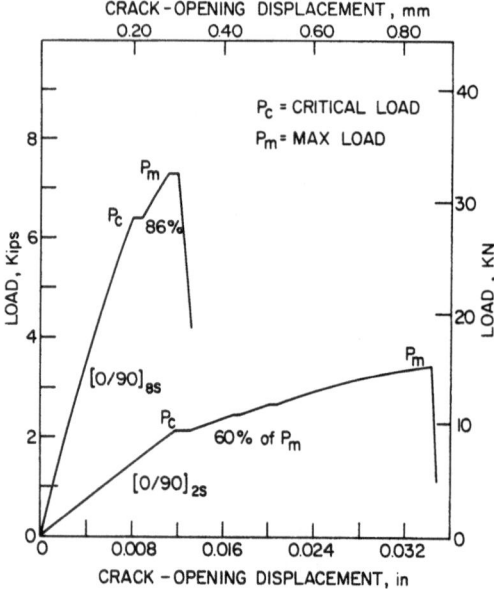

FIG. 2.   Records of typical load versus crack-opening displacement for the $[0/90]_{2s}$ and $[0/90]_{8s}$ laminates.

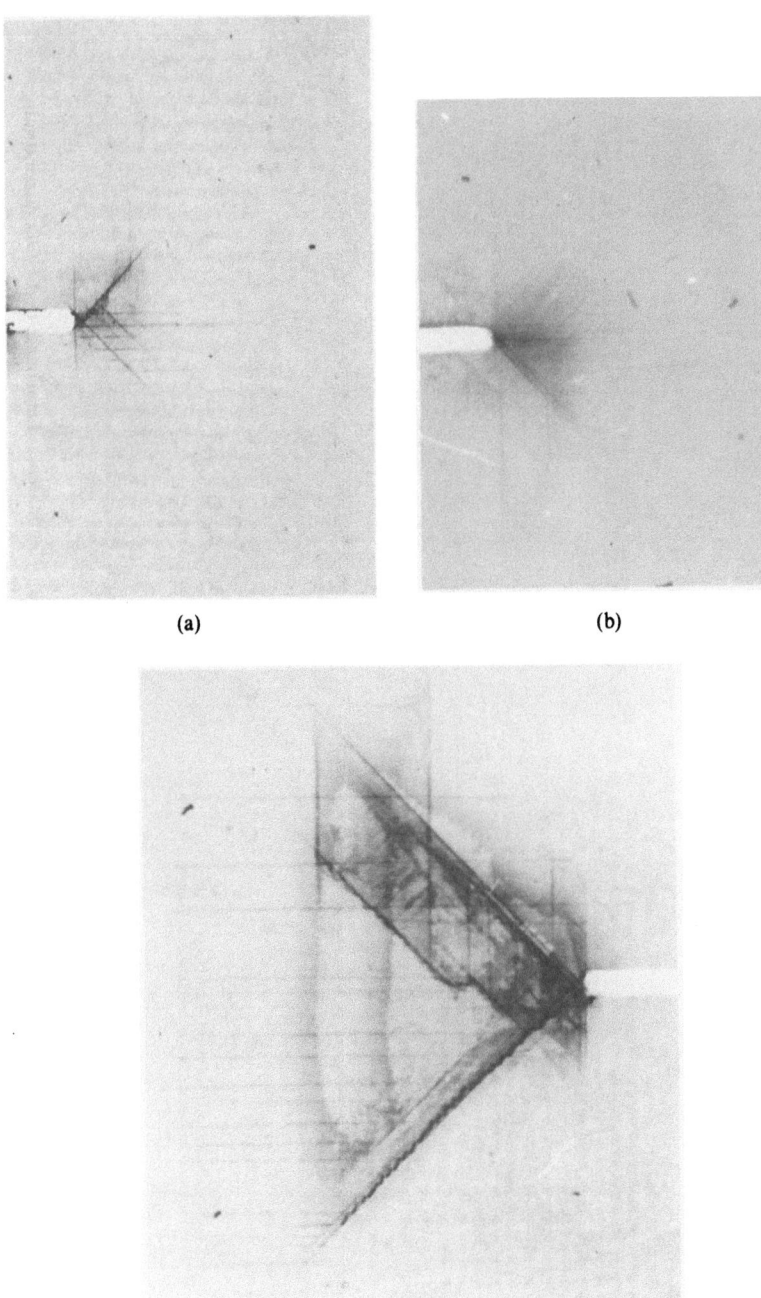

(a)

(b)

(c)

Fig. 3.   Enhanced X-ray photographs (same magnification) of damage in the $[0/\pm 45/90]_{ns}$ laminates. (a) 8 plies at critical load; (b) 64 plies at critical load; (c) 8 plies just prior to final fracture.

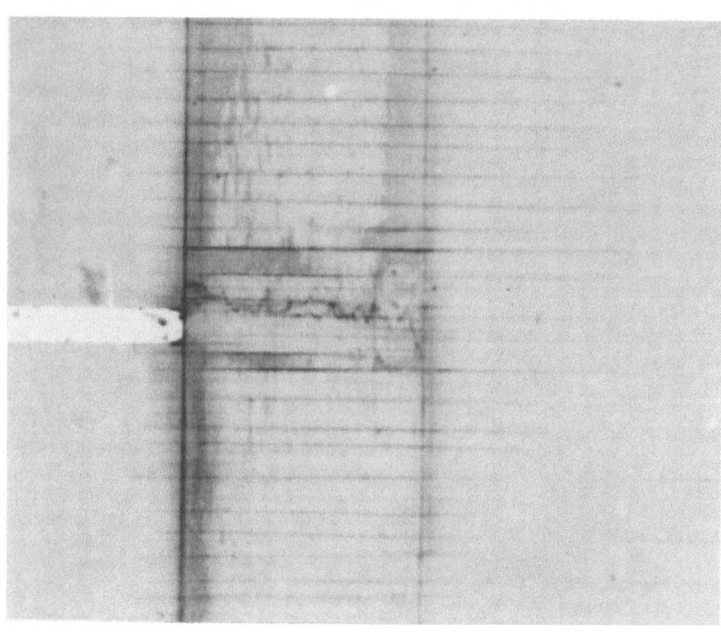

(b)

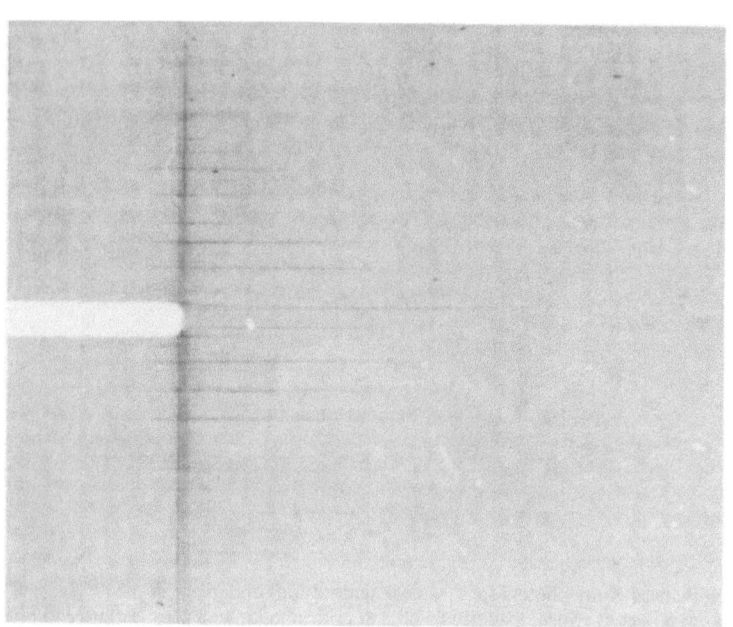

(a)

FIG. 4.   Enhanced X-ray photographs (same magnification) of damage in the $[0/90]_{2s}$ laminate. (a) Damage at first COD discontinuity; (b) damage at a later COD discontinuity.

0·25, 0·375, 0·50 and 0·625. Specimens from the thicker laminates were tested at a crack length-to-width ratio of 0·50. Four replicate tests were conducted at each test condition (laminate type, crack size, laminate thickness). The width of the machined slot simulating the crack was 0·016 in (0·4 mm) and was fabricated by an ultrasonic vibration technique.

The fracture tests were conducted at a constant crosshead displacement rate of 0·05 in/min (0·02 mm/s). The specimen ends were held in 2 in (51 mm) wide wedge-action friction grips such that the specimen length between grip ends was 5 in (127 mm). The thin specimens, 6 or 8 plies, were tested with an antibuckling support to prevent out-of-plane motion.

The recorded test data included a plot of crack-opening displacement (COD) versus load. The COD clip gauge was held directly in the crack (slot) in the composite specimen by machined knife-edge tabs. The inverse of the slope of the initial linear portion of the load versus COD curve, Figs 1 and 2, were taken as the specimen compliance. The critical load, P_c, was defined as the load at which significant damage developed at the crack tip and the specimen compliance was changed. This condition was characterized by a discontinuity or COD jump in the load versus COD records of Figs 1 and 2. X-ray examinations, such as those shown in Figs 3 and 4 were used to verify that significant damage formation occurred with an associated COD jump. In the event of no COD discontinuities the maximum test load, P_m, was taken to be the critical load.

Determination of K by the Stress Analysis Method

A finite element analysis of the center-cracked specimen geometry at the four $2a/\omega$ test values was performed. The computer code developed by the Lockheed-Georgia Company for NASA,[9] utilizes a special crack element that contains the crack tip and performs the singular zone stress analysis. The code performs analyses for either plane stress or plane strain on the basis of linear elastic fracture mechanics for fracture modes I and II. The crack element allows either isotropic material behavior or homogeneous anisotropic material behavior. In this analysis the material was treated as homogeneous and anisotropic. The finite element code requires the matrix relating stress and strain as input. This matrix is a function of the laminate stiffness properties. The computer output included mode I and II stress intensity factors and strain energy release rates for an applied unit load. For a given test condition the fracture toughness at the first COD discontinuity (P_1), critical load (P_c), and maximum load (P_m) were then determined by multiplying the appropriate load by the stress intensity factor for a unit load.

RESULTS AND DISCUSSION

Fracture Toughness of the $[0/\pm45/90]_{ns}$ Laminates

Baseline data for the $[0/\pm45/90]_{ns}$ laminate study were generated from an 8 ply laminate. Center-cracked tension specimens were tested at crack size-to-width ratios of 0·25, 0·375, 0·50 and 0·625. The toughness values experimentally determined by both the stress analysis method and compliance calibration method fell within the range of toughness values reported in the literature[7,10,11] for $[0/\pm45/90]_s$ graphite–epoxy laminates. The load versus crack-opening displacement (COD) record shown in Fig. 1 and the damage illustrated by the enhanced X-ray photographs in Figs 3a and 3c are typical for the 8-ply replicate tests.

There is similarity between the load versus COD records of the 8, 32, 64 and 96 ply $[0/\pm45/90]_{ns}$ laminates of Fig. 1. (Note that in Fig. 1 the load scale is different for each thickness whereas the COD scale is the same for each thickness.) There is one subtle difference in the load records of Fig. 1. The critical load, P_c, where significant damage develops at the crack tip, is a greater percentage of the maximum load as the specimen thickness increases. For example, for 32 plies P_c is 84 % of P_m. However, the type and magnitude of damage associated with P_c is essentially the same. Comparing the enhanced X-rays of Figs 3a, for a typical 8 ply specimen, and Fig. 3b, for a typical 64 ply specimen, it is obvious that both damage zones are comprised of matrix cracks in each fiber direction and delaminations more-or-less confined in a 45° triangle emanating from the crack tip. The damage appears more extensive in the 64 ply specimen but this is deceiving. The X-ray is an integrated through-the-thickness record of damage. Since there are more plies in a 64 ply specimen, one would expect to see more damage in this specimen than in an 8 ply specimen. If the damage zone at each crack tip in Figs 3a and 3b were encircled the size of the zones would be essentially the same.

In addition, there was no difference in the magnitude or type of damage in the specimens of various thicknesses near the maximum load. Figure 3c illustrates the damage revealed by enhanced X-ray examination near the maximum load for an 8 ply specimen. This is typical of all the $[0/\pm45/90]_{ns}$ specimens tested.

The fracture toughnesses of the $[0/\pm45/90]_{ns}$ laminates are shown in Fig. 5 for the 8, 32, 64 and 96 ply thickness. Fracture toughness based on both maximum load and critical load are shown. Both toughness curves show the same trend. Fracture toughness decreases with increasing thickness and appears to approach a lower bound at a thickness of

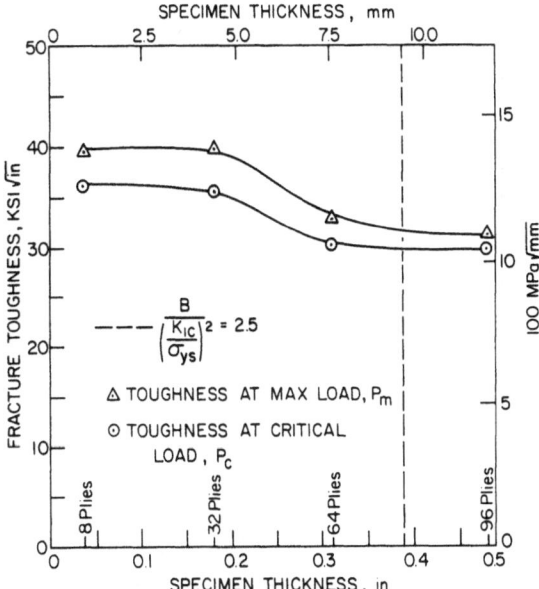

FIG. 5.   Critical and maximum fracture toughness versus specimen thickness for the $[0/\pm45/90]_{ns}$ laminate.

approximately 0·4 in (10 mm). The ASTM isotropic metals specimen thickness requirement for plane strain fracture,[2] $B \geq 2·5(K_{IC}/\sigma_{ys})^2$ where σ_{ys} was approximated by multiplying E_x by the ultimate fiber tensile strain, is indicated by the dashed line in Fig. 5. It appears that this may also be a reasonable indicator of plane strain conditions for the $[0/\pm45/90]_{ns}$ laminate.

Another interesting aspect of Fig. 5 is the relationship between the maximum fracture toughness (K_m) and the critical fracture toughness (K_c). For the 8 ply laminate the difference between K_c and K_m is about 10%. However, both curves appear to approach the same asymptotic value as the thickness increases. This is analogous to the pop-in phenomenon exhibited by some metals. It is believed[1,2] that pop-in is associated with a small increment of crack growth and arrest occurring in the interior region of the crack front; the crack growth appearance resembles plane strain crack growth. Therefore, the fracture toughness at pop-in is usually considered to be the plane strain fracture toughness. Variations in the fracture toughness, at pop-in, with specimen thickness exhibit about the same percentage change in toughness as the change in K_c shown in Fig. 5.[12,13] Thus, even though damage development is different in metals and composites, the fracture toughness–thickness variations are similar.

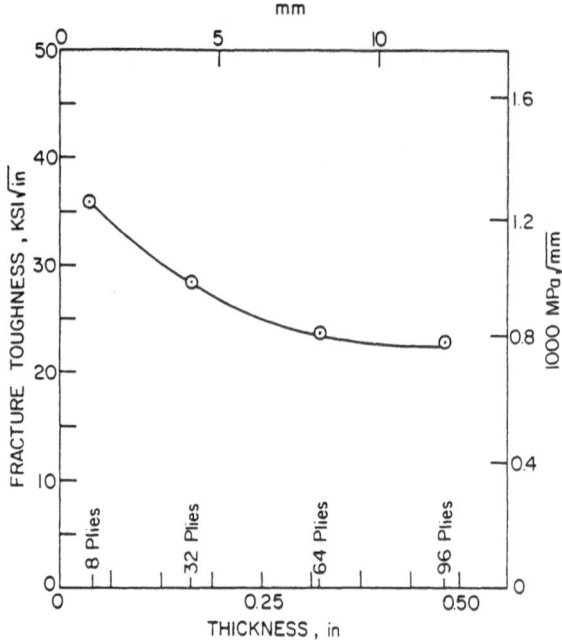

FIG. 6.   Fracture toughness at the first COD discontinuity versus specimen thickness for the $[0/\pm 45/90]_{ns}$ laminate.

One fundamental difference between the behavior of metals and laminated composites is that composites do not generally exhibit macroscopic crack extension colinear to the original crack. This is often the case even when the final fracture is macroscopically colinear, more-or-less, with the original machined crack. The $[0/\pm 45/90]_{ns}$ laminate is an illustration of this. As Fig. 3c clearly illustrates, just before final fracture a zone of damage is present that is anything but colinear with the machined crack. This damage zone is sometimes treated in a fashion similar to the treatment of small plastic zones at the crack tip.[11] The size of the original crack, a, is increased by a measure of the damage zone, ρ, determined from the fracture strength of an unnotched specimen, and the fracture toughness is computed for a crack of length $a + \rho$. This adjustment results in an increase of 1 % to 5 %, typically, in the unadjusted toughness values. This would have little effect on the trends illustrated in Fig. 5. The fracture toughness can be computed at the first perceptible COD discontinuity, P_1 in Fig. 1. X-rays taken prior to load P_1 show almost no detectable damage. Therefore, no damage zone adjustment is necessary at P_1. The fracture toughness, K_1, computed at P_1 is shown in Fig. 6 for the stress analysis method. The fracture toughness decreases with increasing specimen thickness. Note in Fig. 1 that P_1 and P_c are the same for the 8 ply specimens, but P_1 is lower than P_c at the greater thicknesses. X-rays taken of the 32, 64 and 96 ply specimens at P_1 do not show growths in damage formation nor are there changes in specimen compliance at P_1. This led to the definition of P_c as the critical load at which a significant growth in damage occurred and the specimen compliance changed appreciably.

Fracture Toughness of the $[0/90]_{ns}$ Laminates

There are substantial differences between the fracture behavior of the $[0/90]_{2s}$ laminate and the $[0/90]_{8s}$ laminate. Several of these differences are illustrated in Fig. 2 where typical load versus COD records for the two laminates are shown. The first obvious difference is that the COD value of the $[0/90]_{2s}$ laminate at maximum load is several times larger (see Table 1) than the maximum COD of the $[0/90]_{8s}$ laminate. This is in contrast to the load versus COD records for the $[0/\pm 45/90]_{ns}$ laminate, shown in Fig. 1, where the maximum COD is essentially the same regardless of laminate thickness. The second obvious difference is that the critical load, P_c, is a much lower percentage of the maximum load for the $[0/90]_{2s}$ laminate (60 %) than for the $[0/90]_{8s}$ laminate (86 %). These differences can be largely attributed to the dominant role of the axial splits that form at the crack tips in the $[0/90]_{2s}$ laminate. The axial splits are clearly illustrated in the two enhanced X-ray photographs of Fig. 4. Figure 4a shows the crack tip damage immediately after the first discontinuity in the COD plot and Fig. 4b shows the damage after a later COD discontinuity. Also visible in Fig. 4b is the macro crack that has extended colinear with the machined crack to where it is arrested at the second major axial split. These axial splits occur in the outside $0°$ plies and minimize the stress singularity at the crack tip. With the stresses at the crack tip reduced, a higher load is required to break the specimen than would be the case in the absence of the axial splits. The specimen is also more compliant once the axial splits form, so the COD values at final fracture are much higher than otherwise would be the case. Similar results for $[0/90]_{2s}$ graphite/epoxy laminates have been previously reported.[14-16] The 8 ply data tabulated in Table 1 illustrate the role of the axial splits. X-ray examinations reveal that specimen 18-5 did not form major axial splits and it can be seen that the fracture toughness and COD at the maximum load are substantially lower than their counterparts with major axial splits such as those illustrated in Fig. 4. The load versus COD plot for specimen 18-5 is similar to the $[0/90]_{8s}$ record in Fig. 2 except that the maximum load is much lower.

TABLE 1

Test data for the $[0/90]_{2s}$ and $[0/90]_{8s}$ laminates

Specimen ID	Thickness (No. of plies)	Max. load		Critical load	
		K_{max}	COD_{max}	K_c	COD_c
		ksi $\sqrt{in}$ (MPa $\sqrt{mm}$)	in (mm)	ksi $\sqrt{in}$ (MPa $\sqrt{mm}$)	in (mm)
18-3	8	65·2 (2 266)	>0·030 (>0·76)	32·3 (1 122)	0·009 5 (0·24)
18-4	8	51·8 (1 800)	0·022 (0·56)	25·9 (900)	0·012 (0·30)
18-5	8	36·9 (1 282)	0·012 (0·30)	36·9 (1 282)	0·012 (0·30)
18-6	8	60·0 (2 085)	>0·030 (>0·76)	27·8 (966)	0·012 (0·30)
19-1	32	32·5 (1 129)	0·011 (0·28)	29·2 (1 015)	0·009 (0·23)
19-2	32	30·8 (1 070)	0·010 (0·25)	28·2 (980)	0·008 5 (0·22)
19-3	32	32·4	0·010 (0·25)	30·6 (1 063)	0·009 5 (0·24)
19-4	32	32·4 (1 126)	0·011 (0·28)	28·3 (983)	0·009 (0·23)

Referring to Fig. 2 and Table 1, the fracture toughness of the $[0/90]_{8s}$, 32 ply laminate at maximum load is much lower than the toughness of the 8 ply laminate at the maximum load. The fracture toughness differs by almost a factor of two, if one excludes specimen 18-5. This difference in toughness at maximum load for 8 and 32 plies is much greater for the $[0/90]_{ns}$ laminate than the difference exhibited by the 8 and 32 ply $[0/\pm45/90]_{ns}$ laminates (Fig. 5). However, there is less difference in critical fracture toughness, K_c, between the 8 and 32 ply $[0/90]_{ns}$ laminates (Table 1) as was the case for the $[0/\pm45/90]_{ns}$ laminate (Fig. 5). Also as shown in Table 1 there is no appreciable difference in the COD of the 8 and 32 ply $[0/90]_{ns}$ laminates at the critical load contrary to the large difference in COD at the maximum load. This can again be explained by considering the role of the axial splits. The critical load is taken to be the first significant COD discontinuity; prior to this load there are no major axial splits. Apparently the axial splits that form in the 0° plies of the 32 ply $[0/90]_{ns}$ laminate are confined to the outer surface. Therefore, their effect on the fracture behavior would be much less dominant in a thick laminate than in a thin laminate. This seems to be born out by two facts. First, the maximum COD of the 32 ply laminate is much less than for the 8 ply laminate. The COD at maximum load for the 32 ply $[0/90]_{8s}$ laminate is essentially the same as was the maximum COD for the $[0/\pm45/90]_{ns}$ laminates. Second, the critical load, P_c, is a much higher percentage of the maximum load at 32 plies than at 8 plies. At 32 plies P_c is about the same percentage of P_m for both the $[0/90]_{8s}$ and the $[0/\pm45/90]_{4s}$ laminates, 86% and 84% respectively.

CONCLUDING REMARKS

The role of thickness on the fracture behavior of laminated graphite/epoxy composites is continuing to be investigated in an ongoing research program at Virginia Tech. Early program results for the $[0/\pm45/90]_{ns}$ and $[0/90]_{ns}$ laminates are reported in this paper.

The fracture toughness of the $[0/\pm45/90]_{ns}$ laminate at thicknesses ranging between 8 plies and 96 plies seems to exhibit the same variation with thickness as do isotropic metals. The fracture toughness is a decreasing function of increasing specimen thickness and asymptotically approaches a lower bound. A critical fracture toughness (K_c) can be defined for the $[0/\pm45/90]_{ns}$ laminate that exhibits a weaker variation with specimen thickness than the fracture toughness at maximum load and appears to have some analogy to the 'pop-in' phenomenon of metals.

Both the fracture toughness and COD at the maximum load for the 8 ply $[0/90]_{2s}$ laminate are significantly higher than their counterparts for the 32 ply laminate. The role of the axial splits that form in the 0° plies at the crack tip seems to be the explanation for these differences. The critical fracture toughness (K_c) of the 8 ply and 32 ply $[0/90]_{ns}$ laminates are essentially the same (as was the case with the 8 and 32 ply $[0/\pm45/90]_{ns}$ laminates. The fracture toughness of 64 and 96 ply $[0/90]_{ns}$ laminates has yet to be determined.

ACKNOWLEDGEMENTS

The financial support provided for this work by NASA Grant NAG-1-264 from the Fatigue and Fracture Branch of NASA-Langley Research Center is gratefully acknowledged. Further, sincere appreciation is extended to C. C. Poe, Jr of NASA-Langley for his encouragement and helpful discussions.

REFERENCES

1. BROWN, W. F., JR and SRAWLEY, J. E. (eds), *Plane Strain Crack Toughness Testing of High Strength Metallic Materials*, ASTM STP 410, American Society for Testing and Materials, 1966.

2. BROWN, W. F., JR (ed.), *Review of Developments in Plane Strain Fracture Toughness Testing*, ASTM STP 463, American Society for Testing and Materials, 1970.

3. PARKER, A. P., *The Mechanics of Fracture and Fatigue*, London, E. & F. N. Spon Ltd, 1981.

4. CRUSE, T. A. and OSIAS, J. R., *Exploratory Development on Fracture Mechanics of Composite Materials*, Air Force Materials Laboratory, Report No. AFML-TR-74-111, 1974.

5. OWEN, M. J. and CANN, R. J., Fracture toughness and crack-growth measurements in GRP, *J. Mater. Sci.*, **14**, Aug. 1979, 1982–1996.

6. SUN, C. T. and PREWO, K. M., The fracture toughness of boron aluminum composites, *J. Comp. Mater.*, **11**, April 1977, 164–175.

7. HAHN, H. T. and MORRIS, D. H., Fracture resistance characterization of graphite/epoxy composites, in: *Composite Materials: Testing and Design (Fourth Conference)*, ASTM STP 617, American Society for Testing and Materials, 1977, pp. 5–17.

8. JONES, R. M., *Mechanics of Composite Materials*, McGraw-Hill Book Company, New York, 1975.

9. CHU, C. S., ANDERSON, J. M., BATDORF, W. J. and ABERSON, J. A., *Finite Element Computer Program to Analyze Cracked Orthotropic Sheets*, NASA Contractor Report CR-2698, National Aeronautics and Space Administration, 1976.

10. SHIH, T. T. and LOGSDON, W. A., Fracture behavior of a thick-section graphite/epoxy composite, *Fracture Mechanics: Thirteenth Conference*, ASTM STP 743 (Richard Roberts, ed.), American Society for Testing and Materials, 1981, pp. 316–337.

11. POE, JR, C. C., A unifying strain criterion for fracture of fibrous composite laminates, *Engng Fract. Mech.*, **17**, 1983, 153–171.

12. KAUFMAN, J. C., Progress in fracture testing of metallic materials, in: *Review of Developments in Plane Strain Fracture Toughness Testing*, ASTM STP 463, American Society for Testing and Materials, 1970, pp. 3–21.

13. JONES, M. H. and BROWN, W. F., JR, The influence of crack length and thickness in plane strain fracture toughness tests, *Review of Developments in Plane Strain Fracture Toughness Testing*, ASTM STP 463, American Society for Testing and Materials, 1970, pp. 63–101.

14. MANDELL, J. F., WANG, S. S. and MCGARRY, F. J., *Fracture of Graphite Fiber Reinforced Composites*, Air Force Materials Laboratory Report No. AFML-TR-73-142, 1973.

15. SLEPETZ, J. M. and CARLSON, L., Fracture of composite compact tension specimens, in: *Fracture Mechanics of Composites*, ASTM STP 593, American Society for Testing and Materials, 1975, pp. 143–162.

16. YEOW, Y. T., MORRIS, D. H. and BRINSON, H. F., Fracture behavior of graphite/epoxy laminates, *Exp. Mech.*, **19**, Jan. 1979, 1–8.

37

The Suitability of Fibre–Cement Composites for Reinforced Adobe Structures in Earthquake Zones

D. G. SWIFT

Appropriate Technology Centre, Kenyatta University College,
P.O. Box 43844, Nairobi, Kenya

R. B. L. SMITH and K. S. RANGASAMI

Department of Civil Engineering, University of Nairobi,
P.O. Box 30197, Nairobi, Kenya

ABSTRACT

Fibre reinforced composite structures can play a useful role in low-cost housing in developing countries through the use of natural fibres like sisal.[1-3] Such composites appear to be particularly effective at enabling adobe structures to withstand earthquake loading. The paper gives results of six tests conducted on a specially designed test rig, whereby models built on a suspended platform are subjected to known accelerations by means of a load impacting against the platform.[4]

INTRODUCTION

There is need for improvement of the low-cost adobe housing used by rural low-income populations of many developing countries, where earth tremors occur causing collapse of many adobe and masonry structures with resultant loss of life and property. There is very little research devoted to earthquake protection of adobe structures.

Some properties of 'sisal–cement' have been published in other papers.[1-4] A sisal–cement adobe-walling technique developed by the authors and now used in several trial buildings was found in earlier tests to offer remarkably high resistance to impact and vibration. In this paper, the earthquake protection offered by this technique is demonstrated

525

experimentally using a specially designed rig to model the initial acceleration of a seismically loaded adobe structure.

Considering the three failure criteria of adobe structures in tension, flexure and shear, simple model structures made of adobe or adobe covered with sisal–cement, H-shaped in plan, were tested under impact loads in the test rig described below. The structure was covered with sheet metal tray containing suitable loads to simulate light or heavy roof loads.

IMPACT LOADING OF MODEL STRUCTURES (SIMULATING EARTHQUAKE SHOCKS)

The specially designed test rig is shown in Fig. 1, schematically. The purpose of the rig is to impart a single controllable acceleration to the model structure and the platform on which it is built. The platform is a reinforced concrete slab 1·5 m × 1·5 m × 10 cm in size surrounded by steel channels and suspended from a heavy girder by means of four steel cables. A second block of concrete loads tied up in a band of steel channels, weighing 1000 kg, and similarly suspended from the same girder by four cables, is allowed to fall in an arc from a known height, causing impact load on the structure platform, through the buffer springs. The platform supporting the model swings freely and is arrested at its highest point by a cable wrapped round a ratchet drum. One end of this cable is connected to the platform through a spring to reduce the shock; the other end of the cable is connected to a counterweight after wrapping over the ratchet drum. The platform is restored to its initial position by raising the counterweight after each impact.

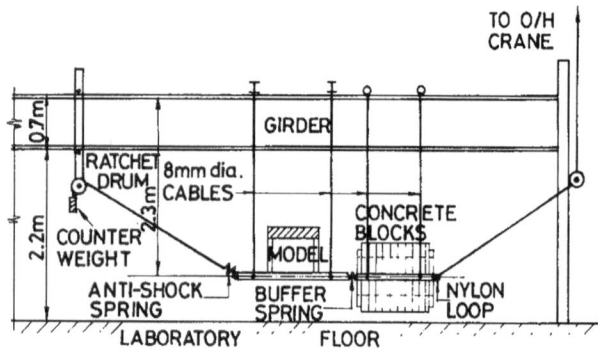

FIG. 1.   Testing rig.

The parameters and accelerations are calculated as described in another paper by the authors.[4]

TEST DATA AND RESULTS

The six test specimens were closely of the same shape and size except for the variations in the incorporation of sisal–cement covering and roof loads as described below.

Adobe bricks used in the specimens were of size $12 \times 6 \times 4.5$ cm; 5% cement was added to the soil for stabilization since it was not convenient to sun-dry the bricks in the laboratory. For sisal–cement covered structures, the bricks were laid without mortar between the joints but sisal fibres were placed transversely across the horizontal courses. The volume of fibres was about 11% of the cement mortar. The length of fibres was about 15 cm and evenly extended outside both faces of the wall. These extensions of fibres were bent down while incorporating into the cement mortar rendering, forming about 1·5 cm thick skins over the adobe structure. Care was taken to keep the fibres in the middle layer of the skin. The height of the structure was made up of eight such courses of bricks. A G.I. sheet tray containing large bricks was fixed to the structure to simulate roof loads. The tray was tied to the structure with steel sheet straps and clamps as shown in Fig. 2. All adobe structures were laid with mud mortar and had no outside rendering. The impact tests and results are given in Table 1.

FIG. 2. Specimen 5: before impact: front view.

TABLE 1

Serial No. of test specimen and description	Serial No. of impact	Horizontal distance of load at start of fall (m)	Corresponding acceleration (m/s²)	Effects/Remarks
1 All adobe structure with small roof load; walls: 74·0 kg roof: 73·5 kg	1	0·3	10·3	Roof and end walls swayed to the right, i.e. towards the impact load
	2	0·3	10·3	Entire structure collapsed; tests were videotaped
2 Part sisal and part adobe with large roof load, i.e. the end walls only were rendered in sisal–cement; middle wall was adobe only; walls: 100·5 kg roof: 112·0 kg	1	0·3	10·3	Small visible vertical cracks in middle wall
	2	0·3	10·3	The vertical cracks widened; two horizontal cracks in the middle wall appeared
	3	0·3	10·3	The middle wall collapsed; but sisal–cement end walls stood well supporting the roof
	4	0·3	10·3	The end walls collapsed like a mechanism since there was no middle wall. The roof fell to the right. Tests were videotaped

Specimen	Description	No.			Remarks
3	Part sisal–cement and part adobe; small roof load; walls: 107.4 kg roof: 57.0 kg	1	0.3	10.3	No effect
		2	0.3	10.3	No effect
		3	0.3	10.3	No effect
		4	0.3	10.3	No effect
		5	0.3	10.3	Small vertical cracks in middle adobe wall at junction with sisal–cement end walls
		6	0.4	14.6	The crack widths increased to 5 mm
		7	0.4	14.6	The crack widths increased to 6 mm
		8	0.4	14.6	The crack widths increased to 8 mm
		9	0.6	22.3	Left end wall broke in the middle height where the roof holding steel straps passed; the structure collapsed like a mechanism to the left away from impact load. This compared to the 2nd specimen shows more than double resistance to shock loads. Light roof is therefore very desirable; photographs failed as camera was not loaded properly
4a	All adobe with large roof load	—	—	—	The end walls collapsed even as the roof tray was attempted to be placed in position
4b	All adobe with large roof load; walls: 74.0 kg roof: 112.0 kg	1	0.3	10.3	Structure collapsed at first impact; this shows that all adobe is very unstable and very weak

(continued)

TABLE 1—contd.

Serial No. of test specimen and description	Serial No. of impact	Horizontal distance of load at start of fall (m)	Corresponding acceleration (m/s²)	Effects/Remarks
5. All sisal–cement with small roof load; walls: 110 kg roof: 49 kg	—	—	—	Figures 2 and 3; before impact
	1	0·3	10·3	No effect
	2	0·3	10·3	No effect
	3	0·3	10·3	No effect
	4	0·4	14·6	No effect
	5	0·4	14·6	No effect
	6	0·5	18·9	Figure 4: visible cracks inside right end wall
	7	0·5	18·9	Figure 5: cracks increased in size
	8	0·5	18·9	Figure 6: crack widths increased
	9	0·5	18·9	Figure 7: cracks inside right end wall
				Figure 8: cracks outside left end wall
	10	0·5	18·9	Figure 9: collapse of end walls
	11	0·5	18·9	Figure 10: collapse of roof to the right; middle wall unaffected
6. All sisal–cement with large roof load; walls: 98 kg roof: 112 kg	—	—	—	Figure 11: before impact
	1	0·4	14·6	No effect
	2	0·5	18·9	Visible crack inside right end wall
	3	0·5	18·9	Crack widths increased
	4	0·5	18·9	Figure 12: crack widths further increased
	5	0·5	18·9	Figure 13: crack widths further increased
	6	0·5	18·9	Figure 14: collapse of end walls; middle wall unaffected

FIG. 4. Specimen 5: after 6th impact.

FIG. 6. Specimen 5: after 8th impact.

FIG. 3. Specimen 5: before impact: diagonal view.

FIG. 5. Specimen 5: after 7th impact.

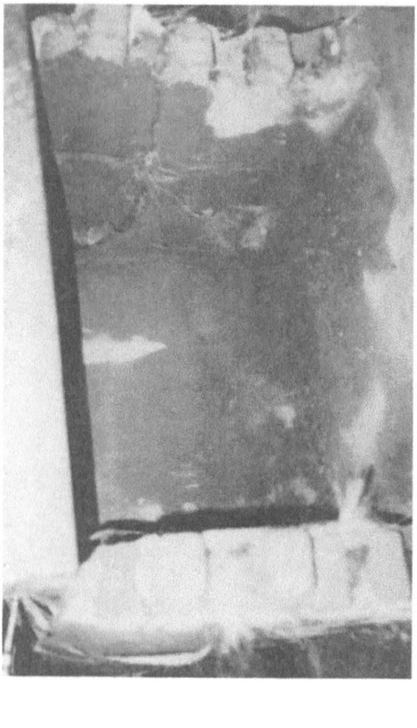

FIG. 8. Specimen 5: after 9th impact: left end wall outside view.

FIG. 7. Specimen 5: after 9th impact.

FIG. 10. Specimen 5: after 11th impact.

FIG. 9. Specimen 5: after 10th impact.

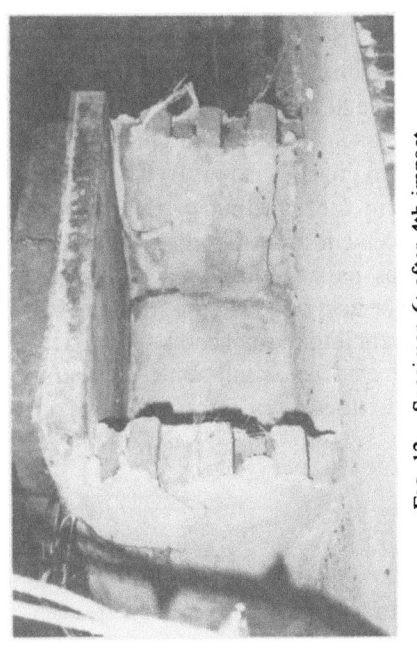

FIG. 12. Specimen 6: after 4th impact.

FIG. 14. Specimen 6: after 6th impact.

FIG. 11. Specimen 6: before impact.

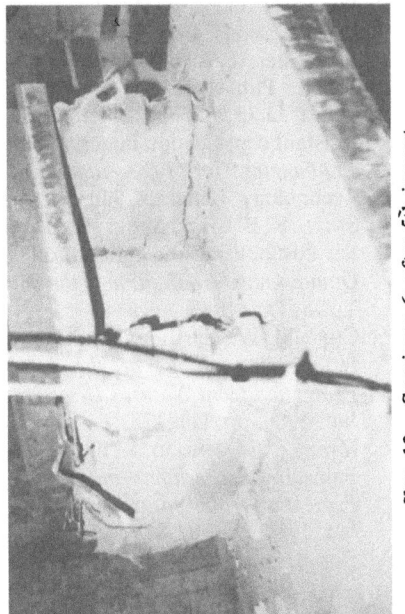

FIG. 13. Specimen 6: after 5th impact.

CONCLUSIONS

Adobe structures without sisal–cement protection have no shock resisting capacity. The sisal–cement walling technique described here has very great toughness against shock loads. Light roof loads are better than heavy roof loads to resist shocks; this is also observed in previous papers by Chin[5] and Mian et al.[6] Interior longitudinal walls in the direction of the impact loading need not be reinforced with sisal–cement as seen in the test on the third specimen. Roof-holding steel straps passing through walls cause initiation of cracks in walls. Roofs should be held or tied to the structure of the walls by more suitable methods to prevent initiation of cracks at impact loads. According to these tests and other evidence, sisal–cement appears to be well suited to protecting adobe structures, even under earthquake loading.

REFERENCES

1. SWIFT, D. G. and SMITH, R. B. L., Sisal fibre reinforcement of cement paste and concrete. Volume 1 of *International Conference on Materials of Construction for Developing Countries*, Bangkok, 1978, pp. 221–234.
2. SWIFT, D. G., The use of natural organic fibres in cement: some structural considerations, *Composite Structures*, International Conference on Composite Structures, Paisley, September 1981 (ed. I. Marshall), pp. 602–617, Applied Science Publishers.
3. SWIFT, D. G. and SMITH, R. B. L., Fibre reinforced concrete as an earthquake-resistant construction material, *Proceedings of the International Conference on Engineering for Protection from Natural Disasters*, Asian Institute of Technology, Bangkok, Jan. 1980, pp. 325–336.
4. SMITH, R. B. L. and SWIFT, D. G., Applications of low modulus fibre concrete to low-cost housing and agricultural buildings, accepted for publication at *Seventh Quinquiennial Convention of the South African Institute of Civil Engineers*, Cape Town, Oct. 1983.
5. CHIN, M. W., Earthquake resistant construction of lower cost housing in West Indies, *Proceedings of the International Conference on Engineering for Protection from Natural Disasters*, Asian Institute of Technology, Bangkok, Jan. 1980, pp. 111–128.
6. MIAN, Z., MAHMOOD, K. and WASTI, S. T., Building in natural hazard areas of Pakistan, *Proceedings of the International Conference on Engineering for Protection from Natural Disasters*, Asian Institute of Technology, Bangkok, Jan. 1980, pp. 907–919.

38

Performance of Banana Fabric–Polyester Resin Composites

K. G. Satyanarayana, K. Sukumaran, A. G. Kulkarni,
S. G. K. Pillai and P. K. Rohatgi*

Regional Research Laboratory
(Council of Scientific and Industrial Research),
Trivandrum—695 019, Kerala, India

ABSTRACT

In this paper techniques of fabrication of laminates and some typical consumer articles (like voltage stabilizer covers, 16 mm projector covers) from banana-cotton (hybrid fabric)/polyester resin composites are reported. Both laminates and consumer articles made out of composites were exposed to indoor environments and their properties and performance were monitored through destructive and non-destructive testing methods. The results demonstrate that these composites have several potential uses.

INTRODUCTION

In recent years, natural fibres like sisal, coir and banana have attracted the attention of scientists and technologists for applications in consumer articles, low cost housing and civil structures where the high cost of synthetic fibres like glass, carbon and boron restrict their use. Other reasons for preferring natural fibres are their low density and low cost. In addition, natural fibres represent a very abundant renewable resource. There have been attempts to fabricate laminates/structures and consumer articles using some of these natural fibres in polymeric matrices,[1-14] where high strength/high modulus requirements are not very demanding with a view to substituting synthetic fibres in such applications. None of these

Present address: Regional Research Laboratory (C.S.I.R.), Bhopal—462 026 (MP), India.

studies report the detailed fabrication process in making natural fibre–polymer composites. Further, most of these studies deal with jute fibre incorporated in either phenol formaldehyde or polyester or epoxy resins.[1–8,10,11,14] In these earlier studies, properties of these composites were evaluated and product development, including the development of roofing sheets for low cost housing applications, is described. These studies have indicated the possibility of using natural fibres like jute as reinforcement or filler in polymeric matrices for common uses like construction of grain storage silos and small fishing boats. One of the studies[10] reports the performance of jute fibre–resin composites after exposing these laminates to weathering for nearly seven years, pointing to the need for surface modification of fibres for increasing the durability of these composites. In our earlier papers[12,13] it is reported that fibres like coir and banana can also be incorporated in polymer matrix to make laminates and consumer articles like roofing, stabilizer cover mirror casings and projector covers. However, the durability of these composites has not been evaluated, which is essential for successful use of these new materials.

This paper reports a detailed fabrication process for preparing banana fabric–polyester resin composite to prepare laminates and a few consumer articles. These new materials were subjected to typical indoor environments and their performance was monitored through destructive and non-destructive testings. Structural observations of the composites have also been reported which throw light on the integrity of the composites after typical indoor exposures.

EXPERIMENTAL PROCEDURE

Banana–cotton fabric with banana fibre as weft and cotton in warp directions was obtained from M/s Khadi & Village Industries Commission, a rural development organization, Trivandrum. Average length of the banana fibre available is about 1 m and therefore by using knots, the fabric was prepared on a usual fabric weaving machine. Characteristics of the fabric are given in Table 1. A general purpose polyester resin (properties are listed in Table 2), which is one of the cheapest resins available in the market, was used. A simple hand layup process was used to fabricate laminates/consumer goods using the banana fabric and polyester resin in an industrial atmosphere.

First the resin was mixed with suitable proportions of MEK peroxide and cobalt naphthenate, as hardener and catalyst respectively. Moulds of simple design were made. After preparing the moulds to sufficient degree of finish, the releasing agent (polyvinyl alcohol) was applied. A gel coat

TABLE 1
Properties of banana cloth

1.	Air permeability ($cm^3/s/cm^2$)	169·7
2.	Bursting strength (MN/m^2)	0·41
3.	*Tensile strength—Warp (kg)	20·3
4.	Elongation at break %—Warp	14·1
5.	Tensile strength—Weft (kg)	27·0
6.	Elongation at break—Weft	3·1
7.	Ends/in	52
8.	Picks/in	52

Suction head area = $2·84\,cm^2$.
Water column pressure difference = $10\,mm$, * $5 \times 20\,cm$ ravelled strip.

consisting of an unthinned normal resin with or without pigment was then applied on the mould so as to obtain resin rich surface. Layer of resin wet fabric was next laid on clean platform to build up the desired thickness. The saturated fabric was then pressed by a roller to remove air bubbles and worked up till the wrinkles and air pockets were removed from the mould at each step. A slight pressure was applied to keep the fabric in position whenever required. This process also enabled one to get smooth, uniform and parallel faces. Number of layers of fabric in each laminate was varied (9 to 18 wt %) to obtain optimum weight percentage that can be incorporated by hand layup process. Fabrication of these laminates was found to be as easy as glass fibre reinforced plastics. This process also eliminates the usual irritation caused to workers when glass fibre is used. The laminate/composite was then allowed to set, the articles were removed from the mould, and then they were subjected to post curing. In all cases, trimming of edges was required, and only one surface of the component had a smooth surface finish, as expected.

TABLE 2
Properties of polyester resin

1.	Density (kg/m^3)	1 300
2.	Strength (MN/m^2): Tensile	41·38
	Flexural	89·69
3.	Percentage elongation	2·65
4.	Modulus of elasticity (GN/m^2)	2·06
5.	Flexural modulus (GN/m^2)	5·10
6.	Impact resistance unnotched (kg/m^2)	77·5
7.	Water absorption	0·21–0·40
8.	Volume resistivity ($\Omega\,cm$)	1 000

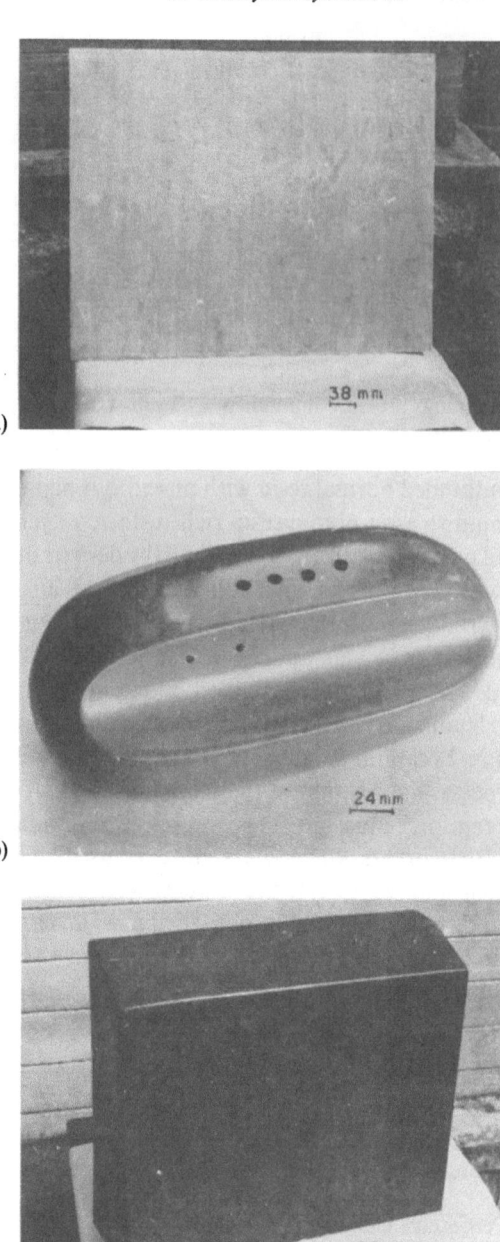

FIG. 1. Components made of banana fabric–polyester resin: (a) laminate; (b) voltage.
stabilizer cover; (c) 16 mm projector cover.

In the case of components, viz. voltage stabilizer cover and 16 mm projector covers, 14 wt % fabric was used as this amount of fabric would give the desired thickness of the component.

Both laminates and components were then subjected to indoor weathering exposure as they would be subjected to during their normal use. Figure 1(a, b and c) show these components after exposure for 242 days.

Strength properties of laminates were measured before, and at various intervals of exposure using a 10 ton Instron testing machine. Testing of composites for ultimate tensile strength (UTS) and percentage elongation was carried out as per ASTM Standard No. D 635, while flexural strength and flexural modulus were measured to per ASTM D 790.

Impact resistance was measured using a pendulum impact testing machine No. PSW 0.4 as per German Standard No. TGL 0-53453. Volume resistivity was measured using a Million Megahommeter Model RM-160 Mk III A with a D.C. voltage of 100–1000 V as described elsewhere.[12] In all cases, specimens were cut in such a way that tests were carried out in the weft direction, i.e. in the direction parallel to the direction of banana fibres.

Non-destructive testing of components and laminates before and at various intervals after exposure was carried out using a Fokker bond tester, with a view to finding any debonding that might occur between the fabric and resin on exposure over the interval of time of testing used in this investigation. The Fokker bond tester works on the principle of measuring the variation in ultrasonic resonance frequency due to the effect of loading on the transducer. The loading is a result of difference in bonding between unbonded and bonded laminates resulting in a certain level of resonance frequency. If there is a change in bonding of a material over the period of time, a change in this resonance frequency level will be observed. If there is no change in this level, it is inferred that no change in bonding of the material has taken place. Thus, this technique will help in predicting the mechanical testing data at various intervals provided correlations between NDT data and strength data by mechanical testing are determined in the initial stages.[15]

All measurements were made at 60–65 % RH and 25–30 °C.

RESULTS AND DISCUSSION

Table 3 lists the UTS, percentage elongation, Young's modulus and flexural strength of polyester resin and banana fabric–polyester resin composite (as

TABLE 3
Mechanical properties of composites

Materials	UTS (MN/m^2)	Elongation percentage	Modulus (GN/m^2)	Flexural strength (MN/m^2)
Polyester resin	41·38	2·65	2·06	89·69
Banana fibre	529–754	1–3·5	7·7–20·0	—
Cotton fabric incorporated	48–62	—	2·76–4·14	89–124
Banana–cotton fabric incorporated				
9 wt %	25·86	1·88	1·36	52·38
14 wt %	30·96	2·68	2·03	61·24
18 wt %	29·50	2·18	1·98	60·40

prepared) studied in this investigation. Properties of banana fibre– and cotton fabric–polyester resin prepared by compression moulding are also included in the table for comparison. It can be seen from the table that incorporation of 9 wt % banana–cotton fabric has resulted in a decrease of up to 50 % in tensile strength and up to 25 % of flexural strength of polyester matrix with relatively smaller decreases in the modulus and percentage elongation. By increasing the fabric content up to 18 wt %, the value of tensile strength and flexural strength have shown marked increases over the 9 % of composite with 14 % fabric showing the best properties. However, it was found that this is the upper limit of weight percentage (i.e. 18 wt %) of the fabric that could be introduced in the polyester matrix by the hand layup technique. The decrease in tensile and flexural strength values of the composite from those of both polyester resin or banana fibre may be due to the poor tensile strength of the fabric itself and/or poor bonding between the fabric and the resin used in this study. The latter is evident from Fig. 2(b) which is a scanning electron micrograph showing debonding of fibres. Poor bonding was even evident in the freshly prepared sample (Fig. 2(a)). Therefore, the fabric only acts as a filler in the present experiments. However, as in the case of coir,[13] surface modification of the fabric may improve the wettability between the fabric and resin.

It may be noted that the strength properties of composites made in this investigation were adequate for intended applications like voltage stabilizer covers and projector covers even though wetting was not obtained and there were voids between the fibres and polyester matrix. The laminates and

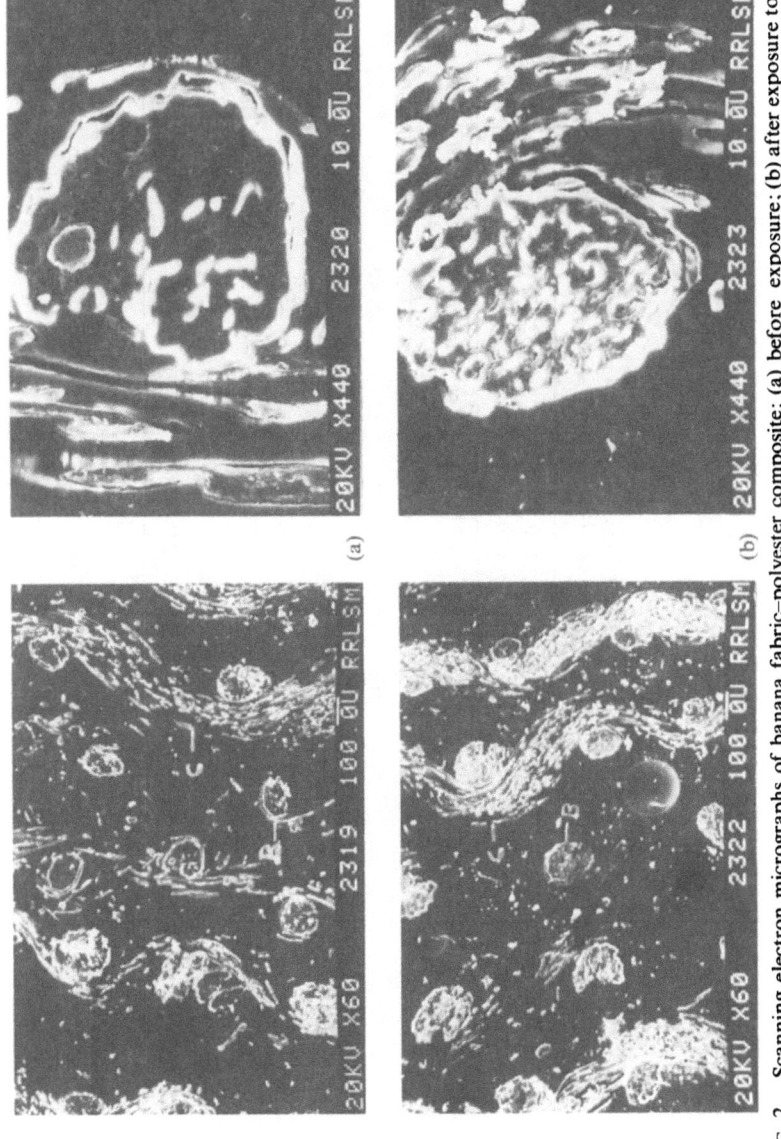

FIG. 2. Scanning electron micrographs of banana fabric–polyester composite: (a) before exposure; (b) after exposure to indoor atmosphere for 242 days (B, banana fibre; C, cotton).

TABLE 4

Physical and mechanical properties of 14 wt % banama fabric–polyester composites after different duration of exposure

Property	0 Exposure	110 Days exposure	152 Days exposure	212 Days exposure	242 Days exposure
1. Density (kg/m³)	1 215	1 215	1 197	1 207	1 219
2. Strength (MN/m²)					
Tensile	27·96	31·15	30·60	26·98	24·28
Flexural	64·00	56·66	58·46	62·79	55·56
3. Percentage elongation	3·40	3·00	2·60	3·9	3·28
4. Modulus of elasticity (GN/m²)	3·34	2·52	2·28	2·11	2·07
5. Flexural modulus (GN/m²)	4·16	2·63	3·3	3·38	3·51
6. Impact resistance un-notched (kg/m²)	748·50	714·64	673·78	650·74	625·00
7. Water absorption (24 h room temp.) %	1·93	2·28	2·32	2·10	2·32
8. Volume resistivity (Ω cm)	400	275–400	—	—	290–400

consumer articles continued to show integrity of structure even after eight months of indoor exposure.

Table 4 lists density, tensile and flexural strength, Young's modulus, flexural modulus, impact strength, percentage elongation, water sorption and volume resistivity of 14 wt % banana fabric–polyester composite after exposing the laminates to indoor atmospheres for various intervals of time up to 8 months.

It can be seen that no significant changes were observed in the properties of composite during the time period of the present investigation. Statistical analysis of the results showed no significant variation in the observed values. Even the case of the impact strength values of composite, which showed a decrease from 748·5 kg/m² at zero exposure to 625 kg/m² after 242 days may not be a significant change.

The above results point to the fact that there is no significant further decrease in the bonding between fabric and the resin as a result of exposure to indoor atmosphere. This is also evident from Fig. 3 which shows typical stress–strain curves of banana fabric–polyester composite

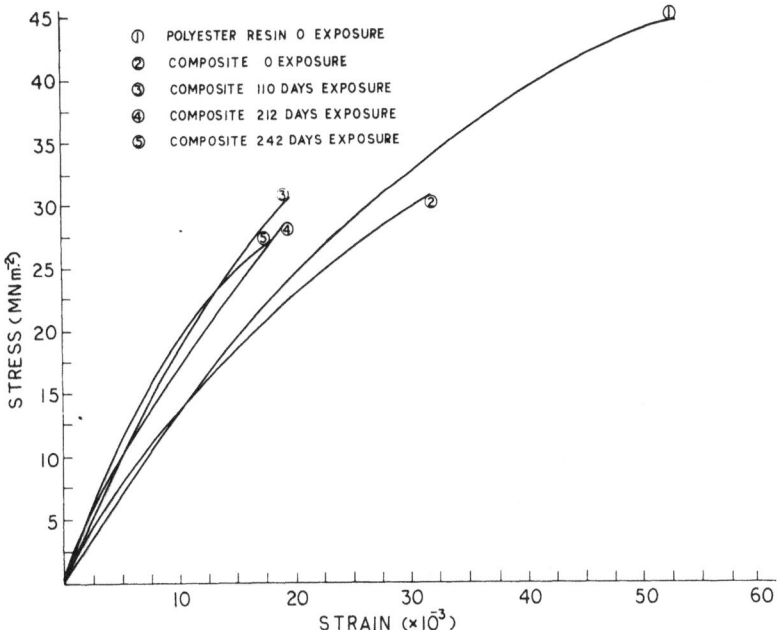

Fig. 3.   Stress–strain curves of banana fabric–polyester composite after various intervals of indoor exposure.

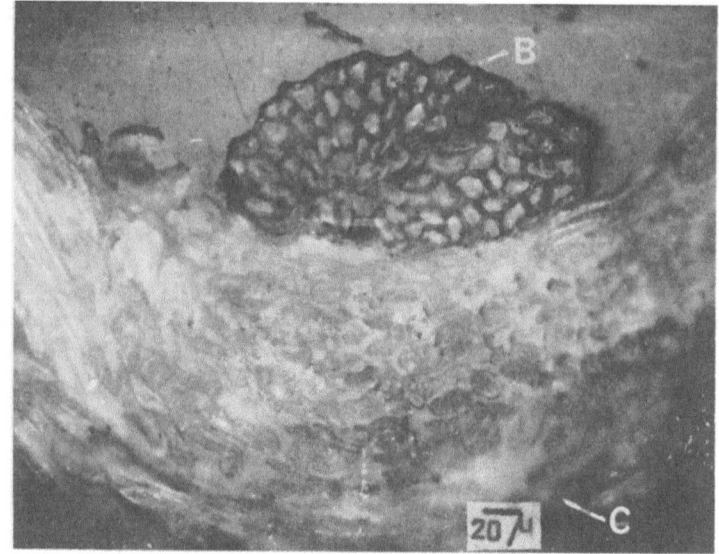

(a)

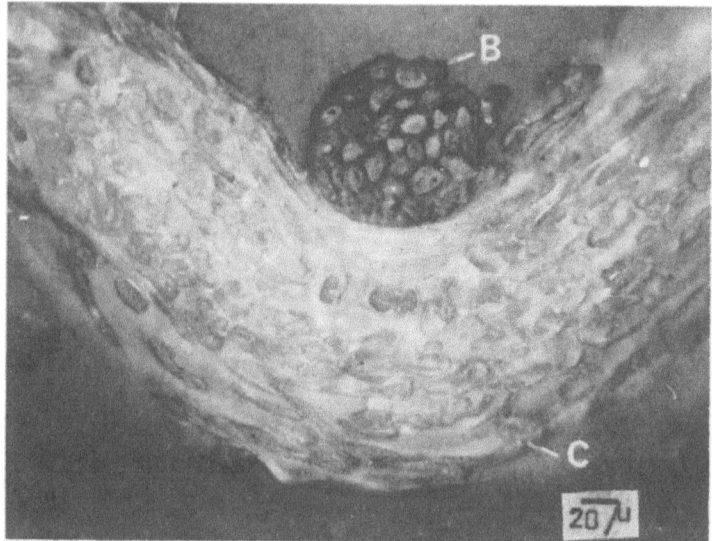

(b)

FIG. 4.   Optical micrographs of banana fabric–polyester composite: (a) before exposure;
(b) after exposure to indoor atmosphere for 242 days (**B**, banana; C, cotton).

exposed to indoor atmospheres at various intervals of time. As can be seen there is no significant change in the pattern of the stress–strain curves. If anything, the modulus of the composites increases with an accompanying decrease in percentage elongation after 100 days of exposure.

No further change in bonding between fabric and polyester resin was also confirmed by scanning electron micrographs shown in Figs 2 and 4. Figure 4(a and b) are photomicrographs of composites before and after exposure showing no change in the close pattern of the fabric and position of the banana fibre even after exposure of 242 days. Similarly, Fig. 5(a and b) are fractographs of the composite before and after 242 days of exposure identical to Fig. 2, showing the same degree of debonding between the fabric and polyester resin.

Further evidence for no change in bonding between fabric and polyester resin over the period of testing used in the present study was collected using the Fokker bond testing. It was found that bond tester values remained unchanged (maintained same level) at identical positions from zero exposure to 242 days of exposure of banana fabric–polyester composite, confirming that no further debonding of layers had occurred due to exposure which otherwise would have resulted in lower mechanical properties.

The present study thus indicates that natural fibres like banana fibre in the form of fabric can be incorporated in a polymeric matrix for fabrication of laminates and consumer articles for low strength engineering applications. After 242 days exposure to indoor atmospheres, tensile and flexural strength values of the composite were $24\cdot28\,\mathrm{MN/m^2}$ and $55\cdot56\,\mathrm{MN/m^2}$ respectively, percentage elongation was $3\cdot28$, and modulus of elasticity was $3\cdot51\,\mathrm{GN/m^2}$. Further, these composites continued to show structural integrity indicating that they are adequate for several applications.

No further change in lack of bonding between the fabric and polyester resin was observed when the composite was exposed to indoor atmospheres up to a period of eight months. However, as indicated elsewhere,[17] further studies are required to overcome the problem of compatibility of fibres (or improving their wettability) with the polymeric matrix through surface modifications. By proper surface modifications, as in the case of jute fibres,[14] it may even be possible to reduce resin consumption by natural fibres and also reduce moisture absorption by polymer–natural fibre composites. Further work in this direction is going on at RRL Trivandrum so that new openings for the abundantly available natural fibres may be developed.

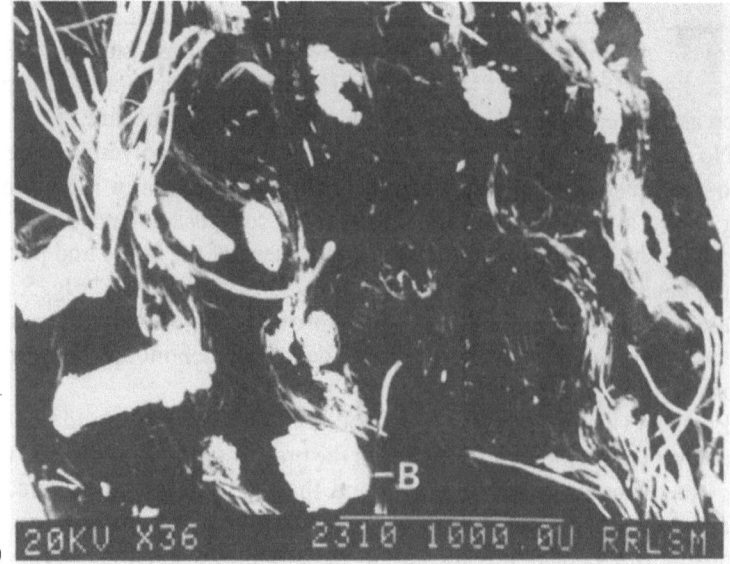

(a)

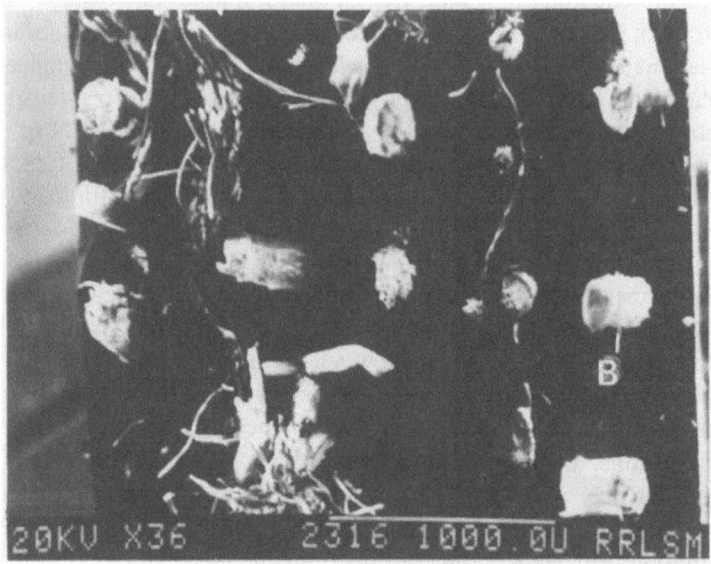

(b)

FIG. 5. Fractograph of composites: (a) before exposure; (b) after indoor exposure for 242 days (B, banana; C, cotton).

CONCLUSIONS

(1) Up to 18 wt % banana–cotton fabric can be incorporated into polyester resin by the hand layup process to get adequate strength properties for certain applications.

(2) It is possible to subject banana fibres for moulding into various shapes in a polyester resin by a proper fabrication technique.

(3) Consumer articles like voltage stabilizer covers and 16 mm projector covers have been made using the 14 wt % banana–cotton fabric and polyester resin. These consumer articles continued to retain structural integrity after 242 days of exposure and showed no signs of degradation and deterioration.

(4) Strength properties of the composites are lower than the matrix resin used indicating lack of bonding between fabric and the resin. The voids between banana fibres and the polyester matrix in the as-prepared composites were confirmed using scanning electron microscope studies.

(5) No significant change in properties of the composites has been observed even after indoor exposure for about eight months.

(6) Periodic non-destructive testing (Fokker Bond Testing) was successfully used to monitor the continued structural integrity of the composites over the time period of study.

ACKNOWLEDGEMENTS

The authors express their sincere thanks to REPLACE Division of the Vikram Sarabhai Space Centre for extending their NDT facility. They are particularly thankful to Mr Vijayan of REPLACE in helping with NDT measurements. Thanks are also due to Mr Appukuttan Nair of Consolidated Fibres for extending facilities for fabrication of composites, Dr C. Pavithran, Scientist, R.R.L. Trivandrum for helpful discussions and Mr K. K. Ravikumar for his help with the experimental work.

REFERENCES

1. BHATTACHARYA, D. N., CHAKRAVARTHI, I. B. and SENGUPTA, S. R., *J. scient. Ind. Res.*, **200** (1961), 193.

2. WINFIELD, A. G. and WINFIELD, B. Z., in: *Fillers and Reinforcements for Plastics*, Advances in Chemistry Series No. 134, American Chemical Society, Ohio, 1974.

3. PARAMASIVAN, T. and ABDUL KALAM, A. P. J., On the study of indigenous natural fibre composites, *Fibre Sci. & Techn.*, **7** (1974), 85–88.

4. NAGABHUSHANAM, T., RADHAKRISHNAN, G., JOSEPH, K. P. and SANTAPPA, M., *Proc. II Symposium on New Fibers and Composites*, Sponsored by Dept of Science and Technology, India and UNIDO, January 10–11, 1977, p. 3.1.

5. WINFIELD, A. G., Jute reinforced polyester project for UNIDO, Govt of India, *ibid.*, p. 18.1.

6. SATHYA, C. R., Progress report from VSSC of the project on newer fibres and composites, *III Int. Symp. on Newer Fibres and Composites*, SASMIRA, Bombay, India, 1978.

7. WINFIELD, A. G., Jute reinforced polyester project for UNIDO, Govt of India, *Plastic and Rubber International*, **4** (1979), 23.

8. Save Energy—Save Money, Composite News, *Composites*, **10** (1979), 61.

9. MCLAUGHLIN, E. C., The strength of bagasse fibre reinforced composites, *J. Mater. Sci.*, **15** (1980), 886–890.

10. SINGH, S. M. and JAIN, S. K., Jute reinforced polyester sheet and its performance, *Plastics & Rubber Materials and Applications*, May 1980, 65–66.

11. SHAW, A. N. and LAKKAD, S. L., Mechanical properties of jute reinforced plastics, *Fibre Sci. & Tech.*, **15** (1981), 41–46.

12. SATYANARAYANA, K. G., KULKARNI, A. G., SUKUMARAN, K., PILLAI, S. G. K., CHERIYAN, K. A. and ROHATGI, P. K., On the possibility of using natural fibre composites, *Composite Structures*, Proc. Int. Conf. on Composite Structures, Paisley, Scotland (ed. I. Marshall), Applied Science Publishers, London, 1981, p. 42.

13. SUKUMARAN, K., KULKARNI, A. G., PAVITHRAN, C., SATYANARAYANA, K. G., PRASAD, S. V., PILLAI, S. G. K. and ROHATGI, P. K., Properties of natural fibre–polyester composites, paper presented at the *National Sem. on Building Materials, Their Science and Technology*, organised by Indian National Science Academy, Institution of Engineers, New Delhi, CBRI, Roorkee, 15–16 April 1982.

14. SRIDHARA, M. K., BASAVARAJAPPA, G., KASTURI, S. G. and BALA-SUBRAMANIAN, N., Evaluation of jute as a reinforcement in composites, *Ind. J. Text. Res.*, **7** (1982), 87–92.

15. THOMAS, K. K., The status of NDT of composites in India, *Proc. Int. Symp. on Fibres and Composites*, New Delhi, Jan. 1976, pp. 34-1–16.

39

Comparative Study on the Incorporation of Composite Material for Tyre Computation*

HEINRICH ROTHERT and BA NGUYEN

Institut für Statik, University of Hannover,
Callinstrasse 32, D-3000 Hannover, West Germany

and

ROLF GALL

Institut für Mechanik, Hochschule der Bundeswehr Hamburg,
Holstenhofweg 85, D-2000 Hamburg 70, West Germany

ABSTRACT

Three-dimensional linear elastic orthotropic finite elements were used for the approximation of reinforced tyre materials. Various authors have expressed analytically the elastic constants. In order to decide which of the different methods to find realistic constants can be recommended, three-dimensional FEM computations have been compared to experimentally obtained results of tension tests. It was found that all the different methods lead to equivalent results.

Finally it is shown that the applied three-dimensional element is best suited to investigate all the effects of tyres due to reinforcement.

NOTATION

E_F, E_M	Young's moduli of fibre/matrix
v_F, v_M	Poisson's ratio of fibre/matrix

* The work reported in this paper has been supported by the West German Minister of Research and Technology. Experiments were carried out by Continental Gummi Werke AG, Hannover, West Germany.

G_F, G_M Shear moduli of fibre/matrix
ϕ Volume ratio
θ Cord angle
E_a, E_b, E_c $a/b/c$—direction modulus in the material axes system
$\nu_{ab}, \nu_{ac}, \nu_{bc}$ Strain ratios in the material axes system
G_{ab}, G_{bc}, G_{ca} Shear moduli in the material axes system
C_{abc} Stress–strain matrix in the material axes system
${}_0^t S_{ij}$ Second Piola–Kirchhoff stress tensor
${}_0^t C_{ijrs}$ Constitutive tensor
${}_0^t \varepsilon_{rs}$ Strain tensor
${}^t \tau_{mn}$ Cauchy stress tensor
${}_t^t C_{mnpq}$ Constitutive tensor
${}_t^t \varepsilon_{pq}$ Strain tensor

1. INTRODUCTION

Due to its heterogeneous, anisotropic, laminated and non-linear nature, the tyre is one of the most complexly loaded structures. In recent years, the finite element method (FEM) became a useful and powerful analytical tool for predicting deformations, strains and stresses in loaded tyres.[1-5] In this report, geometrically non-linear and orthotropic linear-elastic two-dimensional-axisymmetric and three-dimensional finite element structures (see Fig. 1) are used in the inflation analysis of tyres. Great emphasis is being placed on experimental investigations of (reinforced) rubber disks.

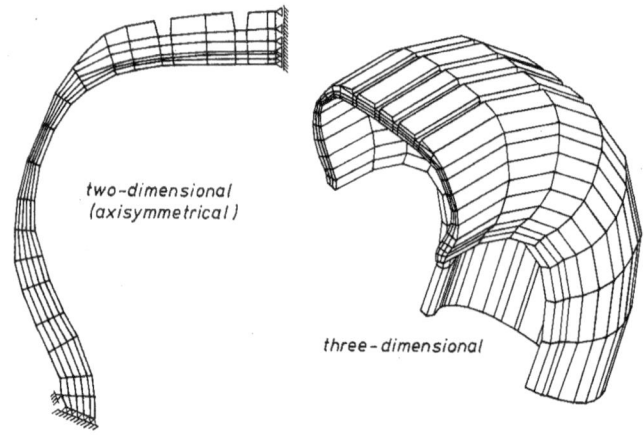

two-dimensional
(axisymmetrical)

three-dimensional

FIG. 1.   FEM models of a passenger car tyre.

REINFORCED LAYER   ORTHOTROPIC ELEMENT

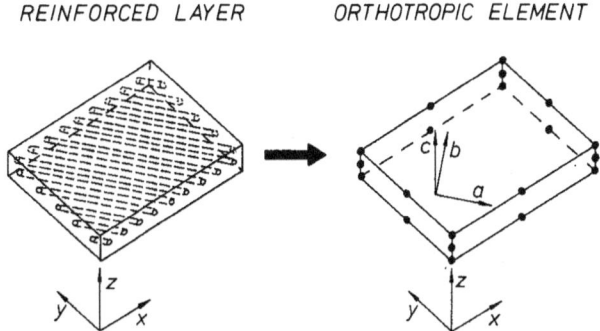

FIG. 2.   Orthotropic linear-elastic 3D element approximating a fibre-reinforced layer.

The authors believe that only with the comparison of numerical and test results can realistic material properties be derived satisfactorily.

Each layer of the tyre can be approximated as being homogeneous and orthotropic. Various authors[5-10] have expressed analytically the elastic constants of orthotropic laminated layers dependent upon the Young's moduli, Poisson's ratio and shear moduli of the cord and rubber according to their volume ratio. One purpose of this paper is to discuss the applicability of these different methods for the approximation of reinforced tyre material by comparing experimentally obtained results to finite element calculations using three-dimensional orthotropic linear-elastic elements.

2.   ELASTIC CONSTANTS OF A FIBRE-REINFORCED LAYER

The behaviour of a unidirectional reinforced layer with a very stiff cord and a very compliant matrix can be described approximately by a linear-elastic orthotropic material model (*see* Fig. 2) as it is used for example in the FEM program ADINA.[11] In the most general case, there are nine elastic constants needed to evaluate the stress–strain matrix C_{abc}:

$$C_{abc}^{-1} = \begin{bmatrix} 1/E_a & -v_{ab}/E_b & -v_{ac}/E_c & 0 & 0 & 0 \\ & 1/E_b & -v_{bc}/E_c & 0 & 0 & 0 \\ & & 1/E_c & 0 & 0 & 0 \\ & & & 1/G_{ab} & 0 & 0 \\ & & & & 1/G_{bc} & 0 \\ \text{symmetric} & & & & & 1/G_{ca} \end{bmatrix}$$

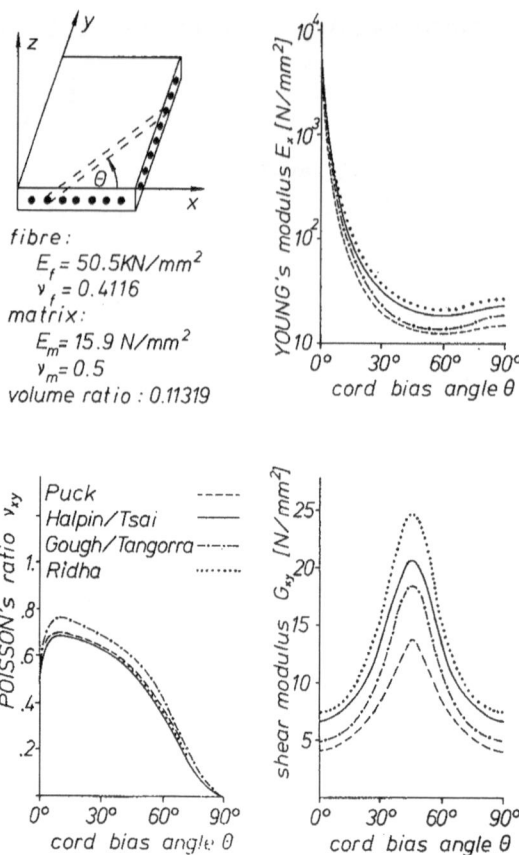

FIG. 3.   Elastic constants for a reinforced layer.

We assume transverse isotropy in the b–c plane

$$E_b = E_c$$
$$\nu_{ab} = \nu_{ac}$$
$$G_{ab} = G_{ca}$$

so that only five independent elastic constants are needed for the evaluation of the stress–strain matrix. Some authors[5–10] have expressed analytically the elastic constants of an orthotropic layer dependent upon the Young's moduli, Poisson's ratio and shear moduli of the cord and rubber according to their volume ratio (*see* Table 1). The transformation of the stress–strain matrix to the global co-ordinates xyz shows a variety of results for the elastic constants dependent on the applied method (*see* Fig. 3).

TABLE 1
Elastic constants of a reinforced layer

Ridha:[5]

$$E_a = E_F \cdot \phi + E_M \cdot (1 - \phi)$$

$$E_b = E_M \cdot \frac{\xi \cdot \eta \cdot \phi + 1}{1 - \eta \cdot \phi}$$

$$v_{ab} = v_F \phi + v_M (1 - \phi)$$

$$G_{ab} = G_M \cdot \frac{1 + \xi \cdot \eta \cdot \phi}{1 - \eta \cdot \phi}$$

Gough/Tangorra:[6]

$$E_a = E_F \phi + E_M (1 - \phi)$$

$$E_b = \frac{4 \cdot E_M \cdot (1 - \phi) \cdot [E_F \cdot \phi + (1 - \phi) \cdot E_M]}{3 E_F \cdot \phi + 4 \cdot E_M \cdot (1 - \phi)}$$

$$G_{ab} = G_M (1 - \phi)$$

$$v_{ab} = 0 \cdot 5$$

Akasaka/Hirano:[6]

$$E_a = E_F \cdot \phi$$

$$E_b = \frac{4}{3} \cdot E_M$$

$$v_{ab} = 0 \cdot 5$$

$$v_{ba} = 0$$

$$G_{ab} = G_M$$

Halpin/Tsai:[6]

$$E_a = E_F \cdot \phi + E_M \cdot (1 - \phi)$$

$$E_b = \frac{E_M \cdot (1 + 2\phi)}{1 - \phi}$$

$$G_{ab} = \frac{G_M \cdot [G_F + G_M + \phi \cdot (G_F - G_M)]}{G_F + G_M - \phi \cdot (G_F - G_M)}$$

$$v_{ab} = v_F \cdot \phi + v_M \cdot (1 - \phi)$$

Jones:[7]

$$E_a = E_F \cdot \phi + E_M \cdot (1 - \phi)$$

$$E_b = \frac{E_F \cdot E_M}{E_F \cdot (1 - \phi) + E_M \cdot \phi}$$

$$v_{ab} = v_F \cdot \phi + v_M \cdot (1 - \phi)$$

$$G_{ab} = \frac{G_F \cdot G_M}{G_F \cdot (1 - \phi) + G_M \cdot \phi}$$

E_F/E_M: Young's moduli of fibre/matrix

v_F/v_M: Poisson's ratio of fibre/matrix

G_F/G_M: shear moduli of fibre/matrix

ϕ: volume ratio

3. COMPARISON OF TENSILE TESTS AND FEM CALCULATIONS

In order to decide which of the different methods to find realistic elastic constants can be recommended for FEM computations of tyres and/or reinforced rubber materials, three-dimensional FEM computations are being compared to experimentally obtained results of tension tests.

3.1. Experiments

Three different kinds of tensile test samples were prepared for the experiments:

—plane rubber samples,
—samples with one reinforced layer
  ($\theta = 0°$, $10°$, $20°$, $30°$),
—samples with two reinforced layers
  ($\theta = \pm 0°$, $\pm 10°$, $\pm 20°$, $\pm 30°$).

The samples' length between the clamps was always 100 mm, further details are given in Fig. 4. The tensile loading versus strain was plotted and photographs of the unloaded and loaded samples were taken so that the deformations could be compared to the FEM computations. For this a 1 × 1 cm net had been marked on the samples.

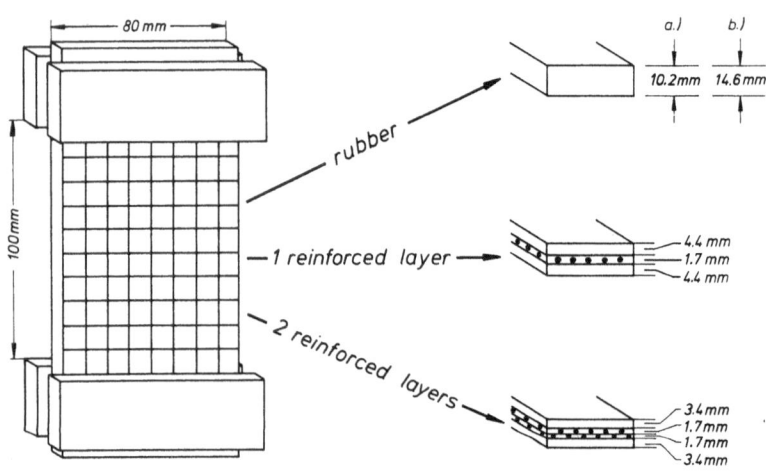

FIG. 4. Samples for tensile tests.

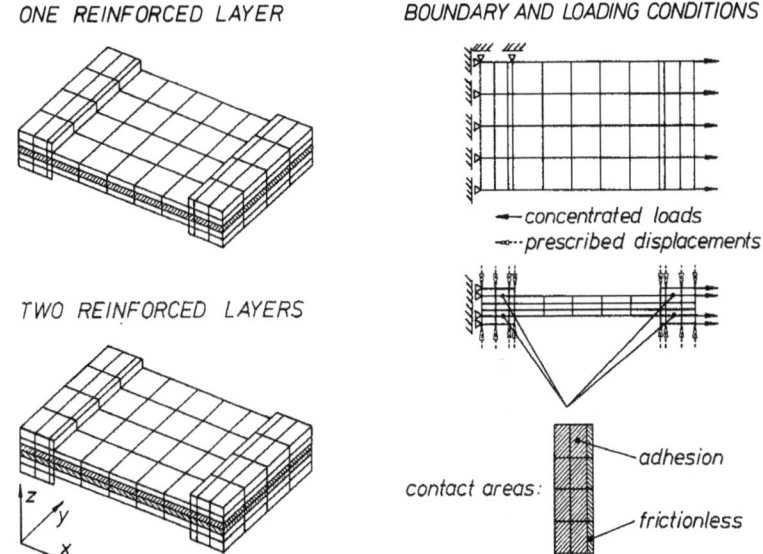

FIG. 5. FEM model of the tensile tests.

3.2. FEM Models

The non-linear finite element computer program ADINA[11] was used to model the tensile test samples including the clamps. Linear-elastic isotropic (for plane rubber layers) and linear-elastic orthotropic (for reinforced layers) three-dimensional 20-node and 8-node isoparametric elements, respectively, were used. The elastic constants for the orthotropic material model were calculated applying the equations listed in Tables 1 and 2. The loading and boundary conditions are illustrated in Fig. 5. The loading conditions at the clamps in y-direction were idealized by prescribing the measured displacements of the clamps. The contact area clamp/sample was assumed to be partly frictionless and partly ideal adhesive. Fixed boundary conditions were specified for the nodes on the negative x-direction face of one pair of clamps. The tensile load is modelled by concentrated loads applied to nodes on the positive x-direction face of the other pair of clamps. The displacement of the clamps is used to determine the analytically obtained tensile load versus strain.

3.3. Plane Rubber Samples

In order to test the boundary and loading conditions of the FEM model, plane rubber samples have been investigated and compared to the finite element calculations.

TABLE 2

Elastic constants of a reinforced layer

Puck:[8]

$$E_a = E_F \cdot \phi + E_M \cdot (1 - \phi)$$

$$E_b = \frac{E_{OM} \cdot (1 + 0.85\phi^2)}{(1 - \phi)^{1.25} + \phi \cdot \dfrac{E_{OM}}{E_F}}$$

$$v_{ab} = v_F \cdot \phi + v_M \cdot (1 - \phi)$$

$$G_{ab} = G_M \frac{1 + 0.6 \cdot \sqrt{\phi}}{(1 - \phi)^{1.25} + \phi \cdot \dfrac{G_M}{G_F}}$$

Förster/Knappe:[9]

$$E_a = E_F \cdot \phi + E_M (1 - \phi)$$

$$E_b = \frac{E_{OM}}{(1 - \phi)^{1.45} + \phi \cdot \dfrac{E_{OM}}{E_F}}$$

$$v_{ab} = v_F \cdot \phi + v_M \cdot (1 - \phi)$$

$$G_{ab} = G_M \frac{1 + 0.4 \cdot \sqrt{\phi}}{(1 - \phi)^{1.45} + \phi \dfrac{G_M}{G_F}}$$

Whitney/Riley:[10]

$$E_a = \phi \cdot E_F + (1 - \phi) \cdot E_M$$

$$+ \frac{2(v_F - v_M)^2 \cdot E_F \cdot E_M \cdot (1 - \phi) \cdot \phi}{E_M \cdot (1 - \phi) \cdot (1 - v_F - 2 \cdot v_F^2) + E_F \cdot \phi \cdot [(1 - v_M - 2 \cdot v_M^2) + (1 + v_M)]}$$

$$E_b = \frac{2 \cdot K_{23} \cdot E_a \cdot (1 - v_{bc})}{E_a + 4v_{ab} \cdot K_{23}}$$

$$v_{ab} = v_M - \frac{2(v_M - v_F) \cdot (1 - v_M^2) \cdot E_M \cdot \phi}{E_M \cdot (1 - \phi)(1 - v_F - 2 \cdot v_F^2) + E_F \cdot [\phi(1 - v_M - 2 \cdot v_M^2) + (1 + v_M)]}$$

$$v_{bc} = v_F \phi + (1 - \phi)v_M$$

$$G_{ab} = G_M \frac{(G_F + G_M) + \phi \cdot (G_F - G_M)}{(G_F + G_M) - \phi \cdot (G_F - G_M)}$$

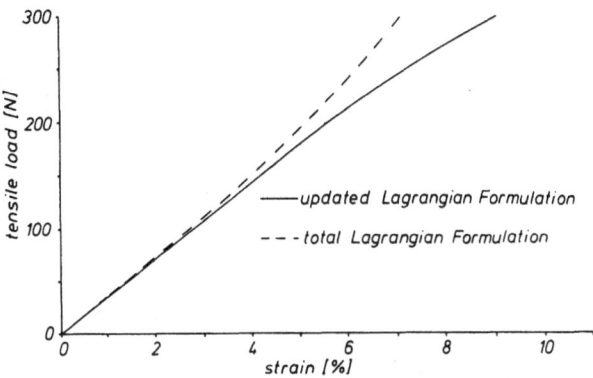

FIG. 6. Calculated tensile load versus strain curves using updated and total Lagrangian formulation.

Due to large strain the Total Lagrangian Formulation

$$ {}_0^t S_{ij} = {}_0^t C_{ijrs}\, {}_0^t \varepsilon_{rs} $$

and the Updated Lagrangian Formulation

$$ {}^t \tau_{mn} = {}_t^t C_{mnpq}\, {}_t^t \varepsilon_{pq} $$

lead to different results (*see* Fig. 6) for the tensile load versus strain.[12]

For the Total Lagrangian Formulation (required for orthotropic material in the FEM program ADINA) the components of ${}_0^t C_{ijrs}$ must be available.

The results illustrated in Figs 7 and 8 show that the calculated tensile load versus strain and the experimentally obtained curve are nearly

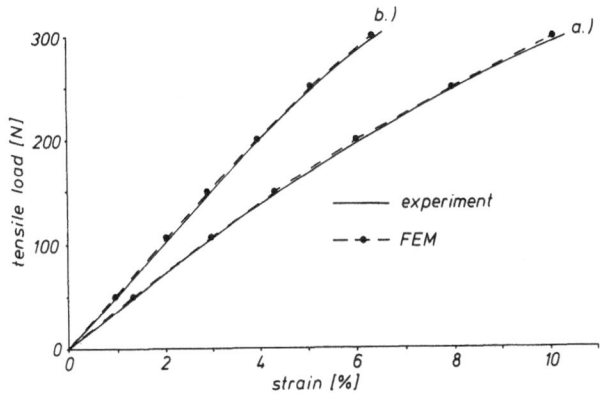

FIG. 7. Calculated tensile load versus strain of an unreinforced sample. Thickness: (a) 10·2 mm; (b) 14·6 mm.

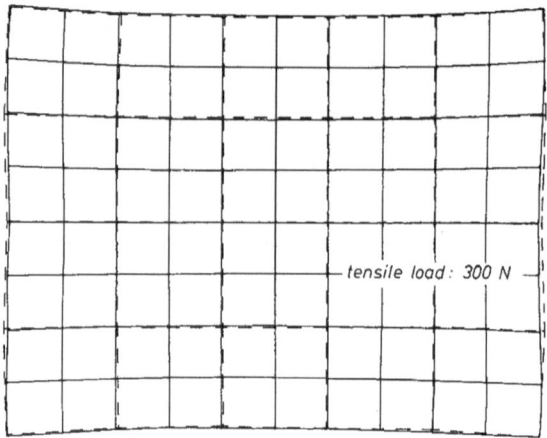

FIG. 8.   Deformations of an unreinforced sample (thickness: 10·2 mm). Experiment: ———;
calculation: – – –.

identical. Furthermore, the calculated deformations, except the one in the clamp area, do agree very well with the measured ones. It is shown that:

—the FEM models illustrated in Fig. 5 are valid to approximate the tensile tests and its boundary and loading conditions;
—isotropic linear-elastic material, that assumes that the compressive moduli are the same as the tensile moduli, can be used to describe the behaviour of rubber in zones of tension. In order to get better results for the deformations in the clamp area, another material description like the non-linear Mooney–Rivlin's approach, should be used.

3.4. Samples with Reinforced Layers

Samples with two symmetrical reinforced layers were modelled by using orthotropic elements for the reinforced layers. The angle between the material co-ordinates a-b-c and the global coordinates x-y-z is the same for all elements of one layer ($\theta_1 = 10°$, $20°$, $30°$) and otherwise identical for the elements of the other layer ($\theta_2 = -\theta_1$). The elastic constants describing the orthotropic material behaviour were calculated according to the equations in Tables 1 and 2.

The deviation among the various theories is very small if comparing just the tensile load versus strain curves, and all the results are in good agreement with the experimentally obtained curves (see Fig. 9). Bigger differences (max. 8%) will be noticed if comparing the calculated deformations to those experimentally obtained (see Fig. 10). For all test

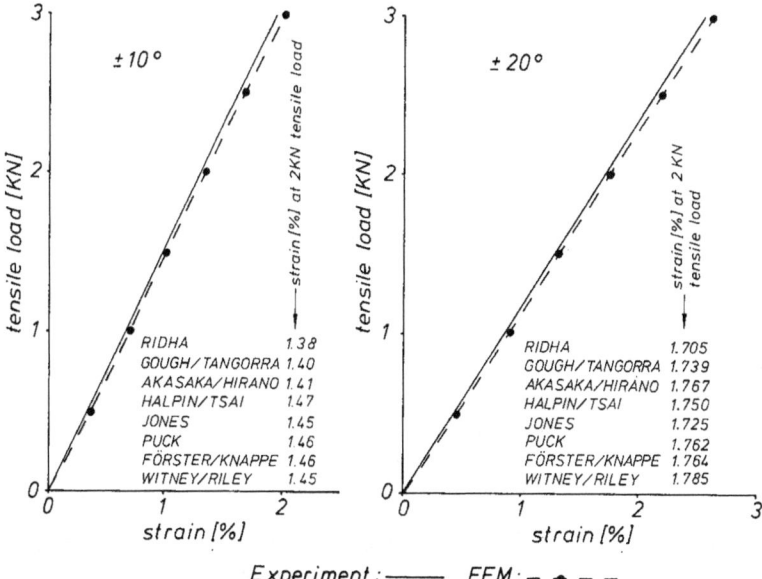

FIG. 9.   Tensile load versus strain curves for samples with two reinforced layers.

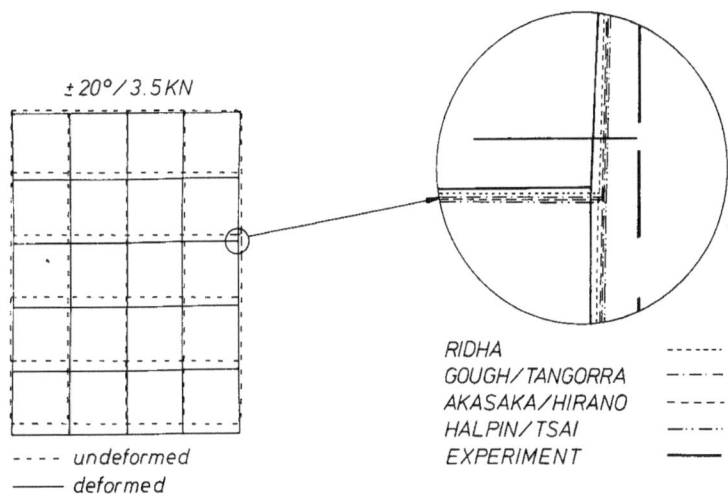

FIG. 10.   Deformations of sample with two reinforced layers.

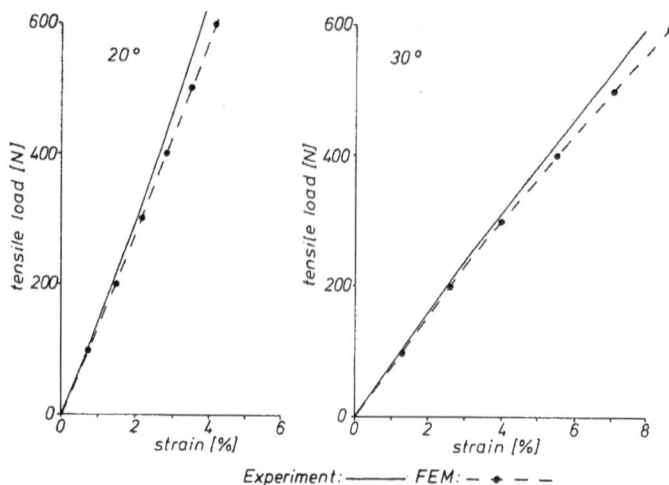

FIG. 11.   Tensile load versus strain curves for samples with one reinforced layer.

samples ($\theta = \pm 10°$, $\pm 20°$, $\pm 30°$) the computations, applying the elastic constants being determined according to the method of Ridha hardly differ from those being obtained experimentally. The method giving less agreement with the experiments, but still good for an approximation, is the method of Halpin/Tsai.

The comparison of FEM computations of tensile tests with samples with one reinforced layer to experiments leads to the same conclusions. The

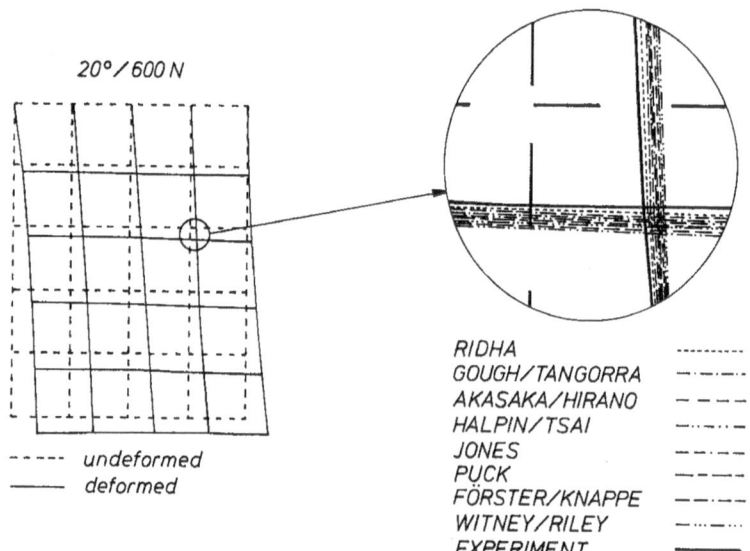

FIG. 12.   Deformations of sample with one reinforced layer.

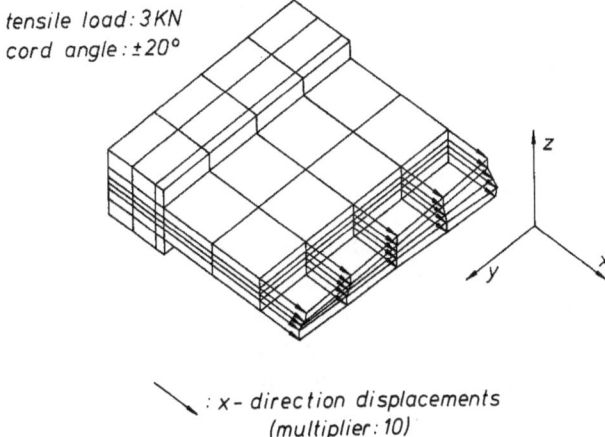

tensile load: 3KN
cord angle: ±20°

  : x- direction displacements
     (multiplier: 10)

FIG. 13.   Interlaminar effect of a sample with two reinforced layers.

agreement of predicted tensile load versus strain curves, using any of the methods to calculate the elastic constants for the orthotropic material, with the experimentally obtained curves are good (Fig. 11) for small strain (4 %). The main reason for the differences between the predicted and measured curves for large strain is the non-updating of the orientation of the material co-ordinates.

The FEM computations, being based on elastic constants due to the method of Ridha are predicting deformations with most agreement with the experiments (Fig. 12).

The models applied can also be used for examining the in-plane shear deformations of each layer and the interlaminar effects when two layers of opposite fibre orientation are bounded together (*see* Fig. 13) as well as the out-of-plane twisting caused by in-plane tractions (*see* Fig. 14).

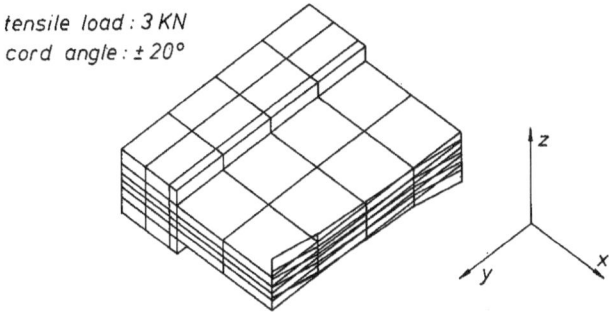

tensile load: 3KN
cord angle: ±20°

z - direction displacements (multiplier: 30)
FIG. 14.   Out-of-plane twisting caused by in-plane traction.

4. TWO-DIMENSIONAL-AXISYMMETRICAL CALCULATION OF A TYRE

Different authors[13-15] are using two-dimensional-axisymmetrical finite element models for inflation analysis of tyres. It is thus assumed that the plane of orthotropy is also the plane of symmetry.

In Fig. 15, an axisymmetric finite element model of the Continental 175 SR14 tyre is shown. Reinforced layers were assumed to be orthotropic linear elastic and the constants were calculated by using the different methods shown in Tables 1 and 2. The stress–strain matrix was transformed from material axis to global axisymmetric co-ordinates and then it was assumed that these co-ordinates are principal axes for the orthotropic material.

Geometrically non-linear computations of the pressurized (2·0 bar) tyre using any of the sets of elastic constants have been carried out; some deformed configurations are plotted in Fig. 16. All the computations lead to nearly the same inflated shape—the results differ less than 6% concerning the deformations.

The computed deformations of the crown region agree very well with experimentally measured deformations while the computed deformations of the sidewall region differ a lot from measured deformations (compare also Ref. 15). Due to this, the bead region has been remodelled in a different way and the boundary conditions have been changed. An axisymmetric

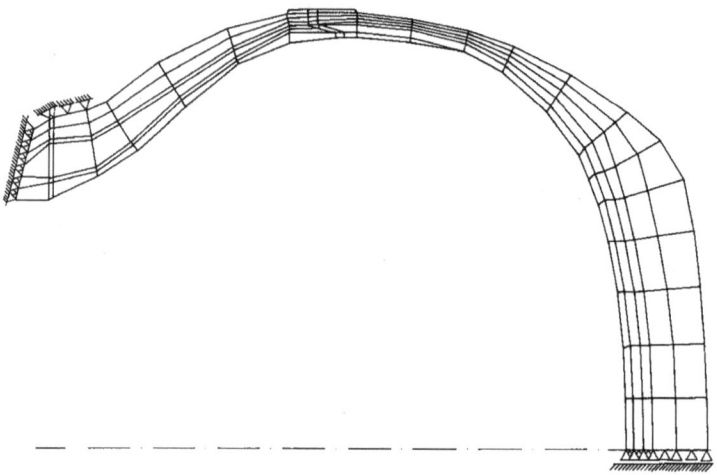

FIG. 15.   Axisymmetric finite element model of the CONTI 175 SR14 tyre.

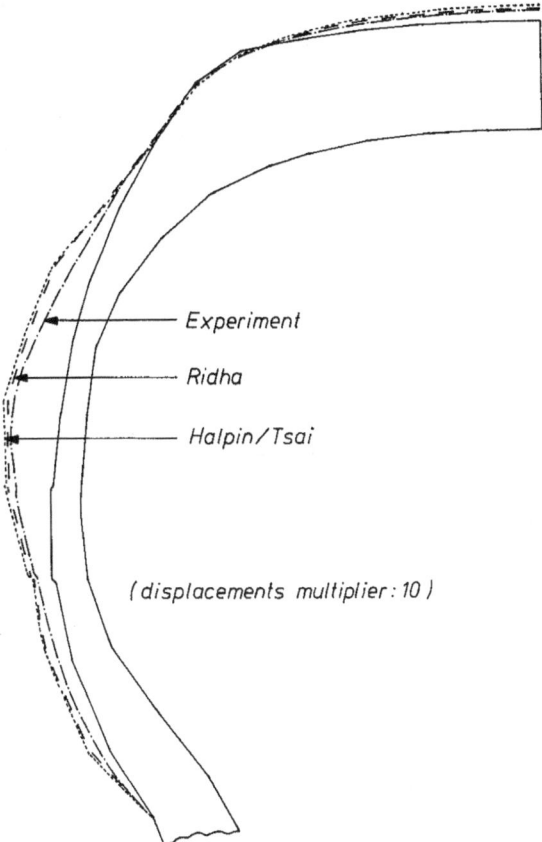

FIG. 16. Deformed shapes of the CONTI 175 SR14 tyre using different methods for calculating the elastic constants.

finite two-components element using displacement functions

$$u_r(r, \theta) = a_1(r) + \sum_{n=1}^{n_0} b_{1n}(r) \sin(n\theta) + \sum_{n=1}^{n_0} c_{1n}(r) \cos(n\theta)$$

$$u_\theta(r, \theta) = a_2(r) + \sum_{n=1}^{n_0} b_{2n}(r) \sin(n\theta) + \sum_{n=1}^{n_0} c_{2n}(r) \cos(n\theta)$$

was used to model the single bead wires and displacements in radial direction were prescribed at the nodes in the contact area between tyre and rim. This new model leads to better results for the sidewall region (Fig. 17).

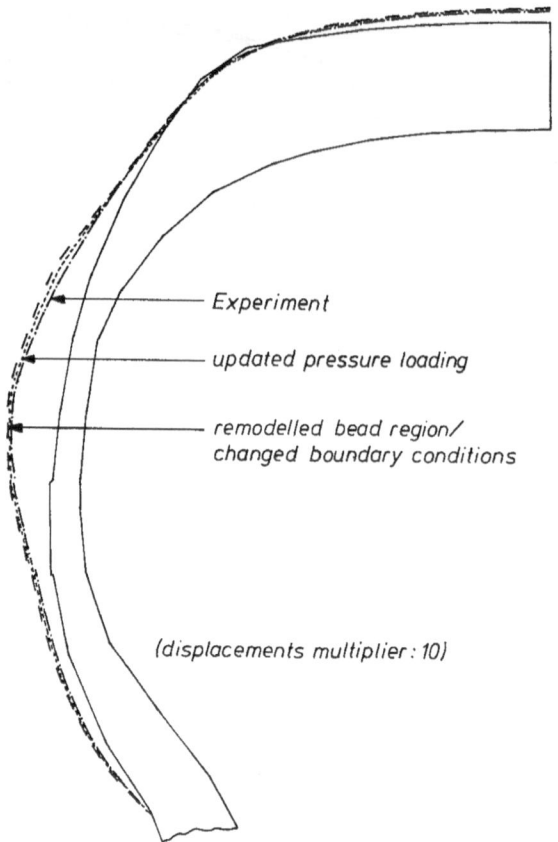

FIG. 17.   Calculated deformed shapes of CONTI 175 SR14 tyre using finite element models
with different boundary and loading conditions.

Another improvement is the result of updated pressure loading due to geometrical changes (Fig. 17). It is shown that boundary and loading conditions have a great influence on the results while the method of calculating the elastic constants of the orthotropic elements, describing the reinforced layers, is of less importance.

5.   THREE-DIMENSIONAL CALCULATION OF A TYRE

All the shown effects of reinforced layers, like in-plane traction, out-of-plane twisting and interply shear effects, cannot be described by an axisymmetrical FEM model presented in Section 4, so that a three-dimensional FEM model, equivalent to the one that was used to model

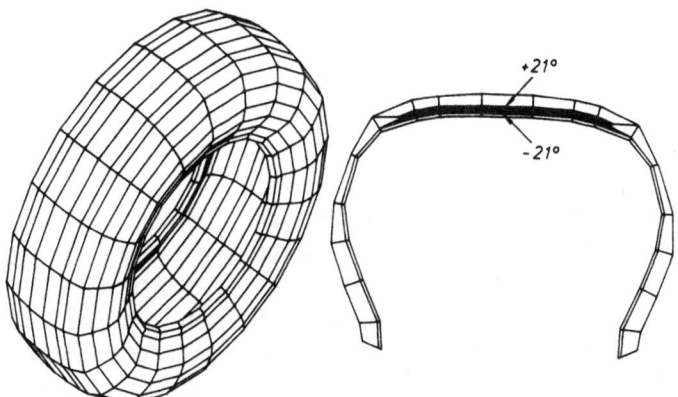

FIG. 18.   Computer plots of a three-dimensional finite element model of a tyre (1520 nodes, 1080 elements).

CROWN REGION ELEMENTS

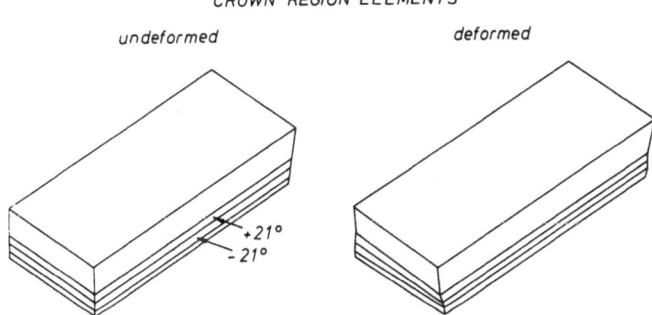

FIG. 19.   Interply shear effect due to inflation.

the tension test, is necessary for tyre computations, even though the loading conditions are axisymmetrical (e.g. inflated tyre). In Fig. 18 a three-dimensional FEM model is illustrated.

This model allows the computation of the three-dimensional effects of the tyre due to reinforced layers, e.g. interply shear strain due to inflation (*see* Fig. 19) as in Ref. 16.

6.   CONCLUSIONS

Three-dimensional orthotropic linear-elastic finite elements can be used for the description of reinforced tyre material.* In contrast to two-dimensional

* Results of tyre computations using three-dimensional orthotropic finite elements will be published in the near future.

elements, as used by other authors, all the three-dimensional effects, like the interlaminar effects and the out-of-plane twisting, can be investigated by using this element type. The different methods used to calculate the elastic constants of the orthotropic element dependent upon the elastic constants and volume ratio of the cord and rubber, lead to equivalent results.

The authors believe that three-dimensional modelling of a tyre with orthotropic elements approximating reinforced layers is best suited when examining effects, due to reinforcement for axisymmetrically and non-axisymmetrically loaded tyres. Isotropic elements can be used for approximating rubber material, but the assumption that the compressive moduli are the same as the tensile moduli will lead to erroneous results in areas with large compressive stresses, like the bead region and crown region (contact area tyre/ground). Another material description like the non-linear Mooney–Rivlin's approach, should be used.

REFERENCES

1. RHIDA, R. A., Analysis for tire mold design, *Tire Sci. Technol.*, **3** (1974), 195–210.
2. DEESKINAZI, J., SOEDEL, W. and YANG, T. Y., Contact of an inflated toroidal membrane with a flat surface as an approach to the tire deflection problem, *Tire Sci. Technol.*, **3** (1975), 43–61.
3. DEESKINAZI, J., YANG, T. Y. and SOEDEL, W., Displacements and stresses resulting from contact of a steel belted radial tire with a flat surface, *Tire Sci. Technol.*, **6** (1978).
4. KENNEDY, R. H., PATEL, H. P. and McMINN, M. S., Radial truck tire inflation analysis: theory and experiment, *Rubber Chem. Technol.*, **54** (1981), 751–766.
5. RIDHA, R. A., Computation of stresses, strains and deformation of tires, *Rubber Chem. Technol.*, **53** (1980), 849–902.
6. WALTER, J. D. and PATEL, H. P., Approximate expressions for the elastic constants of cord-rubber laminates, *Rubber Chem. Technol.*, **52** (1979), 710–724.
7. JONES, R. M., *Mechanics of Composite Materials*, London, McGraw-Hill (1975).
8. PUCK, A., *Zur Beanspruchung und Verformung Mehrschichtiger Verbundstoff-Bauelemente aus Glasseidensträngen und Kunststoff*, Dissertation TH Darmstadt (1967).
9. FÖRSTER, R. and KNAPPE, W., Experimentelle und theoretische Untersuchungen zur Rißbildungsgrenze an zweischichtigen Wickelrohren aus Glasfaser/Kunststoff unter Innendruck, *Kunststoffe*, **6** (1971), 583–588.
10. WHITNEY–RILEY, *Analysis of Structure Composite Materials*, New York (1973).

11. BATHE, K. J., *Static and Dynamic Geometric and Material Nonlinear Analysis Using ADINA*, Report 82339-2, Aeronautics and Vibration Laboratory, Department of Mechanical Engineering, Massachusetts Institute of Technology.

12. BATHE, K. J., *Finite Element Procedures in Engineering Analysis*, Prentice Hall, Englewood Cliffs, New Jersey (1982).

13. DURAND, M. and JANKOVICH, E., *Nonapplicability of Linear Finite Element Programs to the Stress Analysis of Tires*, 2nd NASTRAN User's Colloquium, NASA Langley Research Center, NASA TMX-2637 (1972).

14. KAGA, H., OKAMOTO, K. and TOZAWA, Y., Stress analysis of a tire under vertical load by a finite element method, *Tire Sci. Technol.*, 5 (1977), 102–118.

15. PATEL, H. P. and KENNEDY, R. H., Nonlinear finite element analysis for composite structures of axisymmetric geometry and loading, *Computers & Structures*, 15 (1982), 79–84.

16. TURNER, J. L. and FORD, J. L., Interply behaviour exhibited in compliant filamentary composite laminates, *Rubber Chem. Technol.*, 55 (1982), 1078–1094.

Index